Management of Micro and Nano-plastics in Soil and Biosolids

Sartaj Ahmad Bhat • Vineet Kumar
Fusheng Li • Sunil Kumar
Editors

Management of Micro and Nano-plastics in Soil and Biosolids

Fate, Occurrence, Monitoring, and Remedies

Springer

Editors
Sartaj Ahmad Bhat
Gifu University
Gifu, Japan

Fusheng Li
Gifu University
Gifu, Japan

Vineet Kumar
Department of Microbiology
Central University of Rajasthan
Ajmer, India

Sunil Kumar
CSIR-National Environmental Engineering
Research Institute (CSIR-NEERI)
Nagpur, India

ISBN 978-3-031-51966-6 ISBN 978-3-031-51967-3 (eBook)
https://doi.org/10.1007/978-3-031-51967-3

This Springer imprint is published by the registered company Springer Nature Switzerland AG
The registered company address is: Gewerbestrasse 11, 6330 Cham, Switzerland

Paper in this product is recyclable.

Preface

Management of Micro and Nano-plastics in Soil and Biosolids: Fate, Occurrence, Monitoring, and Remedies is a comprehensive book that addresses the challenges and recent advances in understanding and mitigating the impact of micro- and nano-plastics on soil and biosolids. Micro- and nano-plastics, which are tiny plastic particles measuring less than 5mm and 100nm, respectively, have gained significant attention due to their widespread occurrence and potential environmental and health hazards. This book provides valuable insights into the fate, monitoring, and remediation of these pollutants, featuring contributions from leading experts in the field.

Part I: Micro-Nano-plastics in the Environment: The first part of the book focuses on micro- and nano-plastics in the environment, particularly in aquatic ecosystems. It consists of five chapters that delve into various aspects of this issue. The chapters discuss the impact of micro- and nano-plastics on aquatic environments, the sources of contamination, and the challenges associated with analyzing and assessing the risks they pose. Understanding the dynamics of micro- and nano-plastics in aquatic systems is crucial, as these particles can be transported from stormwater runoffs to water bodies, affecting aquatic life and posing risks to human health.

Part II: Micro-Nano-plastics in the Soil Systems: This explores the presence and effects of micro- and nano-plastics in soil systems. It comprises six chapters that examine the impact of these particles on soil health, plant performance, and soil-dwelling organisms such as earthworms. The toxicological effects of these plastics on plant growth and soil fauna are of particular concern. Understanding how micro- and nano-plastics interact with soil ecosystems is essential, as they can alter the physical, chemical, and biological properties of soil, potentially affecting crop production and soil biodiversity.

Part III: Micro-Nano-plastics in the Biosolids: The third part of the book focuses on micro- and nano-plastics in biosolids, which are organic materials derived from sewage sludge and compost. It includes three chapters that discuss the occurrence and environmental risks associated with biodegradable microplastics in biosolids. As biosolids are often used in agriculture as soil amendments, the presence of micro- and nano-plastics in these materials can potentially lead to their introduction into food chains, raising concerns about human exposure to these contaminants.

Part IV: Micro-Nano-plastics Remedies from Contaminated Areas: The fourth and final part of the book addresses potential remediation strategies for micro- and nano-plastics in contaminated areas. It includes three chapters that examine various techniques for degrading these plastic particles. Microbial degradation and enzyme-assisted biodegradation methods are explored as innovative approaches to reducing micro- and nano-plastics pollution. These methods are crucial in managing the persistence of micro- and nano-plastics in soil and biosolids, which pose a long-term threat to ecological functions and soil biodiversity.

Micro- and nano-plastics are persistent environmental pollutants that can remain in the soil for extended periods, potentially disrupting ecosystems, compromising soil health, affecting global food production, and posing risks to human health. The book emphasizes the importance of adopting a nexus approach, which integrates knowledge and solutions from multiple disciplines, to effectively manage these contaminants. The nexus approach considers the interconnectedness of various environmental and societal factors, seeking to find sustainable solutions to complex problems.

In conclusion, *Management of Micro and Nano-plastics in Soil and Biosolids: Fate, Occurrence, Monitoring, and Remedies* provides a valuable resource for understanding and addressing the challenges posed by micro- and nano-plastics in soil and biosolids. This book gathers the expertise of leading scientists in the field and offers insights into the fate, occurrence, monitoring, and potential remedies for these pollutants. By using a nexus approach and considering various aspects of the issue, it aims to contribute to the development of effective strategies for managing micro- and nano-plastics in our environment.

Gifu, Japan — Sartaj Ahmad Bhat
Ajmer, India — Vineet Kumar
Gifu, Japan — Fusheng Li
Nagpur, India — Sunil Kumar

Contents

Contributors

Deachen Angmo Department of Botanical and Environmental Sciences, Guru Nanak Dev University, Amritsar, India

Prodipto Bishnu Angon Faculty of Agriculture, Bangladesh Agricultural University, Mymensingh, Bangladesh

Derya Ayral-Çınar Department of Environmental Engineering, Gebze Technical University, Kocaeli, Turkey

Gayatri Barabde Department of Environmental Science, The Institute of Science, Dr. Homi Bhabha State University, Mumbai, India

Department of Analytical Chemistry, The Institute of Science, Dr. Homi Bhabha State University, Mumbai, India

Dilara Büşra Bartan Department of Environmental Engineering, Gebze Technical University, Kocaeli, Turkey

Sartaj Ahmad Bhat River Basin Research Center, Gifu University, Gifu, Japan

Akshay Botle Department of Environmental Science, The Institute of Science, Dr. Homi Bhabha State University, Mumbai, India

Eda Ceylan Department of Environmental Engineering, Gebze Technical University, Kocaeli, Turkey

Mary Carolin Kurisingal Cleetus Faculty of Science and Technology, Research Centre for Experimental Marine Biology and Biotechnology (PIE-UPV/EHU), Plentzia, Basque Country, Spain

Faculty of Sciences, University of Liege, Liege, Belgium

Guangyu Cui School of Environment and Energy, Peking University Shenzhen Graduate School, Shenzhen, China

Arpan Das Faculty of Agriculture, Bangladesh Agricultural University, Mymensingh, Bangladesh

Shiamita Kusuma Dewi United Graduate School of Agricultural Science, Gifu University, Gifu, Japan

Arun Dhanasekaran Department of Biomedical Sciences, School of Biosciences and Technology, Vellore Institute of Technology, Vellore, Tamil Nadu, India

Dushyant R. Dudhagara Department of Life Sciences, Bhakta Kavi Narsinh Mehta University, Junagadh, India

Judith Osaretin Ehigie Faculty of Science and Technology, Research Centre for Experimental Marine Biology and Biotechnology (PIE-UPV/EHU), Plentzia-Biskaia, Basque Country, Spain

Helmholtz Centre for Environmental Research – UFZ, Leipzig, Germany

Afsana Ahamed Eva Faculty of Agriculture, Bangladesh Agricultural University, Mymensingh, Bangladesh

Megha S. Gadhvi Department of Life Sciences, Bhakta Kavi Narsinh Mehta University, Junagadh, India

Zaw Min Han Graduate School of Engineering, Gifu University, Gifu, Japan

Pinjing He Institute of Waste Treatment and Reclamation, Tongji University, Shanghai, China

Mihir Herlekar Department of Environmental Science, The Institute of Science, Dr. Homi Bhabha State University, Mumbai, India

Jatinder Kaur Katnoria Department of Botanical and Environmental Sciences, Guru Nanak Dev University, Amritsar, Punjab, India

Kannabiran Krishnan Department of Biomedical Sciences, School of Biosciences & Technology, Vellore Institute of Technology, Vellore, Tamil Nadu, India

Bindhi S. Kumar Research Department of Fisheries and Aquaculture, St. Albert's College (Autonomous), Kochi, Kerala, India

Fusheng Li River Basin Research Center, Gifu University, Gifu, Japan

Shitosri Mondal Faculty of Agriculture, Bangladesh Agricultural University, Mymensingh, Bangladesh

M. Muruganandam Project Prioritization, Monitoring & Evaluation, and Knowledge Management Unit, ICAR-Indian Institute of Soil & Water Conservation (ICAR-IISWC), Dehradun, Uttarakhand, India

P. Nisha Department of Biosciences, MES College, Ernakulam, Kerala, India

M. Nithya Department of Civil Engineering, Yashoda Technical Campus, Wadhe, Maharashtra, India

Cyril Oziegbe Onogbosele Department of Zoology, Faculty of Life Sciences, Ambrose Alli University, Ekpoma, Edo State, Nigeria

İrem Öztürk-Ufuk Department of Environmental Engineering, Gebze Technical University, Kocaeli, Turkey

Elif Pehlivanoğlu-Mantaş Department of Environmental Engineering, Gebze Technical University, Kocaeli, Turkey

Department of Environmental Engineering, Istanbul Technical University, Istanbul, Turkey

M. P. Prabhakaran Department of Aquatic Environment and Management, Kerala University of Fisheries and Ocean Studies, Kochi, Kerala, India

Jyothirmayee Kola Pratap Department of Biomedical Sciences, School of Biosciences & Technology, Vellore Institute of Technology, Vellore, Tamil Nadu, India

A. K. Priya Department of Chemical Engineering, KPR Institute of Engineering and Technology, Uthupalayam, Tamil Nadu, India

Project Prioritization, Monitoring & Evaluation, and Knowledge Management Unit, ICAR-Indian Institute of Soil & Water Conservation (ICAR-IISWC), Dehradun, Uttarakhand, India

Lama Ramadan Department of Environmental Engineering, Gebze Technical University, Kocaeli, Turkey

V. J. Rejish Kumar Faculty of Ocean Science and Technology, Kerala University of Fisheries and Ocean Studies, Kochi, Kerala, India

Department of Aquaculture, Kerala University of Fisheries and Ocean Studies, Kochi, Kerala, India

Sayli Salgaonkar Department of Environmental Science, The Institute of Science, Dr. Homi Bhabha State University, Mumbai, India

Ajay Valiyaveettil Salimkumar Faculty of Science and Technology, Research Centre for Experimental Marine Biology and Biotechnology (PIE-UPV/EHU), Plentzia-Biskaia, Basque Country, Spain

Sirat Sandil Institute of Aquatic Ecology, Centre for Ecological Research, Budapest, Hungary

National Laboratory for Water Science and Water Security, Institute of Aquatic Ecology, Centre for Ecological Research, Budapest, Hungary

Priyanka Sharma Department of Botanical and Environmental Sciences, Guru Nanak Dev University, Amritsar, Punjab, India

Surbhi Sharma Department of Botanical and Environmental Sciences, Guru Nanak Dev University, Amritsar, Punjab, India

Jaswinder Singh Department of Zoology, Khalsa College, Amritsar, Punjab, India

Kuok Ho Daniel Tang Department of Environmental Science, The University of Arizona, Tucson, AZ, USA

Babita Thakur Department of Microbiology, School of Bioengineering and Biosciences, Lovely Professional University, Phagwara, Punjab, India

Emel Topuz Department of Environmental Engineering, Gebze Technical University, Kocaeli, Turkey

Md. Shakil Uddin Faculty of Agriculture, Bangladesh Agricultural University, Mymensingh, Bangladesh

Anjana K. Vala Department of Life Sciences, Maharaja Krishnakumarsinhji Bhavnagar University, Bhavnagar, India

Ayushi Varshney Anantaa GSK Innovations Pvt Ltd., Faridabad, Haryana, India

Meryem Vasef Department of Environmental Engineering, Gebze Technical University, Kocaeli, Turkey

Adarsh Pal Vig Department of Botanical and Environmental Sciences, Guru Nanak Dev University, Amritsar, India

Suhas J. Vyas Department of Life Sciences, Bhakta Kavi Narsinh Mehta University, Junagadh, India

Ashna Waseem Department of Environmental Engineering, Gebze Technical University, Kocaeli, Turkey

Yongfen Wei River Basin Research Center, Gifu University, Gifu, Japan

Xiaoyi Wu School of Environment and Energy, Peking University Shenzhen Graduate School, Shenzhen, China

Qiyong Xu School of Environment and Energy, Peking University Shenzhen Graduate School, Shenzhen, China

Gyula Zaray Institute of Aquatic Ecology, Centre for Ecological Research, Budapest, Hungary

National Laboratory for Water Science and Water Security, Institute of Aquatic Ecology, Centre for Ecological Research, Budapest, Hungary

Institute of Chemistry, Eötvös Loránd University, Budapest, Hungary

About the Editors

Sartaj Ahmad Bhat is working as JSPS Postdoctoral Researcher at the River Basin Research Center, Gifu University, Japan. He received his PhD in Environmental Sciences from Guru Nanak Dev University, Amritsar, India in 2017. His research interests focus on vermicomposting treatment of various solid wastes, especially for investigations on fate and behavior of emerging pollutants during biological treatment of organic wastes. He has published more than 65 papers in peer-reviewed journals and edited over 15 books published by Elsevier, Springer, CRC Press, IWA, and RSC. Dr. Bhat is serving as an Associate/Academic Editor and Editorial Board Member/Advisory Board Member of more than 20 journals published by Frontiers, Springer, Elsevier, PLOS, Wiley, Hindawi, and De Gruyter. Dr. Bhat is a recipient of several prestigious awards such as the JSPS Postdoctoral Fellowship to pursue research at River Basin Research Center, Gifu University, Japan, the Basic Scientific Research Fellowship (BSR JRF, SRF) by the University Grants Commission (UGC) India, the DST-SERB National Postdoctoral Fellowship at CSIR-NEERI, Nagpur, India, and Swachhta Saarthi Fellowship by the Govt. of India. He has also received the 2020 Outstanding Reviewer Award by the *International Journal of Environmental Research and Public Health*, MDPI, and the Top Peer Reviewer 2019 award in *Environment and Ecology* by Web of Science. He has more than 800 verified reviews and 75 editor records to his credit.

Vineet Kumar is working as a National Postdoctoral Fellow in the Department of Microbiology, School of Life Sciences at the Central University of Rajasthan, Rajasthan, India. He earned his PhD (2018) in Environmental Microbiology from Babasaheb Bhimrao Ambedkar (A Central) University, Lucknow, India. Dr. Kumar's research work mainly focuses on wastewater treatment and solid waste management. He has published more than 50 articles in peer-reviewed international journals of repute, 24 books, and 52 book chapters, on various aspects of science and engineering, with more than citations of 2550, and h-index of 31. Dr. Kumar has been serving as a Guest Editor and Reviewer in more than 65 prestigious international journals. He has served on the editorial board of various reputed journals. He has presented several papers relevant to his research areas at national and international conferences. He is also a recipient of various prestigious fellowships and awards, such as the Young Scientist Award, Rajiv Gandhi National Fellowship by UGC, and National Postdoctoral Fellowship by the Science and Engineering Research Board (SERB), Govt. of India. He is an active member of numerous scientific societies including the Microbiology Society (UK), the Indian Science Congress Association (India), the Association of Microbiologists of India (India), etc. He is the founder of the Society for Green Environment, India (website: www.sgeindia.org).

Fusheng Li is a Professor in the Division of Water System Safety and Security Studies and the Graduate School of Engineering at Gifu University, Japan. He received his BS degree in Environmental Engineering from Lanzhou Jiaotong University of China in 1986, MS degree from Kitami Institute of Technology of Japan in 1994, and PhD degree from the Gifu University of Japan in 1998. Dr. Li is directing the Division of Water Quality Studies that covers the fields from water quality to water and wastewater treatment, and recently to resource and energy recovery from organic waste. The ongoing research projects in his lab include adsorption; membrane filtration, enhanced coagulation, disinfection; biological water and wastewater treatment; vermicomposting treatment of vegetable waste and activated sludge; microbial fuel cell; physicochemical water quality assessment; biological water quality assessment. He has over 350 scholarly publications, including more than 200 in peer-reviewed journal papers. As principal supervisor, he has already guided 50 master's and 21 doctorate graduate students to the completion of their degrees. Dr. Li is the recipient of awards from several academic societies and associations for his research work on water treatment and water quality dynamics studies.

Sunil Kumar is a well-rounded researcher with more than 22 years of experience in leading, supervising, and undertaking research in the broader field of Environmental Engineering and Science with a focus on Solid and Hazardous Waste Management. Dr. Kumar is a graduate of Environmental Engineering and Management from the Indian Institute of Technology, Kharagpur, India. He completed his PhD in Environmental Engineering from Jadavpur University, Kolkata, India. His primary area of expertise is solid waste management (Municipal Solid Waste, Electronic waste etc.) over a wide range of environmental topics including contaminated sites, EIA, and wastewater treatment. His contributions in these fields led to a citation of 14972, an h-index of 60, and i10-index of 246 (Google scholar). His contributions since inception at CSIR-National Environmental Engineering Research Institute (NEERI), India, in 2000 include 261 refereed publications, 5 books, 40 book chapters, 10 edited volumes, and numerous project reports to various governmental bodies and private, local, and international academic/research bodies. He is the Associate Editor of peer-reviewed journals of international repute, such as *Environmental Chemistry Letter*, *International Journal of Environmental Science and Technology*, and ASCE *Journal of Hazardous, Toxic and Radioactive Waste*. He also served on the Editorial Board of *Bioresource Technology*, Elsevier. He has completed many research projects as PI with 18 (12 awarded) PhD and 20 MPhil/MTech thesis/dissertations. Dr. Kumar was awarded the most prestigious award Alexander von Humboldt-Stiftung Jean-Paul-Str.12 D-53173 Bonn, Germany, as a Senior Researcher for developing a Global Network and Excellence for more advanced research and technology innovation.

Part I
Micro-Nano-plastics in the Environment

Microplastic and Nanoplastic: A Threat to the Environment

A. K. Priya, M. Muruganandam, and M. Nithya

1 Introduction

Over recent years, an estimated 335 million tons of polymers have been used as plastics worldwide. Plastics have many benefits, such as their adaptability, solidity, and low cost, which have led to trillions of dollars in financial gains worldwide (Ciriminna & Pagliaro, 2020). Microplastic (MP) particles are available all through the environment and are a critical wellspring of concern since they are so tiny (characterized here as 100 nm to 5 mm in size) and nano-estimated particles that they can be ingested by a wide variety of living things, raising the chance of bioaccumulation and biomagnification (CONTAM, 2016). Marine natural organic organisms ingest microplastics, and there is evidence that these organisms pass them from one trophic level to the next after passing through the stomach. This evidence is still in its early stages (Burrows et al., 2020). Nanoplastics may be more dangerous than microplastics due to their ability to penetrate natural layers. Earthbound investigations into microplastic ingestion are emerging for soil organic entities. The sources

A. K. Priya (✉)
Department of Chemical Engineering, KPR Institute of Engineering and Technology, Arasur, Uthupalayam, Tamil Nadu, India

Project Prioritization, Monitoring & Evaluation, and Knowledge Management Unit, ICAR-Indian Institute of Soil & Water Conservation (ICAR-IISWC), Dehradun, Uttarakhand, India
e-mail: a.k.priya@kpriet.ac.in

M. Muruganandam
Project Prioritization, Monitoring & Evaluation, and Knowledge Management Unit, ICAR-Indian Institute of Soil & Water Conservation (ICAR-IISWC), Dehradun, Uttarakhand, India

M. Nithya
Department of Civil Engineering, Yashoda Technical Campus, Wadhe, Maharashtra, India

S. A. Bhat et al. (eds.), *Management of Micro and Nano-plastics in Soil and Biosolids*,
https://doi.org/10.1007/978-3-031-51967-3_1

and predetermination of microplastics in earthly climate are being studied, and we foster their work by investigating how plant and soil standard components in agro-ecosystems could be affected, from the solitary level to the eco-framework level (Horton et al., 2017).

Marine debris is still present in the atmosphere and has either caused or taken care of strongholds in the ocean. Its worldwide extension represents a danger to the sea life's natural framework. Among the different materials associated with portraying marine litter, microplastics and nanoplastics are viewed as arising impurities of concern. Plastics are naturally occurring polymers artificially modified to have high strength, low weight, and affordable cost properties. These characteristics make plastics useful for many applications, from empty plastic jugs to holders for food and consumer goods to transportation, development, media communications, and healthcare. Their extensive use increased their environmental release, whether done purposefully (Priya et al., 2022). From 1.7 million tons in the 1990s to 335 million tons in 2016, the annual production of plastic has increased.

Additionally, 4.8–12.7 million tons of plastic flotsam and jetsam are estimated to enter the ocean annually. The plastic polymers that are most commonly manufactured include polypropylene, polyethylene terephthalate (PET), polyvinyl chloride, polyurethane, polystyrene, and low-thickness polyethylene. These materials are utilized in different creative projects, from equipment to clinical considerations (Worm et al., 2017). For example, in a review zeroed in on the southern Adriatic Sea, 120 samples (water and buildup) were gathered, and 80.6% contained plastic junk, with 38.7% of the examples being made of polystyrene plastic (Šilc et al., 2018).

The production and consumption of plastics are increasing globally, increasing the mass and rate of micro(nano)plastics (MNPs) released into the environment. These micro(nano)plastics can adversely impact global biodiversity, including terrestrial and marine biodiversity (Wang et al., 2020). Human activities, such as the production and disposal of plastics, are significant factors influencing biodiversity. Microplastics and nanoplastics result from these human activities and can be ingested by many marine organisms, especially those that filter feed (Hernandez-Gonzalez et al., 2016). When ingested, micro(nano)plastics can physically block the digestive system, causing damage and reducing nutrient uptake, leading to death. Research has shown that polystyrene nanoplastics, which are very small in size, can pass through cell membranes and enter surrounding tissues, the bloodstream, and even the brains of medaka fish (Kashiwada, 2006). They are challenging to eliminate and can have long-lasting impacts on organisms.

In addition to the environmental effects, micro(nano)plastics pose a potential risk to human health. Studies have found that microplastics and nanoplastics are present in various food sources, such as seafood, tap water, and bottled water. This suggests that humans may be exposed to micro(nano) plastics through diet and drinking water. The health effects of micro(nano)plastics on humans are still largely unknown, but some studies have suggested that ingestion of micro(nano)plastics could lead to inflammation, oxidative stress, and genotoxicity. There is also concern about the potential for microplastics and nanoplastics to act as carriers for harmful chemicals and pathogens, potentially increasing their bioavailability and toxicity (Brewer et al., 2020; Fu et al., 2020). Various strategies have been proposed to mitigate the environmental and health impacts of micro(nano)plastics. These include reducing plastic waste through improved waste management practices, promoting biodegradable and compostable plastics, and reducing overall plastic consumption. In addition, developing effective methods for detecting and quantifying micro(nano)plastics in environmental samples is essential for monitoring and understanding their distribution and impacts. Finally, research on the distribution and fate of micro(nano) plastics in the environment should reveal effective relief techniques (Shiu et al., 2020).

As a result, marine plastic pollution has emerged as a significant ecological problem. Even though MPs and NPs are still having an effect on climate, they may also be having a significant influence on land-based ecosystems. Microplastics in soil and residue have an underappreciated effect on earthbound biological systems. Studies reveal that 98% of the assessed MP and NP defilement in the ocean originates from land. Due to their smaller size, MPs and NPs are easily collected by various species in the marine climate. Analyzing the threat that MPs and NPs pose to marine life and human health is crucial. This chapter discusses MP and NP origins, routes into the ocean, oceanic degradation, and global transportation. Studies are also conducted to determine how MPs/NPs impact human and marine health. This study will offer crucial information to researchers, analysts, and those formulating plans to reduce MP/NP pollution. It is essential that this chapter thoroughly examines and addresses the majority of the perspectives related to small objects and nanoplastics as a whole.

2 Nano- and Microplastics: Their Characteristics and Sources

MPs and NPs are particles of varying sizes, densities, and compound structures in the marine climate. Even though MPs can be categorized as mandatory or optional, there is currently no consensus on what constitutes nanoplastics so research has been focusing on identifying plastic particles (Avila et al., 2020). In late cycles, compelling MPs were used and perceived, whereas discretionary MPs are an unintended result of bigger plastics degrading more than a few cycles. MPs can be found in similar things or created due to other influential MPs contaminating everyday items. Modern abrasives, blower scrubbers, microbeads for personal care products, accidental spills, and pointless plastic powders are the best examples of MPs (Du et al., 2021). The primary sources of microplastic pollution, which account for more than 80% of all naturally occurring microplastic pollution, are materials, tires, and city dust. Some specific micro- and nanoplastic sources include sewage overflow, clothing, the pharmaceutical industry, production, the fishing industry, packaging and transportation, and plastic storage containers. Many countries use sewage flood disengaged by light, temperature, and other environmental factors as manure. As a result, storm drains and the resulting waterways frequently become the final destination for the microplastics from these biosolids. In addition, studies have shown that microplastics pass through some wastewater treatment facilities' filtration processes. Additional sources of MPs and NPs include tires (Magalhães et al., 2020). The movement of microplastics into the atmosphere is significantly accelerated by tire wear. One of the primary sources of income for MPs and NPs is clothing. Table 1 shows the various micro- and nanoplastic sources in the environment and their applications.

Numerous engineered filaments can be shed from clothing that stays in the environment for quite a while. A collection of clothes contains more than 1900 microplastic strands in each item. According to measurements, every piece of clothing emits about 170% more microplastics than the 700,000 strands that can be released from a typical 6 kg wash pile. It is also mentioned that plastic containers could be a source of microplastics. According to research, microplastics contaminate 91% of tests on filtered water. Microplastic was discovered to be twice as common in filtered water as in unfiltered water. This may be made worse by contamination from the packaging system. The drying of clothing, the production of polyethylene foils, and the warm cutting of polystyrene froth are all examples of NP sources (Tiwari et al., 2020). The development of 3D printing, which is currently quickly available for quick models and small projects, has been credited to the development of ultrafine particles. Using biomedical products made with polymeric nanoparticles in clinical settings can also cause ultrafine particles to be present in the air (Begines et al., 2020). As long as people live on and manage the land, it makes perfect sense to dump MPs and NPs there quickly. As a result, the NPs and MPs that are necessary and optional in the sea are obtained from terrestrial sources (Dąbrowska et al., 2021). The leading causes are poor waste management, illegal unloading, and simultaneous releases during construction, handling, horticulture, housekeeping,

Table 1 Source of micro- and nanoplastics in the environment and their applications

S. No.	Source	Route into environment	Applications	References
1	Plastic granules	Drifting and surface runoff	All plastic items are essentially made of crude materials and building blocks	Karkanorachaki et al. (2018)
2	Personal care products	Treated wastewater, sewage sludge	Peeling, film-framing, hydrophilic, scouring experts, and functionalized polymers are used in private consideration products and biomedical applications	Nel et al. (2019)
3	Paint	Surface runoff	Marine, automotive, street stamping paint, and structural coatings	Gaylarde et al. (2021)
4	Textile	Treated wastewater, sewage sludge	An increase in the appeal and utility of synthetic textures	Cai et al. (2020)
5	Sports ground	Drifting/surface runoff	–	van Kleunen et al. (2020)

and relaxation activities. But because many plastic particles are found in treated modern water and waste muck, they are ultimately stored in marine structures. Although ebb and flow wastewater treatment processes have been shown to evacuate microplastics successfully, they have not been replicated for nanoplastics. The advancements for eliminating MPs also influence the treatment cycles for channeling NPs; it has been acknowledged that the NPs were not effectively removed in wastewater treatment facilities (Petroody et al., 2020). The release of cleaning agents and personal care products into domestic wastewater has been identified as a critical starting point for polymeric MPs and NPs found in marine environments. MPs and NPs have been detected in marine regions, including those derived from plastic products and gum spills. These materials are often discarded improperly and can end up in waterways, breaking into smaller particles over time. These microplastics can harm marine life, as they can be mistaken for food and ingested by animals, leading to physical harm or even death (Priya et al., 2023).

In the marine environment, the following are the primary sources of micro- and nano-scale plastic particles: (1) intentionally produced polymer nanoparticles, such as those used in corrective devices, 3D printer ink, and drug delivery systems; (2) plastics breaking down due to UV photodegradation; (3) wastewater treatment plants (biosolids and effluent water); (4) mechanical activity in the environment; (5) hydrolysis of materials; and (6) microorganism activity. The primary sources of micro- and nanoplastics and their conduct through relocation are depicted in Fig. 1.

Plastic pellets and individual consideration items containing microbeads are the fundamental sources of essential MNPs. Other significant sources of necessary MNPs include painted surfaces, washed materials, sewage ooze, plastic track surfaces in schools, and elastic streets, contributing to vehicle tire wear (Smyth et al., 2021). Most of the time, made from polyethylene, polypropylene, and polystyrene, microbeads are extraordinary microplastics that are purposefully added to personal care and beauty products. In research on biomedicine and health and personal consideration items, they expressly serve as experts in peeling and cleaning (Dalili et al., 2019). MNP microbeads are also used as silicones, functionalized polymers, hydrophilic specialists, and film-shaping specialists in personal care and beauty products. Sphericity and molecule consistency, for instance, contribute to a metal ball's impact and result in a luxurious surface and spreadability, two beneficial

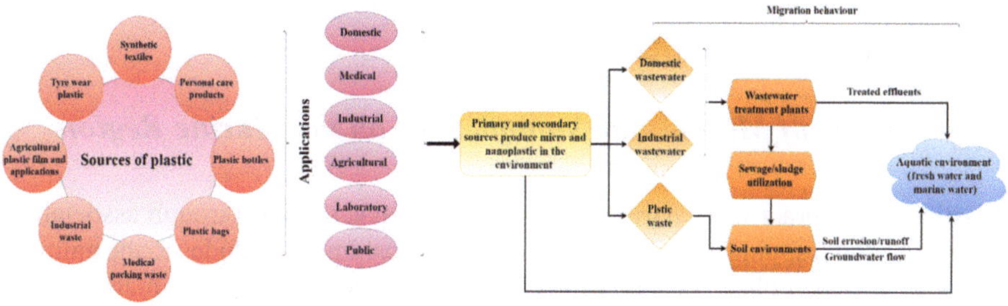

Fig. 1 Sources of micro- and nanoplastics and their migration behavior

properties for cosmetics (Alves et al., 2020). As an alternative to standard materials like actuated carbon and pumice stone, these MNPs can also be used. They have a string-like appearance and are round, bent, and frayed unevenly. Colored micro-beads also increase the visual appeal of things for individualized thought. MNPs are know to be dangerous source of microplastics when they are finished for the redirect and end up in channels, streams, and other streams because they directly pass through sewage treatment plants after being washed down the track (Ding et al., 2020). MNP is not entirely settled to make up 11% of the plastic waste unloaded into the North Sea. Pre-creation pitch pellets (pound), a key source of MNP trash, have also been identified. These pellets are primarily used in the production of modern plastic. These plastic pellets are likewise a result of reused plastic, explicitly during the patterns of cleaning, crushing, dissolving, arranging, and finally managing.

Macroplastic materials are separated using various environmental deterioration cycles, including biodegradation, chemical (use, photooxidation, temperature), and mechanical (scratched spot deterioration, wave action) exercises. Metropolitan wastes such as those generated during the production of movies, plastic bags and containers, fishing equipment, transportation, tire wear, and other large-scale plastic wastes have been considered important sources of assistance for MNPs. According to the ongoing analyses, most MNPs in maritime and terrestrial settings originate from helper wellsprings (An et al., 2020). Due to the increasing number of vehicles on the planet, street stamping, scraped spots, and tire wear are considered among the most common sources of natural MNPs (Kitahara & Nakata, 2020). MNPs are typically discovered in their path into water bodies and wastewater treatment facilities after being shed significantly by deliberate material strands during washing. Depending on the kind of material, about 124–308 mg of microplastics, or 640,000–1,500,000 MNP particles, are sent per kg of washed surface. Countless MNPs are additionally released by the development business from the plastic polymers utilized in cladding, safeguarded materials, and lines. Notwithstanding, it should be noticed that MNPs are often completed on building locales because of reasons or unfortunate ends. MNPs are additionally utilized in specific applications, for example, shooting experts during sandblasting to clean, refine, or roughen surfaces or eliminate paint (Li et al., 2020).

3 Micro- and Nanoplastic Toxicity

3.1 Effects of Micro- and Nanoplastics on the Marine Region

Plastics can interact with marine life in various ways, such as through association, ingestion, and trapping. Catch makes use of ghost fishing and alludes to the animal's yield. It has been observed in various animals, including enormous marine vertebrates and cowardly planktonic animals. Ingesting plastic waste can happen on purpose, accidentally, or indirectly (by eating animals that have consumed the plastic) (Wright et al., 2020). Communication includes either crashing or covering and avoids entangling carelessness with plastic trash. Micro- and nano-scale plastics will comprise some materials used to make plastics. These microscopic and nano-scale plastics will likely impact everything, ranging from sub-organismal to population levels, affecting regular cycles. High concentrations of microplastics can alter the vulnerability and force of buildup absorbance, according to research, which raises concerns about their impact on species that are not determined by temperature and other factors, such as sea turtles and species that live in mud (Ogunola et al., 2018). These micro- and nano-scale plastics will probably affect all that from sub-organismal to populace levels, influencing standard cycles.

Social changes in marine living beings are proportions of ecological pressure. In this manner, people's behavior should be observed and perceived to comprehend the environmental impacts of miniature and nanoplastics. When in contact with an eco-crown, which can alter a nanoparticle's character, size, capacity, and compatibility with natural elements, bio-macromolecules carried by animals may include smaller-than-expected plastics and nanoplastics. For instance, look at how a protein from *Daphnia magna* hailing part covers nanoplastics and makes an eco-crown, empowering nanoplastics to help cell surface confirmation receptors (Jemec et al., 2016). Adopting microplastics and nanoplastics and the potential harm they may cause to marine life and its habitats are concerns raised by the critical view of eco-crown improvement.

Moreover, planned endeavors among these plastic particles, standard substances, and marine fixations repackage micro- and nanoplastics into fertilizer pellets and marine snows (customary materials more significant than 0.5 mm that can tumble down the water fragment), causing micro- and nanoplastics to total on the ocean bottom. It has been demonstrated that these pellets sink at rates different from average pellets, which may impact how quickly carbon builds up in the ocean in areas with higher levels of microplastic pollution (Rios Mendoza et al., 2018). As benthic bivalve species move closer to the bottom of the ocean, they will feel the effects of residual life as they draw in smaller-than-average and nano plastics. Then, through tunneling, bioturbating species will coordinate with smaller-than-anticipated particles of nanoplastics. These ignored plastic materials will impact a few marine organic entities when they eventually reach the benthos. Several studies have focused on the impact of nano- and microplastics on specific living things, such as the damaging effects of plastic particles on biological diversity and the critical natural strategy of bivalves that overpower particular natural environments.

Blue mussels and European-level shellfish are two distinct varieties of bivalves, and both produced high-thickness polyethylene or polylactic disastrous microplastics (2.5 or 25 g/L) in the extra seawater (Ferreira et al., 2019). Mesocosms are enclosed outdoor environments focusing on the natural habitat under controlled conditions. Because of the natural presence of microplastics, there were antagonistic impacts on the two specific kinds of bivalves. *Mytilus edulis* showed diminished filtration following 50 days of openness to 25 g/L of microplastics; yet neither the consistent plan nor invertebrate species were hurt. However, it was discovered that

Ostrea edulis had a prolonged filtration effect after 50 days of exposure to 2.5 or 25 g/L of microplastics. Because of this effect, the amount of ammonium in the pore water and the biomass of benthic cyanobacteria decreased (Green et al., 2017).

3.1.1 Effects of Microplastics on the Marine Region

A few exploratory studies have predicted the typical toxicological risk of microplastics in the organic frameworks of marine life. The presence of polyvinyl chloride (PVC) on the outer layer of the microalga *Skeletonema costatum* and the subsequent restriction of microalgal improvement and photosynthesis were evidence of the adsorption of microplastics and their constituent parts at the lower level of the marine food web. Microplastic use has been demonstrated to reduce zooplankton productivity and uptake (7.3–30.6 m polystyrene dabs). Microplastics consumed by common mussels can move from their stomachs to their circulatory system and persist for at least 48 days. This could have potential implications for the health and survival of the mussels and other organisms in the marine ecosystem and human health if these contaminated organisms are consumed as food. It highlights the importance of reducing the amount of plastic waste that ends up in the ocean to minimize its impact on the environment and living organisms. Microplastics accumulate in marine life, miss the spine, and cause blockages throughout the stomach-related system, which prevents eating. Microplastics will therefore be dispersed throughout the established food chain by the trackers of those vile, land- and water-proficient, spineless animals. Examinations on Norway lobsters (*Nephrops norvegicus*) further demonstrated the presence of plastic in marine organisms' stomachs. Using scanning electron microscopy (SEM), the foregut (abdomen) and midgut were examined, and three different plastic types were identified: balls, strands, and balls strands. Nano- and microscale balls ought to have been visible in the SEM images.

A portion of the microfilaments' polypropylene structure was revealed by Raman spectroscopy. The effects of ingesting microplastic on marine life have been examined in studies carried out in research facilities. Microplastics are typically consumed by fish, copepods, and lugworms, which led to biochemical and cellular changes that accelerated the decomposition of the lugworms and copepods. Fish and aquatic birds can get a lot of their nutrition from lugworms. Microplastics are likely to blame for this effect on copepods and lugworms, as they reduced adapting behaviors, upset, and ultimately caused lower energy reserves than non-uncovered creatures.

M. edulis enjoys consuming particles based on their size, shape, or thickness, paying little attention to the makeup of their atoms. All methods of caring for bivalves will likely suffer if microplastics are not eliminated before consumption. In another study that examined the effects on marine life, female Japanese medaka (a type of fish) produced fewer chemicals when exposed to polyethylene pre-creation pellets. The growth and incubation rates of European roost were slowed by exposure to microplastic pellets, and the development of copepods (zooplankton) was hindered by exposure to polystyrene microplastics. According to research, European perch copepods exposed to polystyrene microplastic produced less substantial eggs than those produced by non-exposed animals, and these fish were likely already conceived. Due to the susceptibility framework's interactions with micro- and nanoplastics which can be ingested by phagocytic cells, adverse effects are anticipated. The safe framework in mussels is the goal of nanoplastics. Their examination uncovered that delivering hemocytes with PS-NH_2 suspensions in hemolymph serum expanded cell harm and responsive oxygen species (ROS), which was then used to build a nanoplastic bio-crown protein in marine living things.

3.1.2 Effects of Nanoplastics on the Marine Region

The fast advancement of nanoplastics and the debasing of plastic materials into nanoparticles represent an immediate and extreme danger to common frameworks. The surfaces of living things can become adhered to by these particles. Atomic force microscopy (AFM) and exhaust analyses of *Pseudokirchneriella subcapitata* cells have shown that nanoplastics can be adsorbed through the cell mass. The external layer of *Artemia franciscana,* a type of saline water shrimp, and the sensory antennules can hold nanoplastics, preventing them from moving freely. *Chlorella* and *Scenedesmus surfaces* had excellent 20 nm nanoplastic adsorption due to their electrostatic communications with the cellulose portion of the cells. The morphology and motility of the algae were critical factors in this adsorption. The electrostatic associations slowed algal photosynthesis because the nanoplastics interfered with light and wind flow (Ferreira et al., 2019).

Additionally, ROS production was accelerated by nanoplastics, endangering the suitability of the marine food chain. When hazardous hydrophobic poisons like phenanthrene become adsorbed at their surface, nanoplastics act as solid adsorbents for these substances, which they helped collect in *D. magna*. According to the analysis presented, the accumulation of toxic hydrophobic substances in the tissues of marine life is likely facilitated by the adsorption of nanoplastics, which affects how these organisms behave and interact with the public. Essential courses (i.e., blended on the micro- or nano-scale) and optional methods (i.e., provided due to the degradation of plastics) can be used to create micro- and nanoplastics. Since the particles are used by microalgae at the base of the natural food chain, they can be consumed by zooplankton, fish, bivalves like channel feeders, and other organisms (Liang et al., 2021).

A portion of the micro- and nanoplastics are ingested by marine life, connected to surfaces, or developed from their outer members. Their presence can dial back real work and providing care, influence the social way of behaving, cause genetic changes in the handling of the liver and muscles, and lead to morphologic changes in the cerebrum or muscles, at last, reassuring passing. Additionally, many nano- and small-scale plastics are dispersed in waste pellets that sink to the ocean floor, primarily affecting benthic natural life (De-la-Torre, 2020).

4 Micro- and Nanoplastics Pose a Risk to Human Health

4.1 Effects of Microplastics on Human Health

The significant amount of cancer-causing and hazardous manufactured materials used in producing and transporting these plastics is the primary source of concern for human health. The typical person is exposed to microplastics through the various foods consumed during daily eating routines. Individuals typically consume at least 50,000 MP particles annually and earn a familiar story amount. A new examination indicated that there were 0.44 MP/g of sugar, 0.03 MP/g of liquor, 0.11 MP/g of salt, and 0.09 MP/g of separated water.

Food (fish, bundling), water consumption, and air inhalation can all result in MP gathering and tainting in humans, leading to cytotoxicity, an intense reaction like hemolysis and touchiness, and unfavorable resistant reactions. The MPs that people consume through marine life (bivalves, fish, and shellfish) are currently exceptional (Galloway et al., 2017). Humans can ingest the microplastics that fish and shellfish consume. A bioaccumulation illustration of the significant structure influencing human receptivity to MPs was used to analyze tissue samples from mussels to check for microplastic content. According to the study, building confirmation in the gathered domain exposes residents to an average of 123 microplastic particles annually. Microplastic receptivity has been estimated to rise to 4620 particles per capita in countries with higher shellfish consumption, given that diets can differ significantly (Catarino et al., 2018). Microplastics have the potential to disrupt neurotoxicity and resistance.

Additionally, it might affect human skin. Eight people from Japan and Europe were examined, and it was discovered that they had microplastics on their skin (Schwabl et al., 2019). Six people in the group consumed fish after consuming plastic-encased food, drinking water from plastic bottles, and testing positive for at least one type of microplastic. Despite this, the analysis was fundamental, insufficient, and unprepared to reveal the trustworthy source of the plastic particles. Each of the three different salt types discovered in stores contained microplastic. According to the test results, sea salt, lake salt, and stone/well salt each have the highest microplastic concentrations.

Additionally, it has been discovered that the rock salt and sea salt commonly used in Spanish table salt contain microplastics. The most widely regarded microplastic in these two tests was polyethylene terephthalate (Iñiguez et al., 2017). Figure 2 shows micro- and nanoplastic‘s sources, routes, and effects on human health.

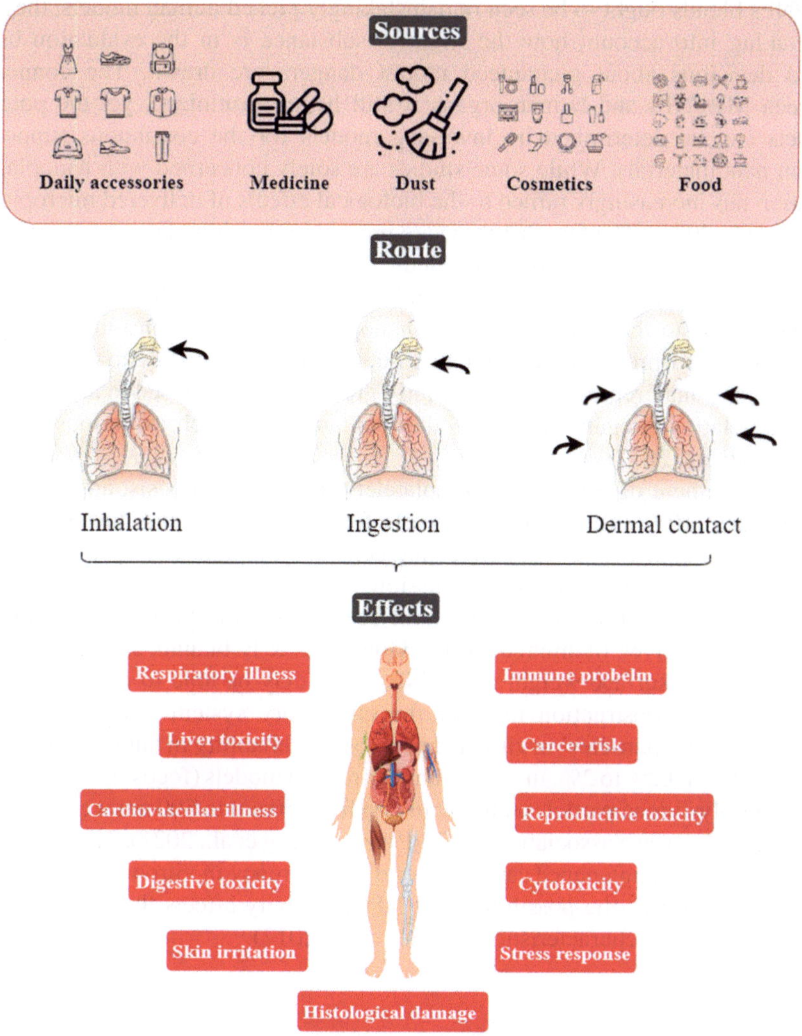

Fig. 2 Source, route, and effects of micro and nanoplastic on human health

4.2 Effects of Nanoplastics on Human Health

The three main pathways for human NP transparency are the gastrointestinal tract, lungs, and skin bundle (Lehner et al., 2019). Nanoplastic can enter the brain, but little is known about the number of molecules that enter the brain or their potential for neurotoxicity (Prüst et al., 2020). There is a severe lack of clarity in the knowledge about the threats to human prosperity due to micro- and nanoplastic exposure. The quality or approach to the investigation is currently the main drawback (Koelmans et al., 2019). The likely harmful quality modes for different size-type

MPs/NPs blends ought to be seen in painstakingly picked human models, they reason, taking into account how the frightful substance is in the evaluation before sound decisions about guaranteed mortal dangers are drawn. The connection between MPs/NPs and human organs is still being scrutinized, yet the potential impacts can be determined by involving models for the continuous support of human nanomaterials. While some studies are solely concerned with nanoplastics, attention has increasingly turned to the biological effects of delivered microplastics (Yang et al., 2021). The fact that these are present in the air and pose a severe threat to human prosperity is widely acknowledged, even though it was difficult to recognize their reality sufficiently due to the inherent difficulties in shielding and assessing them. Due to their tiny size (1 m), nanoparticles can be consumed by people with little to no problem. Additionally, due to their high locale-to-volume ratio, nanoparticles may represent a significantly greater risk if exposed to amplification, bioaccumulation irregularities, and other contaminants, such as various common toxins (Vickers, 2017).

The movement of cells, including platelets and photosynthesis, are affected by nanoplastics (100 nm), which are notable for appearing to pass through organic films. For nanoplastics (NPs), connecting cells and the digestive system is a challenge. For a very long time, oral admissions of polystyrene NPs have been noted, and both in vivo and in vitro studies have been done on gastrointestinal entries of delivered NPs (silver, titanium dioxide) (Goncalves & Bebianno, 2021). Depending on the size and surface weight of the NPs, they may be able to pass through the stomach-related obstruction to enter the circulatory system after development. Polystyrene nanoparticles' in vitro and in vivo bioavailability in humans and rodents increased from 0.2% to 2%. In a few stomach-related models (focusing on 1.5–10%), different polystyrene particles ranging in size from 50 nm to 500 nm were examined using various surface associations and NP sizes (Liang et al., 2021). The unexpected increase in iron maintenance caused by oral transparency in vitro to 50 nm polystyrene particles raises the possibility that NP receptivity affects the gastrointestinal epithelium's block characteristics (Mahler et al., 2012).

5 Nano- and Microplastic Biodegradation

In any case, one of the anticipated alternative strategies for plastics deterioration, the degrading of MNPs using microorganisms, has been the focus of research over the last few years. MNPs are challenging to corrupt because they are quietly present in the climate, requiring the utilization of powerful waste administration innovation, which can immensely assist in combating the terrible impacts (Pathak, 2017). Biodegradation, hydrolysis, photodegradation, and thermooxidative debasement are the four main processes that cause microplastic corruption to complete its cycle. Typically, hydrolysis will be essential in affecting plastic biodegradation, given the presence of water in fluid circumstances. However, in earthbound living environments, heat (thermo oxidative) and light (photodegradation) are thought to impact life forms' ability to degrade plastic polymers significantly. However, it has been observed that the communication of a considerable number of regular variables presents a massive number of complexities in the degradation of plastic polymers under various circumstances. For instance, it has been suggested that the rate of hydrolysis of most plastic polymers in the sea is unimportant. The lower temperature and greater accessibility of oxygen in seawater also significantly reduce the photodegradation impact.

Archaea, microbes, and parasites, among other (miniature) creatures, break down complex polymeric materials into harmless substances that can be reintroduced into the biogeochemical cycles. The specific factors contributing to the microbial debasement of these MNPs are currently poorly understood. Scientists have proposed various mechanisms for the degradation of microplastics in the environment. Biodeterioration refers to the breakdown of plastic materials through the activity of microorganisms, such as bacteria and fungi, that can break down the polymer chains in plastic. Biofragmentation is a similar process in which larger plastic particles are broken down into smaller fragments by microorganisms or physical processes such as UV radiation or wave action. Assimilation refers to the uptake of microplastics by organisms, such as plankton or filter feeders, which can then break down the plastic particles through enzymatic activity in their digestive systems. Mineralization is the final step in the degradation process, in which the breakdown products of plastic are converted into simple organic and inorganic compounds, such as carbon dioxide and water (Amobonye et al., 2021).

Like other types of plastic, the biodeterioration of MNPs typically begins with surface colonization and microbial attachment to the polymer's external layer. In this context, synthetics, light, and temperature, along with the substance and actual activities of the microorganism, are enhanced by the training of other organic specialists to change the properties of the MNPs (Habib et al., 2020). As a result, the extracellular proteins of the life forms depolymerize the decayed plastics and create free extremists. Enzymatic degradation of plastics can produce oligomers or monomers that can be reused in making new plastic materials. For example, the breakdown of polyethylene terephthalate (PET) plastic through enzymatic activity can yield terephthalic acid and ethylene glycol, which can then be used as feedstocks to produce new PET plastic. Numerous systems in various life forms, including active and aloof vehicles, have been shown to work with the digestion of less complex intermediates into microorganisms (Amobonye et al., 2021). The ATP-constricting tape protein assembly, monooxygenases, and porins have all been employed as the assimilative vehicles for the intermediates of plastic degradation. The final step involves intracellular impetuses, which convert ingested metabolites into oxidized ones that contain CH_4, H_2O, CO_2, and N_2. In any case, it has been observed that the final results shaped are a part of the organism's breath state. When oxygen acts as the electron acceptor in stressful situations, CO_2 and H_2O are produced. Different combinations, such as those containing carbon dioxide, iron, manganese, nitrates, and sulfates, may ultimately function as the last electron acceptors in anaerobic conditions (Ahmed et al., 2018).

Different MNPs can be destroyed by microorganisms at the mineralization site, including tiny creatures and developments (Sánchez, 2020). Numerous studies have demonstrated how microbes, especially those from marine environments, can colonize various corrupt MNPs. *Bacillus* and *Pseudomonas* species were typical of the nearby bacterial neighborhood's new colonization of MNPs in an estuary (Wu et al., 2020). In earlier research, these two genera have consistently been found to have the most remarkable potential for plastic biodegradation. Colonizing microplastics in a waterway by various local bacterial area groups was also demonstrated, revealing the possibility of plastic-corrupting organisms (Niu et al., 2021). Recently, there has also been some interest in the potential role of infectious organic entities in the degradation of MNPs based on oil (Sánchez, 2020). For instance, two parasites studied for their ability to degrade MNPs are *Aspergillus flavus*, isolated from a bug's stomach, and *Zalerion maritimum*, a marine organism (Zhang et al., 2020). However, the natural significance of organisms' biodegradative activity on MNPs and plastics is still largely unknown. According to various creators, the rate of microbial corruption is incredibly low, significantly impacting efforts to restore the environment. This has expanded the requirement for using momentary exploratory outcomes to anticipate long-haul contamination pathways and utilizing computational methods to recreate the defilement of MNPs (Chamas et al., 2020).

6 Effects of Micro- and Nanoplastics on the Economy

MPs and NPs significantly impact people's ability to make a living. A decrease in a region's educational, compositional, or sporting standards and risks to human health are examples of the social harm caused by microplastics and nanoplastics. Financial damage includes direct expenses and pay loss caused by MPs and NPs as marine litter, affecting various marine areas, including hydroponics, horticulture, fisheries, transportation, influence age, neighborhood specialists, everyday use, and the travel industry. A financial loss is also caused by the impact of corroding marine litter on organic structure labor and products. Figure 3 shows the socioeconomical impact of micro- and nanoplastics on the environment.

6.1 Effects of Microplastics on the Economy

There is critical worry about gathering tiny bits of MP in light of the sluggish organic and compound debasement rate and their high commonness. A path for synthetics to travel from plastics to the food supply is provided by MP ingestion. MPs pose a severe threat to the fashion sensibilities of nearby beachgoers and tourists, primarily when they transport clean, sewage-related, and medical waste that could also endanger their health (Ström et al., 2018). The diving industry may suffer

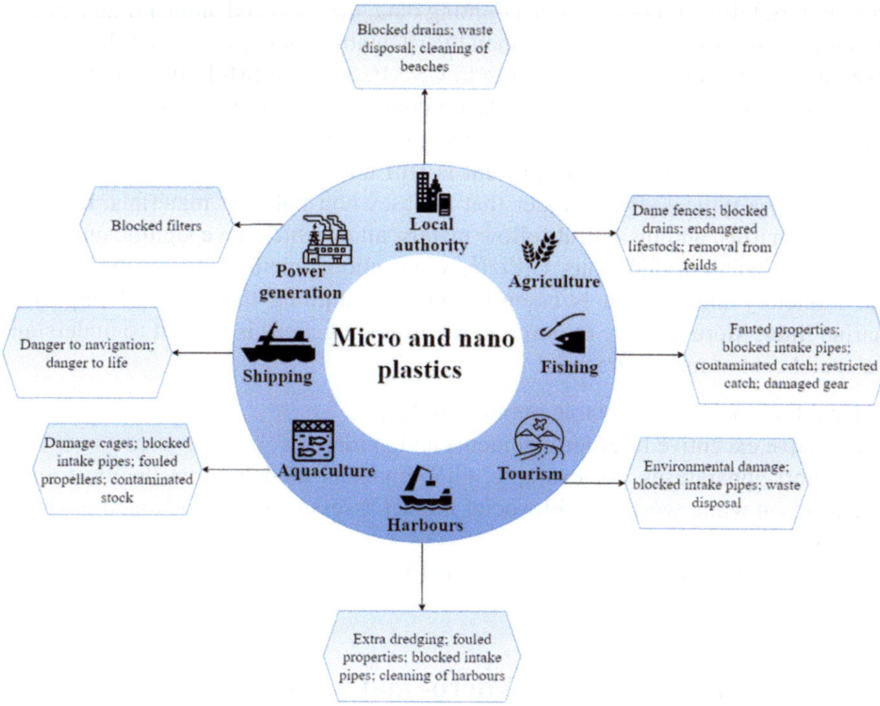

Fig. 3 Impact of micro and nano-scale plastics on the economy

if there is a significant concentration of trash on coral reefs and the ocean floor because divers avoid heavily polluted areas. Expenses related to agriculture and search and rescue efforts to aid clogged transportation engines or hindered affirmations are regarded as legitimate costs. More than 45% of master anglers in the eastern United States had their propellers tangled, more than 30% had their stuff halted, and more than 35% had plastic trash obstructing their engine's cooling structure. Fix costs and missed fishing days, especially for limited, uninsured fishing, can be fundamental (Ström et al., 2018). The financial burden of plastic waste on marine businesses is alleviated in some ways, but spatial inclusion may suffer because information gathering requires intentional actions.

6.2 *Effects of Nanoplastics on the Economy*

Ecological NP comes with some dangers and challenges. Similar to the travel industry, NPs incur high and occasionally elusive costs frequently not covered by polluters or potential producers. NP has a significant impact on temporary vulnerability and supportability. According to accounting data, the surveyed annual financial cost of marine waste to Asia-Pacific Financial Participation increased eight-fold between 2009 and 2015, from US $1.26 billion to US $10.8 billion (McIlgorm et al., 2020). Contamination of soil and air with plastic-containing materials was another financial consequence. It is still difficult to categorize, gather, and identify nanoparticle evidence, and whether NPs are present is still uncertain. Most experts agree that NPs can be a typical stress trigger that releases non-polymer materials, naturally occurring harms, and insightful follow metals, all of which have significant regular and human health implications. Utilizing conventional methods for fusing practical and monetary information is essential when examining the financial impact of marine litter. More specific research in the following areas is needed to understand the social cost of marine trash:

- Establish socially acceptable public standards
- Effective executive litter arrangements and guidelines
- Support apparatuses, for example, GIS, financial models, float demonstrating, and so on while working with social impact assessments and advancing administration endeavors
- Make marine garbage influence models for human wellbeing

7 Systems for Regulating Micro- and Nanoplastics

The quick progression of plastics throughout a pivotal period will probably make the surge of MP and NP particles stay high for quite a while. Understanding the origins and social gatherings of these plastics and NPs is essential to lessen MPs/NPs' responsibility to average marine resources and people. Additionally, raising awareness of the problems with small-scale and nanoplastics will be aided by increasing public mindfulness through training in both private and public settings. Effective management of the marine trash issue can benefit from outside input and collaboration among different stakeholders. The Australian Public Spot for Ocean Resources and Security report highlights the importance of a multi-disciplinary and collaborative approach to tackling the issue of marine debris (McIlgorm et al., 2020):

- Using specialized litter traps.
- Taking care of marine debris "problem areas".
- Methods and activity plans for managing public waste.
- Adopting plans for expanded maker obligation (extended producer responsibility, EPR) that include customers.
- Establishing guidelines to reduce waste, using financial tools to advance reuse, and mobilizing financial resources to draw attention to the secret area.

7.1 *Handling Microplastics in a Controlled Manner*

Many organizations were motivated to create the executive's rules due to concern regarding MPs. For example, the United Nations Environment Programme (UNEP) requested immediate action to reduce the trade of MPs and NPs into oceans after discovering that plastics are consumed by significant marine natural substances, causing harm. UNEP has supported a program involving a sizable population (>40 million) from 120 different nations to raise awareness of the issue and promote decreased plastic use. Marine conditions are heavily impacted by MP pollution, and their reality is generally growing. Due to their small size and reduced visibility compared to large-scale plastics, MPs are very challenging to remove physically. However, biodegradation is a potential method to remove the MPs in a marine environment. Microorganisms play a significant role in biodegradation, leading to corruption in MPs. In this cycle, MPs serve as a carbon and energy source for organisms. The capacity of creatures like *Staphylococcus* sp., *Pseudomonas* sp., and *Bacillus* sp. to sully soil-attracted polyethylene microplastics has made sense. Two or three of the various degraders of polyethylene terephthalate and polystyrene mentioned in the composition include *Staphylococcus aureus*, *Aspergillus Niger*, *Rhodococcus*, *Pseudomonas aeruginosa*, *Bacillus subtilis*, and *Streptococcus pyogenes*. Microorganisms degrade microplastics by forming biofilms that cause the polystyrene to deteriorate and extracellular catalysts that cause a variety of polymers to degrade (Caruso, 2015; Singh et al., 2016).

7.2 *Handling Nanoplastics in a Controlled Manner*

Large-scale research projects are anticipated to more easily understand NP risks. However, the first step is gathering thorough information on marine, freshwater, and terrestrial openness. Environmental NP focuses are currently just gauges because there is limited information due to the lack of explicit logical techniques (Schwaferts et al., 2019). The development of emerging remediation, government strategy, instruction, and care comprise the core components of NP alleviation methods. The development of NP remediation strategies has made less progress, but recent analyses have suggested a few possible directions. Consistent substituting of traditional nonbiodegradable NPs with eco-accommodating biodegradable NPs is made conceivable by headways in biotechnology (Silva et al., 2018). Extraordinary chemicals, organisms, and growths may also be necessary for a successful and efficient removal process (Pico et al., 2019). The effective way to reduce NPs in wastewater treatment facilities is to stop them from growing in marine environments. Although pretreatment techniques (coagulation, thickness, and layer package) effectively remove MPs from consumable water, more research is necessary to determine whether they are also effective in eliminating NPs (Enfrin et al., 2019). According to some theories, altering the soil, such as increasing biochar, can reduce the growth of NP in porous media, lowering the risk associated with the change (Tong et al., 2020). In other studies, the idea that inorganic or natural substances can effectively aid NP assortment in a wastewater treatment process, resulting in compelling NP parcel, has been explored. Extending remediation techniques for various toxic substances to NPs is also encouraged. To immobilize soil, for instance, "green" remediation materials like biochar and soil minerals have proven effective (Wang et al., 2021).

8 Conclusion and Prospective

Plastics are generally utilized and influence our everyday presence; they might perhaps be the most vital material family in the twenty-first century. Even though they offer substantial monetary benefits, plastics are essential wellsprings of natural weakening. The particular mass plastic material, which has specific physical and substance properties, causes the plastic to become consumed in MPs and NPs. Minuscule, splendid particles known as MPs and NPs, are brought into the marine climate through land- and ocean-based exercises. Most MPs and NPs start ashore and travel to the oceans via land, air, or water. Because of the NPs' dim qualities, NPs have just been found in animal tests; in any case, MPs have been tracked down in seawater, improvement, and beaches in critical pieces of the nations of Asia, Africa, America, and Europe. MPs and NPs significantly affect marine life and human government assistance. Aquatic species can easily consume MPs because they are so small and light, resulting in various problems with the tissues, circulatory systems, and mind. Standing apart from MPs, NPs can be distributed into animal bodies significantly more profitably and transferred between different organs.

Additionally, they provide even more special adsorbents and pollution transportation and demonstrate threats to the prosperity of humans and animals. Despite this, NPs have frequently been over-searched in examinations due to their enigmatic existence, testing and examination convention restrictions, and units of fundamental boundaries like particulate matter overflow. MPs and NPs have significant, frequently irrational financial effects on industries like the travel sector, creating costs typically not attributed to producers or potential polluters. MPs and NPs disapprove of the biological system's value for recreation, style, and legacy, and it seems reasonable that these impurities will continue to grow significantly, given that it is impossible to imagine a world without these contaminants. However, it is impossible to prevent the adverse effects of MPs and NPs without addressing general society, the financial sector, the travel industry, and waste management companies. Also, strategies for dealing with studies concentrating on the organisms that may degrade marine plastics are being considered. Once contaminated areas have been cleaned up, such microorganisms can be used. Infecting MPs and NPs with tiny living things offers a practical and generally secure execution system that will enable the association of MPs and NPs with no problematic results.

References

(CONTAM), E.P. on C. in the F.C. (2016). Presence of microplastics and nanoplastics in food, with particular focus on seafood. *EFSA Journal, 14*, e04501.

Ahmed, T., Shahid, M., Azeem, F., Rasul, I., Shah, A. A., Noman, M., Hameed, A., Manzoor, N., Manzoor, I., & Muhammad, S. (2018). Biodegradation of plastics: Current scenario and future prospects for environmental safety. *Environmental Science and Pollution Research, 25*, 7287–7298.

Alves, T. F. R., Morsink, M., Batain, F., Chaud, M. V., Almeida, T., Fernandes, D. A., da Silva, C. F., Souto, E. B., & Severino, P. (2020). Applications of natural, semi-synthetic, and synthetic polymers in cosmetic formulations. *Cosmetics, 7*, 75.

Amobonye, A., Bhagwat, P., Singh, S., & Pillai, S. (2021). Plastic biodegradation: Frontline microbes and their enzymes. *Science of the Total Environment, 759*, 143536.

An, L., Liu, Q., Deng, Y., Wu, W., Gao, Y., & Ling, W. (2020). Sources of microplastic in the environment. In *Microplastics in Terrestrial environments: Emerging contaminants and major challenges* (pp. 143–159). Springer.

Avila, C., Angulo-Preckler, C., Martín-Martín, R. P., Figuerola, B., Griffiths, H. J., & Waller, C. L. (2020). Invasive marine species discovered on non–native kelp rafts in the warmest Antarctic Island. *Scientific Reports, 10*, 1639.

Begines, B., Ortiz, T., Pérez-Aranda, M., Martínez, G., Merinero, M., Argüelles-Arias, F., & Alcudia, A. (2020). Polymeric nanoparticles for drug delivery: Recent developments and future prospects. *Nanomaterials, 10*, 1403.

Brewer, A., Dror, I., & Berkowitz, B. (2020). The mobility of plastic nanoparticles in aqueous and soil environments: A critical review. *ACS ES&T Water, 1*, 48–57.

Burrows, S. D., Frustaci, S., Thomas, K. V., & Galloway, T. (2020). Expanding exploration of dynamic microplastic surface characteristics and interactions. *TrAC Trends in Analytical Chemistry, 130*, 115993.

Cai, Y., Yang, T., Mitrano, D. M., Heuberger, M., Hufenus, R., & Nowack, B. (2020). Systematic study of microplastic fiber release from 12 different polyester textiles during washing. *Environmental Science & Technology, 54*, 4847–4855.

Caruso, G. (2015). Plastic degrading microorganisms as a tool for bioremediation of plastic contamination in aquatic environments. *Journal of Pollution Effects & Control, 3*, 1–2.

Catarino, A. I., Macchia, V., Sanderson, W. G., Thompson, R. C., & Henry, T. B. (2018). Low levels of microplastics (MP) in wild mussels indicate that MP ingestion by humans is minimal compared to exposure via household fibres fallout during a meal. *Environmental Pollution, 237*, 675–684.

Chamas, A., Moon, H., Zheng, J., Qiu, Y., Tabassum, T., Jang, J. H., & Suh, S. (2020). Degradation Rates of Plastics in the Environment. *ACS Sustainable Chemistry & Engineering, 8*, 3494–3511.

Ciriminna, R., & Pagliaro, M. (2020). Biodegradable and compostable plastics: A critical perspective on the dawn of their global adoption. *ChemistryOpen, 9*, 8–13.

Dąbrowska, J., Sobota, M., Świąder, M., Borowski, P., Moryl, A., Stodolak, R., Kucharczak, E., Zięba, Z., & Kazak, J. K. (2021). Marine waste—Sources, fate, risks, challenges and research needs. *International Journal of Environmental Research and Public Health, 18*, 433.

Dalili, A., Samiei, E., & Hoorfar, M. (2019). A review of sorting, separation and isolation of cells and microbeads for biomedical applications: Microfluidic approaches. *Analyst, 144*, 87–113.

De-la-Torre, G. E. (2020). Microplastics: An emerging threat to food security and human health. *Journal of Food Science and Technology, 57*, 1601–1608.

Ding, N., An, D., Yin, X., & Sun, Y. (2020). Detection and evaluation of microbeads and other microplastics in wastewater treatment plant samples. *Environmental Science and Pollution Research, 27*, 15878–15887.

Du, J., Zhou, Q., Li, H., Xu, S., Wang, C., Fu, L., & Tang, J. (2021). Environmental distribution, transport and ecotoxicity of microplastics: A review. *Journal of Applied Toxicology, 41*, 52–64.

Enfrin, M., Dumée, L. F., & Lee, J. (2019). Nano/microplastics in water and wastewater treatment processes–origin, impact and potential solutions. *Water Research, 161*, 621–638.
Ferreira, I., Venâncio, C., Lopes, I., & Oliveira, M. (2019). Nanoplastics and marine organisms: What has been studied? *Environmental Toxicology and Pharmacology, 67*, 1–7. https://doi.org/10.1016/j.etap.2019.01.006
Fu, W., Min, J., Jiang, W., Li, Y., & Zhang, W. (2020). Separation, characterization and identification of microplastics and nanoplastics in the environment. *Science of the Total Environment, 721*, 137561.
Galloway, T., Cole, M., & Lewis, C. (2017). Interactions of microplastic debris throughout the marine ecosystem. *Nature Ecology & Evolution, 1*, 0116.
Gaylarde, C. C., Neto, J. A. B., & da Fonseca, E. M. (2021). Paint fragments as polluting microplastics: A brief review. *Marine Pollution Bulletin, 162*, 111847.
Goncalves, J. M., & Bebianno, M. J. (2021). Nanoplastics impact on marine biota: A review. *Environmental Pollution, 273*, 116426.
Green, D. S., Boots, B., O'Connor, N. E., & Thompson, R. (2017). Microplastics affect the ecological functioning of an important biogenic habitat. *Environmental Science & Technology, 51*, 68–77.
Habib, S., Iruthayam, A., Abd Shukor, M. Y., Alias, S. A., Smykla, J., & Yasid, N. A. (2020). Biodeterioration of untreated polypropylene microplastic particles by Antarctic bacteria. *Polymers (Basel), 12*, 2616.
Hernandez-Gonzalez, A., Saavedra, C., Gago, J., Covelo, P., & Santos, M. B. (2016). Analysis and quantification of microplastics in the stomachs of common dolphin (Delphinus delphis) stranded on the Galician coasts (NW Spain). In *MICRO 2016: Fate and impact of microplastics in marine ecosystems* (p. 123). Elsevier.
Horton, A. A., Walton, A., Spurgeon, D. J., Lahive, E., & Svendsen, C. (2017). Microplastics in freshwater and terrestrial environments: Evaluating the current understanding to identify the knowledge gaps and future research priorities. *Science of the Total Environment, 586*, 127–141.
Iñiguez, M. E., Conesa, J. A., & Fullana, A. (2017). Microplastics in Spanish table salt. *Scientific Reports, 7*, 1–7.
Jemec, A., Horvat, P., Kunej, U., Bele, M., & Kržan, A. (2016). Uptake and effects of microplastic textile fibers on freshwater crustacean Daphnia magna. *Environmental Pollution, 219*, 201–209.
Karkanorachaki, K., Kiparissis, S., Kalogerakis, G. C., Yiantzi, E., Psillakis, E., & Kalogerakis, N. (2018). Plastic pellets, meso-and microplastics on the coastline of Northern Crete: Distribution and organic pollution. *Marine Pollution Bulletin, 133*, 578–589.
Kashiwada, S. (2006). Distribution of nanoparticles in the see-through medaka (*Oryzias latipes*). *Environmental Health Perspectives, 114*, 1697–1702.
Kitahara, K.-I., & Nakata, H. (2020). Plastic additives as tracers of microplastic sources in Japanese road dusts. *Science of the Total Environment, 736*, 139694.
Koelmans, B., Pahl, S., Backhaus, T., Bessa, F., van Calster, G., Contzen, N., Cronin, R., Galloway, T., Hart, A., & Henderson, L. (2019). *A scientific perspective on microplastics in nature and society*. SAPEA.
Lehner, R., Weder, C., Petri-Fink, A., & Rothen-Rutishauser, B. (2019). Emergence of nanoplastic in the environment and possible impact on human health. *Environmental Science & Technology, 53*, 1748–1765.
Li, J., Huang, W., Jiang, R., Han, X., Zhang, D., & Zhang, C. (2020). Are bacterial communities associated with microplastics influenced by marine habitats? *Science of the Total Environment, 733*, 139400.
Liang, B., Zhong, Y., Huang, Y., Lin, X., Liu, J., Lin, L., Hu, M., Jiang, J., Dai, M., & Wang, B. (2021). Underestimated health risks: Polystyrene micro- and nanoplastics jointly induce intestinal barrier dysfunction by ROS-mediated epithelial cell apoptosis. *Particle and Fibre Toxicology, 18*, 20.
Magalhães, S., Alves, L., Medronho, B., Romano, A., & da Rasteiro, M. G. (2020). Microplastics in ecosystems: From current trends to bio-based removal strategies. *Molecules, 25*, 3954.

Mahler, G. J., Esch, M. B., Tako, E., Southard, T. L., Archer, S. D., Glahn, R. P., & Shuler, M. L. (2012). Oral exposure to polystyrene nanoparticles affects iron absorption. *Nature Nanotechnology, 7*, 264–271.

McIlgorm, A., Raubenheimer, K., & McIlgorm, D. E. (2020). *Update of 2009 APEC report on economic costs of marine debris to APEC economies*. University Wollongong Wollongong.

Nel, H. A., Dalu, T., Wasserman, R. J., & Hean, J. W. (2019). Colour and size influences plastic microbead underestimation, regardless of sediment grain size. *Science of the Total Environment, 655*, 567–570.

Niu, L., Li, Y., Li, Y., Hu, Q., Wang, C., Hu, J., Zhang, W., Wang, L., Zhang, C., & Zhang, H. (2021). New insights into the vertical distribution and microbial degradation of microplastics in urban river sediments. *Water Research, 188*, 116449.

Ogunola, O. S., Onada, O. A., & Falaye, A. E. (2018). Mitigation measures to avert the impacts of plastics and microplastics in the marine environment (a review). *Environmental Science and Pollution Research, 25*, 9293–9310.

Pathak, V. M. (2017). Navneet Review on the current status of polymer degradation: a microbial approach. Bioresour. *Bioprocess. 4*, 15. https://doi.org/10.1186/s40643-017-0145-9

Petroody, S. S. A., Hashemi, S. H., & van Gestel, C. A. M. (2020). Factors affecting microplastic retention and emission by a wastewater treatment plant on the southern coast of Caspian Sea. *Chemosphere, 261*, 128179.

Pico, Y., Alfarhan, A., & Barcelo, D. (2019). Nano-and microplastic analysis: Focus on their occurrence in freshwater ecosystems and remediation technologies. *TrAC Trends in Analytical Chemistry, 113*, 409–425.

Priya, A. K., Jalil, A. A., Dutta, K., Rajendran, S., Vasseghian, Y., Karimi-Maleh, H., & Soto-Moscoso, M. (2022). Algal degradation of microplastic from the environment: Mechanism, challenges, and future prospects. *Algal Research, 67*, 102848. https://doi.org/10.1016/j.algal.2022.102848

Priya, A., Anusha, G., Thanigaivel, S., Karthick, A., Mohanavel, V., Velmurugan, P., Balasubramanian, B., Ravichandran, M., Kamyab, H., Kirpichnikova, I. M., & Chelliapan, S. (2023). Removing microplastics from wastewater using leading-edge treatment technologies: A solution to microplastic pollution—A review. *Bioprocess and Biosystems Engineering, 46*, 309–321. https://doi.org/10.1007/s00449-022-02715-x

Prüst, M., Meijer, J., & Westerink, R. H. S. (2020). The plastic brain: Neurotoxicity of micro- and nanoplastics. *Particle and Fibre Toxicology, 17*, 1–16.

Rios Mendoza, L. M., Karapanagioti, H., & Álvarez, N. R. (2018). Micro(nanoplastics) in the marine environment: Current knowledge and gaps. *Current Opinion in Environmental Science & Health, 1*, 47–51. https://doi.org/10.1016/j.coesh.2017.11.004

Sánchez, C. (2020). Fungal potential for the degradation of petroleum-based polymers: An overview of macro-and microplastics biodegradation. *Biotechnology Advances, 40*, 107501.

Schwabl, P., Köppel, S., Königshofer, P., Bucsics, T., Trauner, M., Reiberger, T., & Liebmann, B. (2019). Detection of various microplastics in human stool: A prospective case series. *Annals of Internal Medicine, 171*, 453–457.

Schwaferts, C., Niessner, R., Elsner, M., & Ivleva, N. P. (2019). Methods for the analysis of submicrometer-and nanoplastic particles in the environment. *TrAC Trends in Analytical Chemistry, 112*, 52–65.

Shiu, R.-F., Vazquez, C. I., Tsai, Y.-Y., Torres, G. V., Chen, C.-S., Santschi, P. H., Quigg, A., & Chin, W.-C. (2020). Nano-plastics induce aquatic particulate organic matter (microgels) formation. *Science of the Total Environment, 706*, 135681.

Šilc, U., Küzmič, F., Caković, D., & Stešević, D. (2018). Beach litter along various sand dune habitats in the southern Adriatic (E Mediterranean). *Marine Pollution Bulletin, 128*, 353–360.

Silva, A. B., Costa, M. F., & Duarte, A. C. (2018). Biotechnology advances for dealing with environmental pollution by micro (nano) plastics: Lessons on theory and practices. *Current Opinion in Environmental Science & Health, 1*, 30–35.

Singh, G., Singh, A. K., & Bhatt, K. (2016). Biodegradation of polythenes by bacteria isolated from soil. *International Journal of Research and Development in Pharmacy & Life Sciences, 5*, 2056–2062.

Smyth, K., Drake, J., Li, Y., Rochman, C., Van Seters, T., & Passeport, E. (2021). Bioretention cells remove microplastics from urban stormwater. *Water Research, 191*, 116785.

Ström, G., Albihn, A., Jinnerot, T., Boqvist, S., Andersson-Djurfeldt, A., Sokerya, S., Osbjer, K., San, S., Davun, H., & Magnusson, U. (2018). Manure management and public health: Sanitary and socio-economic aspects among urban livestock-keepers in Cambodia. *Science of the Total Environment, 621*, 193–200.

Tiwari, N., Santhiya, D., & Sharma, J. G. (2020). Microbial remediation of micro-nano plastics: Current knowledge and future trends. *Environmental Pollution, 265*, 115044.

Tong, M., He, L., Rong, H., Li, M., & Kim, H. (2020). Transport behaviors of plastic particles in saturated quartz sand without and with biochar/Fe_3O_4-biochar amendment. *Water Research, 169*, 115284.

van Kleunen, M., Brumer, A., Gutbrod, L., & Zhang, Z. (2020). A microplastic used as infill material in artificial sport turfs reduces plant growth. *Plants, People, Planet, 2*, 157–166.

Vickers, N. J. (2017). Animal communication: When i'm calling you, will you answer too? *Current Biology, 27*, R713–R715.

Wang, F., Wang, B., Qu, H., Zhao, W., Duan, L., Zhang, Y., Zhou, Y., & Yu, G. (2020). The influence of nanoplastics on the toxic effects, bioaccumulation, biodegradation and enantioselectivity of ibuprofen in freshwater algae *Chlorella pyrenoidosa*. *Environmental Pollution, 263*, 114593.

Wang, L., Wu, W.-M., Bolan, N. S., Tsang, D. C. W., Li, Y., Qin, M., & Hou, D. (2021). Environmental fate, toxicity and risk management strategies of nanoplastics in the environment: Current status and future perspectives. *Journal of Hazardous Materials, 401*, 123415.

Worm, B., Lotze, H. K., Jubinville, I., Wilcox, C., & Jambeck, J. (2017). Plastic as a persistent marine pollutant. *Annual Review of Environment and Resources, 42*, 1–26.

Wright, S. L., Ulke, J., Font, A., Chan, K. L. A., & Kelly, F. J. (2020). Atmospheric microplastic deposition in an urban environment and an evaluation of transport. *Environment International, 136*, 105411.

Wu, N., Zhang, Y., Zhao, Z., He, J., Li, W., Li, J., Ma, Y., & Niu, Z. (2020). Colonization characteristics of bacterial communities on microplastics compared with ambient environments (water and sediment) in Haihe Estuary. *Science of the Total Environment, 708*, 134876.

Yang, L., Zhang, Y., Kang, S., Wang, Z., & Wu, C. (2021). Microplastics in soil: A review on methods, occurrence, sources, and potential risk. *Science of the Total Environment, 780*, 146546.

Zhang, J., Gao, D., Li, Q., Zhao, Y., Li, L., Lin, H., Bi, Q., & Zhao, Y. (2020). Biodegradation of polyethylene microplastic particles by the fungus Aspergillus flavus from the guts of wax moth Galleria mellonella. *Science of the Total Environment, 704*, 135931.

Impact of Microplastics and Nanoplastics in the Aquatic Environment

Sirat Sandil and Gyula Zaray

1 Introduction

Over the last 70 years, human population expansion and the emergence of industrialized civilizations have resulted in a significant increase in plastic consumption. Plastics have an unparalleled demand in domestic and industrial applications ascribable to their characteristics, including lightweight, low cost, durability, and resistance to shock, corrosion, and chemicals (Yan et al., 2019; Tien et al., 2020; Dalu et al., 2021). Simultaneously, plastic pollution has surged steadily, with global plastic manufacturing growing from 1.7 to 335 million tons (Karthik et al., 2018) with over 350 million tons of plastics manufactured on an annual basis (Zhang et al., 2022). Nevertheless, the rate of plastic recycling remains insufficient, with only 10–30% of plastics being recycled globally. The bulk ends up in landfills and various environmental locations, and approximately 10% of these plastics wind up in the oceans annually (Free et al., 2014; Zhang et al., 2022). Plastics are considered to be the most prevalent and persistent pollutants in aquatic environments such as rivers, lakes, and oceans, arriving via an array of land-based channels, notably plastic litter, wastewater effluent, industries, agricultural runoff, stormwater runoff,

S. Sandil (✉)
Institute of Aquatic Ecology, Centre for Ecological Research, Budapest, Hungary

National Laboratory for Water Science and Water Security, Institute of Aquatic Ecology, Centre for Ecological Research, Budapest, Hungary
e-mail: sirat.sandil@ecolres.hu

G. Zaray
Institute of Aquatic Ecology, Centre for Ecological Research, Budapest, Hungary

National Laboratory for Water Science and Water Security, Institute of Aquatic Ecology, Centre for Ecological Research, Budapest, Hungary

Institute of Chemistry, Eötvös Loránd University, Budapest, Hungary

S. A. Bhat et al. (eds.), *Management of Micro and Nano-plastics in Soil and Biosolids*,
https://doi.org/10.1007/978-3-031-51967-3_2

riverine transport, atmospheric deposition, in addition to sea-based ones including fishing and shipping operations (Karthik et al., 2018; Crew et al., 2020; Schrank et al., 2022).

In the natural environment, plastics are extremely resilient and do not degrade directly (Rios et al., 2007); instead, they are fragmented into smaller pieces by physical, chemical, and biological factors (Binelli et al., 2020). The resultant microscopic particles are termed microplastics (MPs) and nanoplastics (NPs). While all plastic fragments <5 mm are considered MPs, there is ongoing debate concerning the size range of NPs, with some classifying those between 1 and 100 nm as NPs and others between 1 and 999 nm as NPs (Desforges et al., 2014; Ter Halle et al., 2017; Davranche et al., 2020). When MPs or NPs are purposely produced in the micro- or nano-size range for usage (abrasive beads in industrial applications and scrubbers in cosmetics), they are referred to as primary MPs or NPs, whereas when they are produced by the disintegration of bigger plastic items, such as bottles, packaging material, and fishing nets, as well as the deterioration of synthetic fabrics, they are cataloged as secondary MPs or NPs (Song et al., 2015; Maes et al., 2017). In aquatic ecosystems, MPs and NPs are derived largely from terrestrial anthropogenic sources that include textile manufacturing, industrial production, personal care goods and cosmetics, everyday plastic consumption, agriculture, and wastewater treatment plants (WWTPs). Maritime activities like shipping, fishing, and transportation are additional contributors to these particles within aquatic environments (Karthik et al., 2018; Eo et al., 2021). MPs and NPs do not biodegrade easily and can linger in the environment for ages, making them a major predicament in aquatic ecosystems due to their potentially detrimental impacts on ecosystems and organisms. They can traverse vast distances via water or wind, causing pollution in remote as well as pristine regions far from their sources (Eo et al., 2021). Microplastics have been reported in lakes, rivers, and oceans, as well as in far-flung mountainous and polar regions (Goldstein et al., 2013; Pan et al., 2019; Eo et al., 2021). The world's oceans act as a significant repository for plastic debris (Eriksen et al., 2013), receiving plastic from atmospheric deposition, marine activities, riverine runoff, and wastewater effluent (Cai et al., 2018). An estimated 1.15–2.41 million tons of plastic is introduced into the oceans yearly through rivers (Schrank et al., 2022). In recent times, MP presence has been documented in drinking water, salt, food, soil, and even human bodies, potentially affecting ecological functions and human health (Zhang et al., 2022).

MPs and NPs can subsist in aquatic ecosystems for extended periods courtesy of their stability and durability (Andrady, 2011). Recent investigations have discovered large quantities of MPs and NPs in aquatic habitats, prompting focus on their probable toxicity and ecological risks (Dalu et al., 2021). The environmental impact of MPs and NPs includes ingestion, accumulation of contaminants, and alteration of habitat by the introduction of hard substrate (Goldstein et al., 2013). Owing to their minute size and color, MPs and NPs resemble natural food and can be consumed by numerous aquatic organisms, such as zooplankton, mollusks, fish, and other marine creatures (Andrady, 2011; Nelms et al., 2018; Fraser et al., 2020). Ingestion of MPs might result in blockage of digestive tracts and diminished dietary intake efficiency, thereby resulting in poor nutrient intake, malnutrition, impaired growth and reproduction, and even death (Du et al., 2021; Zhang et al., 2021). Besides ingestion, MPs and NPs can cause physical harm to aquatic organisms by entangling and damaging delicate tissues and altering behavior. MPs and NPs possess the capacity to migrate and amass in the food chain; when smaller species consume MPs and NPs, these microscopic particles can ascend the food chain as predators consume their prey. This can result in the build-up of plastics in higher trophic levels, including fish, and eventually impact human health when people consume tainted seafood (Nelms et al., 2018). MPs and NPs are known to carry chemicals either through addition (chemicals added during manufacture) or sorption (Mai et al., 2018). They have a high surface area-to-volume ratio and substantial adsorption capacity, enabling them to adsorb and concentrate toxic contaminants from their immediate environment. They serve as carriers for contaminants, including heavy metals (HMs), endocrine-disrupting chemicals, and persistent organic pollutants (POPs), including polychlorinated biphenyls (PCBs) and polycyclic aromatic hydrocarbons (PAHs), accumulating contaminants at concentrations significantly greater than those found in the aqueous phase (Tan et al., 2019; Yin et al., 2023). These MPs and NPs upon ingestion by organisms can release contaminants within the bodies of organisms, leading to adverse physical and chemical effects in the organisms along with the bioaccumulation of these contaminants in the food web (Mai et al., 2018; Sarkar et al., 2021; Tien et al., 2020).

Additionally, MP and NP buildup in sediments can change sediment structure and composition, as well as the physical properties of habitats, disrupting the ecosystem's equilibrium. This might have an impact on the behavior and survival of benthic creatures that rely on these environments for shelter, feeding, and reproduction, as well as variations in population dynamics, community structure, and ecosystem functioning (Seeley et al., 2020; Bowman et al., 2021). For instance, alterations in feeding habits or reproductive success of key species can have a cascading effect across the food web, affecting overall ecosystem health and stability (Cedervall et al., 2012). In this chapter, the authors address the abundance of MPs and the occurrence of NPs in aquatic environments across the world, along with their source, fate, and potential impact on organisms. The impact of MPs and NPs on aquatic species is examined in-depth along with their synergistic effects with other contaminants, the trophic transfer of these particles, and variations in the toxicity between MPs and NPs.

2 Sources of Microplastics and Nanoplastics in the Aquatic Environment

Plastics in aquatic environments are derived from a multitude of terrestrial sources (Yin et al., 2023), such as industries, consumer plastic debris, agriculture, laundry, abrasions from tires and road marking, plastic-based paint (Maes et al., 2017), tourism (Free et al., 2014; Karthik et al., 2018), wastewater treatment plant (WWTP) effluent, and landfills leachate, and marine-based sources like fishing, shipping, and aquaculture (Bhattacharya et al., 2010; Song et al., 2015). Additional origins of plastic waste in aquatic bodies comprise runoff, storm water (Maes et al., 2017), and atmospheric deposition (Eo et al., 2021). A large number of MPs and NPs are contributed by WWTPs; WWTPs have been found to be capable of removing up to 98% of MPs, but even then, because of the large volumes of effluent discharged a significant amount of MPs enter the rivers. WWTPs with tertiary treatments can remove up to 91–96% of NPs, but they are still the source of 25% of environmental NPs (Tien et al., 2020; Mohana et al., 2021). Larger plastics undergo photo-oxidative degradation by ultraviolet radiation, mechanical abrasion, wave impact, corrosion, weathering, and biodegradation, and are broken down into tiny MPs and NPs (Baldwin et al., 2016; Rodrigues et al., 2018). The disintegration of MPs in the aquatic setting is extremely sluggish as opposed to on land on account of reduced UV exposure and lower temperatures in water (Munari et al., 2017; Karthik et al., 2018). MPs from terrestrial sources are discharged directly or indirectly into rivers and transported downstream to estuaries, seas, and oceans. MPs are classified into two categories contingent upon their source: primary MPs are consciously manufactured for particular applications, e.g., scrubbers and exfoliants in personal hygiene goods (toothpaste, exfoliating soaps, cosmetics), pellets and abrasives in detergents, medicine, raw material for plastic production, etc., whereas secondary MPs are formed through the decomposition of bigger plastic items as well as the disintegration of tiny fibers from textiles during washing (Pan et al., 2019; Tan et al., 2019; Fraser et al., 2020). MPs in the environment can be found in a variety of morphologies, including fragment, film, sheet, angular, foam, spherical, pellet, and fiber, which aid in pinpointing their possible source. Fragments are derived from the degradation of thick plastic materials like bottles and containers; films stem from bags; microbeads/pellets arise from personal care items; foam arises from expanded polystyrene (PS) products; and fibers from textiles, clothing, laundry, fishing ropes, and nets (Tan et al., 2019; Fraser et al., 2020; Almeshal et al., 2022).

Plastics in the environment encompass an assortment of sizes, colors, and polymer types. Plastic coloration is carried out to improve the attractiveness and lifespan of plastic materials (Fraser et al., 2020; Ta & Babel, 2020; Bowman et al., 2021). The most frequently utilized plastic polymers are polyethylene (PE), polypropylene (PP), polystyrene (PS), polyvinyl chloride (PVC), and polyethylene terephthalate (PET), which comprise 90% of global plastic production, and are also dominant in aquatic environments (Munari et al., 2017; Zhang et al., 2022). PE is used in packaging material, water bottles, and personal care products like scrubs, and has an annual global production of 80 million tonnes, whereas, PP is used in packaging material, reusable containers, stationery, textiles, furnishings, pipes, fishing nets, and ropes, with a yearly output of 55 million tons (Munari et al., 2017; Pan et al., 2019). Polyester (PES) is used extensively in the clothing industry; PS likely arises from fishing lines or nets, and food containers, while PET is utilized in the manufacture of synthetic garments (Mai et al., 2018; Pan et al., 2019). Millions of tons of plastic end up in the oceans, and the abundance of certain plastics is determined by

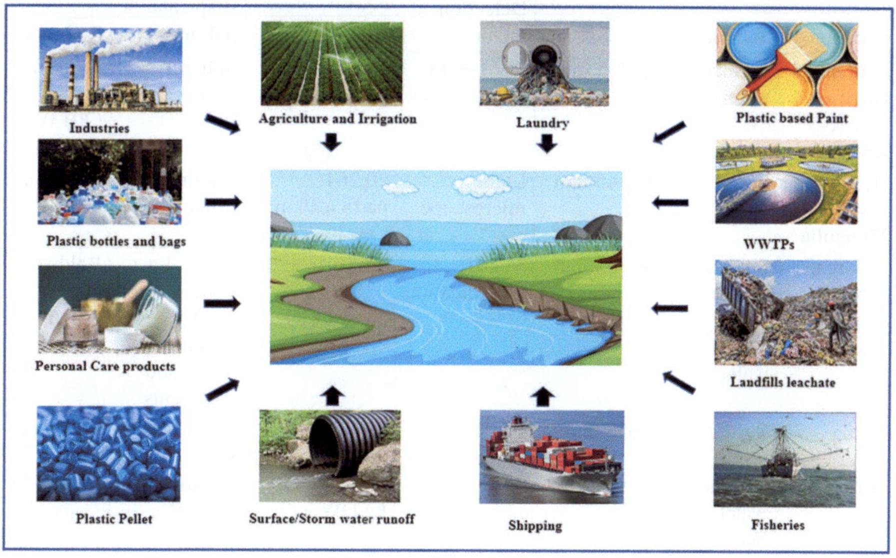

Fig. 1 Sources of microplastics and nanoplastics in the aquatic environment

the amount of plastic manufactured, used, and discarded (Munari et al., 2017; Zhang et al., 2022).

Similar to MPs, NPs are also of two types: primary NPs are produced for industrial usage and released by cosmetics, pharmaceuticals, paints, and electronics, while secondary NPs arise from the fragmentation or degradation of MPs (Gangadoo et al., 2020; Wan et al., 2021; Zhang et al., 2021). Because of their smaller sizes and larger surface areas, NPs have more complex surface properties and can exhibit greater toxicity than MPs (Mohana et al., 2021). Some sources of MPs and NPs are illustrated in Fig. 1.

3 Distribution of Microplastics and Nanoplastics in Freshwater and Marine Environments

3.1 Microplastics in the Freshwater Environment

MPs have been discovered in abundance in the water and sediments of rivers, lakes, and estuaries (Table 1). Depending on the MP concentration, rivers like the Rhine and Danube are less contaminated as compared to the moderately contaminated Antua River or the highly contaminated Pearl River (Table 1). Lakes and fluvial networks (rivers, streams) can function as provisional or long-standing repositories for MPs, and rivers can additionally act as conduits for MPs from the terrestrial ecosystem into the aquatic ecosystem (Dalu et al., 2021; Zhang et al., 2022). Rivers

Table 1 Abundance of microplastics in the freshwater environment

Location	Sample	Size	Detection method	Abundance	Dominant shape and polymer	References
Laurentian Great Lakes, USA	Surface water	0.33–5 mm	SEM-EDX spectroscopy	43,157 particles/km^2	Spheres/ beads	Eriksen et al. (2013)
Hovsgol Lake, Mongolia	Surface water	0.33–5 mm	Light microscope	20,264 particles/km^2	Fragments, films, fibers/ lines	Free et al. (2014)
29 Great Lakes tributaries, USA	Surface and upper 20–35 cm layer water	0.33–5 mm	Visual identification	4.2 particle/m^3	Fibers/lines, pellets/ beads, foams, films, and fragments	Baldwin et al. (2016)
Antua River, Portugal	Surface and bottom water Sediment (up to 12 cm depth)	0.055–5 mm	FTIR spectroscopy	58–1265 items/m^3 2.6–71.4 mg/ kg DW	Foams, fibers, fragments; PE, PP	Rodrigues et al. (2018)
Pearl river and estuary	Surface water	<5 mm	μ-Raman spectroscopy	19,860 items/ m^3 and 8902 items/m^3	Films, fragments, fibers; Polyamide (PA), cellophane	Yan et al. (2019)
St. Lawrence River, Canada	Sediment (up to 15 cm)	<5 mm	Fluorescent microscopy	832 particles/ kg Dry weight (DW)	Micro-beads, fragments, fibers	Crew et al. (2020)
Qiantang River, China	Sediment (0–5 cm)	<5 mm	μ-FTIR spectroscopy	0.23 particles/g DW	Fragments, fibers, microbeads; PE	Fraser et al. (2020)
Lake Maggiore, Como, Iseo, and Garda, Italy	Water (20 cm below surface)	0.33–5 mm	FTIR spectroscopy	0.057–0.500 MP/m^3	Films; PE, PP, PS	Binelli et al. (2020)
Ganga River, India	Surface water Sediment	0.3–5 mm	ATR-FTIR spectroscopy	380–684 items/1000 m^3 17–36 items/ kg DW	Films, foams, fragments, filaments; PE	Singh et al. (2021)
Rhine and Danube River, Germany	Surface water (upper 15 cm)	0.3-5 mm	ATR-FTIR spectroscopy	11.5 particles/ m^3 and 48.7 particles/m^3	Fragments; PE, PP	Schrank et al. (2022)

ATR-FTIR spectroscopy attenuated total reflection-FTIR spectroscopy; *FTIR spectroscopy* Fourier-transform infrared (FTIR) spectroscopy; *PE* polyethylene, *PP* polypropylene, *PS* polystyrene, *SEM-EDX spectroscopy* scanning electron microscopy-energy dispersive x-ray spectroscopy can also act as producers of MPs by exposing larger plastics to sun, wind, and other mechanical stresses found in rivers and enabling their breakdown via mechanical and/or UV weathering (Maes et al., 2017). Multiple studies have reported the MP concentration in rivers to be higher than that of ocean gyres (Baldwin et al., 2016). This is because, upon entering the ocean, MPs will scatter and dilute owing to hydraulic action and the enormous water volume (Zhang et al., 2022).

MP particles are not homogeneously distributed in rivers and their abundance and distribution are dependent on prevailing surface circulation, wind, waves, and propinquity to cities (Free et al., 2014; Schrank et al., 2022). Other influencing factors include flow velocity, discharge rate, storms, rainfall, and floods (Eriksen et al., 2013; Rodrigues et al., 2018). In Lake Hovsgol, the MP density declined as the distance from the developed and populated shore increased, and this decrease was guided by the prevailing southwesterly winds and the drainage of the lake into Eg River (Free et al., 2014). The seasonal disparity in MP abundance was noticed by Dalu et al. (2021), wherein MP concentration was highest in the hot-wet season with high rainfall compared to the cool/hot-dry seasons. Elevated concentrations of MPs in rivers ascribable to anthropogenic activities and urbanization have been reported in Antua River (Rodrigues et al., 2018), St. Lawrence River (Crew et al., 2020), and Pearl River (Yan et al., 2019). The impact of anthropogenic activities in terms of municipal discharge and coal-fired power plants was observed in the Laurentian Great Lakes where a site downstream of the urban center contained over 466,000 particles/km^2 (Eriksen et al., 2013).

Another prominent source of MPs in rivers is effluent from WWTPs (Yan et al., 2019; Fraser et al., 2020). Emscher River, a tributary of the Rhine River, was found to contain a higher concentration of MPs (213 particles/m^3) because it receives high amounts of wastewater from WWTPs (Schrank et al., 2022). In the Great Lakes tributaries, the concentration of MPs (fragments, pellets/beads, foam, films) correlated favorably with the proportion of watershed associated with urban usage, the population size, and the extent of impermeable surface, and elevated amount of MPs were detected during higher runoff situations (Baldwin et al., 2016). Similarly, in the St. Lawrence River sediments, the high MP concentration was correlated with urban land use and sediment particle characteristics (Crew et al., 2020). Plastic debris is more prevalent in metropolitan settings than in other land covers because the extensive tracts of impermeable surfaces and stormwater drains enable the transportation of plastics to recipient bodies of water under runoff events (Baldwin et al., 2016). However, even in isolated regions with low population density, for instance, Lake Hovsgol, Mongolia, a high MP density of 20,264 particles/km^2 was observed. The MP abundance occurred as a result of inadequate waste management of consumer plastic items (Free et al., 2014).

In the freshwater environment, small-sized MPs were dominant. In the Great Lakes tributaries, 72% of MPs were 0.355–0.99 mm in size and 26% were 1.0–4.75 mm in size (Baldwin et al., 2016). In the Rhine and Danube rivers, MPs <1000 μm constituted 87% of the total MPs of which 61% were in the size class 20–300 μm (Schrank et al., 2022). Even in sediments, small microbeads (<400 μm) were dominants (Crew et al., 2020).

In aquatic ecosystems, secondary MPs significantly outnumber primary MPs (Tables 1 and 2). In the Rhine and Danube catchment, fragments were the prevailing shape (90.7%), followed by fibers (6.5%), beads (1.7%), films (1.0%), and pellets (0.02%) (Schrank et al., 2022). In the Great Lakes tributaries, the MPs were in the form of fibers, beads, foams, films, and fragments, with fibers and fragments comprising 71% and 17%, respectively, of each sample (Baldwin et al., 2016). In the Ganga River, films were dominant, with a lesser percentage of foam, fragments, and filaments (Singh et al., 2021). In sediments of the St. Lawrence River, microbeads were the most abundant type (489 MP/kg DW), followed by fragments (220 MP/kg DW) and fiber (122 MP/kg DW) (Crew et al., 2020). Fragments form as a result of the disintegration of bigger plastic items through photooxidation, thermal degradation, or biodegradation. Fibers are derived from wastewater effluent contaminated with textile fibers or fishing and boating activities or are deposited by atmospheric input, while films arise from packaging materials (Munari et al., 2017; Desforges et al., 2014). Conversely, pellets representing primary MPs arise from accidental spills of plastic pellets during production, packaging, and transport (Pan et al., 2019). In the Rhine and Danube catchment, PE (49.2%) and PP (33.2%) were the most frequent polymer types probably because of their low density and high production rates (Schrank et al., 2022). A similar abundance of PE and PP was noted in the Antua River (Rodrigues et al., 2018) and Italian lakes (Binelli et al., 2020), and PP in the Ganga River (Singh et al., 2021).

3.2 *Microplastics in the Marine Environment*

The oceans, where rivers congregate, have been identified as a repository of MPs. MPs are ubiquitous in the marine realm, being discovered at the surface as well as at depth in seas, oceans, and along coastlines (Desforges et al., 2014). Researchers have discovered the presence of MPs in surface water and sediment of numerous seas and the deep ocean and continental shelf of the Pacific and Atlantic Oceans (Table 2). As can be comprehended from Table 2, the amount of MPs in marine environments varies widely around the globe. A higher number of MPs have been recorded in investigations employing smaller mesh sizes for sample collection. In the South China Sea, the abundance of MPs with size >333 μm was 4.5×10^{-2} particles/m^3 and with size 0.02–0.3 mm was 2.57×10^3 particles/m^3 (Cai et al., 2018). Smaller particles were predominant with more than 50% having a size between 0 and 100 μm (Cai et al., 2018). The concentration of MPs changes across the depth of the water column, with respective mean abundances of MPs in the surface, thermocline, and bottom layers being 266, 228, and 136 particles/m^3 in the continental shelf area, and 289, 260, and 1136 particles/m^3 in the deep sea area (Eo et al., 2021).

MP abundance and spatial heterogeneity in the marine domain indicate the effect of land-based inputs, ocean currents, wave- and wind-driven turbulence, or hydrodynamic effects (e.g., downwelling and upwelling, gyres or fronts). The primary driver of MP pollution in the marine domain is anthropogenic plastic waste from

Table 2 Abundance of microplastics in the marine environment

Location	Sample	Size	Detection method	Abundance	Dominant shape and polymer	References
Northeast Pacific Ocean	Surface water	0.33–5 mm	FTIR spectroscopy	0.448 particles/m^2 in summer and 0.021 particles/m^2 in fall	–	Goldstein et al. (2013)
Northeast Pacific Ocean	Sub-surface water (4.5 m below surface)	<100 μm-5 mm	Light microscopy	8 to 9200 particles/m^3	Fibers	Desforges et al. (2014)
Jinhae Bay, South Korea	Sea surface microlayer (150–400 μm depth)	<5 mm	FTIR spectroscopy	88 particles/L (12,600 particles/m^3)	Fragments, spheres, fibers, sheets; PP, PE, phenoxy resin, PS, PES	Song et al. (2015)
Seas of Northwest Europe Southern North Sea	Surface water Sediment (top 5 cm)	0.33–5 mm	Light microscopy	0–1.5 MP/m^3 0–3146 particles/kg DW	Fragments Fibers and spheres	Maes et al. (2017)
Northern Adriatic coast	Sediment (top 5 cm)	0.33–5 mm	FTIR spectroscopy	5.99–21.6 items/kg DW	Fragments, films, and fibers; PE, PP, nylon, PS, PET, PVC, thermoplastic polyurethane	Munari et al. (2017)
Southeast coast of India	Sediment (top 5 cm)	0.33-5 mm	FTIR spectroscopy	44.6 particles/m^2	Fragments, line/fibers, foams; PE, PP, PS, nylon	Karthik et al. (2018)
Bohai Sea, China	Surface water	0.33-5 mm	Visual sorting and ATR-FTIR spectroscopy	0.650 items/m^3	Foams, films, lines, and pellets; PS and PE	Mai et al. (2018)
Huanghai Sea, China	Surface water	0.33-5 mm	Visual sorting and ATR-FTIR spectroscopy	0.0900 items/m^3	Foams, films, lines, and pellets; PS and PE	Mai et al. (2018)

(continued)

Table 2 (continued)

Location	Sample	Size	Detection method	Abundance	Dominant shape and polymer	References
South China Sea	Surface water	0.02–0.3 mm and 0.33–5 mm	Microscope and μ-FTIR spectroscopy	2.57×10^3 particles/m^3 (0.02–0.3 mm) and 4.5×10^{-2} particles/m^3 (>0.333 mm)	21 polymer types—alkyd resin, polycaprolactone, polyethyl acrylate, PS, Polytetrafluoroethylene (PTFE), etc., in the 0.02–0.3 mm size range, PES, PE, PP-PE copolymer in >0.333 mm size range	Cai et al. (2018)
North-Western Pacific Ocean	Surface water	0.33–5 mm	Stereo light microscope, Raman spectroscopy, and SEM	6.2×10^4 items/km^2	Fragments, pellets, fibers, lines, films, foams; PE, PP, PES, PS, PE-PP copolymer, PET, PA	Pan et al. (2019)
Yellow sea, East Sea, South Sea, West Pacific Ocean	Surface water	0.33–5 mm	μ-FTIR spectroscopy	15–9400 particles/m^3	35 polymer types, mainly high-density polymers	Eo et al. (2021)

ATR-FTIR spectroscopy attenuated total reflection-FTIR spectroscopy, *FTIR spectroscopy* Fourier-transform infrared (FTIR) spectroscopy, *PE* polyethylene, *PET* polyethylene terephthalate, *PP* polypropylene, *PS* polystyrene, *PVC* polyvinyl chloride, *SEM-EDX spectroscopy* scanning electron microscopy-energy dispersive x-ray spectroscopy inland or coastal activities (Mai et al., 2018) which is conveyed into the ocean via rivers (Cai et al., 2018). Maes et al. (2017) observed higher concentrations of MPs near the coast and estuaries indicating the importance of inputs through rivers. Similarly, a greater concentration of MPs was reported for beaches in close proximity to the river mouths because the rivers with high water flow pass through densely populated cities (Karthik et al., 2018). The MP abundance in rivers, estuaries, and oceans is closely related to watershed demographics and urban growth, as well as local hydrodynamic conditions.

Munari et al. (2017) found a greater concentration of MPs in Volano and Rosolina, two beaches impacted by Po and Adige River outflow, respectively, than at Bellocchio or Bevano, which are affected by considerably lesser riverine runoff. A larger prevalence of tiny plastic debris would be predicted on beaches subjected to heavier riverine runoff, since increased flow may favorably affect the higher plastic density. Karthik et al. (2018) recorded the presence of a greater abundance of MPs from the high tide line (9–178 No/m^2) compared to the low tide line (2–64 No/m^2). MP particles were less abundant in the low tide line, probably because of their inundation for the majority of the day and susceptibility to easy removal and transit to the open ocean by wave action. On regional scales, plastic debris movement is influenced by the interaction of large-scale atmospheric and oceanic circulation patterns (Goldstein et al., 2013; Desforges et al., 2014). Eo et al. (2021) noted the transfer of small MPs to far-off offshore locations about 210 km from land. An elevated concentration of MPs was witnessed in Queen Charlotte Sound in the North East Pacific Ocean, which was probably a result of oceanographic conditions (Pacific currents, northward coastal currents) that trap and concentrate plastic debris (Desforges et al., 2014). The transport and dispersal of MPs on coastal beaches in South India were regulated by the southwest trade winds blowing toward the Northern Hemisphere during monsoon and the East Indian Coastal Current (Karthik et al., 2018). Plastic debris in oceans is transported to a place far away from its source area and may converge in a place where local circulation is able to retain such plastics. For instance, surface plastic garbage in the oceans accumulates in gyres, which drives and accrues plastics in the oceanic regions, such as the Great Pacific Garbage Patch (Cai et al., 2018; Pan et al., 2019; Davranche et al., 2020).

Another factor is the proximity to urban settlements or human population centers (Goldstein et al., 2013). In the NE Pacific Ocean, the MP concentration at nearshore sites was 4–27 times higher than offshore sites, confirming that terrestrial anthropogenic activities constitute the primary source of MPs in the marine environment (Desforges et al., 2014). The MP abundance was several times higher in an inner sea, the Bohai Sea, compared to an open sea, the Huanghai Sea (Mai et al., 2018). Similarly, in Jinhae Bay, which is surrounded by a coastal city, MP was significantly higher ($p < 0.05$) than the east coast of Geoje, which is a relatively open sea (Song et al., 2015). The Jinhae Bay is surrounded by coastal cities with high population and industrial complexes, which release a high plastic burden into the Nakdong River, which empties into the Jinhae Bay (Song et al., 2015).

Smaller MPs usually contribute a larger percentage of marine plastic remnants (Goldstein et al., 2013). Munari et al. (2017) stated that MPs made up 61% of the total plastic debris in Adriatic sediments and small plastic debris (MPs and mesoplastics, 5–20 mm) accounted for 89.9% of the total plastic. The highest MP abundance was contributed by MPs of the size range 1.18–2.36 mm on the southern coast of India (Karthik et al., 2018) and of 50–150 μm range in the Yellow Sea, South Sea, and East Sea (Eo et al., 2021). Further, there was an increase in the abundance of small MPs (<1 mm) with decreasing particle size, indicating that smaller-sized fractions contributed to the total plastic burden (Cai et al., 2018). Multiple studies have reported a generalized decrease in the mean size of plastic fragments coupled with a rising abundance of smaller particles owing to continuous degradation (Munari et al., 2017). Notwithstanding the multiple reports of microplastic presence in aquatic bodies, any comparison is difficult in view of the absence of a unified sampling, treatment, and identification method.

The MPs shapes commonly observed in surface water are fragments, films, and fibers, with a much smaller fraction of pellets (Munari et al., 2017; Karthik et al., 2018; Maes et al., 2017). In the surface water in the Yellow, East, and South Sea in the western Pacific (Eo et al., 2021) and North West European seas (Maes et al., 2017) fragments accounted for more than 70% of the MPs followed by fibers. In the Northwest Pacific, fragments and pellets were dominant, followed by a smaller percentage of fiber, line, film, and foam (Pan et al., 2019). Contrastingly, in the North East Pacific Ocean, fibers/filaments formed the dominant fraction followed by fragments (Desforges et al., 2014). In marine sediments, a similar distribution of MP shape has been documented (Munari et al., 2017; Karthik et al., 2018). However, Maes et al. (2017) noted a larger percentage of MPs in the form of fibers and spheres in the sediments of the North West European seas, with a lower amount of fragments. The spheres and fibers made of PS and polyacrylamide are heavier and can readily sink (Maes et al., 2017). The presence of different shapes and sizes of MPs might be determined by variables controlling plastic distribution, that include source, quality, rate of degradation, wind drift, etc. (Karthik et al., 2018). A prominence of PE was observed in the Adriatic sea sediments, along with the presence of other plastics such as PP, Nylon, PS, PET, and PVC (Munari et al., 2017). In the Northwest Pacific Ocean, PE and PP accounted for >60% of the MPs, while the rest were made up of PES, PS, PE-PP, PET, and PA (Pan et al., 2019). In the Bohai and Huanghai Seas, the main MP polymers identified were PS and PE (Mai et al., 2018).

3.3 Occurrence of Nanoplastics

Nanoplastics are expected to be ubiquitous in the environmental compartments (Davranche et al., 2020), but have not been quantified yet (Andrady, 2011). At present, studies on established analytical methods for detecting NPs in the aquatic environment are lacking and NP analysis is still indeterminate (Koelmans et al., 2015). Although the separation, concentration, and identification of NPs are difficult, several studies have tried to establish the presence of NPs in environmental specimens. The first observation of NPs (1–999 nm) in environmental samples was reported by

Ter Halle et al. (2017) in water samples of the North Atlantic Subtropical Gyre. They isolated the colloidal fraction of the water, ultra-filtered it, and demonstrated the presence of NPs with dynamic light scattering (DLS). Additionally, they obtained a chemical fingerprint of the colloids with pyrolysis coupled with gas chromatography-mass spectrometry (Py-GCMS) and noted that the NPs were mainly PVC, PET, PS, and PE. However, the authors could not quantify the NPs or showcase an accurate size distribution. Davranche et al. (2020) proved the existence of NPs in sand water extracts from the coast exposed to the North Atlantic Gyre using a combination of TEM, DLS, and Py-GCMS. They expected the presence of NPs on the coast due to a large quantity of MPs, but were unable to quantify the NPs or identify their origin.

4 Migration and Fate of Microplastics and Nanoplastics

In the aquatic environment, the abundance and distribution of MPs are governed by an assortment of factors including their physicochemical properties (shape, size, color, polymer type, density), prevailing surface circulation, and wind (Desforges et al., 2014; Maes et al., 2017). MPs in the aquatic environment might remain floating in the water column or descend to the sediment based on their polymer type, density, and shape (Tien et al., 2020). Denser MPs (>1.2 g/cm^3) such as PET and PTFE sink through the water column, while lighter MPs (<1.2 g/cm^3) like PE and PP float on the water's surface and are dominant in the water column (Tien et al., 2020; Eo et al., 2021). Denser MPs might however remain suspended in water as a consequence of greater surface area to volume ratio and be resuspended because of water circulation aided by riverine inflow, wind, evaporation, and tidal currents (Tan et al., 2019). Low-density MPs recirculate more extensively between sediments and water due to wind and water currents compared to high-density MPs which are deposited in sediments (Karthik et al., 2018). However, low-density polymers have also been reported in sediments on account of changes in their buoyancy and density brought on by biological interactions and physical mixing (Maes et al., 2017). Plastic particles may sink by reason of density changes induced by interaction with aggregates, fecal matter of marine organisms, organic matter, clay minerals, and biofouling (Goldstein et al., 2013; Maes et al., 2017; Eo et al., 2021). Biofouling increases the density of MPs and facilitates deposition in addition to aiding in their environmental aging and degradation (Pan et al., 2019). MPs aggregate with marine biogenic particles which accelerate their sinking velocity and contribute to their vertical transport (Eo et al., 2021). In estuarine sediments and organic sediments, Maes et al. (2017) noted a greater quantity of MPs with a strong correlation between the number of plastic particles/kg dry sediment and total organic carbon ($p = 0.001$). Fine muds with higher organic materials and smaller grain sizes are thus expected to be hotspots of MPs. This further points out that benthic organisms feeding and burrowing in the mud will likely be exposed to greater concentrations of MPs than benthic creatures in regions with bigger grain sizes and diminished organic matter

(Maes et al., 2017). Additionally, low-density polymers may contain added chemicals, which alter their specific gravity and enable their deposition (Eo et al., 2021). In Korean offshore waters, Eo et al. (2021) observed low levels of aggregate fraction in the continental shelf as well as deep seawater and dominance of high-density plastic polymers. They confirmed that the aggregates were responsible for vertical and long-range transport of high-density MPs as the sinking rate of high-density polymers decreases when in aggregation with less dense materials enabling their transportation to farther distances (Eo et al., 2021). Further, physical forces such as wind-driven mixing can aid in the vertical transportation of MPs (Eo et al., 2021).

The horizontal movement is locally governed by wind patterns which control the distribution of the debris by mixing particles of different densities (Goldstein et al., 2013), while regional plastic debris movement is influenced by the interaction of large-scale atmospheric and oceanic circulation patterns (Goldstein et al., 2013; Desforges et al., 2014). Atmospheric transportation is another important pathway for the transport of MPs to far-flung regions. Eo et al. (2021) demonstrated the atmospheric transportation of MPs to far offshore water in the East Sea, Korea.

The behavior of NPs in aquatic domain is regulated by properties such as aggregation, agglomeration, dissolution, and redox reactions, which are further controlled by surface properties, pH, aging, ionic strength, and the presence of natural organic matter (Gangadoo et al., 2020). Aggregation reflects the environmental destiny of NPs in terms of stability, fluidity, and bioavailability, while the subsequent step is deposition, suggesting that when NPs agglomerate, they descend due to gravity and are deposited (Zhang et al., 2021). In aqueous environments, NPs exhibit Brownian motion and can form agglomerates of sizes 1–100 nm in the presence of organic matter or colloidal materials (Davranche et al., 2020). The accretion and sedimentation of NP aggregates in the aquatic system enhances their propensity for entering organisms (Zhang et al., 2021).

5 Impact of Microplastics and Nanoplastics in the Aquatic Environment

Since the presence of MPs has been documented in numerous aquatic bodies and the presence of smaller nonquantifiable MPs and NPs is foreseeable, it is pertinent that the aquatic primary producers and organisms will encounter MPs and NPs.

5.1 Toxicity Effect of Microplastics and Nanoplastics on Aquatic Primary Producers

Numerous studies have explored the interaction and effect of MPs/NPs in algae, cyanobacteria, and higher aquatic plants and published contradictory discoveries (Table 3). Algae occupy the foundational tier of the food chain, and their

Table 3 Effects of microplastics and nanoplastics on aquatic primary producers

Aquatic plant	Polymer type; size; concentration of microplastics/nanoplastics	Effects on plant	References
Microplastics			
Chlamydomas reinhardtii	PP and HDPE; 400–1000 μm; 100 mg/L	MPs >400 μm did not directly affect microalgal growth; formation of hetero-aggregates of microalgae, MPs, and exopolysaccharides in case of PP; over-expression of genes in the sugar biosynthesis pathways after 70 day	Lagarde et al. (2016)
Chlorella vulgaris (microalgae), *Dunaliella tertiolecta* (marine flagellate), *Thalassiosira pseudonana* (diatom)	PS; uncharged (0.05, 0.5, 6 μm) and charged (0.5 μm); all at 25 and 250 mg/L	No effect on photosynthesis; negative effect (up to 45%) on microalgal growth by uncharged PS MPs at 250 mg/L concentration	Sjollema et al. (2016)
Lemna minor	PE; 30–600 μm; 0, 10, 50, 100 mg/L	No significant effect on leaves; reduced root length	Kalčíková et al. (2017)
Lemna minor	PE; 10–45 μm	No significant effect on photosynthesis or plant growth	Mateos-Cárdenas et al. (2019)
Spirodela polyrhiza	PS beads; 50 and 500 nm; 10^2, 10^3, 10^4, 10^5, 10^6 particles/mL	No significant effect on plant growth or chlorophyll production; attachment of NPs to plant roots	Dovidat et al. (2020)
Chlamydomas reinhardtii	PS; 300–600 nm; 5, 25, 50 and 100 mg/L	Decreased density of microalgae with increasing MP concentration (highest inhibitory rate was 45.8% on the seventh day at 100 mg/L); inhibition of EPS content; reduction of chlorophyll-a fluorescence and photosynthesis at all dosages	Li et al. (2020b)
Chlorella pyrenoidosa, *Microscystis flosaquae*	PVC, PP; 5, 10, 50, 100, 250, 500 mg/L	Negative effect on chlorophyll-a content in both algae; PVC at 250 mg/L reduced chlorophyll-a content of *C. pyrenoidosa* by 55.23%; higher concentrations of PVC and PP stimulated lower values of Fv/Fm, $Fv/F0$, and Fv'/Fm'	Wu et al. (2019a)

(continued)

Table 3 (continued)

Aquatic plant	Polymer type; size; concentration of microplastics/nanoplastics	Effects on plant	References
Myriophyllum spicatum *Elodea*	PS; 50–190 nm, 20–500 µm; 0.1, 0.3, 1, 3, and 10% sediment DW	MPs reduced the main shoot length in *M. Spicatum*; NPs reduced shoot/root ratios of both plants and decreased *M. spicatum* main shoot length	van Weert et al. (2019)
Nanoplastics			
Chlorella spp., *Scenedesmus* spp.	PS beads positively and negatively charged; 20 nm	Increased adsorption of NP beads, decreased photosynthesis, increased reactive oxygen species (ROS) generation	Bhattacharya et al. (2010)
Platymonas helgolandica	PS; 70 nm; 20, 200, 2000 µg/L	Growth inhibition during the first 4 day; elevated heterocyst frequency in 200 and 2000 µg/L treatment; augmented membrane permeability, mitochondrial membrane potential, lessened light energy usage in photochemical processes after exposure to 200 and 2000 µg/L NPs for 4 day; morphological changes (surface folds, fragmentation, aggregation, and rupture)	Wang et al. (2020)
Chlorella pyrenoidosa	PS NPs; 80 nm; 5, 10, 20, 30, 40, and 50 mg/L	Maximum inhibition rates of 27.73%, 29.64%, and 11.76% on algal growth, chlorophyll-a, and *Fv/Fm*, respectively, at 50 mg/L	Yang et al. (2021)
Amphidinium carterae	PS; 50 nm; 10, 20, 50, 80, and 100 mg/L	Chlorophyll synthesis was inhibited on day 4; significant growth inhibition at high concentrations (50 mg/L); high reactive oxygen species (ROS) level and malondialdehyde (MDA) content at 50 mg/L; significantly higher contents of extracellular and intracellular hemolytic toxins in all groups, except 100 mg/L	Wang et al. (2021a, b)

photosynthetic activity plays a crucial role in generating a significant portion of the earth's oxygen. These primary producers are vital for the maintenance of biogeochemical cycles and the overall operation of aquatic ecosystems (Bhattacharya et al., 2010; Li et al., 2020b). Hence, any fluctuations in algae in the freshwater environment have a cascading effect on the makeup and operation of aquatic ecosystems (Wu et al., 2019a). Generally, toxicity seems to be dependent on parameters such as size, concentration, surface charge, and polymer type (Lagarde et al., 2016; Sjollema et al., 2016; van Weert et al., 2019; Li et al., 2020b).

Primary producers have the potential to adsorb and accumulate floating MPs, thus acting as both a sink and a source of MPs in the aquatic ecosystem (Grgić et al., 2023). Studies have documented the attachment of MPs and NPs to the surfaces of algae and aquatic plants (Bhattacharya et al., 2010; Mateos-Cárdenas et al., 2019; Dovidat et al., 2020) without any discernible impact on plant development or photosynthesis. Mateos-Cárdenas et al. (2019) demonstrated the quick adsorption of PE MPs on *Lemna minor*, with the highest adsorption occurring after 72 h of exposure. The MPs exhibited no influence on plant growth indices; however, the authors demonstrated MP trophic transfer from the freshwater primary producer *L. minor* to the consumer amphipod *Gammarus duebeni*. A similar trophic transfer in marine environment was reported by Goss et al. (2018), wherein the leaves of the seagrass *Thalassia testudinum* were encrusted with MPs (microfibers and microbeads in a ratio of 59:14). Seagrasses serve an assorted assemblage of fish and invertebrates and are consumed by grazers, thus transferring MPs in the marine food webs.

On the contrary, Bhattacharya et al. (2010) stated that adsorbed NPs induced phytotoxic effects in the algae *Chlorella* spp. and *Scenedesmus* spp., by blocking light from reaching photosynthetic centers, damaging algal cell walls to form pores, obstructing airflow, and hindering nutrient uptake. PS NPs were found to adsorb on the algal surface, increasing the formation of reactive oxygen species (ROS) and decreasing photosynthesis. The inhibition of photosynthesis probably occurred due to either the shadowing effect caused by the attached PS beads or the obstruction of pathways for carbon dioxide flow and nutrient uptake (Bhattacharya et al., 2010). Another study on *Thalassiosira pseudonana, Dunaliella tertiolecta,* and *Chorella vulgaris* employed negatively charged and uncharged PS MPs of sizes 0.05, 0.5, and 6 μm. They did not observe any effect on microalgal photosynthesis, except for the treatment with the smallest size of uncharged MP at 250 mg/L, which reduced photosynthesis by 45% (Sjollema et al., 2016). Similarly, Kalčíková et al. (2017) did not witness any negative effect of PE microbeads on leaf growth and content of photosynthetic pigments in *L. minor*. However, they reported a significant negative effect on root growth and reduction in root length, probably because of the adsorption of the microbeads on the root surface and subsequent mechanical damage. However, opposite results were published by Li et al. (2020b) in *Chlamydomonas reinhardtii*, wherein MP concentration of 5–100 mg/L reduced chlorophyll-a fluorescence and photosynthetic activities. Additionally, the density of microalgae decreased with increasing MP concentrations, and the highest inhibition of 45.8% was noted under 100 mg/L together with the decrease in the quantity of extracellular polymeric substances. The authors observed that MP beads were wrapped on the surface of microalgae causing membrane damage, which could be the reason for the reduction of photosynthetic activities and the increase of soluble proteins and malondialdehydes (MDA) content. A similar reduction of photosynthetic pigments and photosynthetic activity was reported by Wu et al. (2019a) in *Chlorella pyrenoidosa* and *Microcystis flosaquae*.

NPs can produce different effects on aquatic plants compared to MPs. In *Myriophyllum spicatum* and *Elodea* sp., van Weert et al. (2019) determined the consequence of MP and NP exposure on multiple plant growth indices. MPs did not result in reliable dose-effect interactions on the endpoints studied, except for the reduction of main shoot length in *M. spicatum.* However, significant correlations were obtained between NPs and augmented root biomass and diminished primary shoot length in *M. spicatum,* and surge in root and shoot biomasses, RGR, and lateral shoot length in *Elodea* sp. NPs significantly reduced the shoot/root ratio in the macrophytes because of increased root biomass, indicating that nutrient accumulation may be inhibited upon NP exposure. NPs were also reported to inhibit growth, chlorophyll-a pigment production, and photosynthesis, and trigger oxidative stress in the microalgae *Chlorella pyrenoidosa, Platymonas helgolandica,* and *Amphidinium carterae.* Additionally, lipid peroxidation, change in membrane permeability, and morphology were also reported (Wang et al., 2020, 2021a, b; Yang et al., 2021).

5.2 *Effect of Microplastics on Aquatic Organisms*

Microplastics have been spotted in aquatic biota worldwide, ranging from small organisms like fish larvae to large mammals (Table 4). Plastic debris at environmentally-relevant concentrations is harmful to organisms (Urbina et al., 2023) and affects a large variety of aquatic species, including turtles, birds, and mammals (Desforges et al., 2014; Winkler et al., 2020). MPs in the water as well as the sediments can be consumed by an extensive spectrum of aquatic species, ranging from zooplankton to bigger organisms (Von Moos et al., 2012; Almeshal et al., 2022). Aquatic organisms often confuse them for food because of their minute sizes, coloration, and floating properties (Tien et al., 2020). In coastal waters of India, Karthik et al. (2018) noted that 10% of the studied fishes ingested MPs, with fragments (87.5%) and fibers (12.5%) being the major fractions. The fishes ingested particles <0.5 mm and fishes collected close to the river mouths displayed the highest abundance of MPs. As seen in the previous sections, MPs in the environment exist in an assortment of sizes, shapes, polymer compositions, densities, and colors. These inherent characteristics of MPs affect their distribution, transport, accessibility to organisms, ingestion, and trophic transfer in the aquatic ecosystem:

(i.) Density: The plastic density plays a vital role in whether MP particles remain suspended in the water or settle in the sediment. This, in turn, determines the potential organisms that might consume these plastic particles. Plastics with lower densities, like PE, are prone to ingestion by filter and suspension feeders found in the upper water column. Meanwhile, benthic organisms are more likely to ingest MPs that have settled in sediments or become resuspended due to bioturbation (Tien et al., 2020).

(ii.) Size: The MP size variation is a critical aspect for ingestion in aquatic ecosystems (Klein et al., 2021). Aquatic organisms have been found to ingest MPs of the smaller size range. In the Fengshan River, fishes were found to ingest greater amounts of smaller MPs (43–69%) compared to those of larger MPs (31–57%) (Tien et al., 2020). Research has documented that smaller MPs can

Table 4 Effect of microplastics on aquatic organisms

Organism	Type, size, concentration of microplastic, and duration of experiment	Toxicity effect	References
Mussel (*Mytilus edulis* L.)	PS MPs; 3.0 and 9.6 μm; 15,000 particles/350 mL of seawater; 3 h	Abundance of smaller MPs; MPs presence in haemolymph; MPs transfer from gut to circulatory system within 3 h and persistence for 48 day	Browne et al. (2008)
Blue mussel (*Mytilus edulis*)	HDPE MPs; >0–80 μm; 96 h	MPs uptake in gills, stomach, and digestive gland; inflammation and histological changes; impact of MPs at tissue and cellular level	Von Moos et al. (2012)
Marine copepod (*Tigriopus japonicus*)	PS MPs; 0.05, 0.5, 6 μm; 1.25, 12.5, 25 μg/L; 96 h	No mortality after 96 h; 0.5 and 6 μm MPs decreased fecundity; 0.05 μm beads caused mortality in F0 as well as F1 generations	Lee et al. (2013)
Common goby (*Pomatoschistus microps*)	PE MPs; 420–500 μm; 96 h	Lowered predatory activity; reduction of food intake; alteration in prey selection capability	de Sá et al. (2015)
Daphnia magna	MPs; 1–5 μm; 10^2–10^5 particles/mL; 3 weeks	Significant reduction of total population (number of individuals); 21% reduction in total biomass	Bosker et al. (2019)
Adult zebrafish (*Danio rerio*)	PS MPs; 5 μm; 0.001–20 mg/L	MPs accumulation in gastrointestinal tract and gills; 32% reduction in thickness of muscularis layer in foregut; hyperactivity; scarcity of energy-giving material	Chen et al. (2020)
Detritivores caddisfly larvae (*Sericostoma Pyrenaicum*)	PS MPs; 10 μm; 0–10^3 particles/mL	Increased detrivore mortality (9 times at highest concentration); reduced leaf litter decomposition	López-Rojo et al. (2020)
Diptera larvae (*Chironomus riparius*)	PE MPs; Size class (SC) I: 32–63 μm; II: 63–250 μm; III: 125–500 μm); 1.25, 5, 20 g/kg; 48 h	SC I: High amounts of MPs (~2400 particles/organism) in guts; SC II and III: Visible gut obstruction, diminution in total lipid reserves due to lack of nutrients; SC III: Decreased detoxification capacity; all SC: Oxidative damage and lowered aerobic energy generation	Silva et al. (2021)
Zebrafish (*Danio rerio*)	PS MPs; 1 μm; 10, 100, 1000 μg/L; 21 day	Amplified ROS in male and female fish gonads; molecular responses, histological alterations (basement membrane thinning in testes), and apoptosis in fish gonads indicating adverse effects in reproductive organs	Qiang and Cheng (2021)

(continued)

Table 4 (continued)

Organism	Type, size, concentration of microplastic, and duration of experiment	Toxicity effect	References
Giant river prawn (*Macrobrachium rosenbergii*)	PS (0.5–1 μm) and PE (30–150 μm) MPs; 1, 5, 10 mg/100 g; 60 day	Lower rate of survival; sluggish length and weight gain; increased levels of antioxidant enzymes; inhibited metabolic enzymes activity; positive regulation of HSP70 gene and negative regulation of MSTN gene	Jaikumar et al. (2021)
Filter-feeding crab (*Petrolisthes laevigatus*)	PE MPs; 3–30 μm; 250 particles/L for 5 months and 20,800 particles/L for 48 h	No significant effect of acute exposure (48 h); chronic exposure (5 months) impacted ingestion rate and caused weight loss and nutritional damage	Urbina et al. (2023)

readily permeate tissues and induce toxicity effects in organisms (Fraser et al., 2020; Cole et al., 2020). In the marine copepod *Tigriopus*, smaller MPs (0.05 μm) caused mortality in the first generation at higher concentrations; however, in the offspring, mortality was triggered by MPs at even lower concentrations (Lee et al., 2013).

(iii.) Shape: MP shape affects their uptake by aquatic organisms. Fibers have been observed to be the most prevalent MPs in water, resulting in their hegemony in organisms as well (Winkler et al., 2020). All MPs in the top predator, the kingfisher bird, were fibers with one fragment, most likely acquired by secondary ingestion from fish (Winkler et al., 2020).

(iv.) Concentration: Higher amounts of MPs in marine and freshwater habitats enhance the risk of organisms encountering and ingesting them (Klein et al., 2021). In the Giant River Prawn, Jaikumar et al. (2021) discerned dose-dependent changes in survival rate, weight and length gain, and activities of antioxidant enzymes. Similarly, in zebrafish, Qiang and Cheng (2021) noticed a concentration-dependent stress response. MPs at 10 μg/L did not produce any significant changes while at 1000 μg/L concentration severely affected the reproductive system.

(v.) Color: Plastic particle color plays a crucial role in their ingestion by predators. MPs with a similar hue to the organism's diet have a greater probability of being ingested by mistake. Colored plastics increase the likelihood of ingestion and have a greater proclivity to be found at higher levels in the food chain (Fraser et al., 2020; Sarkar et al., 2021). Fishes in one experiment were found to ingest a higher amount of white, light yellow, and green-colored MPs (Sarkar et al., 2021) while another study published the preferential consumption of red, blue, green, and black-colored MPs (Ta and Babel, 2020).

Moreover, MP ingestion by organisms is predisposed by additional parameters, including hydrodynamic conditions, temperature, habitat, feeding behavior, feed abundance, and exposure route (Nelms et al., 2018; Klein et al., 2021).

MPs can be ingested by organisms in two ways: direct ingestion from the environment due to indiscriminate feeding strategies such as filter-feeding, or through active selection due to misidentification of diet (Nelms et al., 2018). An additional pathway for MPs to enter the food chain is through the consumption of MP-contaminated prey by predators (Nelms et al., 2018). The consumption of MPs by aquatic creatures could culminate in lethal effects in fishes (de Sá et al., 2015; Chen et al., 2020; Qiang & Cheng, 2021), mussels (Browne et al., 2008; Von Moos et al., 2012), copepods (Lee et al., 2013), detrivores (López-Rojo et al., 2020), and crustaceans (Bosker et al., 2019; Jaikumar et al., 2021; Urbina et al., 2023). The effects of MPs on aquatic creatures may manifest as physical, biological, or chemical consequences. Physical impressions of MPs arise from the ingestion of MPs and have been linked to various harmful effects, including gut blockage (Silva et al., 2021), internal injuries, reduced feeding efficiency, malnutrition, increased buoyancy, and dietary dilution (Nelms et al., 2018; Silva et al., 2021; Urbina et al., 2023). In the *Chironomus* larvae, Silva et al. (2021) explored the effect of different MP sizes and reported that larger sizes caused gut obstruction while smaller ones accumulated in the gut in high amounts. Additionally, these MPs caused oxidative damage, reduced energy production, depleted lipid reserves, and decreased the organism's detoxification capacity.

The ingestion or adherence of MPs results in biological responses including inflammatory responses (von Moos et al., 2012), oxidative stress (Silva et al., 2021; Jaikumar et al., 2021), altered immune responses (Von Moos et al., 2012), deviations in energy metabolism (Chen et al., 2020; Silva et al., 2021; Urbina et al., 2023), modified genetic makeup, immunological, neurological, and histological damage, reduced feeding, diminished body weight (Urbina et al., 2023), inhibited growth or length gain, adverse effect on reproduction (Qiang & Cheng, 2021), particle transfer in the circulatory system (Browne et al., 2008), impairment of respiratory system, gastrointestinal tracts, and liver (Von Moos et al., 2012; Silva et al., 2021), and mortality (Bosker et al., 2019; López-Rojo et al., 2020; Jaikumar et al., 2021). MPs can induce significant detrimental impacts on tissues, cells, and subcellular structures, ultimately affecting the overall health of aquatic populations (Von Moos et al., 2012; de Sá et al., 2015).

In prawn, Jaikumar et al. (2021) witnessed copious toxicity effects of PE (0.5–1 μm) and PS (30–150 μm) MPs, including lower survival rates, reduced weight and length, heightened levels of antioxidant enzymes (superoxide dismutase, catalase, glutathione s-transferase, and glutathione peroxidase [GPx]), elevated levels of reduced glutathione (GSH) and MDA, and diminished levels of metabolic enzymes (glutamic oxaloacetic transaminase and glutamic pyruvic transaminase). The MPs influenced the prawn's feeding behavior and digestive processes, disrupting growth and survival, while the prawn attempted to deal with oxidative stress by enhancing the production of antioxidant enzymes. MPs also increased the expression of the heat shock protein (HSP70) gene while decreasing the expression of the myostatin (MSTN) gene. *HSP70* genes are markers of additional biological mechanisms for dealing with stresses like apoptosis, oxidative stress, and inflammation (Jaikumar et al., 2021). In zebrafish, MPs were found to significantly enhance ROS in male and female fish gonads, along with apoptosis in testes. This led to histological alterations in the form of thinning of the testis basement membrane and overall adverse effects on reproductive organs (Qiang & Cheng, 2021). Decreased fecundity on exposure to MPs was also noted by Lee et al. (2013), probably as a result of insufficient nutrition which led to reduced egg production or because of physical inhibition of copepod fertilization by MPs. Enhanced mortality (up to ninefold higher) due to MP exposure was seen in the detrivore caddisfly (López-Rojo et al., 2020). Moreover, MPs affect other ecosystem processes like the decomposition of leaf litter by inducing toxic effects in the organisms concerned with the decomposition procedure (López-Rojo et al., 2020).

The impact of MPs extends beyond physical and chemical effects. Studies have provided evidence of behavioral changes in aquatic organisms exposed to MP particles (Chen et al., 2020). For example, the exposure of the common goby fish to MPs along with food confused the fish and led to the ingestion of all MP particles. It led to a significant reduction in predatory behavior, efficiency, and prey selection capability of fishes (de Sá et al., 2015). In adult zebrafish, Chen et al. (2020) noticed hyperactivity upon exposure to MPs. The fish stayed in manic and active states for extended durations and its swimming distance increased by 1.3–2.4 times compared to the control. These behavioral changes could have cascading effects on their survival, fitness, and interactions with other species and can subsequently impact entire food chains and ecosystem dynamics (de Sá et al., 2015).

Although numerous studies have determined the effect of MPs on organisms, none have investigated the effect of a plethora of irregularly shaped MP polymers at concentrations that are present in the environment. This is important considering that Urbina et al. (2023) observed no significant effect of short-term exposure to extremely high MP concentrations, but long-term exposure to even low concentrations caused severe nutritional damage, reduced feeding, affected energy uptake, increased filtration rates, utilized lipid and carbohydrate reserves, and resulted in weight loss. Binelli et al. (2020) sampled the MPs in four Italian lakes and applied the MP cocktail (mainly PP and PE) to zebra mussels (*Dreissena polymorpha*) for a week to determine the effect of MPs on these mussels. The significant modifications in the antioxidant and detoxifying enzyme activity and decrease in cell viability indicated that the organisms suffered due to the MPs rather than having a defensive response.

5.3 Impact of Nanoplastics on Aquatic Organisms

As mentioned earlier, there is a lack of studies on extracting and detecting NPs in environmental samples. Hence, the potential toxicity effects of NPs on organisms are confined mostly to laboratory studies. NPs are very small and have a large surface area compared to MPs; hence, they can have different ecological effects on aquatic organisms. Owing to their smaller sizes, NPs are more bioavailable to organisms and potentially more lethal (Brandts et al., 2018). Their unique properties, attributable to their nano-size, make them readily consumable by organisms, capable of penetrating tissues, and prone to accumulation within an organism's body (Kelpsiene et al., 2020). The uptake of NPs by aquatic organisms can occur through direct ingestion, trophic transfer into the food chain, and adsorption on the surfaces of living organisms (Brandts et al., 2018; Cole et al., 2020). Another less explored pathway involves the transference of NPs from the mother to offspring (Pitt et al., 2018; Zhang et al., 2020). Nonetheless, different uptake pathways exhibit varying toxicity effects, as witnessed by Zhang et al., 2020 in zebrafish, where waterborne NP exposure was more hazardous than injection exposure (maternal transfer). The studies on the harmful effects of NPs on freshwater and marine organisms are cataloged in Table 5.

The toxicity effects of ingested NPs have been explored in fish (Cedervall et al., 2012; Chen et al., 2017; Brandts et al., 2018; Gu et al., 2020), crustaceans (Li et al., 2021a; Pikuda et al., 2022), and mussels (Cole et al., 2020; Sendra et al., 2020). In the fish *Dicentrarchus labrax*, provisional exposure to NPs altered molecular signaling pathways related to lipid metabolism and decreased esterase activity in the plasma, pointing out the negation of the immune system (Brandts et al., 2018). The disturbance of fat metabolism was also reported by Cedervall et al. (2012) in the Crucian carp, wherein NP ingestion restrained the fish from utilizing its fat reserves by binding to apolipoprotein A-I. They also noted a change in the fish's feeding habits, with the time required for food consumption doubling when exposed to NPs. In the yellow croaker fish, NP exposure produced a detrimental effect on the intestine (Gu et al., 2020). They impeded the function of digestive enzymes and caused significant changes in the amount of three major bacterial groups (*Bacteroidetes, Proteobacteria*, and *Firmicutes*) in the gut, together with significant growth of potentially harmful bacteria. In addition, NPs significantly relegated lysozyme activity and specific growth rate (SGR) while augmenting mortality. The fish attempted to enhance gut probiotics to deal with the proliferation of harmful bacteria, but lower digestive enzyme activity and immunity along with lower growth and heightened mortality insinuated a breakdown of the self-regulating system of the fish (Gu et al., 2020). Exposure to high NP concentrations has been found to produce a negative effect on energy metabolism. In prawn, Li et al. (2021a) noticed the inhibition of the expression of genes associated with energy metabolism, causing a decrease in related enzyme activity and thereby energy metabolism.

In the zebrafish larvae, Chen et al. (2017) mentioned that MPs produced no effect on larval locomotion while NPs inhibited larval locomotion by 22%. Additionally, NPs significantly reduced larvae body length by 6%, inhibited acetylcholinesterase (AChE) (a crucial enzyme in the nervous system) activity by 40%, and upregulated the expression of the genes linked to nervous system development (*gfap* and *α1-tubulin*) and visual activity (*zfrho* and *zfblue*). Likewise, Cole et al. (2020) noted that particle size influenced the toxicity in mussels; while MP exposure did not induce any toxic effects, NP exposure increased the amount of phagocytic

Table 5 Effect of nanoplastics on aquatic organisms

Organism	Type, size, concentration of nanoplastics, and duration of experiment	Toxicity effect	References
Crucian carp (*Carassius carassius*)	PS NPs; 24 nm; 0.01% (w/v); 14, 21, 27, 39 day	Behavioral changes; disturbance in lipid metabolism; NPs transfer in food chain	Cedervall et al. (2012)
Crucian carp (*Carassius carassius*)	PS NPs; 24 and 27 nm; 0.01% (w/v); 24, 61 day	Alterations in behavior in terms of reduced activity and feeding rate; genetic changes in liver and muscle metabolism; morphological changes in brain and muscles	Mattsson et al. (2015)
Zebrafish larvae (*Danio rerio*)	PS NPs; 50 nm; 1 mg/L; 120 h	Significant developmental neurotoxicity; significantly reduced larvae body length; inhibition of larval locomotion; inhibition of acetylcholinesterase activity and upregulation of genes involved in nervous system and visual system	Chen et al. (2017)
Marine fish (*Dicentrarchus labrax*)	Polymethylmethacrylate (PMMA) NPs; 45 nm; 0.02, 0.2, 2, 20 mg/L; 96 h	Significant increase in abundance of mRNA transcripts linked to lipid metabolism; lessened esterase activity levels in plasma denoting compromised immune system; significantly lower levels of alkaline phosphatase in skin mucus	Brandts et al. (2018)
Zebrafish (*Danio rerio*)	PS NP; 42 nm; 10% of food by mass; 7 day	Transfer of NPs from mother to offspring; accumulation of NPs in yolk sac, gastrointestinal tract, liver, and pancreas of offspring; impact on offspring physiology; modified antioxidant system in adults and larvae	Pitt et al. (2018)
Daphnia magna	Positively charged aminated PS NPs; 53 nm; 0.32 mg/L; negatively charged carboxylated PS-NPs; 26 and 62 nm; 25 and 50 mg/L; 103 day	Decrease in duration of life (3 times lesser); reduced number of offspring	Kelpsiene et al. (2020)

(continued)

Table 5 (continued)

Organism	Type, size, concentration of nanoplastics, and duration of experiment	Toxicity effect	References
Large yellow croaker fish (*Larimichthys crocea*)	PS NPs; 100 nm; 0, 10, 104, 106 items/L; 14 day	Reduced digestive enzyme activities; significant changes in the ratio of three major bacterial groups in the gut; significant increase in pathogenic bacteria; significantly reduced lysozyme activity and specific growth rate (SGR), and enhanced mortality	Gu et al. (2020)
Zebrafish (*Danio rerio*)	PS NPs; 42 nm; 96 hpf	Both waterborne and injection exposure to NPs caused accumulation of NPs in the yolk sac, brain, eyes, gut, and swim bladder; injection caused NP accumulation mainly in the trunk area; aqueous exposure caused higher NP concentrations in the brain and eyes, reduced antioxidant gene expression, increased frequency of developmental abnormalities, resulted in lower growth rates, and hypoactivity; waterborne exposure caused higher NP transfer to the brain and higher toxicity	Zhang et al. (2020)
Mussels (*Mytilus* spp.)	PS NPs; 50 nm; 500 ng/mL; 24 h or 7 day	Significant effect on hyalinocyte granulocyte ratios, suggestive of an amplified immune response; 24-h exposure increased superoxide dismutase activity	Cole et al. (2020)
Mussels (*Mytilus galloprovincialis*)	PS NPs; 50 nm, 100 nm, 1 μm; 10 mg/L; 3 h, 24 h	Size-dependent transfer of NPs to hemolymph after 3 h of exposure; NPs in digestive gland and muscles; NPs exposed hemocytes experienced motility changes, apoptosis, production of ROS, phagocytosis	Sendra et al. (2020)
Prawn (*Macrobrachium nipponense*)	PS NPs; 75 nm; 0, 5, 10, 20, 40 mg/L; 7, 14, 21, 28 day	Decreased survival rate; fluctuation in weight gain rate and molting number; negative impact on energy metabolism	Li et al. (2021a)
Daphnia magna	PS-NPs; 20 and 200 nm; 0.1 and 50 mg/L; 21 day	Impact on growth, molting, and reproduction	Pikuda et al. (2022)

hemocytes, indicative of an intensified immune response, and caused a noticeable rise in micronuclei formation. This is presumably because NPs can easily enter the circulatory tissues and fluid, eliciting an immunological response and inducing cytotoxicity (Sendra et al., 2020). NPs at even environmentally relevant concentrations (0.1 mg/L) were found to affect growth, molting, and reproduction in *Daphnia* (Pikuda et al., 2022). However, higher concentrations (50 mg/L) were found to induce additional toxicity symptoms such as altered hopping frequency and modifications in swimming behavior, probably a result of increased NP particle count in the medium (Pikuda et al., 2022).

5.4 Combined Effect of Micro-/Nanoplastics and Other Contaminants

MP and NPs can serve as potential carriers for numerous environmental pollutants, such as POPs, HMs, and additives, enabling their widespread transport (Tan et al., 2019; Fraser et al., 2020; Sarkar et al., 2021) (Table 6 and Fig. 2). Because of their small size, elevated surface area relative to volume, and hydrophobic properties, MPs have a high affinity for contaminants (Tan et al., 2019; Fraser et al., 2020). In the surface sediments of Qiantang River, China, Fraser et al. (2020) reported a considerable correlation between microbeads and films with polychlorinated biphenyls (PCBs) based on a cluster analysis. Microbeads and films exhibit a greater ratio of surface area to volume when compared to other MP shapes, making them advantageous for adsorbing organic pollutants like PCBs. Research has documented the association of MPs with PCBs (Rios et al., 2007; Lo et al., 2019), PAHs (Rios et al., 2007; Tan et al., 2019; Lo et al., 2019), organochlorinated pesticides (Rios et al., 2007; Lo et al., 2019), polybrominated diphenyl ethers (PBDEs) (Rochman et al., 2014), and pharmaceuticals (Klavins et al., 2022). Several studies have detailed the association of MPs with HMs in both water and sediments (Brennecke et al., 2016; Li et al., 2020a, b; Ta & Babel, 2020; Sarkar et al., 2021). The modifications in environmental parameters will influence the relationship between contaminants and MPs, altering the bioavailability as well as accumulation of contaminants. The attributes of MPs, such as their size and polymer composition, influence their capacity to adsorb or leach the environmental contaminants and inherent plasticizer additives. MPs with smaller sizes and large surface area per unit mass exhibit a greater potential for surface chemical interaction and binding of contaminants and display heightened toxicity effects (Tang et al., 2020; Yuan et al., 2020). Moreover, weathering and degradation of MPs play a role as weathered particles have cracks, abrasions, and flakes on their surface, resulting in enhanced surface area and thereby increased adsorption (Tan et al., 2019).

Plastics also contain certain chemical additives or plasticizers, such as bisphenol-A, HMs, phthalate esters, alkylphenols, and polybrominated diphenyl ethers, which are incorporated to improve the elasticity, life, color, and heat resistance of

Table 6 Association of micro–/nanoplastics with contaminants/microorganisms in the aquatic environment

Study	Location	Observations
Organic pollutants		
Rios et al. (2007)	North Pacific Gyre, selected sites in California, Hawaii, and from Guadalupe Island, Mexico	Plastic debris is a POPs trap The total concentration of PCBs was 27–980 ng/g, DDTs was 22–7100 ng/g, PAHs was 39–1200 ng/g, and aliphatic hydrocarbons was 1.1–8600 μg/g
Rochman et al. (2014)	South Atlantic Ocean	A positive relationship was established between the density of plastic debris and the chemical body burden of BDE#s 183–209 in mesopelagic lanternfishes PBDEs can transfer from plastic to organisms, including fish, upon ingestion Higher brominated PBDEs act as an indicator of the exposure of plastic debris to marine organisms
Lo et al. (2019)	Hong Kong sandy beaches, China	In sedimenary MPs, the concentration of ΣPAHs was 70.8–1509 ng/g, ΣPCBs was 13–1083 ng/g, DDT and DDX was 1.96–626 ng/g
Tan et al. (2019)	Feilaixia reservoir, Beijang River, China	The total concentration of 16 priority PAHs on PE, PP, and expanded PS MPs ranged from 282.4 to 427.3 ng/g Chrysene, benzo [ghi] perylene, and phenanthrene were abundant at concentrations 39.5–89.6 ng/g, 34.6–56.8 ng/g, and 25.6–45.6 ng/g, respectively
Fraser et al. (2020)	Qiantang River and Hangzhou Bay, China	The total mean PCBs concentration in the surface sediment was between 1.47 and 1.65 ng/g DW Considerable correlation exists between films and microbeads to both highly chlorinated and less chlorinated PCB congeners
Klavins et al. (2022)	Laboratory study	Sorption of pharmaceuticals was based on the salinity of the aquatic environment; lower salinity increases the sorption ability of MPs and the formation of MP-pharmaceutical complexes In rivers with high organic matter, the MP-pharmaceutical complex absorbs humic substances, forming a large complex that can be ingested by aquatic animals
Heavy metals		
Brennecke et al. (2016)	Laboratory study	There is a significant interaction between MPs and HMs Adsorption of Cu was significantly greater on PVC fragments than on PS beads, likely due to higher surface area and polarity of PVC Concentration of Cu and Zn increased significantly on PVC and PS MPs during the duration of the experiment (except Zn on PS)
He et al. (2020)	Brisbane River, Australia	The concentrations of metals (Al, As, Cr, Co, Cu, Fe, Mg, Mn, Ni, Cd, Se, Sr, Zn, Pb) in river sediments were mainly dependent on their source, rather than the concentration of MPs The risk associated with MPs is based on their monomer composition rather than the quantities present

(continued)

Table 6 (continued)

Study	Location	Observations
Ta and Babel (2020)	Chao Phraya River Estuary, Thailand	Toxic metals (Cr, Cu, Ni, Pb, Cd, and Zn) were found in MPs from surface water samples
Li et al. (2020a)	Hong Kong sandy beaches, China	On MPs, Fe had the highest median concentration (302 mg/kg) followed by Zn (19.6 mg/kg) and Mn (18.6 mg/kg) Very low concentrations of Cu (0.89 mg/kg), Ni (0.15 mg/kg), As (<LOD) and Cd (<LOD) were measured
Sarkar et al. (2021)	Freshwater wetland, eastern India	High content (2.98–439.32 μg/g) of toxic metals (As, Cd, Cr, Cu, Ni, Pb, Zn) was found on PET and PE MPs
Microorganisms/pathogens/antibiotic resistance genes		
Viršek et al. (2017)	Sea surface, North Adriatic	The study identified 28 bacterial species on MPs, including *Aeromonas* spp. and hydrocarbon-degrading bacterial species The study identified the pathogenic fish bacteria *Aeromonas salmonicida* on MPs for the first time
Eckert et al. (2018)	Lake Maggiore and WWTP in Verbania (Italy)	MPs promoted the persistence of typical indicators of microbial anthropogenic pollution in natural waters With increasing MP concentration, the abundance of integrase 1 (int1; a marker associated with mobile genetic elements known as a proxy for anthropogenic effects on the spread of antimicrobial resistance genes) increased in the plastisphere, but not in the surrounding water The microbial community on MPs was more similar to the original wastewater community
Wu et al. (2019b)	Haihe River, China	MP biofilm had a distinctive community structure with the highest node-connected community, lower average path length, network diameter, and modularity compared with natural particles biofilm MP biofilm had broad spectrum and distinctive resistome; it hosted two opportunistic human pathogens (*Pseudomonas monteilii*, *Pseudomonas mendocina*) and one plant pathogen (*Pseudomonas syringae*) MP is a novel microbial niche and may act as a vector for antibiotic resistance genes (ARGs) and pathogens
Wu et al. (2020)	Haihe estuary, China	MPs supplemented the bacteria (e.g., Halobacteriaceae and Pseudoalteromonadaceae) and depreciated the effect of environmental variation. MPs contained significantly higher concentrations of potentially pathogenic bacteria (e.g., *Pseudomonas* and *Bacillus*) than the ambient environment Minor effect of polymer types on the abundance and structure of the bacterial communities Compared with surface water and sediments, MPs were a good habitat for bacterial communities

(continued)

Table 6 (continued)

Study	Location	Observations
Xue et al. (2020)	Urumqi River, China	The diversity of bacteria on MPs was higher than in the surrounding water Plastic debris could accumulate a higher abundance of some potential pathogens than surrounding water even at high antibiotic concentrations About 92.3–99.7% of bacteria communities in downstream water were not from upstream but from the input of downstream human activities; anthropogenic activities had great effects on microbial community structure and functional groups Plastisphere possessed higher ecological functional diversity than the planktonic bacteria and plastisphere communities were much more resistant to human disturbance than planktonic bacteria
Li et al. (2021b)	Beilun River, Sino-Vietnamese border	The antibiotic resistance genes (ARGs) on MPs showed a clear spatial distribution with the abundance increasing by 2 orders of magnitude from rural to urban regions Urbanization predominantly contributed to both the abundance and potential mobile genetic elements (MGEs)-mediated dissemination of ARGs in riverine MPs MP types significantly affected the resistome and dissemination risk of ARGs; PP was the preferred substrate to acquire and spread ARGs

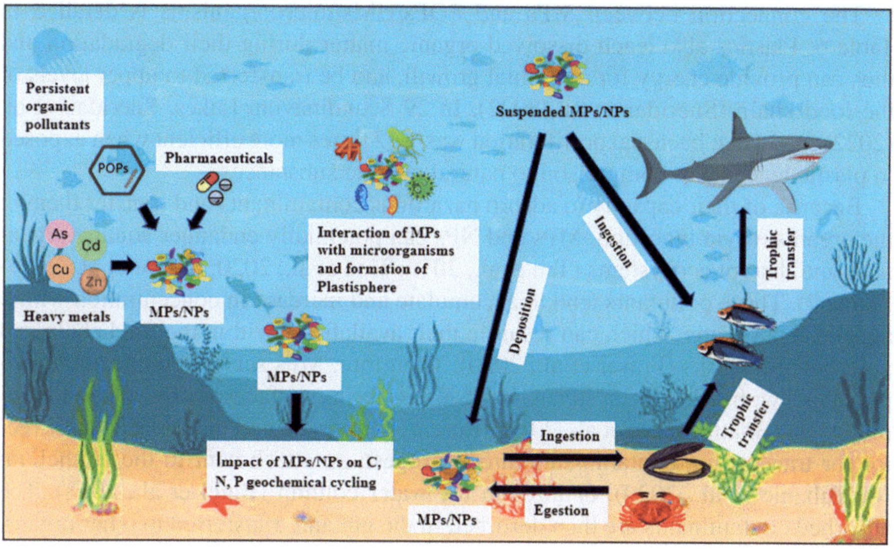

Fig. 2 Interaction of MPs and NPs with POPs/HMs, microorganisms, and their trophic transfer

plastic products. Many of these additives are not tightly sorbed to the plastic polymer and are leached into the tissues of organisms or the surrounding environment (Rochman et al., 2014; He et al., 2020; Ta and Babel, 2020; Sheridan et al., 2022). Bandow et al. (2017) exposed high-density polyethylene (HDPE), PVC, PS, PP, and PET MPs to artificial aging circumstances (thermo- and photo-oxidation) for 2000 h to determine their leaching potential. They observed the leaching of substances incorporated in plastic materials (Cu, Zn, Cl, Ca, and organic carbon) into the water either directly or as metabolites after degradation.

The degradation of MPs and NPs produces persistent carbon-based substrates for microbial growth and biofilm formation. The colonization of plastic debris by various microbial communities results in the formation of a unique niche called "plastisphere." It changes the physical features of MPs, such as increasing their density, which leads them to sink, making them available for ingestion by benthic creatures, and lowering their hydrophobicity, which makes them less susceptible to contamination by organic contaminants. The microbial communities proliferating on plastic fragments are distinct from those found in the surrounding environment, indicating that the functions of microbes on the plastisphere may differ considerably from those in the surrounding environment (Xue et al., 2020; Li et al., 2021a, b). The plastisphere enables increased transfer of plasmids and transports attached microorganisms over long distances, thus altering their natural range and causing them to become invasive species or disease vectors. It also raises the likelihood of MPs being misinterpreted for food by organisms at higher trophic levels (Viršek et al., 2017; Eckert et al., 2018; Wu et al., 2019b, 2020).

The connection between MPs and POPs/HMs/microorganisms is detailed in Table 6. Plastics also leach dissolved organic matter during their degradation and thus can provide energy for microbial growth, and be transferred to upper levels of the food chain (Sheridan et al., 2022). In 29 Scandinavian Lakes, Sheridan et al. (2022) found that bacterial proliferation was 1.72 times more efficient when exposed to plastic leachate in comparison to natural organic substances.

Because of their capacity to adsorb hazardous contaminants and transfer them to the organisms via ingestion, MPs and NPs can potentially endanger aquatic organisms and disrupt ecosystems (Tan et al., 2019; Tien et al., 2020; Sarkar et al., 2021) (Table 7). These pollutants tend to accumulate and increase in concentration within aquatic ecosystems, which can result in their availability within the food chain and pose potential risks (Fraser et al., 2020). Therefore, MPs serve as point sources of pollutants within aquatic systems, while also offering surfaces for the adsorption of pollutants (Bhattacharya et al., 2010). Research has indicated that MPs act as a vector for transferring contaminants through ingestion, attachment to the epithelium (Le Bihanic et al., 2020), or through the water column (Batel et al., 2018). The adsorbed contaminants are then desorbed in the gut and transferred to other organs. Tien et al. (2020) stated that the occurrence of PAHs in the muscles of three fish species in the Fengshan River was significantly positively correlated to the MPs in fish, water, and sediments. Sarkar et al. (2021) stated that HM abundance (As, Cd, Cr, Cu, Ni, Pb) in the fish muscles was associated with the HM content of the MPs, pointing out that MPs acted as HM carriers. Furthermore, MPs increase the toxicity of other chemical pollutants (Rainieri et al., 2018) even when they do not produce any toxicity themselves (Jinhui et al., 2019). In zebrafish, Rainieri et al. (2018) demonstrated that MPs with sorbed contaminants (PCBs, PBDEs, multiple perfluoro compounds, methylmercury) affected the liver to a greater extent than contaminants alone. This was also confirmed by Banaee et al. (2019), wherein MPs and Cd had a synergistic effect on each other, and MPs-Cd mixtures triggered maximum toxicity. Several studies have provided insights into the effects of organic pollutants in aquatic organisms (Table 7). Avio et al. (2015) studied pyrene-contaminated MPs in mussels (*Mytilus galloprovincialis*) and reported that after 7 days of exposure the amount of pyrene in the digestive glands was more than on the MP particles, indicating enhanced bioavailability. The MPs were detected in the digestive glands, intestine, gills, and haemolymph, and toxicity effects such as changes in immunological responses, lysosomal compartment, peroxisomal proliferation, antioxidant system, neurotoxicity, genotoxicity, and deviations in gene expression were observed. Likewise, exposure to HMs has been observed to produce similar toxicity effects (Table 7). In the common carp, Banaee et al. (2019) detected alterations in the biochemical parameters and intrinsic immunological factors, including decreased plasma activities of AChE and gammaglutamyl-transferase (GGT), increased aspartate aminotransferase (AST), alanine aminotransferase (ALT), lactate dehydrogenase (LDH), and alkaline phosphatase (ALP), decreased plasma total protein, albumin, and globulin levels, and increased levels of glucose, triglyceride, and cholesterol, lower lysozyme activity, and low levels of total immunoglobulins on exposure to MPs-Cd.

5.5 Trophic Transfer of MPs and NPs

Although the trophic transfer of MPs/NPs is anticipated, a constrained amount of literature exists on this aspect. Trophic transfer denotes an indirect but significant route through which MPs are incorporated into species that are part of a prey-predator relationship (Nelms et al., 2018) (Fig. 2). The first study showcasing the natural trophic transfer of MPs was conducted by Farrell & Nelson, 2013. The mussels (*Mytilus edulis*) were exposed to PS MPs before being fed to crabs (*Carcinus maenas*). The authors discovered MPs in the crab's stomach, hepatopancreas, ovaries, and gills, roughly about 0.04% of the quantity to which the mussels were exposed remaining in the crab's hemolymph. In another study, Setälä et al. (2014) depicted the potential transfer of MPs through planktonic organisms from one trophic level (mesozooplankton) to a higher trophic level (macrozooplankton). They determined the potential of different Baltic Sea zooplankton taxa (Mysid shrimps, copepods, cladocerans, rotifers, polychaete larvae, and ciliates) to ingest MPs and identified the pelagic polychaete larvae, *Marenzelleria* spp., as having the highest proportion of individuals containing ingested MP spheres. These larvae were fed to mysid shrimps and the presence of the zooplankton prey and MPs in the mysid intestine was noted. Nelms et al. (2018) provided proof of the transference of MPs

Table 7 Effect of micro–/nanoplastics-associated contaminants on aquatic organisms/primary producers

Organism/Primary producers	Type and concentration of MP/NP	Contaminant	Duration	Toxicity effect	References
Organic pollutants					
Mussel (*Mytilus galloprovincialis*)	PE and PS (<100 μm, 1.5 g/L)	PAHs pyrene 200–260 ng/g	7 day	Tissue localization of MPs and pyrene in digestive tissues, hemolymph, gills; antagonistic cellular effects (e.g., alteration in immunological responses, lysosomal compartment, peroxisomal proliferation, antioxidant defenses, neurotoxicity, genotoxicity); altered gene expression	Avio et al. (2015)
Zebrafish (*Danio rerio*)	Fluorescent PE MPs; 1–5 μm and 10–20 μm	PAHs (Resorufin, resorufin ethyl ether, Benzo[a]pyrene); 252 μg/3 mg MPs	6 and 24 h	No permanent accumulation in gills; high number of smaller and heavier MPs on zebrafish egg chorion; BaP spiked MP in gill-induced cytochrome P450 1A; MPs can transfer POPs by attachment to epithelia	Batel et al. (2018)
Red tilapia (*Oreochromis niloticus*)	Fluorescent PS MPs; 0.1 μm; 1, 10, 100 μg/L	Roxithromycin; 50 μg/L	14 day	Augmented bioaccumulation of roxithromycin in fish gut, gills, brain, and liver due to MPs; MPs reduced roxithromycin neurotoxicity; altered cytochrome P450 enzyme activity in liver; MPs-roxithromycin exposure mitigated oxidative damage	Zhang et al. (2019)
Marine medaka fish (*Oryzias melastigma*)	PE MPs; 4–6 μm	Treatments: MPs + Benzo(a)pyrene (BaP); 0.01 and 16.64 μg BaP/g of MP; MPs + Perfluorooctanesulfonic acid (PFOS); 0.12 and 55.65 μg PFOS/g of MP; MPs + Benzophenone-3 (BP3); 0.14 and 24 ng BP3/g of MP	12 day	No ingestion, only agglomeration on surface; MPs-PFOS decreased embryonic survival and hatching; MPs-BaP and MPs-BP3 lessened growth, amplified developmental anomalies, and abnormal behavior	Le Bihanic et al. (2020)

Blood clam (*Tegillarca granosa*)	PS MPs; 500 nm and 30 μm; 1 mg/L	Benzo[*a*]pyrene (5, 50 μg/L) and 17β-estradiol (E2) (0.1, 1 μg/L)	96 h	Immunotoxicity indicated by alterations in haemocyte count, blood cell composition, phagocytic activity, amount of ROS, Ca and lysozyme concentration, lysozyme activity; alteration of Ca signaling and apoptosis-related genes; MPs size dependence of PAHs and E2 toxicity	Tang et al. (2020)
Swamp ghost crab (*Ucides cordatus*)	PE MPs; 100–250 μm; 250 mg/L	Triclosan with dimethylsulfoxide (DMSO) in ratio 1 mg: 0.5 ml	3 day, 7 day	In gills, altered activity of the biomarkers ethoxyresorufin O-deethylase (EROD), dibenzylfluorescein dealkylase (DBF), glutathione S-transferase (GST), and glutathione peroxidase (GPx), reduced glutathione (GSH), and lipid peroxidation (LPO), DNA damage	Nobre et al. (2022)
Moss and caddisflies	HDPE, LDPE, PS, PP; <500 μm; 2000 MPs/L	Pharmaceuticals and endocrine disrupting compounds (1H-benzotriazole, bisphenol-A, caffeine, gemfibrozil, ketoprofen, methylparaben, estriol, diphenhydramine, tris (1-chloro-2-propyl) phosphate); 500 ng/L for each compound	0, 15, 30, 60 day	Transfer of MPs from moss to caddisflies; MPs and pharmaceuticals produced stress in caddisflies comparable to that produced by direct exposure to pharmaceuticals; alterations in phenotypic response	Grgić et al. (2023)
Heavy metals					
Zebrafish (*Danio rerio*)	PS MPs; 5 μm; 20 and 200 μg/L	Cd; 10 μg/L	21 day	Enhanced bioaccumulation and toxicity of Cd in zebrafish tissues, gills>guts>liver; oxidative damage and inflammation	Lu et al. (2018)
Common carp (*Cyprinus carpio*)	PE microbeads from face scrub	Cd Treatments: 100 and 200 μg/L cadmium chloride; 250 and 500 μg/L MPs; 100 μg/L Cd and 250 μg/L MPs; 100 μg/L Cd and 500 μg/L MPs	30 day	Changes in blood biochemical and immunological parameters by MPs or Cd alone; synergistic effect of MP and Cd on each other and enhanced toxicity of MPs-Cd mixture	Banaee et al. (2019)

(continued)

Table 7 (continued)

Organism/Primary producers	Type and concentration of MP/NP	Contaminant	Duration	Toxicity effect	References
Yellow seahorse (*Hippocampus kuda* Bleeker)	HDPE MPs; 15–80 μm	Cu (0.05 mg/L) + cd (0.01 mg/L) + Pb (0.05 mg/L)	45 day	Accumulation of MPs-HMs caused adverse effect on seahorse growth (e.g., body length, body weight, specific growth rate, survival rate), oxidative damage	Jinhui et al. (2019)
Daphnia magna	PS MPs; 10 μm and 50 μm	Pb, Cu, Cd, Ni; 50 mg/L	48 h	Smaller MPs exhibited higher adsorption capacity for metal ions and higher toxicological risks	Yuan et al. (2020)
Asian Clam (*Corbicula fluminea*)	PS MPs; up to 200 μm; 2 mg/L	Cd; 10 μg/L	7 day	Oxidative stress, damage and neurotoxicity; synergetic effect in gill and gonad, and antagonism response in digestive gland	Parra et al. (2021)
Chinese mitten crab (*Eriocheir sinensis*)	PS MPs; 0 and 400 μg/L	Pb; 0, 5, 50 μg/L	21 day	Increased bioaccumulation of Pb in hepatopancreas and liver injury; aggravated oxidative damage; effect on fatty acid synthesis	Yang et al. (2022)
Primary producers					
Chlorella sp. TJ6-5 *Pseudokirchneriella subcapitata* NIES-35	PS MPs (10 μm) and PS NPs (500 nm)	Cu (0, 65.5, 81.9, 102.4, 128, 160, and 200 μM) Cu (0, 72.3, 86.8, 104.2, 125, and 150 μM)	–	MP/NP increased the toxicity of Cu at EC50 in both microalgae; single and combined treatment of MP/NP + Cu caused higher oxidative stress and morphological and ultrastructural changes in both microalgae	Wan et al. (2021)
Microcystis aeruginosa	PS MPs (50 μm) and PS NPs (80 nm); 0.5, 1.0, 2.0, 5.0 mg/L	Cd (0.2, 0.5, 1.0, 1.5, 2.0 mg/L)	–	Negative effect on growth; enhanced toxicity due to MPs + Cd/NPs + Cd	Wang et al. (2023)

across trophic levels, from the Atlantic mackerel (*Scomber scombrus*) to the top marine predator mammal grey seals (*Halichoerus grypus*). Other investigations have described the transfer of MPs from marine aggregates to mussels (*Mytulis edulis*), highlighting the role of aggregates in removing MPs from the ocean surface and enabling their transfer into marine food webs (Zhao et al., 2018). They also established a positive association between the abundance of MPs within organisms and their trophic level, indicating that species belonging to higher trophic levels contained more MPs (Wang et al., 2021b).

The transfer of MPs to aquatic food webs through herbivory, an understudied pathway, has also been established (Grgić et al., 2023). In the freshwater ecosystem, Mateos-Cárdenas et al. (2019) demonstrated the trophic transfer of MPs from *Lemna minor* to *Gammarus duebeni*. They found that only 28.57% of the amphipods could retain one or two MPs in the gut, but the MPs did not cause any deleterious effects in the amphipod. In the marine ecosystem, Goss et al. (2018) reported the presence of MPs on blades of the seagrass *Thalassia testudinum*, a primary producer. This seagrass loaded with epibionts and MPs is consumed by *Gastropoda*, *Echinoidea*, and parrotfishes, thereby enabling MPs transmission to higher trophic levels.

A similar transmission of NPs through the trophic chain has been observed. Cedervall et al. (2012) demonstrated the uptake and transfer of NPs in an aquatic food chain, starting from algae (*Scenedesmus* sp.) to zooplankton (*Daphnia magna*) and finally to the Crucian carp fish (*Carassius carassius*). The ingestion of NPs-containing zooplankton altered the feeding patterns of the fish, decreased lipid metabolism, and caused weight loss. In another study, Mattsson et al. (2015) established the transference of NPs across three trophic levels, starting with algae, followed by zooplankton, and ending with carp fish. They observed harsh effects on the metabolism, feeding, and shoaling behavior of the fish.

5.6 Impact on Ecosystem Services

The occurrence of MPs and NPs in aquatic settings can also disturb ecosystems at a broader level. They can alter water quality, affect light penetration, and nutrient and contaminant cycling. Moreover, the adsorption of HMs/POPs and the accumulation of microbial communities on plastic surfaces can have both direct and indirect ecological consequences (Bowman et al., 2021; Yin et al., 2023). Sediments are repositories of nutrients such as phosphorous (P), nitrogen (N), and carbon (C) in aquatic ecosystems, playing a critical role in the geochemical cycling of these nutrients (Yin et al., 2023). Phosphorus may be liberated into the water by desorption, dissolution, and mineralization, whereas nitrogen can be emitted via nitrification, ammonification, and denitrification, and multiple microorganisms act as drivers of these cycling processes. Additionally, sediments act as repositories for MPs (see Sect. 4) (He et al., 2020; Yin et al., 2023). Seeley et al. (2020) studied the effect of MPs (PE, PVC, polylactic acid [PLA], polyurethane foam [PUF]) on microbial community and nitrogen cycling in a microcosm using salt marsh sediment. They observed that MPs altered the N cycling process and microbial community composition by serving as organic carbon substrates for microbes. PUF and PLA MPs boosted nitrification and denitrification processes, while PE MPs could only slightly promote the process, and PVC hindered both processes. The effect of PP, PVC, and PLA MPs on microbial communities and N and P cycling in freshwater sediments was inspected by Yin et al., 2023. They saw that MPs had a considerable impact on the microbial community composition and nutrient cycling. PP and PVC MPs could stimulate microbial nitrification and nitrite oxidation, while PP could substantially boost alkaline phosphatase (ALP) activity and the presence of phosphorus-regulation gene. On the other hand, PLA MPs demonstrated the capacity to increase microbial phosphorus transporter, nitrogen fixation, and denitrification genes while inhibiting nitrification, resulting in enormous ammonia nitrogen buildup and release. The PLA MPs could boost sedimentary phosphorus release by enhancing dissimilatory iron and sulfite reduction despite suppressing ALP activity and the abundance of the organophosphorus mineralization gene. MPs were also found to alter the biogeochemical cycling of mercury (Hg) in surface waters of Lake Erie, Sacramento River, and Sand Francisco Bay by harboring Hg methylating microbes, adsorbing inorganic Hg and monomethyl mercury (MMHg), sheltering *mer* genes, increasing rates of rapid Hg methylation, and interfering with the indirect photochemical decomposition of MMHg (Bowman et al., 2021). MPs have also been documented to alter nutrient fluxes by affecting organisms. A recent study reported that MPs affected the bioturbator worm *Tubifex tubifex*, also known as ecosystem engineers, and thereby altered biogeochemical processes. The exposure of the tubificid worms to medium MPs concentration (7000 particles/kg sediment DW) enhanced the oxidative stress in the worms, which led to decreased bioturbation activities, and hence inhibited nutrient fluxes and organic matter mineralization (Wazne et al., 2023). Alterations in nutrient cycling, sediment transport, and light penetration can have cascading effects on ecosystem dynamics, ultimately impacting biodiversity and ecosystem services.

6 Conclusion

The damaging effect of MPs and NPs on the aquatic environment is an intricate issue with far-reaching consequences for both ecosystems and human health. Research has shown that these minute plastic particles, derived from the disintegration of bigger plastic items or purposefully manufactured at small scales, have permeated a variety of water bodies, ranging from rivers and lakes to seas and oceans. The influences of MPs and NPs on aquatic ecosystems are multifaceted, encompassing a wide range of biological, ecological, and environmental repercussions. Addressing the issue of MPs and NPs in aquatic habitats necessitates a holistic strategy that includes decreasing plastic waste and improving waste management, creating alternative materials, implementing regulations, and raising public awareness. Urgent and coordinated efforts at the local, regional, and global levels are necessary to reduce plastic pollution and its probable effect on aquatic ecosystems and human health. We can endeavor to preserve the health of aquatic ecosystems and protect both the environment and human well-being by comprehending the potential effects and adopting appropriate mitigation techniques.

Acknowledgments This research was supported by the National Laboratory for Water Science and Water Security (RRF-2.3.1-21-2022-00008).

References

Almeshal, W., Tak, A., Sandil, S., Dobosy, P., & Gyula, Z. (2022). Comparison of freshwater mussels Unio tumidus and Unio crassus as biomonitors of microplastic contamination of Tisza River (Hungary). *Environments, 9*, 122.

Andrady, A. L. (2011). Microplastics in the marine environment. *Marine Pollution Bulletin, 62*, 1596–1605. https://doi.org/10.1016/j.marpolbul.2011.05.030

Avio, C. G., Gorbi, S., Milan, M., Benedetti, M., Fattorini, D., D'Errico, G., Pauletto, M., Bargelloni, L., & Regoli, F. (2015). Pollutants bioavailability and toxicological risk from microplastics to marine mussels. *Environmental Pollution, 198*, 211–222. https://doi.org/10.1016/j.envpol.2014.12.021

Baldwin, A. K., Corsi, S. R., & Mason, S. A. (2016). Plastic debris in 29 Great Lakes tributaries: Relations to watershed attributes and hydrology. *Environmental Science & Technology, 50*, 10377–10385. https://doi.org/10.1021/acs.est.6b02917

Banaee, M., Soltanian, S., Sureda, A., Gholamhosseini, A., Haghi, B. N., Akhlaghi, M., & Derikvandy, A. (2019). Evaluation of single and combined effects of cadmium and micro-plastic particles on biochemical and immunological parameters of common carp (Cyprinus carpio). *Chemosphere, 236*, 124335. https://doi.org/10.1016/j.chemosphere.2019.07.066

Bandow, N., Will, V., Wachtendorf, V., & Simon, F.-G. (2017). Contaminant release from aged microplastic. *Environment and Chemistry, 14*, 394. https://doi.org/10.1071/EN17064

Batel, A., Borchert, F., Reinwald, H., Erdinger, L., & Braunbeck, T. (2018). Microplastic accumulation patterns and transfer of benzo[a]pyrene to adult zebrafish (Danio rerio) gills and zebrafish embryos. *Environmental Pollution, 235*, 918–930. https://doi.org/10.1016/j.envpol.2018.01.028

Bhattacharya, P., Lin, S., Turner, J. P., & Ke, P. C. (2010). Physical adsorption of charged plastic nanoparticles affects algal photosynthesis. *Journal of Physical Chemistry C, 114*, 16556–16561. https://doi.org/10.1021/jp1054759

Binelli, A., Pietrelli, L., Di Vito, S., Coscia, L., Sighicelli, M., Torre, C. D., Parenti, C. C., & Magni, S. (2020). Hazard evaluation of plastic mixtures from four Italian subalpine great lakes on the basis of laboratory exposures of zebra mussels. *Science of the Total Environment, 699*, 134366. https://doi.org/10.1016/j.scitotenv.2019.134366

Bosker, T., Olthof, G., Vijver, M. G., Baas, J., & Barmentlo, S. H. (2019). Significant decline of *Daphnia magna* population biomass due to microplastic exposure. *Environmental Pollution, 250*, 669–675. https://doi.org/10.1016/j.envpol.2019.04.067

Bowman, K. L., Lamborg, C. H., Agather, A. M., & Hammerschmidt, C. R. (2021). The role of plastic debris in the biogeochemical cycle of mercury in Lake Erie and San Francisco Bay. *Marine Pollution Bulletin, 171*, 112768. https://doi.org/10.1016/j.marpolbul.2021.112768

Brandts, I., Teles, M., Tvarijonaviciute, A., Pereira, M. L., Martins, M. A., Tort, L., & Oliveira, M. (2018). Effects of polymethylmethacrylate nanoplastics on *Dicentrarchus labrax*. *Genomics, 110*, 435–441. https://doi.org/10.1016/j.ygeno.2018.10.006

Brennecke, D., Duarte, B., Paiva, F., Caçador, I., & Canning-Clode, J. (2016). Microplastics as vector for heavy metal contamination from the marine environment. *Estuarine, Coastal and Shelf Science, 178*, 189–195. https://doi.org/10.1016/j.ecss.2015.12.003

Browne, M. A., Dissanayake, A., Galloway, T. S., Lowe, D. M., & Thompson, R. C. (2008). Ingested microscopic plastic Translocates to the circulatory system of the mussel, *Mytilus edulis* (L.). *Environmental Science & Technology, 42*, 5026–5031. https://doi.org/10.1021/es800249a

Cai, M., He, H., Liu, M., Li, S., Tang, G., Wang, W., Huang, P., Wei, G., Lin, Y., Chen, B., Hu, J., & Cen, Z. (2018). Lost but can't be neglected: Huge quantities of small microplastics hide in the South China Sea. *Science of the Total Environment, 633*, 1206–1216. https://doi.org/10.1016/j.scitotenv.2018.03.197

Cedervall, T., Hansson, L. A., Lard, M., Frohm, B., & Linse, S. (2012). Food chain transport of nanoparticles affects behaviour and fat metabolism in fish. *PLoS One, 7*, 1–6. https://doi.org/10.1371/journal.pone.0032254

Chen, Q., Gundlach, M., Yang, S., Jiang, J., Velki, M., Yin, D., & Hollert, H. (2017). Quantitative investigation of the mechanisms of microplastics and nanoplastics toward zebrafish larvae locomotor activity. *Science of the Total Environment, 584–585*, 1022–1031. https://doi.org/10.1016/j.scitotenv.2017.01.156

Chen, Q., Lackmann, C., Wang, W., Seiler, T. B., Hollert, H., & Shi, H. (2020). Microplastics Lead to hyperactive swimming behaviour in adult zebrafish. *Aquatic Toxicology, 224*, 105521. https://doi.org/10.1016/j.aquatox.2020.105521

Cole, M., Liddle, C., Consolandi, G., Drago, C., Hird, C., Lindeque, P. K., & Galloway, T. S. (2020). Microplastics, microfibres and nanoplastics cause variable sub-lethal responses in mussels (Mytilus spp.). *Marine Pollution Bulletin, 160*, 111552. https://doi.org/10.1016/j.marpolbul.2020.111552

Crew, A., Gregory-Eaves, I., & Ricciardi, A. (2020). Distribution, abundance, and diversity of microplastics in the upper St. Lawrence River. *Environmental Pollution, 260*, 113994. https://doi.org/10.1016/j.envpol.2020.113994

Dalu, T., Banda, T., Mutshekwa, T., Munyai, L. F., & Cuthbert, R. N. (2021). Effects of urbanisation and a wastewater treatment plant on microplastic densities along a subtropical river system. *Environmental Science and Pollution Research, 28*, 36102–36111. https://doi.org/10.1007/s11356-021-13185-1

Davranche, M., Lory, C., Le Juge, C., Blancho, F., Dia, A., Grassl, B., El Hadri, H., Pascal, P. Y., & Gigault, J. (2020). Nanoplastics on the coast exposed to the North Atlantic Gyre: Evidence and traceability. *NanoImpact, 20*, 100262. https://doi.org/10.1016/j.impact.2020.100262

de Sá, L. C., Luís, L. G., & Guilhermino, L. (2015). Effects of microplastics on juveniles of the common goby (Pomatoschistus microps): Confusion with prey, reduction of the predatory performance and efficiency, and possible influence of developmental conditions. *Environmental Pollution, 196*, 359–362. https://doi.org/10.1016/j.envpol.2014.10.026

Desforges, J. P. W., Galbraith, M., Dangerfield, N., & Ross, P. S. (2014). Widespread distribution of microplastics in subsurface seawater in the NE Pacific Ocean. *Marine Pollution Bulletin, 79*, 94–99. https://doi.org/10.1016/j.marpolbul.2013.12.035

Dovidat, L. C., Brinkmann, B. W., Vijver, M. G., & Bosker, T. (2020). Plastic particles adsorb to the roots of freshwater vascular plant *Spirodela polyrhiza* but do not impair growth. *Limnology and Oceanography Letters, 5*, 37–45. https://doi.org/10.1002/lol2.10118

Du, S., Zhu, R., Cai, Y., Xu, N., Yap, P. S., Zhang, Y., He, Y., & Zhang, Y. (2021). Environmental fate and impacts of microplastics in aquatic ecosystems: A review. *RSC Advances, 11*, 15762–15784. https://doi.org/10.1039/d1ra00880c

Eckert, E. M., Di Cesare, A., Kettner, M. T., Arias-Andres, M., Fontaneto, D., Grossart, H. P., & Corno, G. (2018). Microplastics increase impact of treated wastewater on freshwater microbial community. *Environmental Pollution, 234*, 495–502. https://doi.org/10.1016/j.envpol.2017.11.070

Eo, S., Hong, S. H., Song, Y. K., Han, G. M., Seo, S., & Shim, W. J. (2021). Prevalence of small high-density microplastics in the continental shelf and deep sea waters of East Asia. *Water Research, 200*, 117238. https://doi.org/10.1016/j.watres.2021.117238

Eriksen, M., Mason, S., Wilson, S., Box, C., Zellers, A., Edwards, W., Farley, H., & Amato, S. (2013). Microplastic pollution in the surface waters of the Laurentian Great Lakes. *Marine Pollution Bulletin, 77*, 177–182. https://doi.org/10.1016/j.marpolbul.2013.10.007

Farrell, P., & Nelson, K. (2013). Trophic level transfer of microplastic: *Mytilus edulis* (L.) to *Carcinus maenas* (L.). *Environmental Pollution, 177*, 1–3. https://doi.org/10.1016/j.envpol.2013.01.046

Fraser, M. A., Chen, L., Ashar, M., Huang, W., Zeng, J., Zhang, C., & Zhang, D. (2020). Occurrence and distribution of microplastics and polychlorinated biphenyls in sediments from the Qiantang River and Hangzhou Bay, China. *Ecotoxicology and Environmental Safety, 196*, 110536. https://doi.org/10.1016/j.ecoenv.2020.110536

Free, C. M., Jensen, O. P., Mason, S. A., Eriksen, M., Williamson, N. J., & Boldgiv, B. (2014). High-levels of microplastic pollution in a large, remote, mountain lake. *Marine Pollution Bulletin, 85*, 156–163. https://doi.org/10.1016/j.marpolbul.2014.06.001

Gangadoo, S., Owen, S., Rajapaksha, P., Plaisted, K., Cheeseman, S., Haddara, H., Truong, V. K., Ngo, S. T., Vu, V. V., Cozzolino, D., Elbourne, A., Crawford, R., Latham, K., & Chapman, J. (2020). Nano-plastics and their analytical characterisation and fate in the marine environment: From source to sea. *Science of the Total Environment, 732*, 138792. https://doi.org/10.1016/j.scitotenv.2020.138792

Goldstein, M. C., Titmus, A. J., & Ford, M. (2013). Scales of spatial heterogeneity of plastic marine debris in the Northeast Pacific Ocean. *PLoS One, 8*, e80020. https://doi.org/10.1371/journal.pone.0080020

Goss, H., Jaskiel, J., & Rotjan, R. (2018). *Thalassia testudinum* as a potential vector for incorporating microplastics into benthic marine food webs. *Marine Pollution Bulletin, 135*, 1085–1089. https://doi.org/10.1016/j.marpolbul.2018.08.024

Grgić, I., Cetinić, K. A., Karačić, Z., Previšić, A., & Rožman, M. (2023). Fate and effects of microplastics in combination with pharmaceuticals and endocrine disruptors in freshwaters: Insights from a microcosm experiment. *Science of the Total Environment, 859*, 160387. https://doi.org/10.1016/j.scitotenv.2022.160387

Gu, H., Wang, S., Wang, X., Yu, X., Hu, M., Huang, W., & Wang, Y. (2020). Nanoplastics impair the intestinal health of the juvenile large yellow croaker *Larimichthys crocea*. *Journal of Hazardous Materials, 397*, 122773. https://doi.org/10.1016/j.jhazmat.2020.122773

He, B., Duodu, G. O., Rintoul, L., Ayoko, G. A., & Goonetilleke, A. (2020). Influence of microplastics on nutrients and metal concentrations in river sediments. *Environmental Pollution, 263*, 114490. https://doi.org/10.1016/j.envpol.2020.114490

Jaikumar, I. M., Periyakali, S. B., Rajendran, U., Joen-Rong, S., Thanasekaran, J., & Tsorng-Harn, F. (2021). Effects of microplastics, polystyrene, and polyethylene on antioxidants, metabolic enzymes, HSP-70, and Myostatin expressions in the Giant River Prawn *Macrobrachium rosenbergii*: Impact on survival and growth. *Archives of Environmental Contamination and Toxicology, 80*, 645–658. https://doi.org/10.1007/s00244-021-00833-3

Jinhui, S., Sudong, X., Yan, N., Xia, P., Jiahao, Q., & Yongjian, X. (2019). Effects of microplastics and attached heavy metals on growth, immunity, and heavy metal accumulation in the yellow seahorse, Hippocampus kuda Bleeker. *Marine Pollution Bulletin, 149*, 110510. https://doi.org/10.1016/j.marpolbul.2019.110510

Kalčíková, G., Žgajnar Gotvajn, A., Kladnik, A., & Jemec, A. (2017). Impact of polyethylene microbeads on the floating freshwater plant duckweed *Lemna minor*. *Environmental Pollution, 230*, 1108–1115. https://doi.org/10.1016/j.envpol.2017.07.050

Karthik, R., Robin, R. S., Purvaja, R., Ganguly, D., Anandavelu, I., Raghuraman, R., Hariharan, G., Ramakrishna, A., & Ramesh, R. (2018). Microplastics along the beaches of southeast coast of India. *Science of the Total Environment, 645*, 1388–1399. https://doi.org/10.1016/j.scitotenv.2018.07.242

Kelpsiene, E., Torstensson, O., Ekvall, M. T., Hansson, L. A., & Cedervall, T. (2020). Long-term exposure to nanoplastics reduces life-time in Daphnia magna. *Scientific Reports, 10*, 1–7. https://doi.org/10.1038/s41598-020-63028-1

Klavins, M., Klavins, L., Stabnikova, O., Stabnikov, V., Marynin, A., Ansone-Bertina, L., Mezulis, M., & Vaseashta, A. (2022). Interaction between microplastics and pharmaceuticals depending on the composition of aquatic environment. *Microplastics, 1*, 520–535. https://doi.org/10.3390/microplastics1030037

Klein, K., Heß, S., Nungeß, S., Schulte-Oehlmann, U., & Oehlmann, J. (2021). Particle shape does not affect ingestion and egestion of microplastics by the freshwater shrimp Neocaridina palmata. *Environmental Science and Pollution Research, 28*, 62246–62254. https://doi.org/10.1007/s11356-021-15068-x

Koelmans, A. A., Besseling, E., & Shim, W. J. (2015). Nanoplastics in the aquatic environment. Critical review. In M. Bergmann, L. Gutow, & M. Klages (Eds.), *Marine anthropogenic litter* (pp. 325–340). Springer International Publishing. https://doi.org/10.1007/978-3-319-16510-3_12

Lagarde, F., Olivier, O., Zanella, M., Daniel, P., Hiard, S., & Caruso, A. (2016). Microplastic interactions with freshwater microalgae: Hetero-aggregation and changes in plastic density appear strongly dependent on polymer type. *Environmental Pollution, 215*, 331–339. https://doi.org/10.1016/j.envpol.2016.05.006

Le Bihanic, F., Clérandeau, C., Cormier, B., Crebassa, J. C., Keiter, S. H., Beiras, R., Morin, B., Bégout, M. L., Cousin, X., & Cachot, J. (2020). Organic contaminants sorbed to microplastics affect marine medaka fish early life stages development. *Marine Pollution Bulletin, 154*, 111059. https://doi.org/10.1016/j.marpolbul.2020.111059

Lee, K. W., Shim, W. J., Kwon, O. Y., & Kang, J. H. (2013). Size-dependent effects of micro polystyrene particles in the marine copepod *Tigriopus japonicus*. *Environmental Science & Technology, 47*, 11278–11283. https://doi.org/10.1021/es401932b

Li, W., Lo, H. S., Wong, H. M., Zhou, M., Wong, C. Y., Tam, N. F. Y., & Cheung, S. G. (2020a). Heavy metals contamination of sedimentary microplastics in Hong Kong. *Marine Pollution Bulletin, 153*, 110977. https://doi.org/10.1016/j.marpolbul.2020.110977

Li, S., Wang, P., Zhang, C., Zhou, X., Yin, Z., Hu, T., Hu, D., Liu, C., & Zhu, L. (2020b). Influence of polystyrene microplastics on the growth, photosynthetic efficiency and aggregation of freshwater microalgae *Chlamydomonas reinhardtii*. *Science of the Total Environment, 714*, 136767. https://doi.org/10.1016/j.scitotenv.2020.136767

Li, Y., Liu, Z., Yang, Y., Jiang, Q., Wu, D., Huang, Y., Jiao, Y., Chen, Q., Huang, Y., & Zhao, Y. (2021a). Effects of nanoplastics on energy metabolism in the oriental river prawn (*Macrobrachium nipponense*). *Environmental Pollution, 268*, 115890. https://doi.org/10.1016/j.envpol.2020.115890

Li, R., Zhu, L., Yang, K., Li, H., Zhu, Y. G., & Cui, L. (2021b). Impact of urbanization on antibiotic Resistome in different microplastics: Evidence from a large-scale whole river analysis. *Environmental Science & Technology, 55*, 8760–8770. https://doi.org/10.1021/acs.est.1c01395

Lo, H. S., Wong, C. Y., Tam, N. F. Y., & Cheung, S. G. (2019). Spatial distribution and source identification of hydrophobic organic compounds (HOCs) on sedimentary microplastic in Hong Kong. *Chemosphere, 219*, 418–426. https://doi.org/10.1016/j.chemosphere.2018.12.032

López-Rojo, N., Pérez, J., Alonso, A., Correa-Araneda, F., & Boyero, L. (2020). Microplastics have lethal and sublethal effects on stream invertebrates and affect stream ecosystem functioning. *Environmental Pollution, 259*, 113898. https://doi.org/10.1016/j.envpol.2019.113898

Lu, K., Qiao, R., An, H., & Zhang, Y. (2018). Influence of microplastics on the accumulation and chronic toxic effects of cadmium in zebrafish (*Danio rerio*). *Chemosphere, 202*, 514–520. https://doi.org/10.1016/j.chemosphere.2018.03.145

Maes, T., Van der Meulen, M. D., Devriese, L. I., Leslie, H. A., Huvet, A., Frère, L., Robbens, J., & Vethaak, A. D. (2017). Microplastics baseline surveys at the water surface and in sediments of the North-East Atlantic. *Frontiers in Marine Science, 4*, 1–13. https://doi.org/10.3389/fmars.2017.00135

Mai, L., Bao, L. J., Shi, L., Liu, L. Y., & Zeng, E. Y. (2018). Polycyclic aromatic hydrocarbons affiliated with microplastics in surface waters of Bohai and Huanghai Seas, China. *Environmental Pollution, 241*, 834–840. https://doi.org/10.1016/j.envpol.2018.06.012

Mateos-Cárdenas, A., Scott, D. T., Seitmaganbetova, G., van Pelt Frank, N. A. M., O'Halloran, J., & Marcel, A. K. J. (2019). Polyethylene microplastics adhere to Lemna minor (L.), yet have no effects on plant growth or feeding by *Gammarus duebeni* (Lillj.). *Science of the Total Environment, 689*, 413–421. https://doi.org/10.1016/j.scitotenv.2019.06.359

Mattsson, K., Ekvall, M. T., Hansson, L. A., Linse, S., Malmendal, A., & Cedervall, T. (2015). Altered behavior, physiology, and metabolism in fish exposed to polystyrene nanoparticles. *Environmental Science & Technology, 49*, 553–561. https://doi.org/10.1021/es5053655

Mohana, A. A., Farhad, S. M., Haque, N., & Pramanik, B. K. (2021). Understanding the fate of nano-plastics in wastewater treatment plants and their removal using membrane processes. *Chemosphere, 284*, 131430. https://doi.org/10.1016/j.chemosphere.2021.131430

Munari, C., Scoponi, M., & Mistri, M. (2017). Plastic debris in the Mediterranean Sea: Types, occurrence and distribution along Adriatic shorelines. *Waste Management, 67*, 385–391. https://doi.org/10.1016/j.wasman.2017.05.020

Nelms, S. E., Galloway, T. S., Godley, B. J., Jarvis, D. S., & Lindeque, P. K. (2018). Investigating microplastic trophic transfer in marine top predators. *Environmental Pollution, 238*, 999–1007. https://doi.org/10.1016/j.envpol.2018.02.016

Nobre, C. R., Moreno, B. B., Alves, A. V., de Lima Rosa, J., Fontes, M. K., de Campos, B. G., da Silva, L. F., de Almeida Duarte, L. F., de Abessa, D. M. S., Choueri, R. B., Gusso-Choueri, P. K., & Pereira, C. D. S. (2022). Combined effects of polyethylene spiked with the antimicrobial triclosan on the swamp ghost crab (Ucides cordatus; Linnaeus, 1763). *Chemosphere, 304*, 135169. https://doi.org/10.1016/j.chemosphere.2022.135169

Pan, Z., Liu, Q., Sun, Y., Sun, X., & Lin, H. (2019). Environmental implications of microplastic pollution in the Northwestern Pacific Ocean. *Marine Pollution Bulletin, 146*, 215–224. https://doi.org/10.1016/j.marpolbul.2019.06.031

Parra, S., Varandas, S., Santos, D., Félix, L., Fernandes, L., Cabecinha, E., Gago, J., & Monteiro, S. M. (2021). Multi-biomarker responses of Asian clam Corbicula fluminea (bivalvia, corbiculidea) to cadmium and microplastics pollutants. *Water (Switzerland), 13*. https://doi.org/10.3390/w13040394

Pikuda, O., Dumont, E. R., Matthews, S., Xu, E. G., Berk, D., & Tufenkji, N. (2022). Sub-lethal effects of nanoplastics upon chronic exposure to Daphnia magna. *Journal of Hazardous Materials Advances, 7*, 100136. https://doi.org/10.1016/j.hazadv.2022.100136

Pitt, J. A., Trevisan, R., Massarsky, A., Kozal, J. S., Levin, E. D., & Di Giulio, R. T. (2018). Maternal transfer of nanoplastics to offspring in zebrafish (Danio rerio): A case study with nanopolystyrene. *Science of the Total Environment, 643*, 324–334. https://doi.org/10.1016/j.scitotenv.2018.06.186

Qiang, L., & Cheng, J. (2021). Exposure to polystyrene microplastics impairs gonads of zebrafish (Danio rerio). *Chemosphere, 263*, 128161. https://doi.org/10.1016/j.chemosphere.2020.128161

Rainieri, S., Conlledo, N., Larsen, B. K., Granby, K., & Barranco, A. (2018). Combined effects of microplastics and chemical contaminants on the organ toxicity of zebrafish (Danio rerio). *Environmental Research, 162*, 135–143. https://doi.org/10.1016/j.envres.2017.12.019

Rios, L. M., Moore, C., & Jones, P. R. (2007). Persistent organic pollutants carried by synthetic polymers in the ocean environment. *Marine Pollution Bulletin, 54*, 1230–1237. https://doi.org/10.1016/j.marpolbul.2007.03.022

Rochman, C. M., Lewison, R. L., Eriksen, M., Allen, H., Cook, A. M., & Teh, S. J. (2014). Polybrominated diphenyl ethers (PBDEs) in fish tissue may be an indicator of plastic contamination in marine habitats. *Science of the Total Environment, 476–477*, 622–633. https://doi.org/10.1016/j.scitotenv.2014.01.058

Rodrigues, M. O., Abrantes, N., Gonçalves, F. J. M., Nogueira, H., Marques, J. C., & Gonçalves, A. M. M. (2018). Spatial and temporal distribution of microplastics in water and sediments of a freshwater system (Antuã River, Portugal). *Science of the Total Environment, 633*, 1549–1559. https://doi.org/10.1016/j.scitotenv.2018.03.233

Sarkar, D. J., Das Sarkar, S., Das, B. K., Sahoo, B. K., Das, A., Nag, S. K., Manna, R. K., Behera, B. K., & Samanta, S. (2021). Occurrence, fate and removal of microplastics as heavy metal vector in natural wastewater treatment wetland system. *Water Research, 192*, 116853. https://doi.org/10.1016/j.watres.2021.116853

Schrank, I., Löder, M. G. J., Imhof, H. K., Moses, S. R., Heß, M., Schwaiger, J., & Laforsch, C. (2022). Riverine microplastic contamination in Southwest Germany: A large-scale survey. *Frontiers in Earth Science, 10*, 1–17. https://doi.org/10.3389/feart.2022.794250

Seeley, M. E., Song, B., Passie, R., & Hale, R. C. (2020). Communities and nitrogen cycling. *Nature Communications, 11*, 1–10. https://doi.org/10.1038/s41467-020-16235-3

Sendra, M., Saco, A., Yeste, M. P., Romero, A., Novoa, B., & Figueras, A. (2020). Nanoplastics: From tissue accumulation to cell translocation into *Mytilus galloprovincialis* hemocytes. resilience of immune cells exposed to nanoplastics and nanoplastics plus *Vibrio splendidus* combination. *Journal of Hazardous Materials, 388*, 121788. https://doi.org/10.1016/j.jhazmat.2019.121788

Setälä, O., Fleming-Lehtinen, V., & Lehtiniemi, M. (2014). Ingestion and transfer of microplastics in the planktonic food web. *Environmental Pollution, 185*, 77–83. https://doi.org/10.1016/j.envpol.2013.10.013

Sheridan, E. A., Fonvielle, J. A., Cottingham, S., Zhang, Y., Dittmar, T., Aldridge, D. C., & Tanentzap, A. J. (2022). Plastic pollution fosters more microbial growth in lakes than natural organic matter. *Nature Communications, 13*, 1–9. https://doi.org/10.1038/s41467-022-31691-9

Silva, C. J. M., Patrício Silva, A. L., Campos, D., Machado, A. L., Pestana, J. L. T., & Gravato, C. (2021). Oxidative damage and decreased aerobic energy production due to ingestion of polyethylene microplastics by *Chironomus riparius* (Diptera) larvae. *Journal of Hazardous Materials, 402*, 123775. https://doi.org/10.1016/j.jhazmat.2020.123775

Singh, N., Mondal, A., Bagri, A., Tiwari, E., Khandelwal, N., Monikh, F. A., & Darbha, G. K. (2021). Characteristics and spatial distribution of microplastics in the lower Ganga River water and sediment. *Marine Pollution Bulletin, 163*, 111960. https://doi.org/10.1016/j.marpolbul.2020.111960

Sjollema, S. B., Redondo-Hasselerharm, P., Leslie, H. A., Kraak, M. H. S., & Vethaak, A. D. (2016). Do plastic particles affect microalgal photosynthesis and growth? *Aquatic Toxicology, 170*, 259–261. https://doi.org/10.1016/j.aquatox.2015.12.002

Song, Y. K., Hong, S. H., Jang, M., Han, G. M., & Shim, W. J. (2015). Occurrence and distribution of microplastics in the sea surface microlayer in Jinhae Bay, South Korea. *Archives of Environmental Contamination and Toxicology, 69*, 279–287. https://doi.org/10.1007/s00244-015-0209-9

Ta, A. T., & Babel, S. (2020). Microplastics pollution with heavy metals in the aquaculture zone of the Chao Phraya River Estuary, Thailand. *Marine Pollution Bulletin, 161*, 111747. https://doi.org/10.1016/j.marpolbul.2020.111747

Tan, X., Yu, X., Cai, L., Wang, J., & Peng, J. (2019). Microplastics and associated PAHs in surface water from the Feilaixia Reservoir in the Beijiang River, China. *Chemosphere, 221*, 834–840. https://doi.org/10.1016/j.chemosphere.2019.01.022

Tang, Y., Rong, J., Guan, X., Zha, S., Shi, W., Han, Y., Du, X., Wu, F., Huang, W., & Liu, G. (2020). Immunotoxicity of microplastics and two persistent organic pollutants alone or in combination to a bivalve species. *Environmental Pollution, 258*, 113845. https://doi.org/10.1016/j.envpol.2019.113845

Ter Halle, A., Jeanneau, L., Martignac, M., Jardé, E., Pedrono, B., Brach, L., & Gigault, J. (2017). Nanoplastic in the North Atlantic Subtropical Gyre. *Environmental Science & Technology, 51*, 13689–13697. https://doi.org/10.1021/acs.est.7b03667

Tien, C. J., Wang, Z. X., & Chen, C. S. (2020). Microplastics in water, sediment and fish from the Fengshan River system: Relationship to aquatic factors and accumulation of polycyclic aromatic hydrocarbons by fish. *Environmental Pollution*. https://doi.org/10.1016/j.envpol.2020.114962

Urbina, M. A., da Silva Montes, C., Schäfer, A., Castillo, N., Urzúa, Á., & Lagos, M. E. (2023). Slow and steady hurts the crab: Effects of chronic and acute microplastic exposures on a filter feeder crab. *Science of the Total Environment, 857*, 159135. https://doi.org/10.1016/j.scitotenv.2022.159135

van Weert, S., Redondo-Hasselerharm, P. E., Diepens, N. J., & Koelmans, A. A. (2019). Effects of nanoplastics and microplastics on the growth of sediment-rooted macrophytes. *Science of the Total Environment, 654*, 1040–1047. https://doi.org/10.1016/j.scitotenv.2018.11.183

Viršek, M. K., Lovšin, M. N., Koren, Š., Kržan, A., & Peterlin, M. (2017). Microplastics as a vector for the transport of the bacterial fish pathogen species Aeromonas salmonicida. *Marine Pollution Bulletin, 125*, 301–309. https://doi.org/10.1016/j.marpolbul.2017.08.024

Von Moos, N., Burkhardt-Holm, P., & Köhler, A. (2012). Uptake and effects of microplastics on cells and tissue of the blue mussel Mytilus edulis L. after an experimental exposure. *Environmental Science & Technology, 46*, 11327–11335. https://doi.org/10.1021/es302332w

Wan, J. K., Chu, W. L., Kok, Y. Y., & Lee, C. S. (2021). Influence of polystyrene microplastic and nanoplastic on copper toxicity in two freshwater microalgae. *Environmental Science and Pollution Research, 28*, 33649–33668. https://doi.org/10.1007/s11356-021-12983-x

Wang, S., Liu, M., Wang, J., Huang, J., & Wang, J. (2020). Polystyrene nanoplastics cause growth inhibition, morphological damage and physiological disturbance in the marine microalga Platymonas helgolandica. *Marine Pollution Bulletin, 158*, 111403. https://doi.org/10.1016/j.marpolbul.2020.111403

Wang, S. C., Liu, F. F., Huang, T. Y., Fan, J. L., Gao, Z. Y., & Liu, G. Z. (2021a). Effects of nanoplastics on the dinoflagellate amphidinium carterae hulburt from the perspectives of algal growth, oxidative stress and hemolysin production. *Nanomaterials, 11*. https://doi.org/10.3390/nano11102471

Wang, F., Wu, H., Wu, W., Wang, L., Liu, J., An, L., & Xu, Q. (2021b). Microplastic characteristics in organisms of different trophic levels from Liaohe Estuary, China. *Science of the Total Environment, 789*, 148027. https://doi.org/10.1016/j.scitotenv.2021.148027

Wang, Q., Wang, J., Chen, H., & Zhang, Y. (2023). Toxicity effects of microplastics and nanoplastics with cadmium on the alga Microcystis aeruginosa. *Environmental Science and Pollution Research, 30*, 17360–17373. https://doi.org/10.1007/s11356-022-23278-0

Wazne, M., Mermillod-Blondin, F., Vallier, M., Hervant, F., Dumet, A., Nel, H. A., Kukkola, A., Krause, S., & Simon, L. (2023). Microplastics in freshwater sediments impact the role of a Main Bioturbator in ecosystem functioning. *Environmental Science & Technology, 57*, 3042–3052. https://doi.org/10.1021/acs.est.2c05662

Winkler, A., Nessi, A., Antonioli, D., Laus, M., Santo, N., Parolini, M., & Tremolada, P. (2020). Occurrence of microplastics in pellets from the common kingfisher (*Alcedo atthis*) along the Ticino River, North Italy. *Environmental Science and Pollution Research, 27*, 41731–41739. https://doi.org/10.1007/s11356-020-10163-x

Wu, Y., Guo, P., Zhang, X., Zhang, Y., Xie, S., & Deng, J. (2019a). Effect of microplastics exposure on the photosynthesis system of freshwater algae. *Journal of Hazardous Materials, 374*, 219–227. https://doi.org/10.1016/j.jhazmat.2019.04.039

Wu, X., Pan, J., Li, M., Li, Y., Bartlam, M., & Wang, Y. (2019b). Selective enrichment of bacterial pathogens by microplastic biofilm. *Water Research, 165*, 114979. https://doi.org/10.1016/j.watres.2019.114979

Wu, N., Zhang, Y., Zhao, Z., He, J., Li, W., Li, J., Xu, W., Ma, Y., & Niu, Z. (2020). Colonization characteristics of bacterial communities on microplastics compared with ambient environments (water and sediment) in Haihe Estuary. *Science of the Total Environment, 708*, 134876. https://doi.org/10.1016/j.scitotenv.2019.134876

Xue, N., Wang, L., Li, W., Wang, S., Pan, X., & Zhang, D. (2020). Increased inheritance of structure and function of bacterial communities and pathogen propagation in plastisphere along a river with increasing antibiotics pollution gradient. *Environmental Pollution, 265*, 114641. https://doi.org/10.1016/j.envpol.2020.114641

Yan, M., Nie, H., Xu, K., He, Y., Hu, Y., Huang, Y., & Wang, J. (2019). Microplastic abundance, distribution and composition in the Pearl River along Guangzhou city and Pearl River estuary, China. *Chemosphere, 217*, 879–886. https://doi.org/10.1016/j.chemosphere.2018.11.093

Yang, W., Gao, P., Li, H., Huang, J., Zhang, Y., Ding, H., & Zhang, W. (2021). Mechanism of the inhibition and detoxification effects of the interaction between nanoplastics and microalgae *Chlorella pyrenoidosa. Science of the Total Environment, 783*, 146919. https://doi.org/10.1016/j.scitotenv.2021.146919

Yang, Z., Zhu, L., Liu, J., Cheng, Y., Waiho, K., Chen, A., & Wang, Y. (2022). Polystyrene microplastics increase Pb bioaccumulation and health damage in the Chinese mitten crab Eriocheir sinensis. *Science of the Total Environment, 829*, 154586. https://doi.org/10.1016/j.scitotenv.2022.154586

Yin, M., Yan, B., Wang, H., Wu, Y., Wang, X., Wang, J., Zhu, Z., Yan, X., Liu, Y., Liu, M., & Fu, C. (2023). Effects of microplastics on nitrogen and phosphorus cycles and microbial communities in sediments. *Environmental Pollution, 318*, 120852. https://doi.org/10.1016/j.envpol.2022.120852

Yuan, W., Zhou, Y., Chen, Y., Liu, X., & Wang, J. (2020). Toxicological effects of microplastics and heavy metals on the *Daphnia magna. Science of the Total Environment, 746*, 141254. https://doi.org/10.1016/j.scitotenv.2020.141254

Zhang, S., Ding, J., Razanajatovo, R. M., Jiang, H., Zou, H., & Zhu, W. (2019). Interactive effects of polystyrene microplastics and roxithromycin on bioaccumulation and biochemical status in the freshwater fish red tilapia (*Oreochromis niloticus*). *Science of the Total Environment, 648*, 1431–1439. https://doi.org/10.1016/j.scitotenv.2018.08.266

Zhang, R., Silic, M. R., Schaber, A., Wasel, O., Freeman, J. L., & Sepúlveda, M. S. (2020). Exposure route affects the distribution and toxicity of polystyrene nanoplastics in zebrafish. *Science of the Total Environment, 724*, 138065. https://doi.org/10.1016/j.scitotenv.2020.138065

Zhang, B., Chao, J., Chen, L., Liu, L., Yang, X., & Wang, Q. (2021). Research progress of nanoplastics in freshwater. *Science of the Total Environment, 757*, 143791. https://doi.org/10.1016/j.scitotenv.2020.143791

Zhang, Z., Gao, S. H., Luo, G., Kang, Y., Zhang, L., Pan, Y., Zhou, X., Fan, L., Liang, B., & Wang, A. (2022). The contamination of microplastics in China's aquatic environment: Occurrence, detection and implications for ecological risk. *Environmental Pollution, 296*, 118737. https://doi.org/10.1016/j.envpol.2021.118737

Zhao, S., Ward, J. E., Danley, M., & Mincer, T. J. (2018). Field-based evidence for microplastic in marine aggregates and mussels: Implications for trophic transfer. *Environmental Science & Technology, 52*, 11038–11048. https://doi.org/10.1021/acs.est.8b03467

Microplastics: An Emerging Environmental Issue—Its Bioremediation, Challenges, and a Future Perspective

Megha S. Gadhvi, Suhas J. Vyas, Anjana K. Vala, and Dushyant R. Dudhagara

1 Introduction

A synthetic polymer known as plastic is now produced from fossil fuels because of the Industrial Revolution and the subsequent development of chemical and material sciences. The Latin word "plasticus"—from which the English word "plastic" is derived—originates from the Greek word "plastikos," which means "able to be shaped or fit." A significant development in the field of material sciences is the formation of plastic. The advantages of plastics in terms of economy, efficiency, and substitutability have grown in prominence with improvements in their production technique (Yu et al., 2022). Today, plastics have become an essential component of our daily lives in the form of household items, packaging, clothing, toys, electrical gadgets, and insulating materials, in addition to the pharmaceutical, aerospace, cosmetic industries and in the field of agriculture. Numerous kinds of plastic products have become a significant part of daily life and production. As per statistics, increasing global production of plastics has touched 8.3 billion tons. In 2015, 2016, 2017, and 2018, it was 381, 335, 349 and 359 million tons respectively (Geyer et al., 2017). The recycling of disposed plastic products should have increased as a result of the production rate's excessive growth, but that did not happen. Only 20% of plastics are actually recycled, whereas the other 80% ends up accumulating in the soil and aquatic bodies.

Plastic waste releases environmental toxins which cannot be decomposed in the environment even after a century. According to a report and assessment by Brahney et al., 2020, by the year 2025, the environment will probably include 11 billion tons

M. S. Gadhvi · S. J. Vyas · D. R. Dudhagara (✉)
Department of Life Sciences, Bhakta Kavi Narsinh Mehta University, Junagadh, India

A. K. Vala
Department of Life Sciences, Maharaja Krishnakumarsinhji Bhavnagar University, Bhavnagar, India

S. A. Bhat et al. (eds.), *Management of Micro and Nano-plastics in Soil and Biosolids*,
https://doi.org/10.1007/978-3-031-51967-3_3

of plastic trash. In addition, the plastics that we refer to as "white pollution" fragment into smaller pieces because of the numerous chemical, physical, and biological activities. As per their origin, microplastics can be distinguished into primary and secondary types. Primary microplastics are smaller than 5 mm in size, whereas secondary microplastics are produced by the breakdown and fragmentation of larger primary microplastics. Microplastics can aggregate, migrate, and disseminate in the environment due to their strong hydrophobicity, small particle size, high specific surface area, stable chemical properties, and capacity to transport other environmental pollutants (like heavy metals and antibiotics).

Microplastics and nanoplastics pollution has emerged as a novel category of environmental issue on a worldwide scale and rising plastic pollution has created numerous difficulties for ecosystems. Microplastics are plastics which are less than 5 mm in particle size while nanoplastics are less than 100 nm in particle size. It has been reported that nanoplastics can be produced during the degradation process of microplastics. Thus, the biological technique of removing microplastics pollution will spontaneously decrease the nanoplastics pollution as well (Lu et al., 2023).

Compared to marine environments, terrestrial habitats get around 4–23 times more plastic garbage each year. The biodiversity of aquatic habitats, such as endangered polychaetes, crustaceans, zooplankton, are suffering greatly as a result of plastic trash and microplastics. It also affects soil biodiversity, plant growth, plant production, and seed germination in terrestrial ecosystems. Plastic residue used in paint, cosmetics, coatings, and detergents are the main sources of microplastics; discarded plastics, tire abrasion, synthetic fibers, and urban dust are thought to be minor sources. The plastics used in agriculture, industry, and households are additional sources of microplastics. Both synthetic and biological polymers are responsible for the contamination of land and water by microplastics. Most particles in terrestrial, marine, and aquatic environments are 1 mm in size or smaller. These little particles are referred to as hazardous elements and it is not fully understood how they affect human health (Smith et al., 2018). Degraded plastics impact aquatic and terrestrial ecosystems by affecting the soil, water, and air environments. This seems to be relevant to all ecosystems. Simultaneous to this, plastic particles are gradually contaminating the food and feed chains (Microplastics, Nanoplastics, etc.) (Orlando et al., 2023).

As a result, scientists are quite concerned about preventing the aggregation of synthetic microplastics in the environment. Recently, interest in the bioremediation of microplastics has increased. The ability of some microorganisms, such as fungi and bacteria, to degrade microplastics rapidly and remediate them under stressful conditions by secreting exoenzymes and their by-products has also been studied. Lipases, cutinases, and proteases are a few examples of microbial enzymes that have been found to degrade microplastics. Therefore, nonbiodegradable microplastics might easily be replaced by microbe-assisted bioremediation of microplastics to assure the safety of the environment and ecosystem.

This chapter emphasizes the different sources of microplastics, effect of microplastics in the environment, and the role of various microorganisms and their mechanisms involved in bioremediation of Microplastics present in the environment.

This chapter also discusses the effects of microplastics on ecosystems and their bioremediation technique for a more accurate evaluation of the ecological and environmental concerns of microplastics along with their probable challenges and future directions for research.

2 Sources of Microplastics

There are two main sources of microplastics in the environment: primary and secondary. Microbeads and microfibers are examples of primary microplastics that are deliberately made to be small, while secondary microplastics are produced when larger plastic particles break down (Fig. 1) (Hidalgo-Ruz et al., 2012). Both the sources of microplastics are described in the following sections with some examples.

2.1 Primary Sources of Microplastics

2.1.1 Pellets of Plastics

In order to create various plastic items, plastic pellets, which are granular polymers with a typical thickness of 2–5 mm and a consistent shape, are utilized. Plastic pellets are frequently transported, processed, and stored as unfinished goods. Propylene, styrene, ethylene, polyvinyl chloride, and other materials are made mostly from coal and petroleum, which are also used to make other compounds. Thermoset plastics and thermoplastics are two categories of polymers that are used in

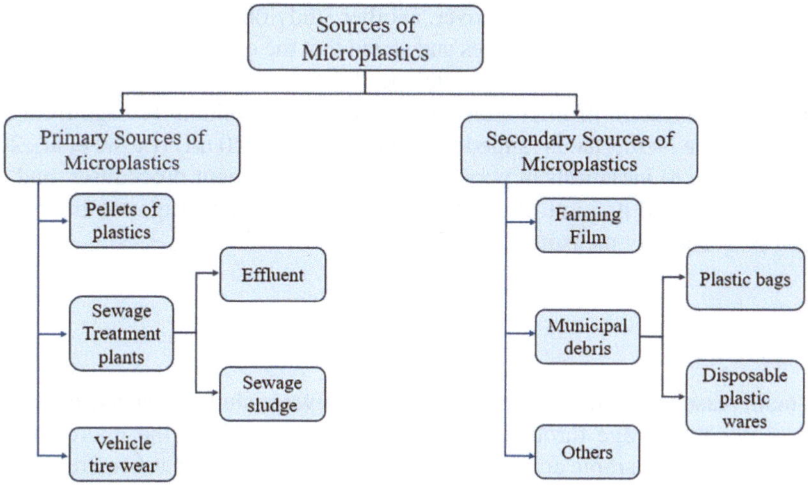

Fig. 1 Different sources of microplastics

manufacturing. The most common thermoplastics are prepared from pure or unused pellets of plastics, often called preproduction beads or pellets of plastics. Generally, these pellets are prepared in recycling facilities or during polymer production. Plastic is widely used in everyday life, including clothing, building supplies, appliances, agriculture, and the chemical industry (Patil et al., 2021). They are also widely used in the production of electrical, electronic, telecommunication, and medical equipment. Plastic pellets are a persistent material, which means that once they are in the environment, they decompose very slowly. A further risk to human health is that because of their small size, they are easily consumed by fish or birds and spread throughout the food chain. Therefore, it is important for researchers to develop plastic substitutes as well as a mechanism for managing and eradicating environmental microplastic pollution (An et al., 2020).

2.1.2 Sewage Treatment Plants

Effluent

All the major effluents that are discharged into sewage treatment plants are from domestic sewage, industrial sewage, and runoff that can contain various types of microplastics from industrial plastic production, personal care products, chemical laundry products, tire wear, and other activities. Although the sewage treatment process may remove more than 90% of the microplastics in wastewater, sewage treatment plant effluents are one of the major sources of microplastics in natural waters because they are discharged directly into surface waters in large quantities (Murphy et al., 2016).

According to the available information, sewage treatment plant effluents contain more than 30 different forms of microplastics, including polyethylene (PE), polypropylene, and polystyrene. Moreover, another study observed 17 different sewage treatment plants in the United States and stated that the effluent of sewage contained an average of 0.05–0.024 microplastics per liter. A different study found that the average concentration of microplastics in effluent and influent sewage from seven locations in the Netherlands ranged from 51–81 to 68–910 n/L (Leslie et al., 2017). Moreover, due to variations in population, sewage treatment procedures, and other factors, the overall amount of microplastics released by sewage treatment plants cannot be accurately measured.

Sewage Sludge

The main reason why there are microplastics in sewage sludge is because they physically enter the sludge through the treatment process. Thus, the microplastics in sludge have a comparable composition to those in effluents. It has been reported that more than 98% of microplastics in influents might dissolve into the sewage sludge (Magnusson et al., 2016). At present, no particular process is used to remove microplastics from sewage sludge, when sludge is composted or used in agriculture it leads to the release of microplastics into the soil.

In May and June 2018, the Gaobeidian wastewater treatment plant, Beijing's largest sewage treatment plant that utilizes the conventional activated sludge process, provided samples for investigation of microplastics in sludge. There were 25 different kinds of plastic polymers identified in the sludge, the majority of which were polybutylene adpareterephthalate (PBA), rayon, polyamide, polybutyl methacrylate (PBMA), polyethylene (PE), and polyethylene terephthalate (PET). The most common type of microplastic in sludge was found to be granular microplastics, which made up 34.07% of the total amount of microplastics and had an average particle size of 621.6 μm. These were followed by microplastic fragments (24.72%), film-like microplastics (22.53%), and fibrous microplastics, which contributed the least amount of microplastics (18.68%). It was found that there were 2933 ± 611, 3867 ± 611, and 5333 ± 3501 particles/kg of dry sludge in May and June 2018 and the amount of microplastics in the sludge amplified every month. The overall emissions of microplastics in sludge from the plant were estimated to be roughly 1.52×10^8 n/day based on the Gaobeidian wastewater treatment plant's sludge production (37.5 t/day dry weight) (Corradini et al., 2019).

2.1.3 Vehicle Tire Wear

One of the primary sources of microplastics in road dust is thought to be the release of microplastics from tires by tire wear. There are three stages in the life of a motor vehicle tire: manufacturing, use, and disposal. The use stage from these three stages has the maximum carbon emissions.

There are two categories of tires: natural rubber (NR) tires and synthetic rubber tires (SN). Styrene butadiene rubber (SBR), or butadiene rubber form, is the majority of synthetic rubber (BR). Advances automotive tires are often made of natural rubber since it is more durable than synthetic rubber. Currently, the *Hevea brasiliensis* (Brazilian rubber tree) is the main source of natural rubber. It is relatively expensive to create rubber from this tree, while the petroleum industry produces synthetic rubber. To achieve the performance requirements of motor vehicle tires, producers often chemically infuse rubber in a range of forms and concentrations. Rubber tires have complex chemical additions.

In Denmark and Norway, rubber tires have been found to be the main sources of environmental microplastics (Sundt et al., 2016). Rubber particles have a density of 1.2–1.3 g/cm^3; therefore, when they enter an aquatic environment, they may sink into the sediment. Surface water, sewage, soil, and air are the various environments where the released rubber particles from motor vehicle tires can be deposited. Increasing tire friction is often thought to shorten a vehicle's stopping distance and enhance driver safety. However, when friction increases, it also increases the rate at which rubber particles are released from tires. To reduce the amount of microplastics produced by tire friction, drivers should avoid using winter tires in summer. Winter tires seem to be softer and have higher grip than summer tires, which causes more tire particles to be discharged on dry winter roads. Additionally, drivers should try to use low-emission tires, repair tires in a timely manner, and choose public transportation to reduce the amount of microplastics produced by tire friction (NIVA, 2020).

2.2 *Secondary Source of Microplastics*

Small plastic fragments that derive from bigger plastic particles that have not been accurately disposed of are known as secondary sources of microplastic. Microplastics are created over time as plastic structures degrade due to physical, biological, and chemical processes like mechanical grinding, biological crushing, and light irradiation aging (Pettipas et al., 2016). A large amount of environmental microplastic comes from secondary sources. Some examples of secondary sources are mentioned below.

2.2.1 Municipal Debris

Plastic Bags

Plastic bags refer to bags made of various plastic raw materials mixed with additives which are processed by heat-sealing or bonding. Many different materials are used in the manufacturing of plastic bags. Plastic bags are widely used due to their low cost, extremely lightweight, large capacity, and easy storage. Globally, up to 5 trillion plastic bags are used per year. In the United States, 100 million plastic bags are used by consumers every year; however, less than 10% of these are recycled (United Nations Environment Programme, 2022).

Since the extensive commercial use of plastic bags appeared in the 1990s, a huge number of plastic bags have been disposed of in the environment, including roads, river banks, and land around cities. Because of their extremely long degradation cycle, the production and use of plastic bags are slowly being banned in some countries. The range and scope of interventions for reducing the use of plastic bags vary between countries; these include prohibiting the sale of lightweight bags, charging customers for lightweight bags, and taxing stores that sell lightweight bags.

After January 2017, the total annual consumption of plastic bags from supermarkets in Israel was 2.7 billion. In Israel, one-quarter of all bags are thrown away immediately after use, and the total weight of plastic bags in waste was estimated to be 15,000 tons per year. However, in 2018, the State of Israel introduced a law that required supermarkets to charge customers at least 0.10 Israeli NIS (about US$0.03) per plastic bag used. Since the introduction of this law, the use of plastic bags has dropped by 80%, and the amount of plastic bag waste found in the sea was half. However, many countries and regions have not implemented charges or bans on the use of plastic bags, such as New Zealand and some states of Australia. Although plastic bag pollution has been reduced by the statement of various bans on plastic bags, there are still a large number of plastic bags left in the environment, which are expected to be a great source of environmental microplastics (An et al., 2020).

Disposable Plasticware

Disposable plastic tableware refers to disposable utensils intended for use during meals or similar purposes which are produced by the thermoplastic molding of resin or other thermoplastic materials. Plastic tableware is extensively used globally due to its low cost, lightness, waterproofing, and durability. The survey suggests that the United States uses up to 500 million plastic drinking straws per year and more than 100 million plastic containers, the large majority of which are not subsequently recycled. Plastic tableware is mainly constructed from polyethylene, polypropylene, polystyrene, and so on. The United States produces about three million tons of disposable polystyrene utensils per year. Polystyrene is mainly used for the manufacture of packaging materials and food service items, such as foam cups, instant noodle boxes, and fast food boxes. If not properly disposed of, plastic tableware may enter sewers, soil, oceans, etc.; over time, it becomes partially degraded and therefore represents a source of environmental microplastics. Globally, it is estimated that plastic tableware might be responsible for 269,000 tons of plastic pollution into waterways and oceans each year (An et al., 2020).

2.2.2 Farming Film

Another prominent source of environmental microplastic is agricultural plastic film. One kind of agricultural mulch is a thin layer made of polyvinylchloride, polyethylene, and other chemicals. About 3.4% of the world's total plastic production is used in agriculture. Early in the 1950s, plastic mulch films were first utilized in agriculture. Such films are extremely important for ensuring food security since they increase soil temperature, which boosts agricultural productivity and income. Due to the high consumption of plastic mulch film and its short lifespan, it is difficult to recover the film, it has a low recycling efficiency, and it readily releases microplastics into the soil. Low-density polyethylene, polyethylene, and linear low-density polyethylene are the most common microplastics released by agricultural film. Large volumes of polyethylene have been discovered. More than any other nation, China uses a lot of plastic film mulch. In 2017, China used 1865.72 hm^2 of plastic film, or 143.37 million tonnes of plastic film mulch. When plastic film is maintained in the soil, it can be absorbed by crops, migrate through the food chain, and eventually harm human health. This eventually reduces the land's ability to produce crops in addition to polluting the soil (An et al., 2020).

2.2.3 Others

Along with the abovementioned areas, the aircraft, building, and medical sectors also use a substantial quantity of plastic. Plastic is widely used in a variety of products, including hubcaps, hinge covers and other automotive parts, pipes and fittings, touch screens, foundations, roofs, panels, floors, and roads, as well as insulation, window frames, profiles, flooring, cable jackets, wall coverings, building insulation, valves, seals, and other items. Because of their resistance to environmental variables, plastic items are frequently utilized outside; they require less maintenance because they do not decay or rust. However, plastics may end up in the solid waste stream. Undisposed plastic trash reaches the ecosystem and contributes significantly to the microplastic pollution of the environment. Depending on how well solid waste is managed, plastic items can contribute to environmental microplastic pollution.

3 Effects of Microplastics on Ecosystems

Microplastics are widely dispersed in ecosystem, and it can affect biota and diversity of ecosystem. To enhance the functionality and applications of plastic products, a variety of additives (including plasticizers, flame-retardants, and stabilizers) are employed during the manufacturing and processing processes. These additives eventually enter soils after prolonged exposure to the environment, where they have a negative impact on the variety and operations of the soil microbial community. Additionally, microplastics can adsorb various pollutants, including polychlorinated biphenyls (PCBs), polycyclic aromatic hydrocarbons (PAHs), dichlorodiphenyl trichloroethane (DDTs), h*exachlorocyclohexanes* (HCHs), *pharmaceuticals and personal care products*, *polyfluoroalkyl substances,* and heavy metals. They act as vehicles for their environmental movement, having an additional negative impact on the health of terrestrial ecosystems. Microplastics may affect soil physicochemical qualities, enzyme activity, microbial populations, soil animals, and plant growth in a positive, negative, or inconsequential manner. These effects can be brought on by variations in soil properties, exposure period, and microplastic features (such as polymer type, content, size, and form). The effects of microplastics on aquatic ecosystems have received some attention from research, while the implications of microplastics on terrestrial ecosystems have received less attention due to a lack of knowledge (Fig. 2) (Hahladakis et al., 2018).

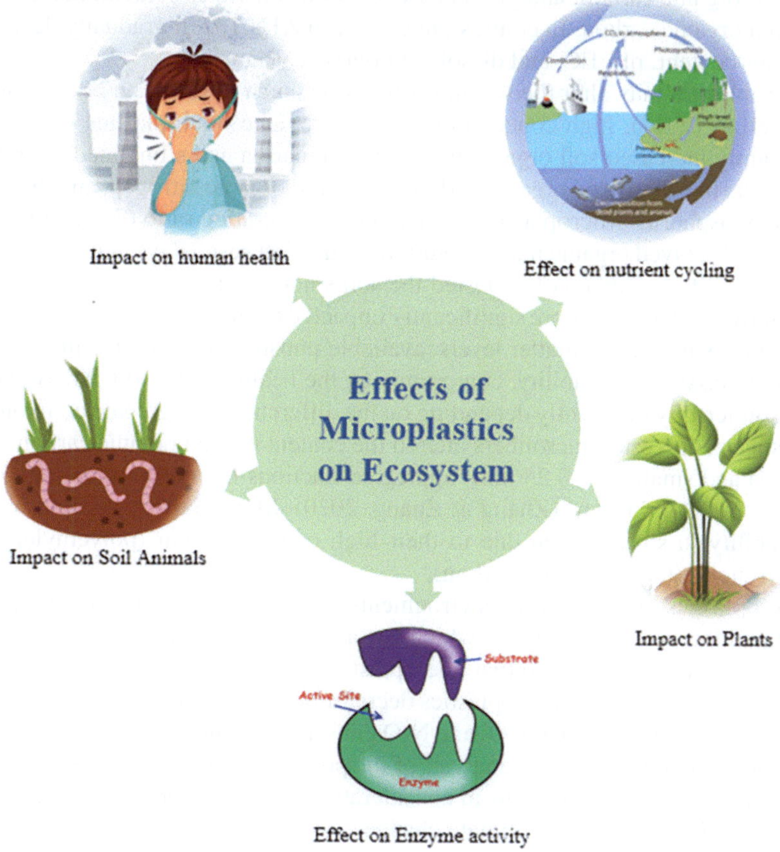

Fig. 2 Effects of microplastic pollution on ecosystem

3.1 Effects of Microplastics on the Nutrient Cycle and Physiochemical Properties of Soil

Microplastics enter soil, aggregate with organic matter and microbial secretions, and become stable in the soil's microstructure. As a result, soil physicochemical parameters are affected by an increase in porosity, a decrease in bulk density, and a reduction in moisture permeability, destroying its structural integrity and changing its structure (Zhang et al., 2020a). Microplastics can also decrease or increase soil pH. For example, polylactic acid microplastics and high-density polyethylene microplastics can increase soil pH, while polystyrene microplastics and polytetrafluorethylene (PTFE) can decrease it. Plant species have an impact on how microplastics affect soil characteristics. For instance, research has found that polymer-treated fertilizer microplastics in corn crop ZNT 488 gradually increased soil pH and electric conductivity (EC), had no effect on soil water content and dissolved organic carbon, and gradually decreased soil water content, pH, EC, and dissolved organic carbon. In contrast, in corn crop ZTN 182, it gradually decreased soil water content, pH, EC, and dissolved organic carbon.

Microplastics may alter the soil nutrition cycle (such as C, P, and N). According to Liu et al. (2017), high-concentration polypropylene microplastics (28% w/w) significantly increased soil organic matter accumulation and stimulated the release of nutrients like dissolved organic carbon, dissolved organic nitrogen, and dissolved organic phosphorus. Microplastics with a low concentration ratio (7% w/w) had no impact on dissolved organic matter solutions in the first seven days, but between day 14 and day 30, they markedly boosted the soil's nutritional content. The dynamics of soil nitrogen cycle are more significantly impacted by microplastics. Microplastics may reduce soil organic matter levels, available phosphorus, alkaline nitrogen levels, and potassium availability, compromising the health of the soil. Microplastics affect soil fertility differently depending on the different soil aggregate components. For instance, polyester microfibers altered the content of total organic carbon in big (>2 mm) and small (2.0–0.25 mm) aggregate fractions but not in micro-aggregate fractions (0.25–0.05 mm) (Zhang & Zhang, 2020). Microplastics can improve the soil's ability to store carbon due to their high carbon content (polyethylene and polystyrene contain over 90% carbon).

Microplastics in agricultural environments alter the physicochemical properties of soil and the cycle of nutrients, which has an uncertain effect on greenhouse gas emissions. It has been studied that microplastics have different effects on soil CO_2, N_2O, and CH_4 emissions. Microplastics decrease soil organic carbon mineralization and decomposition as well as CO_2 and N_2O emissions. Future soil ecological impact assessments should include an analysis of the possible effects of various microplastics on greenhouse gas emissions in consideration of the significance of greenhouse gases to global warming and climate change.

3.2 *Effects of Microplastics on Enzyme Activities*

The cycle of soil nutrients like carbon, nitrogen, and phosphorus is greatly influenced by soil enzymes, which are directly associated with a number of soil biological and biochemical activities. These can also be used to evaluate the soil's fertility, which can provide early indicators of changes in the soil ecosystem. Microplastics may either hinder or stimulate enzyme activity, or they may have no noticeable impact. According to Yu et al. (2020), microplastics significantly decreased the activities of dehydrogenase, FDAse, and urease as well as dehydrogenase activity and the activities of enzymes involved in the P, N, and C cycles of the soil (betaglucosidase and cellobiohydrolase). The concentration of microplastics has an impact on the enzyme activity in soil. According to Liu et al. (2017), a high concentration of polypropylene-microplastics (28% w/w) may stimulate the activities of phenoloxidase and FDAse, whereas a low concentration (7% w/w) had little to no impact. Low concentration of microplastics (0.2% w/w) decreased the phosphatase activity but had no effect on the activities of phosphatase, urease, or catalase. High-dose polylactic acid microplastics (2% w/w), polybutylene succinate microplastics, and polyhydroxybutyrate microplastics increased the phosphatase, urease, and catalase activities. Thus, the effects of microplastics also depend on the polymer type. Although sucrase activity was not significantly affected, polyethylene microplastics can activate the catalase and urease activities. Both polyvinylchloride and polyethylene microplastics had inhibitory effects on the activity of the enzyme FDAse, whereas they had activating effects on the activities of the enzyme's urease and acid phosphatase.

Additionally, the impact of microplastics on enzyme activity depends on their structure. For instance, microsphere polypropylene microplastics have a reverse effect on dehydrogenase while fiber polypropylene microplastics have a higher influence on alkaline phosphatase and urease activities (Yi et al., 2021). With increasing exposure period, microplastics have varying effects on enzyme activity. According to Chen et al. (2021), polyamide-microplastics significantly decreased soil enzyme activity in the first 20 days but had no apparent impact on urease, catalase, or glucosidase activities over the course of a 70-day incubation period (Chen et al., 2020). In a brief, the environment, other variables, and the concentration, size, type, and form of microplastics all have an impact on the degree and direction of the effects of microplastics on enzyme activities, and their mode of action is complex.

3.3 Impacts of Microplastics on Plants

Microplastics may have a wide range of toxic effects on various physiological and biochemical processes in plants, such as reducing or delaying seed germination, hindering plant growth, changing root traits, reducing biomass, reducing, and delaying fruit yield, disrupting with photosynthesis, causing oxidative damage, and producing genotoxicity. Direct toxicity of contaminants (like flame retardants, plasticizers, antioxidants, and colorings) present in microplastics can negatively impact plant growth and performance through a variety of mechanisms, including direct toxicity to plants, indirect effects on plant growth because of changes in soil characteristics, microbial communities, and direct toxicity of contaminants (Rillig et al., 2019).

3.3.1 Direct Toxicity of Microplastics to Plants

Through adsorption, microplastics can have an impact on plants. Microplastics, for instance, may adhere to plant roots and alter their characteristics, making it more difficult for plants to absorb water and nutrients. Smaller-particle microplastics are more harmful to plants (Jiang et al., 2021). Instead, submicrometer or micron-sized microplastics could enter the plant's body and harm it by altering intracellular molecules and cell membranes, leading to oxidative stress (Yin et al., 2021). Interspace tissues of plant roots are sensitive to microplastic accumulation, which can subsequently transfer to flowers, stems, leaves, and fruits of the plant. Microplastics can also interfere with the plant's ability to absorb other substances including Fe, Mn, Cu, and Zn, and their phytotoxicity depends on the type of plant.

3.3.2 Changes in Soil Properties and Microbial Communities: Indirect Effects on Plants

The characteristics of the soil, its biological community, and its diversity all play a significant role in the growth of plants. Microplastics change the physical, chemical, and microbiological composition of soil, that may change the plant growth conditions, rhizosphere, and the availability of nutrients, indirectly influencing plants. According to Wan et al. (2019), microplastics, for instance, sharply increase the rate of soil water evaporation, which may result in soil dryness and potentially have a negative impact on plant performance. Microplastic contamination may lower soil fertility and result in nutrient loss in plants. Furthermore, the diversity of soil bacteria and the number of rhizosphere fungal symbionts may fall due to microplastics, which will reduce plant diversity.

3.3.3 Direct Toxicity of Contaminants Found in Microplastics

Plants may be affected by additives used in plastic, such as flame-retardants and plasticizers, as well as by other heavy metals and organic pollutants adsorbing on the surface of microplastics. These chemical additives rapidly contaminate the soil and have a negative impact on plant development. Depending on the polymer structure, form, degradation, concentration, additives, and location of the microplastics, these negative impacts intensify as the adsorption capacity of the microplastics rises (Zhou et al., 2021).

3.4 Impact of Microplastics on Soil Animals

Although it is generally recognized that flora, including earthworms, springtails, and mites, are crucial for preserving soil quality, the effects of microplastics on these important species could seriously jeopardize the efficiency of the agroecosystem. Consumption of microplastics may cause immunological issues, intestinal damage, oxidative stress, neurotoxicity, DNA damage, and improperly expressed genes in earthworms, springtails, ciliates, amoebas, and other terrestrial species. These effects may even be passed along the food chain. The unfavorable impacts of microplastics are primarily brought about by their significant build-up in soil organisms'guts and stomachs, which can impair their immune systems and have an impact on their feeding habits and development. In addition, microplastics impair invertebrate sperm and the gut cells and DNA of earthworms. They also have the potential to change the diversity and richness of the gut microbiomes of soil animals, which could influence their capacity to take part in the cycle of essential elements and the decomposition of organic matter in the soil. Animal reactions to microplastics are strongly influenced by their exposure concentration, shape, size, kind, and additions (Chen et al., 2020).

3.5 Impact of Microplastics on Human Health

Microplastics move from one trophic level to the next until they reach the top of the food chain for humans, either directly (through contaminated beverages and food, especially those made from edible animals and plastic-made drinks) or indirectly (via contaminated water and air) or through the environment (through contaminated edible animals and plants). Numerous microplastics have been found in foods and beverages that are consumed on a regular basis, including honey and sugar, beer, and drinking water (Pivokonsky et al., 2018). Additionally, wheat and lettuce roots and leaves can also be penetrated by 0.2 and 2 m microplastics (Li et al., 2020). According to these studies, microplastics can get into crops through their seeds or fruits and can even get into people's bodies through their diet. Microplastics can have a negative impact on health if they reach the human body through the transmission of viruses, parasites, and chemical and particulate toxins. According to studies, after entering the human body, microplastics may gather in the intestines. These could potentially alter the diversity and makeup of the intestinal bacterial community and cause anomalies, which would impair normal gastrointestinal function, causing localized inflammation, interfering with endocrine function, and be harmful to human health. Additionally, microplastics have been shown to cross the intestinal barrier and go to the liver and spleen via the circulatory system. The endocrine abnormalities and several health issues, including diabetes, cancer, and obesity, are linked to microplastic additives like bisphenols and phthalates (Teles et al., 2020). The presence of microplastics in agricultural soil poses a major hazard to human health since they can be ingested, inhaled, or come into touch with humans, causing respiratory tract irritation, obesity, cardiovascular disease, and gastrointestinal disorders. Therefore, it is important to eliminate these microplastics from the environment (Yu et al., 2022).

4 Microplastic Remediation

Two ideas that are currently being used for microplastics decomposition are biological degradation and advanced oxidation processes (AOPs). These two procedures can disrupt the chemical connections between polymer microplastics and tiny molecules, allowing them to be further transformed into valuable chemicals or even

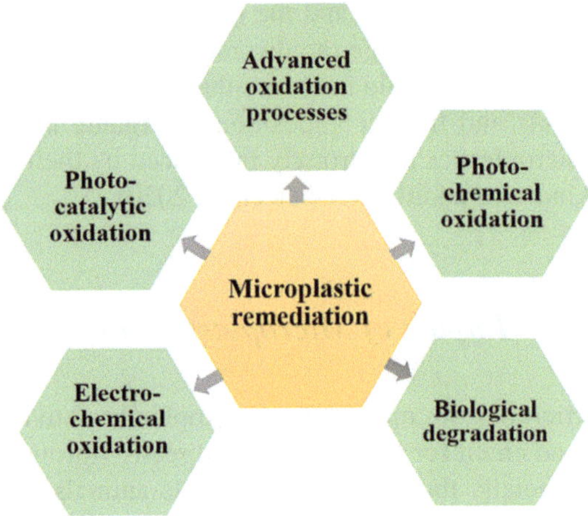

Fig. 3 Different techniques for microplastic remediation

fully mineralized into CO_2 and H_2O (Silva et al., 2018). The polymer chain can break at any monomer, irregularly converting microplastics into readily available organic or inorganic compounds. In some microplastics, the monomer repeating unit ends in a chain fracture, and the remaining monomer then breaks sequentially by chain depolymerization. Environmental conditions have a significant impact on polymer types, which in turn affects the produced products and degradation mechanisms. The authors have explored these processes in the following sections based on the studies that have already been done on the microplastic degradation by advanced oxidation processes (AOPs) (photocatalytic oxidation, direct photodegradation, and electrochemical oxidation), as well as by microbes (Fig. 3).

4.1 Advanced Oxidation Processes (AOPs)

Currently, advanced oxidation processes (AOPs), based on the formation of various reactive oxygen species (ROS), such as hydroxyl radicals (normal hydrogen electrode [NHE]) versus •OH, E0 = 2.7 V for Fenton and sulfate radicals (SR) (SO_4•-, E0 = 3.1 V versus NHE), have demonstrated excellent performance in degrading persistent organic contaminants in water bodies (Liu et al., 2020). Sulfate radical (SR) based on Advanced oxidation processes (AOPs) was considered as a stick-down system to breakdown a broad variety of refractory organic pollutants in water environments, and some research suggested that Fenton treatment may efficiently decompose plastic trash into usable products. According to reports, sulfate radical-based AOPs have demonstrated excellent catalytic degrading potential for decomposing cosmetic microplastics, which primarily consist of polyethylene (Kang et al., 2019). The high redox potential of microplastics is likely to facilitate the oxidation process. This immediately induced the disintegration of microplastics, which causes the breakage of the chain, the synthesis of beneficial products, or even the complete mineralization of microplastics. The detail discussion of advanced oxidation processes is mentioned below in the techniques entitled as, photocatalytic oxidation, including direct photodegradation, electrochemical oxidation processes, photocatalytic oxidation, and the mechanisms of their degradation.

4.2 Photochemical Oxidation

The process of photodegradation is thought to be essential for the degradation of polymers. Considering microplastics, prolonged exposure to sunlight enriched by UV light may cause the synthesis of free radicals that are harmful to the environment, the removal of hydrogen, the addition of oxygen, the cross-linking or scission of chemical chains, as well as morphological changes such as cracks and flakes. It was believed that the primary component affecting this process was UV light. Natural photodegradation of microplastics was unregulated. Therefore, it is essential to research the connection between microplastics' aging traits and their degree of aging. Song et al. (2017) studied the aging process of microplastics under lab-accelerated photodegradation settings and discovered that oxygen-containing groups and cracks were generated on the surfaces of microplastics. However, due to the relatively low concentration of ROS in the aquatic environment, nothing was known about the effects of ROS on the aging processes of microplastics. Additionally, in nature, especially in aquatic habitats, the photodecomposition of microplastics occurs very slowly. Polystyrene microplastics removal is a pressing subject because it is one of the most often found plastics products in aquatic locations. However, earlier studies on the polystyrene's photodegradation have mainly focused on the big plastic film rather than microplastics or nanoplastics. Feng et al. (2011) examined the disintegration of polystyrene microspheres using the photo-mediated Fenton process. The primary limiting element in a Fenton-based reaction is the oxidant. Additionally, because H_2O_2 has a free radical scavenging effect, the secondary pollution brought on by iron ions leaching and significant sludge development prevents their effective employment. Recently, Kang et al. (2019) investigated the catalytic activation of peroxymonosulfate to generate active radicals based on sturdy carbon nanosprings, which can degrade cosmetic microplastics. Despite these developments, oxidation techniques were unable to directly degrade cosmetic microplastics into compounds with value additions. The nontoxic organic byproducts of microplastic breakdown may be digested by microbes, which might then convert them into other useful products like protein, biofuels, and sugar that could be used by people in a sustainable and environmentally acceptable manner. The carbon cycle in nature can thus be realized through this mechanism. Below is a discussion of the photocatalytic oxidation of microplastics on catalysts.

4.3 Photocatalytic Oxidation

Photocatalysis is considered as an AOP for removing contaminants. It is an established green technology, and because it makes use of limitless, cost-free solar energy. It has the potential to be both environmentally friendly and economically viable. Photocatalytic destruction of organic contaminants is primarily based on semiconductor materials. According to Hodges et al. (2018), when the absorbed photon energy (*E*) exceeds the semiconductor's band gap energy (*Eg*) ($E > Eg$), the electrons in the valence band (e^-) will be transported to the conduction band (CB), resulting in the production of positive holes (h^+), which separates electron-hole pairs. Both species (h^+ and e^-) react with O_2, OH^-, or H_2O to form highly reactive oxygen species (ROS), which directly speeds up the decomposition of organic pollutants and plastics. Numerous studies have previously demonstrated that reactive oxygen species can degrade massive plastic films or particles (Jiang et al., 2021). Additionally, Zhang et al. (2020b) described the photodegradation mechanisms of plastics in depth. There are also some studies reporting the photodegradation of microplastics in aquatic environments using semiconductor materials as photocatalysts. According to Hodges et al. (2018), TiO_2-based nanomaterials have been widely used as standard photocatalysts for the breakdown of large-sized plastics and microplastics under visible light. This is most likely because of their superior ability to oxidize organic contaminants. In other investigations, ZnO-based materials were also used to degrade plastics and microplastics due to their high redox potential and strong catalytic activity (Qi et al., 2017). Even though some partial degradation was visible in these constructed photocatalytic systems, the majority of them did not explore the last degradation products using liquid chromatography-mass spectrometry (LC–MS) and other detection methods. Volatile organic molecules may also be released during these processes. Because the catalyst is floating in water, it is difficult to recycle, which leads to some secondary contamination. If the catalyst is immobilized on the conductive substrate surface, these issues can be prevented. So, below, we have covered the recently discovered electrochemical oxidation for microplastics breakdown.

4.4 Electrochemical Oxidation

Indirect cathode oxidation and anodic oxidation (AO) are two types of electrochemical oxidation. The most well-known method is anodic oxidation, which is the direct oxidation of pollutants by organic pollutants on the anode surface through charge transfer or indirect oxidation of pollutants by active H_2O_2, O_3, chlorine species, and peroxymonosulfate in aqueous solution. An electro-Fenton technology is regarded as being indirect cathode oxidation. In electro-Fenton, the Fenton's reaction resulted in the production of •OH and other reactive oxygen radicals from the breakdown of H_2O_2, and the resulting free radicals were in charge of oxidizing organic contaminants. The treatment of refractory organics can thus benefit greatly from this technique. According to Kang et al. (2019), sulfate radical-based AOPs demonstrated superior decomposition performance of microplastics, which are primarily made of polyethylene. This was accomplished by catalytically activating peroxymonosulfate to create active radicals, and an analysis of their toxicity showed that the organic byproducts of the degradation of microplastics were beneficial to aquatic microorganisms. The investigation, however, was unable to classify the many intermediate products. Because of its strong controllability, ease of operation, and lack of secondary pollution, powerful electro-Fenton-like processes have been widely used to generate strong oxidizing •OH for the removal of organic pollutants (Chen et al., 2020). Because the cathodic processes dominate the electro-Fenton-like system, the nature of the cathode elements significantly influenced the complete performance of the degradation. Therefore, it is vital to use catalyst materials that are highly effective. According to the analysis above, it is highly important to research the use of an electro-Fenton-like approach to reduce microplastic pollution. There are not many publications on the use of electrochemistry to cure the microplastic pollution in aquatic environments. A study investigated the electro-Fenton-like system's ability to degrade polyvinyl chloride microplastics using a TiO_2/graphite cathode (Miao et al., 2020). This system demonstrated remarkable polyvinyl chloride decomposition activity through •OH oxidation, cathodic reduction dichlorination, and some desired and practical compounds were generated during the process. This study points the way toward investigating an effective electrochemical oxidation technique for the conversion of microplastics into useful products.

4.5 Biological Degradation

Researchers have discovered that although microplastics are stable in natural environments for long periods of time, particular bacteria are still required for their decomposition. Microbials are extremely adaptable and can live in practically any environment. They can also degrade many organic contaminants, including microplastics. Microplastics were used as substrates for the development of biofilms during the microplastics breakdown process (Zurier & Goddard, 2020). The structure of the microplastics is weakened as a result of pitting and cracking caused by the formation of biofilm. The step that determines the rate is when bacterial enzymes attack the fragments of broken microplastics both selectively and randomly. When microplastic fragments weigh less than 600 kDa, bacteria in the biofilm can easily absorb them during assimilation. The assimilation pieces are further broken down by enzymes into smaller molecules (CO_2, N_2, CH_4, H_2O, H_2S), which are then used as energy sources by microorganisms before being released back into the environment. This process converts the small molecules into products that are now available. To improve the biodegradation efficacy of microplastics and make it a practical method for reducing microplastic pollution, bottlenecks at each stage of the process must be eliminated. For instance, pure bacterial cultures of a single bacterium that biodegrades chemical substances frequently secrete toxic byproducts which prevent the growth of microorganisms. Enzymes that are produced might not be perfectly suited to the plastic substrate. The objective of destroying the microplastics particles is challenging to accomplish. Utilizing particular bacteria improves the biodegradation performance of microplastics, offering a sustainable and promising method to boost natural biodegradation of microplastics with minimal negative effects on natural ecosystems. Therefore, bioremediation is likewise regarded as the most desired method for eliminating microplastics pollution, with the exception of AOPs. Some studies have demonstrated how bacteria can be used to break down microplastics. The utilization of bacterial cultures for the in vitro microbial-mediated breakdown of microplastics was the main area of concern. The majority of the wastewater, sludge, and silt were removed from these pure cultures. Two pure bacterial cultures from mangrove soil were investigated by Auta et al. (2018) for the breakdown of polypropylene microplastics. After 40 days of testing, *Rhodococcus* sp. strain 36 and *Bacillus* sp. strain 27 accelerated weight loss of polypropylene by 4.0% and 6.4%, respectively. Diverse abnormalities were discovered on the polypropylene surface, indicating that bacterial cultures isolated from the environment may stick to, colonize, and harm polypropylene. Fungi, in addition to bacteria, can use and break down microplastics. Recent years have seen the discovery of novel fungus with enhanced microplastic breakdown abilities. While studies on the fungal-associated degradation of microplastics have been rarely described, indicating that it may be challenging to find fungal strains that perform better at degrading microplastics by ectopic screening. The biodegradation of microplastics by fungi under varied environmental circumstances is still an active study area; however, some advancements have been made. The successful use of fungus to break down microplastics and other environmental contaminants has steadily gained attention. Table 1 details the microbial biodegradation of various forms of microplastics (bacteria and fungi). Although it is clear that each of these microbes has the potential to degrade microplastics to some level, it took at least 21 days for them to achieve the desired biological decomposition outcomes. Despite this, these works still provide a framework for the creation of fresh, cutting-edge approaches that attempt to lessen the negative environmental effects of microplastic pollution (Du et al., 2021).

5 Challenges and Future Perspectives

This chapter summarized the source of microplastics pollution in the environment. In addition, the impacts of microplastics on ecosystems are also discussed, particularly the risks of microplastics to human health. For a better understanding of the sources and ecological risks of microplastics in ecosystems, there are some challenges that we need to solve:

Table 1 Plastics degradation by microorganisms, assay conditions for the biodegradation, and methods used for detection

Types of plastic	Strain	Assay conditions for biodegradation (media, time, rpm, T °C)	Methods for detection of biodegradation	Biodegradation of polymers	References
Polyethylene, polyethylene Terephthalate, polystyrene (pre-treated with UV)	*Bacillus cereus*	Mineral salt media with 0.5 g polymer, for 38 days, 150 rpm, at room temperature	Weight loss, Fourier transform infrared spectroscopy, scanning electron microscopy	1.6%, 6.6%, 7.4%, respectively	Auta et al. (2017)
Polyethylene, polypropylene, terephthalate, polystyrene (pre-treated with UV)	*Bacillus gottheilli*	Mineral salt media with 0.5 g polymer, for 38 days, 150 rpm, at room temperature	Weight loss, Fourier transform infrared spectroscopy, scanning electron microscopy	6.2%, 3.0%, 3.6%, 5.8%, respectively	Auta et al. (2017)
Polyethylene	Marine microbial bioaugmented indigenous community with *Salinibacterium* sp. and *Lysinibacillus* sp.	Nutrient broth media supplemented with Saline water, for 180 days, 120 rpm, at 25 °C	Weight loss, Fourier transform infrared spectroscopy, scanning electron microscopy	19%	Syranidou et al. (2017)
Polyethylene	Mixed mesophilic bacterial culture (*Bacillus* sp. and *Paenibacillus* sp.)	Basal media with 0.1 mg of polymer, for 60 days, at 30 °C	Weight loss, Fourier transform infrared spectroscopy, scanning electron microscopy	14.7%	Park and Kim (2019)

(continued)

Table 1 (continued)

Types of plastic	Strain	Assay conditions for biodegradation (media, time, rpm, T °C)	Methods for detection of biodegradation	Biodegradation of polymers	References
Polyethylene and polypropylene	Biofilm composed of *Pirellulaceae, Phycisphaerales, Cyclobacteriaceae*, and *Roseococcus*	Dechlorinated tap water, woods hole media, for 0.7 months (incubation in tanks in a greenhouse exposed to natural light)	DNA extraction, amplification, and Sequencing (evaluation of the effects of substrate type on microbial communities)	Information not available	Miao et al. (2019)
Polyethylene	*Aspergillus oryzae*	Synthetic medium containing ampicillin, for 120 days, at 28 °C in an incubator with shaker	Weight loss; Fourier transform infrared spectroscopy, gas chromatography–mass spectrometry	36.4%	Muhonja et al. (2018)
Polyethylene	*Zalerion maritimum*	Minimum growth media with 0.130 g of polymer, for 28 days, 120 rpm, at 25 °C	Weight loss, Fourier transform infrared spectroscopy, nuclear magnetic resonance (NMR) spectroscopy	43%	Paço et al. (2017)
Polypropylene	*Bacillus* sp. strain 27	Bushell Haas media with 0.5 g polymer, for 38 days, at room temperature	Weight loss; Fourier transform infrared spectroscopy, scanning electron microscopy, pH variation	4%	Auta et al. (2018)
Polypropylene	*Rhodococcus* sp. strain 36*	Bushnell Haas (BH) media, for 38 days, at 29 °C	Weight loss, Fourier transform infrared spectroscopy, scanning electron microscopy	6.4%	Auta et al. (2018)

Polystyrene	*Cephalosporium* sp. (NCIM 1251)	Mineral salt media, for 60 days, at 120 rpm, at 28 °C	Weight loss, Fourier transform infrared spectroscopy, scanning electron microscopy, pH variation, gel permeation, chromatography	2.17%	Chaudhary and Vijayakumar (2020)
Polystyrene	*Mucor* sp. (NCIM 881)	Mineral salt media, for 60 days, at 120 rpm, at 28 °C	Weight loss, Fourier transform infrared spectroscopy, scanning electron microscopy, pH variation; gel permeation, chromatography	1.81%	Chaudhary and Vijayakumar (2020)
Polystyrene (styrofoam)	*Pestalotiopsis* sp.	Malt extract broth, for 30 days, at 25 °C	Weight loss, Fourier transform infrared spectroscopy, scanning electron microscopy	74.43%	Yanto et al. (2019)

1. There are few regional and global data sources for microplastics pollution in agroecosystems, necessitating further research. Second, there is no way to compare the results of previous studies due to the absence of consistent evaluation criteria for microplastic pollution. To establish the baseline concentration and calculate the possible impact on biodiversity and biota exposure risk, it is required to evaluate the microplastic contamination in various habitats and practices.
2. Microplastics are available in a variety of shapes, sizes, and sorts, and, depending on the product they are used in, they may have varying effects on ecosystems. Therefore, investigations to explore the impact of diverse forms of microplastics on the ecological environment should include microplastics from a range of applications and sources. It is vital to investigate the comprehensive impacts of various types of microplastic mixes on the terrestrial ecosystem because the majority of recent studies have concentrated on the ecotoxicity of a single microplastic to individual animals.
3. Organic contaminants that are difficult to degrade and persist for a long time in soil are microplastics. Long-term identification of the microplastics' influencing mechanism and process is necessary. However, the majority of pertinent investigations have relatively short research timelines and only provide a preliminary impact profile. In the future, extensive field-scale studies will be required to assess the biological and environmental impacts of microplastics in order to more accurately and comprehensively show their long-term consequences on the terrestrial environment.
4. In agricultural soils, microplastics commonly exist with environmental toxins such as heavy metals and organics. Microplastics and soil contaminants existing in the soil might have additive and antagonistic effects that will have additional complex effects and unclear environmental concerns on the soil, particularly on soil ecosystems. Future studies should focus more on the combined impacts of microplastics and other contaminants coexisting.
5. Soil microbes are necessary for the cycling of soil nutrients and the preservation of soil functions. There are currently limited investigations on specific bacterial communities and their roles, particularly microbiological methods including biogeochemical cycles of N, C, and P such as greenhouse gas emissions. Instead, most studies concentrate on microbial populations and their activities. The effect of microplastics on important bacterial species which are essential to fundamental soil activities has to be further studied.

Further recommendations and future aspects for microplastics degradation are:

1. In order to selectively degrade microplastics into usable and nontoxic products through the control of the breaking of chemical bonds, high activity and selectivity catalysts should be developed. For biodegradation, particular functional microorganisms should be grown. Additionally, to increase the efficiency of microbes-mediated microplastics breakdown, future research should integrate genomes and proteomics engineering methods.

2. High-performance liquid chromatography coupled high-resolution mass spectrometry (HPLC-HRMS), positron annihilation spectroscopy, free radical trapping and diffuse reflection, Fourier transform, and infrared spectroscopy technique experiments can reveal the selective degrading mechanism of microplastics. Thus, to extract highly purified degradation products, suitable techniques must be investigated.
3. To evaluate the toxic effects of the degraded intermediates that have been produced during the breakdown of microplastics, toxicity research should be carried out.
4. It is challenging to bring together the degraded products from the soil or other ecosystems, in contrast to the aquatic environment. Therefore, the improvement of an extraction method for degradation products needs to receive much more attention.
5. To determine the effects of the interactions among several functional microbial species and their numerous enzymes on the degradation of microplastics, it is essential to observe the degradation products and efficiency of microplastics.

6 Conclusion

Due to their strength and slow rate of deterioration, plastic materials persist in the environment. Natural factors have changed it into microplastics, that are easily absorbed into all ecosystems and then into living things where they build up, increase, and have adverse effects. This chapter summarized the sources, effects, remediation, and future aspects of microplastics degradation. Biodegradation is eco-friendly and is also a promising approach for the removal of microplastics pollution. However, it usually takes a longer period, and it is unclear how it degrades. Therefore, choosing the appropriate decomposition techniques for the various types of microplastics treatment is important. This chapter offers a clearer grasp of the decomposition of microplastics through advanced oxidation processes and microbial degradation as well as an understanding of challenges and future prospects for the degradation of microplastics.

References

An, L., Liu, Q., Deng, Y., Wu, W., Gao, Y., & Ling, W. (2020). Sources of microplastic in the environment. In D. He & Y. Luo (Eds.), *Microplastics in terrestrial environments. The handbook of environmental chemistry* (Vol. 95). Springer.

Auta, H. S., Emenike, C. U., & Fauziah, S. H. (2017). Screening of bacillus strains isolated from mangrove ecosystems in peninsular Malaysia for microplastic degradation. *Environmental Pollution, 231*, 1552–1559.

Auta, H. S., Emenike, C. U., Jayanthi, B., & Fauziah, S. H. (2018). Growth kinetics and biodeterioration of polypropylene microplastics by Bacillus sp. and Rhodococcus sp. isolated from mangrove sediment. *Marine Pollution Bulletin, 127*, 15–21.

Brahney, J., Hallerud, M., Heim, E., Hahnenberger, M., & Sukumaran, S. (2020). Plastic rain in protected areas of the United States. *Science (New York, N.Y.), 368*(6496), 1257–1260.

Chaudhary, A. K., & Vijayakumar, R. P. (2020). Studies on biological degradation of polystyrene by pure fungal cultures. *Environment, Development and Sustainability, 22*(5), 4495–4508.

Chen, H., Wang, Y., Sun, X., Peng, Y., & Xiao, L. (2020). Mixing effect of polylactic acid microplastic and straw residue on soil property and ecological function. *Chemosphere, 243*, 125271.

Chen, J., Wan, J., Gong, Y., Xu, K., Zhang, H., Chen, L., Liu, J., & Liu, C. (2021). Effective electro-Fenton-like process for phenol degradation on cerium oxide hollow spheres encapsulated in porous carbon cathode derived from skimmed cotton. *Chemosphere, 270*, 128661.

Corradini, F., Meza, P., Eguiluz, R., Casado, F., Huerta-Lwanga, E., & Geissen, V. (2019). Evidence of microplastic accumulation in agricultural soils from sewage sludge disposal. *The Science of the Total Environment, 671*, 411–420.

Du, H., Xie, Y., & Wang, J. (2021). Microplastic degradation methods and corresponding degradation mechanism: Research status and future perspectives. *Journal of Hazardous Materials, 418*, 126377.

Feng, H. M., Zheng, J. C., Lei, N. Y., Yu, L., Kong, K. H., Yu, H. Q., Lau, T. C., & Lam, M. H. (2011). Photoassisted Fenton degradation of polystyrene. *Environmental Science & Technology, 45*(2), 744–750.

Geyer, R., Jambeck, J. R., & Law, K. L. (2017). Production, use, and fate of all plastics ever made. *Science Advances, 3*(7), e1700782.

Hahladakis, J. N., Velis, C. A., Weber, R., Iacovidou, E., & Purnell, P. (2018). An overview of chemical additives present in plastics: Migration, release, fate and environmental impact during their use, disposal and recycling. *Journal of Hazardous Materials, 344*, 179–199.

Hidalgo-Ruz, V., Gutow, L., Thompson, R. C., & Thiel, M. (2012). Microplastics in the marine environment: A review of the methods used for identification and quantification. *Environmental Science & Technology, 46*(6), 3060–3075.

Hodges, B. C., Cates, E. L., & Kim, J. H. (2018). Challenges and prospects of advanced oxidation water treatment processes using catalytic nanomaterials. *Nature Nanotechnology, 13*(8), 642–650.

Jiang, R., Lu, G., Yan, Z., Liu, J., Wu, D., & Wang, Y. (2021). Microplastic degradation by hydroxy-rich bismuth oxychloride. *Journal of Hazardous Materials, 405*, 124247.

Kang, J., Zhou, L., Duan, X., Sun, H., Ao, Z., & Wang, S. (2019). Degradation of cosmetic microplastics via functionalized carbon Nanosprings. *Matter, 1*(3), 745–758.

Leslie, H. A., Brandsma, S. H., van Velzen, M. J., & Vethaak, A. D. (2017). Microplastics en route: Field measurements in the Dutch river delta and Amsterdam canals, wastewater treatment plants, North Sea sediments and biota. *Environment International, 101*, 133–142.

Li, L., Luo, Y., Li, R., Zhou, Q., Peijnenburg, W. J. G. M., Yin, N., Yang, J., Tu, C., & Zhang, Y. (2020). Effective uptake of submicrometre plastics by crop plants via a crack-entry mode. *Nature Sustainability, 3*(11), 929–937.

Liu, H., Yang, X., Liu, G., Liang, C., Xue, S., Chen, H., Ritsema, C. J., & Geissen, V. (2017). Response of soil dissolved organic matter to microplastic addition in Chinese loess soil. *Chemosphere, 185*, 907–917.

Liu, F., Xu, Y., Zhang, B., Liu, Y., & Zhang, H. (2020). Heterogeneous degradation of organic contaminant by peroxydisulfate catalyzed by activated carbon cloth. *Chemosphere, 238*, 124611.

Lu, Y., Li, M. C., Lee, J., Liu, C., & Mei, C. (2023). Microplastic remediation technologies in water and wastewater treatment processes: Current status and future perspectives. *The Science of the Total Environment, 868*, 161618.

Magnusson, K., Eliasson, K., Fråne, A., Haikonen, K., Hultén, J., Olshammar, M., Stadmark, J., & Voisin, A. (2016). Swedish sources and pathways for microplastics to the marine environment. *Swedish Environmental Protection Agency, 183*(C 183), 1–89.

Miao, L., et al. (2019). Distinct community structure and microbial functions of biofilms colonizing microplastics. *Science of the Total Environment, 650*, 2395–2402.

Miao, F., Liu, Y., Gao, M., Yu, X., Xiao, P., Wang, M., Wang, S., & Wang, X. (2020). Degradation of polyvinyl chloride microplastics via an electro-Fenton-like system with a TiO_2/graphite cathode. *Journal of Hazardous Materials, 399*, 123023.

Muhonja, C. N., Makonde, H., Magoma, G., & Imbuga, M. (2018). Biodegradability of polyethylene by bacteria and fungi from Dandora dumpsite Nairobi-Kenya. *PLoS One, 13*(7), 1–17.

Murphy, F., Ewins, C., Carbonnier, F., & Quinn, B. (2016). Wastewater treatment works (WwTW) as a source of microplastics in the aquatic environment. *Environmental Science & Technology, 50*(11), 5800–5808.

NIVA. (2020). Integrated environmental assessment and management. *Microplastics in road dust – Characteristics, pathways and measures REPORT SNO. 7526-2020.*

Orlando, M., Molla, G., Castellani, P., Pirillo, V., Torretta, V., & Ferronato, N. (2023). Microbial enzyme biotechnology to reach plastic waste circularity: Current status, problems and perspectives. *International Journal of Molecular Sciences, 24*(4), 3877.

Paço, A., et al. (2017). Biodegradation of polyethylene microplastics by the marine fungus Zalerion maritimum. *Science of the Total Environment, 586*, 10–15.

Park, S. Y., & Kim, C. G. (2019). Biodegradation of micro-polyethylene particles by bacterial colonization of a mixed microbial consortium isolated from a landfill site. *Chemosphere, 222*, 527–533.

Patil, S., Bafana, A., Naoghare, P. K., Krishnamurthi, K., & Sivanesan, S. (2021). Environmental prevalence, fate, impacts, and mitigation of microplastics-a critical review on present understanding and future research scope. *Environmental Science and Pollution Research International, 28*(5), 4951–4974.

Pettipas, S., Bernier, M., & Walker, T. R. (2016). A Canadian policy framework to mitigate plastic marine pollution. *Marine Policy, 68*, 117–122.

Pivokonsky, M., Cermakova, L., Novotna, K., Peer, P., Cajthaml, T., & Janda, V. (2018). Occurrence of microplastics in raw and treated drinking water. *The Science of the Total Environment, 643*, 1644–1651.

Qi, K., Cheng, B., Jiaguo, Y., & Ho, W. (2017). Review on the improvement of the photocatalytic and antibacterial activities of ZnO. *Journal of Alloys and Compounds, 727*, 792–820.

Rillig, M. C., Lehmann, A., de Souza Machado, A. A., & Yang, G. (2019). Microplastic effects on plants. *The New Phytologist, 223*(3), 1066–1070.

Silva, A. B., Costa, M. F., & Duarte, A. C. (2018). Biotechnology advances for dealing with environmental pollution by micro(nano)plastics: Lessons on theory and practices. *Current Opinion in Environmental Science & Health, 1*, 30–35.

Smith, M., Love, D. C., Rochman, C. M., & Neff, R. A. (2018). Microplastics in seafood and the implications for human health. *Current Environmental Health Reports, 5*(3), 375–386.

Song, Y. K., Hong, S. H., Jang, M., Han, G. M., Jung, S. W., & Shim, W. J. (2017). Combined effects of UV exposure duration and mechanical abrasion on microplastic fragmentation by polymer type. *Environmental Science & Technology, 51*(8), 4368–4376.

Sundt, P., Syversen, F., Skogesal, O., & Schulze, P.-E. (2016). Primary microplastic-pollution: Measures and reduction potentials in Norway. *Mepex*, (April), 117.

Syranidou, E., et al. (2017). Development of tailored indigenous marine consortia for the degradation of naturally weathered polyethylene films. *PLoS One, 12*(8), 1–21.

Teles, M., Balasch, J. C., Oliveira, M., Sardans, J., & Peñuelas, J. (2020). Insights into nanoplastics effects on human health. *Science Bulletin, 65*(23), 1966–1969.

United Nations Environment Programme. (2022). *Visual feature | beat plastic pollution*. UNEP.

Wan, Y., Wu, C., Xue, Q., & Hui, X. (2019). Effects of plastic contamination on water evaporation and desiccation cracking in soil. *The Science of the Total Environment, 654*, 576–582.

Yanto, D. H. Y., et al. (2019). Biodegradation of styrofoam waste by ligninolytic fungi and bacteria. *IOP Conference Series: Earth and Environmental Science, 308*(1).

Yi, M., Zhou, S., Zhang, L., & Ding, S. (2021). The effects of three different microplastics on enzyme activities and microbial communities in soil. *Water Environment Research: A Research Publication of the Water Environment Federation, 93*(1), 24–32.

Yin, L., Wen, X., Huang, D., Du, C., Deng, R., Zhou, Z., Tao, J., Li, R., Zhou, W., Wang, Z., & Chen, H. (2021). Interactions between microplastics/nanoplastics and vascular plants. *Environmental Pollution (Barking, Essex: 1987), 290*, 117999.

Yu, H., Fan, P., Hou, J., Dang, Q., Cui, D., Xi, B., & Tan, W. (2020). Inhibitory effect of microplastics on soil extracellular enzymatic activities by changing soil properties and direct adsorption: An investigation at the aggregate-fraction level. *Environmental Pollution (Barking, Essex: 1987), 267*, 115544.

Yu, H., Ying, Z., Wenbing, T., & Zheng, Z. (2022). Microplastics as an emerging environmental pollutant in agricultural soils: Effects on ecosystems and human health. *Frontiers in Environmental Science, 10*(March), 1–18.

Zhang, G. S., & Zhang, F. X. (2020). Variations in aggregate-associated organic carbon and polyester microfibers resulting from polyester microfibers addition in a clayey soil. *Environmental Pollution (Barking, Essex: 1987), 258*, 113716.

Zhang, D., Ng, E. L., Hu, W., Wang, H., Galaviz, P., Yang, H., Sun, W., Li, C., Ma, X., Fu, B., Zhao, P., Zhang, F., Jin, S., Zhou, M., Du, L., Peng, C., Zhang, X., Xu, Z., Xi, B., Liu, X., & Liu, H. (2020a). Plastic pollution in croplands threatens long-term food security. *Global Change Biology, 26*(6), 3356–3367.

Zhang, F., Zhao, Y., Wang, D., Yan, M., Zhang, J., Zhang, P., Ding, T., Chen, L., & Chen, C. (2020b). Current technologies for plastic waste treatment: A review. *Journal of Cleaner Production, 282*, 124523.

Zhou, J., Yuan, W., Miles, R. M., Zhao, J., Gui, H., Yang, Y., Zeng, Z., Jones, D. L., & Zang, H. (2021). Microplastics as an emerging threat to plant and soil health in agroecosystems. *Science of the Total Environment, 787*, 147444.

Zurier, H. S., & Goddard, J. M. (2020). Biodegradation of microplastics in food and agriculture. *Current Opinion in Food Science, 37*, 37–44.

Micro-Nanoplastics from Stormwater Runoffs to Water Bodies: An In-Depth Investigation

Sayli Salgaonkar, Akshay Botle, Gayatri Barabde, and Mihir Herlekar

1 Introduction

The global problem of waste plastic pollution, due to the detrimental effects of widely dispersed micro-nanoplastics (MnPs) on the ecosystem, has reached worrisome proportions (Sarkar et al., 2022). Plastic makes up between 60% and 80% of all marine trash, making it the major marine litter in the world. Approximately 80% of the plastic trash in the ocean derives from land, whereas only 18% is the result of the fishing sector. As a result, when plastic is discharged into the environment, a sizeable portion will wind up in aquatic environments where it is broken down into tiny bits by ultraviolet (UV) radiation, mechanical erosion, biological breakdown, and disintegration. Between 7000 and 35,000 tons of plastic debris are thought to be produced globally, with open ocean surface waters accounting for the majority. Most particles are <10 mm in size (Mattsson et al., 2018). Nanoplastics (NPs) (0.2 mm) are the category for the ultra-fine fraction of plastics, while microplastics (MPs) and macroplastics are often differentiated by their size ranges of 5 mm and >5 mm, respectively (Sarkar et al., 2022). Plastic breaks down into smaller bits, altering its chemical and physical properties, making it more accessible to aquatic species and perhaps having a negative biological impact. In all likelihood, the breakdown processes will keep on producing nanoscale plastic, which will be

S. Salgaonkar · A. Botle · M. Herlekar (✉)
Department of Environmental Science, The Institute of Science, Dr. Homi Bhabha State University, Mumbai, India
e-mail: tevs.mihir@iscm.ac.in

G. Barabde
Department of Environmental Science, The Institute of Science, Dr. Homi Bhabha State University, Mumbai, India

Department of Analytical Chemistry, The Institute of Science, Dr. Homi Bhabha State University, Mumbai, India

S. A. Bhat et al. (eds.), *Management of Micro and Nano-plastics in Soil and Biosolids*,
https://doi.org/10.1007/978-3-031-51967-3_4

distinct from both the original substance and microplastic (MP) (Mattsson et al., 2018). MnPs are divided into primary and secondary MnPs, depending on the source. Primary MnPs, including microfibers/beads and plastic pellets, develop as a result of direct contaminant discharge from personal care items and other industries. These anthropogenic MnPs are typically found in home, industrial, and wastewater treatment plants (WWTPs). Secondary MnPs are produced when bigger plastic particles deteriorate due to UV exposure and mechanical weathering. Macroplastic decomposition may lead to MnP buildup in soil and aquatic environments. In comparison to primary MnP sources, marine and terrestrial ecosystems present more challenges for the tracing, management, and quantification of MnPs (Sarkar et al., 2022).

Due to their widespread use and the great diversity of applications, MPs are a problem. Aquatic species have the ability to consume small plastic waste fragments, which can then build up as they move up the food chain and eventually reach people. They can have a number of negative impacts on organisms, including stunted growth, behavioral issues, reproductive problems, nutrition issues, reduced viability, and even mortality (Chae & An, 2017). Since the early 1970s, there have been reports of MPs in marine waters and their harmful consequences, which has sparked research on MPs in aquatic environments. MPs in the soil have been the subject of significant research, whereas NPs in the environment have received less attention. Horton et al. (2017) found MPs in the basin of the River Thames, with the source of MPs identified as paints used in road marking; whereas a study conducted by Ziajahromi et al. (2020) in Gold Coast, Queensland, found synthetic rubber tires as the source of MPs in stormwater, with runoff from roads being the most likely pathway. Tire and bitumen, generated due to traffic, were identified as major sources of MPs by Järlskog et al. (2020) contributing to stormwater pollution by MPs.

The use of items containing MnPs and their release into the ecosystem must be controlled, even if managing MnPs is difficult. This can be done by enacting new laws and regulations. Policymakers, including the United Nations Environment Programme (UNEP), have focused on MnPs by implementing a number of international and national legislation and policies, in addition to the issue of global plastic debris. It has been determined that the bulk of MnPs in the environment are mostly caused by the careless disposal of plastic trash (Sarkar et al., 2022).

The objectives of this chapter are to (1) offer a complete review of the many sources of micro-nanoplastics in water bodies, with a particular emphasis on plastics emerging from stormwater runoff; (2) analyze the negative environmental consequences of micro-nanoplastics, comprising aquatic ecosystems and on ecology; (3) assess the possible health effects of micro-nanoplastic contamination; and (4) present readers with a complete set of mitigating solutions for stormwater runoff-derived micro-nanoplastics contamination. As a result, this chapter will help to provide further scientific knowledge of MnP contamination in aquatic bodies produced by stormwater runoff. Furthermore, it will provide vital information to lawmakers, academics, and environmental practitioners in order to design focused and effective approaches to address this global environmental issue followed by future trends in this area.

2 Stormwater Runoff

2.1 *Introduction of Stormwater Runoff*

The runoff from rain or snowstorm that is accumulated in the pipe, ditch, stream, or river after being precipitated and reaching the ground is known as urban stormwater (National Research Council, 2008). The runoff from stormwater is a nonpoint pollution source comprising debris, litter, and sediment, which are the visible components, and coliforms, toxic chemicals, and heavy metals, which are unseen elements. These constituents present in stormwater pose a serious threat to animals, humans, and plants.

2.2 *Sources of Stormwater Runoff*

The sources of stormwater pollution can be classified and grouped in many ways but the source categories are not well defined. Different sources contribute to pollution under ambient conditions or at different times (Müller et al., 2020). The sources are divided into intrinsic and extrinsic types by Loganathan et al. (2013). Intrinsic sources comprise pesticide usage, littering by pedestrians, and vehicular-related sources, and extrinsic sources comprise road structures such as industries, buildings, and vegetation and soil. Classification based on the type of land use and activity performed was done by Lundy et al. (2012) which included highway surfaces, rooftops, construction sites, misconnections, etc. The sources of stormwater runoff are grouped into the following types: drainage, atmospheric deposition (AD), and anthropogenic activities. Anthropogenic activities were described as those activities that are carried out by people in urban settings or are produced as a result of such acts, with varying frequency and intensity. Vehicle movement, construction work, or winter road maintenance are a few examples of activity-related sources (Müller et al., 2020). Other sources of pollution in the built-up setting are stationary and mainly related to surfaces or land covers; these can give rise to pollution in damp weather despite man-made activity. The anthropogenic sources comprise things like parkland, building exteriors, and infrastructural components. However, a number of variables, like the age of the material, the rainfall intensity, and others, may affect how much pollution is released from surfaces (Müller et al., 2020).

2.2.1 Atmospheric Deposition

Atmospheric deposition (AD) that occurs due to deposition or precipitation in dry weather allows easier transportation of pollutants and other materials to the urban drainage surface from the atmosphere. Thus, AD consists of both components, wet and dry; the wet components potentially contribute after the wash-off of materials and chemicals from catchment surfaces and the dry components directly contribute to the contamination of surface runoff (Müller et al., 2020). Regarding runoff pollution, metropolitan areas are especially relevant as they have many sources that contribute to air pollution and a high possibility that pollutants may enter stormwater runoff due to the existence of impermeable areas as well as runoff transportation networks (Hobbie et al., 2017). In terms of pollution caused by stormwater, the atmosphere mostly functions as a channel for the transmission of pollutants rather than a direct cause of the pollution that is observed. Emissions and atmospheric transport from various sources introduce atmospheric contaminants in the urban environment from natural as well as manmade sources (Müller et al., 2020).

2.2.2 Drainage Surfaces

Paved surfaces cause stormwater pollution in two main ways: mechanical wear caused by vehicle tires and chemical elution that runs off of the pavement surface into water (Müller et al., 2020). Any surface coming into contact with surface runoff from rainwater or stormwater impacts the quality of those waters. In an urban set-up, many different structures are exposed to rainwater or stormwater, including buildings, sculptures, and lampposts, to mention a few. Their surfaces are made of a variety of materials, several of which were examined thoroughly to understand the quality of stormwater. These surfaces can contribute to thermal pollution like other resistant surfaces found in the built environment (Müller et al., 2020). A variety of substances (inorganic and organic) are employed in structuring envelopes to shield construction and other structures from the consequences of extreme weather, stormwater, and AD. The majority of these coatings, which include paints and biocides, have dangerous ingredients that end up in stormwater. New potential sources of developing contaminants in urban runoff are represented by manufacturing paint formulas, surface coatings, new materials, etc. with their use in urban settings. Modern urban designing understands the worth and advantages of preserving or growing green spaces in cities. Streets with grass and trees, neighborhood gardens, regional or central parks, and a variety of sports fields, including golf courses, are just a few examples of these spaces in varied shapes and sizes (Müller et al., 2020). The existence of green spaces always causes interaction with "gray" drainage systems, resulting in dissolved organic matter (DOM) releasing, transferring, and mixing with runoff from various street drains and storm sewer sources. The fate and transit of contaminants are both impacted by DOM, in addition to playing a significant influence in the formation of aquatic life systems (Müller et al., 2020).

2.2.3 Anthropogenic Activities

When it comes to pollution of urban stormwater, the number of substances contributed by specific transportation-related sources is significantly higher than that of other source categories. This is probably because runoff from roads was first identified as a significant pollution source, based on the traffic. For wintertime maintenance of road in chilly temperate climates, the use of grit and salt is necessary to maintain a safe driving surface. Both of these compounds have the ability to impact arriving waters by compromising water quality processes and, in the event of salts, causing toxicity. The cleaning of traffic tunnels is an additional maintenance task that contributes to pollution from runoff by polycyclic aromatic hydrocarbons (PAHs) (typical for road runoff) and trace metals (Müller et al., 2020). It is known that industrial operations are contaminant source that can go into stormwater in one of two ways: through atmospheric emissions, following dry and wet AD, or directly through surface overflow from industrial land. Evidently, the different types of industrial activities that contribute to stormwater pollution will vary as those activities evolve in response to demand in the market and the development of novel technologies for both industrial production and preventing pollution (Müller et al., 2020). Stormwater discharge is contaminated by both traditional pollutants and micropollutants from industrial activity. Even though the latter group is more significant and reflects the compounds used in certain industrial operations, the former group is linked to sources that are common in many industrial activities (Müller et al., 2020). It is well recognized that construction sites and related operations add a significant amount of soil particles and sediments to surface runoff, which then ends up in the receiving waters. Due to the removal of the soil's natural vegetation cover during building, the increased runoff volumes that cause erosion, and the transport of sediment in unlined channels, soil erosion in urbanizing areas increases during catchment development (Müller et al., 2020). Construction activities contribute to PAHs, trace metals, polychlorinated biphenyls (PCBs), trace metals, and nutrients (P and N) attached to solids, construction chemicals, pesticides, and MPs. Construction activities are unquestionably the biggest source of stormwater solids production (Müller et al., 2020). Both purposeful and accidental littering causes litter to be deposited either directly into stormwater inlets or on surfaces of the catchment where it can be removed and carried with runoff (Müller et al., 2020). Accidental spills are a substantial source of illegal releases to stormwater gutters and are more likely to happen at commercial, industrial, or sites of vehicle transportation. These sources with unpredictable character contributing to stormwater pollution are made worse by the nature of littering, illegal dumping, and home waste disposal. Concerns about these sources, which are known for contributing plastics to urban runoff, are exacerbated by the recent discovery that MPs are pollutants in marine ecosystems (Müller et al., 2020). Because relatively little to no runoff is predicted from gardens, there have not been many studies published on gardening practices as sources of stormwater pollution. Runoff from pervious surfaces may happen during intense thunderstorms or when the ground is already moist, adding suspended solids or nutrients from, for example, fertilizers used in grassed areas to stormwater (Gromaire et al., 2001).

2.3 Types of Pollutants in Stormwater Runoff

Volatile organic compounds (VOCs), a separate class of chemicals, are carried through the atmosphere. VOCs are used in a variety of products (paints, fuels, solvents, fumigants, precursors in the manufacture of chemicals, and pesticides (Müller et al., 2020)) or they are created as derivatives of processes taking place in cities (such as trihalomethanes created as byproduct of water disinfection). As a result, VOCs are pervasive in metropolitan areas and may penetrate groundwater as well as surface waters, including stormwater. Vehicle fuels (from exhaust gases or gas stations, fuel spills), industrial releases, and atmospheric transport and deposition are the significant sources of VOCs in stormwater (Müller et al., 2020). Compared with mobile emission sources (road dust and traffic), several studies emphasized the significance of regional stationary sources of AD. Other investigations noted the existence of trace organics in urban stormwater, the majority of which were priority pollutants. They generally occurred in wet AD and were present in low or trace levels (Müller et al., 2020). In Minnesota, United States, it was discovered from urban stormwater pond sediments that 67% of the detected PAHs were caused by sealants from coal tar, which are used in improving the structural integrity and aesthetic appeal of asphalt roadways, pavement, parking lots, walking trails, and some playgrounds (Müller et al., 2020). Bitumen, an asphalt component containing PAHs, could be a PAH source in the runoff. The Portland cement used in the manufacturing of concrete was the source of the chromium that was discovered to leach from concrete pavement materials (Kayhanian et al., 2009). According to Zhang et al. (2017), pedestrian routes made of natural stone, due to their high total solids content, created the maximum PAH surface load, which served as a carrier for pollutants. Roofing materials (concrete and clay) raise the pH of runoff, and corrosion of metal buildings, frequently seen on things like roofs, contribute to copper and zinc in runoff (Müller et al., 2020). In a populated residential area of central Paris, the primary source (>80%) of cadmium, lead, and zinc, contributing to the sewer system, was the runoff from roofs owing to corrosion of roof cover materials (Gromaire et al., 2001). The chief sources of Pb and Zn in historical structures were window frames, ventilation holes, slate roofs, etc.; whereas the chief sources of Cd and Zn, minor components of zinc goods, were Zn sheets used in roofs and gutters. Another significant Zn source is galvanized steel that is employed on urban-built environment surfaces, including crash barriers, building envelopes, and light poles (Müller et al., 2020). Many studies have recorded the discharge of numerous chemicals from building materials. For example, Burkhardt et al. (2011), in their study, demonstrated that significant levels of pesticides used in building envelopes (diuron, terbutryn, irgarol, and carbendazim) infiltrated runoff. Pb, pesticides, and PCBs can all be found in paints used on various types of surfaces (Müller et al., 2020). It was established that turf areas (parks, sports fields, golf courses, etc.) and residential lawns in urban environments are "hotspots" of nutrient input into runoff (Hobbie et al., 2017). Hobbie et al. (2017) examined lawns and their soil as a substantial source of dissolved and total phosphorus in runoff from urban places. In addition, fallen leaves accounting for more than 50% of the annual phosphorus output are a substantial source of nutrients, particularly phosphorus, in urban runoff (Müller et al., 2020). Pesticides were discovered as a source of runoff at golf courses. Artificial turf and recycled rubber mulch are widely utilized in playgrounds. According to Magnusson et al.'s study (2016), rubber mulch is a clear source of MPs and also of PAHs, Zn, and phthalates discharged into runoff. Nickel was found to be a component of diesel engine exhausts (Duong & Lee, 2011). According to Liu et al. (2018), gas stations were identified as "hotspots" of benzene series (BTEX) pollutants and as such, they were considered possible sources of BTEX pollutants in stormwater. Tire and brake wear were considered the most significant nonexhaust pollution sources, with the most significant contaminants being Zn, Cu, Cd, PAHs, and MPs (Horton et al., 2017; Müller et al., 2020). According to the study by Björklund (2010), phthalates, nonylphenols, and nonylphenol ethoxylates (NPEOs) were commonly found in urban runoff from vehicle washing and wear.

When Pitt et al. (1995) looked at the source of the stormwater pollutants, they concluded that parking lots, storage facilities, and vehicle service sites were likely traditional pollution hotspots in industrial and commercial sectors. When compared with other land uses, like residential areas, Liu et al. (2018) observed that commercial and industrial areas produced greater concentrations of metals in stormwater runoff. This was particularly true for Zn, Cu, and Pb. Perfluorooctane sulfonate was found to be significantly influenced by industrial operations; studies by Xiao et al. (2012) and Liu et al. (2018) also demonstrated that industrial land usage increased stormwater BTEX loads. Manufacturing processes, like those used to make rubber, chemicals, paints, and varnishes, as well as the production of MP pellets for use in plastic media for abrasive blasting, personal care products, and plastic pellets, all signify MP sources from industrial areas in stormwater (Magnusson et al., 2016; Horton et al., 2017).

The US EPA (2005) identified additional significant sources of pollution in construction regions like solid waste and debris, fertilizers (source of nitrogen and phosphorus), application of insecticides and other chemicals used in construction, spills or littering, and washout from concrete trucks. Old building façade plaster was mentioned as a potential source of PCB during demolition. Through the use of building materials made of plastic-like polyvinyl chloride and polyethylene, as well as through maintenance or construction work (drilling and sawing) using plastic materials, MPs could be dispersed in the environment (Magnusson et al., 2016). Such characteristics of material might also make stormwater transportation more likely.

According to recent research (Magnusson et al., 2016), the degradation and disintegration of plastic litter, like bags used to collect pet waste, bottles, and other items, is an important source of MPs for the marine ecosystem. According to research by Hahladakis et al. (2018), plastic litter can also transport and absorb adsorbed organic contaminants and trace metals that can be discharged into the environment. Many organic pollutants in trace amounts, including polybrominated diphenyl ethers, phthalates, and bisphenol A (BPA), along with metals used as colorings have been observed for being emitted from plastic trash (Horton et al., 2017; Hahladakis et al., 2018). Stormwater quality may be negatively affected by illegal or unintentional disposal and spills of garbage or chemicals from, among other places, houses, businesses, and other sources. These incidents can also be sources of a variety of pollutants. The quality of stormwater discharge may be impacted by pesticides used in gardening to manage weeds, fungi, and insects (Müller et al., 2020).

2.4 *Effects of Stormwater Runoff on Water Bodies*

The quantity and quality of water are impacted by stormwater runoff which has a detrimental effect on the biological integrity of receiving waterbodies (Feist et al., 2017). Urban stormwater has become a serious hazard to neighboring rivers, lakes, and estuaries because of the increased fertilizer and contamination loads it carries. Although the chemical combinations in urban stormwater runoff are complex and unexpected (McCarthy et al., 2008), hydrocarbons and heavy metals from commercial and industrial land use, along with medications and pesticides, are pervasive in urban catchments. Urban stormwater exposure can occasionally have immediate, fatal consequences. For instance, compared with Coho salmon (*Onchorhynchus kisutch*) that return to nonurban streams, adult Coho salmon that return back to urban creeks face substantially higher mortality rates (>50%) before spawning (Spromberg et al., 2016). According to experimental findings, chemicals in leachates from tire-wear particles are the most likely culprits for this prespawn mortality (Peter et al., 2018). In fact, a number of urbanization-related factors, such as the amount of imperviousness in the watershed and the density of roads, are correlated with Coho pre-spawn mortality (Feist et al., 2017). Stormwater can change the physiology of a single organism, leading to conditions such as pericardial edema and sensory deprivation in young fish (McCarthy et al., 2008). The physiological change can then have long-lasting, multigenerational effects by altering behavior or reducing survival or fertility rates (Saaristo et al., 2018). According to the study by Saaristo et al. (2018), such effects can spread to population and community levels. These effects may then affect processes occurring at the ecosystem level, such as the movement of nutrients, ecosystem resilience, water quality, and carbon sequestration. Some organic substances, like dichlorodiphenyltrichloroethane (DDT) and PCBs, can be transported over vast geographical scales by highly migratory species. These substances also remain in the ecosystem and accumulate in the tissues of animals for a long duration (Pesiakova et al., 2018).

3 Micro-Nanoplastics in Water Bodies

3.1 Introduction to Micro-Nanoplastics

The daily lives of people have significantly improved over the past 50 years thanks to the use of plastics (Oliveira & Almeida, 2019). Recent years have seen increased attention given to plastic pollution, a new sort of emerging environmental contamination (Shen et al., 2019). The attention of the public and scientific community has recently been focused on the aquatic pollutants generated by human and industrial activities, and importantly the pollution of marine and freshwater habitats with plastic waste. Synthetic polymers have evolved into a serious environmental hazard due to their large manufacturing, physical-chemical properties, and thoughtless disposal and recycling (Strungaru et al., 2019).

Plastic is an easily available, affordable, and practical material that is widely used in businesses and everyday life across the globe and provides a great deal of convenience for people (Zhou et al., 2022). Owing to their low cost, strength, and flexibility, they are used in a variety of commercially accessible items, which has resulted in a geometric growth in manufacturing (Benson et al., 2022). With its practical use, it has lately come to light as a danger to the environment worldwide (Zhou et al., 2022). Over 0.3 billion tonnes of plastic is manufactured each year, with half of that amount utilized as items that can only be used once (Allouzi et al., 2021). Plastic wastes are continuously accumulating in the environment because they are chemically stable, resistant to corrosion, and difficult to degrade through biodegradation (Shen et al., 2019). These properties cause plastic to build up in the aquatic, atmospheric, and terrestrial environments as well as in various environmental media, including marine ecosystems, landfills, agrarian ecosystems, like mulching film, as well as the air (Allouzi et al., 2021). This contributes to plastic pollution, a problem that the world is currently facing on the rise.

Plastics are organic polymers of synthetic origin with structures that are stable, lipophilic, and water-resistant, and are created through condensation or addition processes of substituted or unsubstituted hydrocarbons (Benson et al., 2022). Large polymers would break down into smaller species when exposed to sunshine, erosion, weathering, and immersion (Shen et al., 2019). Brahney et al. (2020) predicted that by 2025, 11,000 million tons of plastics are anticipated to amass in the environment. The recovery rate for plastics, however, is only around 5%. In addition, the plastics referred to as "white pollution" will break down into minute particles known as micro-nanoplastics due to numerous chemical, physical, and biological activities (Zhou et al., 2022). Plastic trash can break down into debris made of plastic or even smaller fragments that have a significant negative influence on the terrestrial and marine environments. According to Allouzi et al. (2021), these particles are classified as microplastics (MPs) (0.1 μm to 5 mm) and nanoplastics (NPs) (0.001–0.1 μm).

Micro-nanoplastic (MnP) research has undergone a paradigm shift in the past 10 years as a result of recent revelations about the scope and severity of MnP contamination, even in the most remote regions of the globe. Initially, it was believed that MnP pollution was primarily a marine issue, with impacts that mostly affected marine biota and were primarily sourced from land-based sources. The seas were considered the last sink for plastic pollution, where it all flowed and eventually became stuck. However, due to recent findings in soil, groundwater, snow, freshwater, terrestrial, sediment, aquatic biota, air, and oceans, MnP research has advanced quickly (Allen et al., 2022).

3.2 *Sources of Micro-Nanoplastics in Water Bodies*

Little is known about this issue for freshwater ecosystems despite the extensive and important research explaining the origin and consequences of MnPs in the ocean (Strungaru et al., 2019). Few investigations are carried out on MnPs in freshwater ecosystems compared to marine habitats, even though freshwater ecosystems can amass tremendous amounts of MnPs (Allen et al., 2022). Wastewater treatment plant effluent, AD, and stormwater runoff are just a few ways that MPs can get into freshwater bodies of water (Koutnik et al., 2021). The most significant sources are those that are on land, such as storm sewers, currents in water systems, wastewater treatment effluents, domestic and industrial drainage systems, wind, and runoff soils from agricultural activities like mulching of plastic or treatment with sewage sludge, which is frequently used as fertilizer. The marine-based sources are connected to activities including product transportation, leisure cruises, oil extraction, unlawful dumping, and fishing. MnPs are also used in various products, like abrasives in cleaning products, biomedical applications, personal care products (shower/bath gels, toothpaste, scrubs, makeup, peelings, facial cleansers, insect repellents, and sunscreen), and personal care products with the goal of being released into sewage systems and/or the environment (Oliveira & Almeida, 2019). According to research on six different facial scrub brands, each usage of the items might result in the discharge of 4594 to 94,500 microbeads into the wastewater stream (Napper et al., 2015). Moreover, pellets are used to make plastic consumer goods, and MPs are incorporated into commercial items for cleaning like rust or paint scrubbers (Anderson et al., 2016).

Through wastewater treatment plants (WWTPs), enormous amounts of MnPs can be released into open waterways (Allen et al., 2022). After analyzing wastewater samples from 17 US municipal WWTPs, Mason et al. (2016), in a study, estimated the average release of microbeads to be 13 billion particles per day. In addition, surface runoff from nearby landscapes of urban and rural areas, events of stormwater overflow, AD, incidental release, and plastic films used in agrarian soils for cultivation are other sources of MnPs in river ecosystems (Allen et al., 2022). The existence of MP in a freshwater system in the UK, which included both urban and rural areas, was described by Horton et al. (2017). Most of the MPs found in sediments comprised artificial fibers that were brought by sewage effluent and secondary MP that was locally produced (e.g., broken down from larger objects).

Due to fishing activities, tourism and recreation have also become important sources of MnPs. For instance, freshwater ecosystems have to deal with the degradation of fishing nets (Talbot & Chang, 2022). In addition, commercial fisheries that discard fishing gear and equipment such ropes, nets, plastic lures, lines, and occasionally abandoned trawlers may introduce plastic trash from anthropogenic sources into freshwater and marine habitats. MPs are now more prevalent in freshwater environments, like lakes, rivers, and sand beaches, due to inappropriate storage and treatment (Benson et al., 2022).

MnPs are broadly used in the beauty and hygiene care industries, particularly exfoliating products and toothpaste, as well as in the electronics, textile, plastics, automotive, and 3D-printing industries. The industries that produce 3-D printer ink, household goods, beauty products, and raw materials for the plastics industry are good sources of primary MnPs. When huge plastic items are broken down by a variety of plastic particles due to abiotic causes like temperature, reactive species (ozone), microbial decay, UV radiation, atmospheric pressure, and physical abrasion, secondary MnPs are created (Allouzi et al., 2021). Figure 1 lists various sources of MnPs in water bodies.

The types of MnPs found in water bodies are described in Fig. 2.

4 Impacts of Micro-Nanoplastics

The various impacts of MnPs on humans, environment, and flora and fauna are illustrated in Fig. 3.

4.1 On Environment

The environmental effects of MPs or NPs in freshwater habitats were taken into account in some cases (Sana et al., 2020). Owing to their minute size, MPs or NPs are more likely than larger particles to be taken directly and indirectly by aquatic organisms, sometimes as food, which has a negative physical impact. For example, data from maritime research indicates that MP or NP ingestion results in digestive

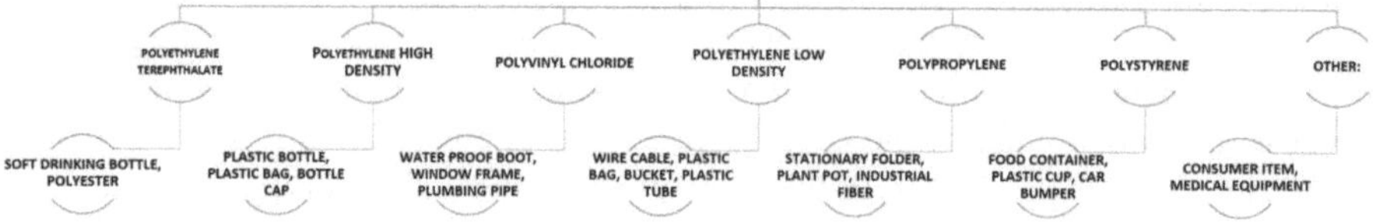

Fig. 1 Sources of MnPs

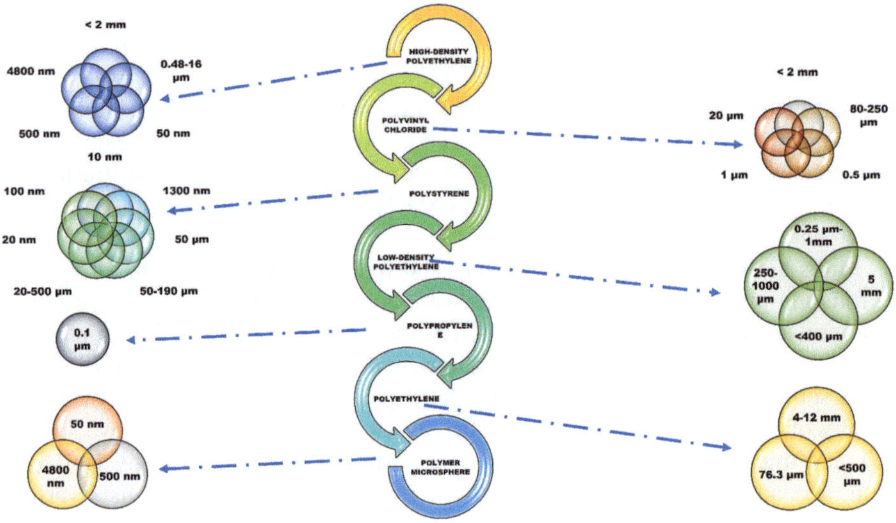

Fig. 2 Types and sizes of MnPs in waterbodies

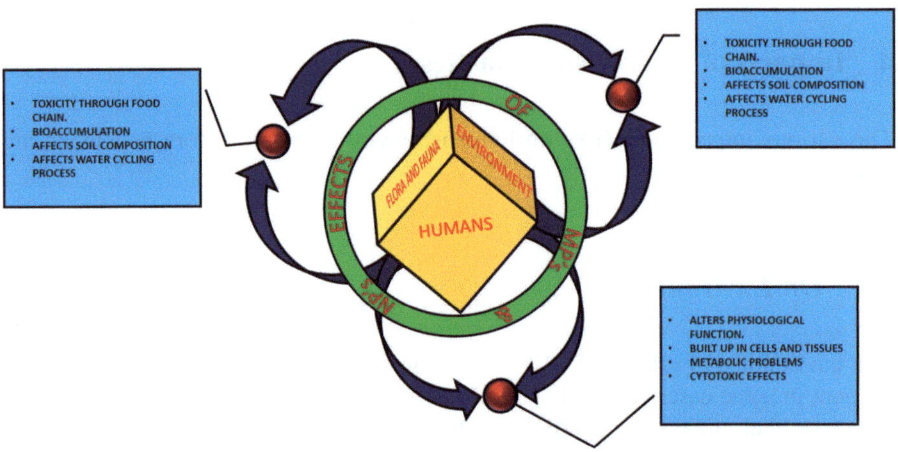

Fig. 3 Effects of MnPs on environment, humans, flora, and fauna

tract obstruction, shock, weakened state, organ damage, and ultimately death. In addition, persistent organic pollutants (POPs) absorbed by MPs or NPs may cause toxicity throughout the food chain. Through bioaccumulation, people may finally be exposed to this. The combined effects of MPs can affect plants by causing physico-chemical changes in soil composition and appearance, which affect water cycling process, the functioning of the ecology in terrestrial environments, and other plant responses (Sana et al., 2020). In this way, changes in soil hydrology brought on by MPs may have an impact on microbial development of soil, with implications on symbiotic relationships in terrestrial ecosystems like mycorrhizal groups and N-fixing organizations (Sana et al., 2020).

The ability of MnPs to alter the physiochemical characteristics of sediments is another problem associated with their presence in the environment. Both the permeability of sediment and the maximum temperature are decreased by MPs. MP-contaminated sediments lose their ability to absorb heat, which can have a drastic impact on the biota (Boyle & Örmeci, 2020).

Hydrophobic chemicals are also known to be used in plastic hydrophobic surfaces and their eco-corona associates (Sana et al., 2020). When compounds adsorbing onto MPs were linked to interaction processes of the intra- or interspecies in the marine environment, trophic implications and other ecological effects were seen (Sana et al., 2020). Numerous amphiphilic and hydrophobic chemicals regulate species interactions and ecosystem functions in soil. For instance, the fungus naturally produces hydrophobins, a widespread amphiphilic protein found in soil. These cysteine-rich polypeptides have immediate effects on erosion of soil and biogeochemical cycles due to their important roles in hydrophobicity and aggregation stabilization. Research laboratory results show that in toxicity caused by NPs, hydrophobins help by shielding filamentous fungus (Nomura et al., 2016), and MPs had different sorption capabilities for inorganic components of the soil (Sana et al., 2020). Accumulated MnPs interact with other dangerous substances and organic pollutants, enhancing their potential and adversely altering the diverse terrestrial biota (Chai et al., 2020), causing irreparable harm to soil systems.

Particularly, it has been seen that MnPs that might last for hundreds of years interact with soil organic matter, changing physiochemical parameters of soil and contaminating groundwater, hence lowering plant growth and general productivity. MnPs also significantly impair the reproduction, lifespan, growth, and survival of soil fauna, particularly earthworms and nematodes, through a variety of toxic mechanisms, including bioaccumulation, gut microbiota dysbiosis, genotoxicity, histopathological damage, neurotoxicity, oxidative stress, metabolic problems, DNA damage, and reproductive toxicity (Amobonye et al., 2021).

4.2 On Humans

Since humans are at the apex of many food chains, it is critical to emphasize any potential effects that MnPs may have on them. Numerous studies have shown that smaller NPs are generally more likely to enter and build up in various cells and tissues, where they would then alter the physiological functions of the tissues and cells. It was demonstrated in vitro that the unaltered plastic polymers directly affected the expression of inflammatory genes, viability, and morphology of stomach cell lines (Sana et al., 2020). The MnPs can cause hazardous impacts on the cellular environment by inducing oxidative stress, which sets off a chain reaction of unfavorable cellular functions and eventual destruction (Guo et al., 2020). As a result, it is thought that MnPs might build up in cells and tissues, leading to local inflammation and metabolic problems. On a variety of human cell lines, like lung cells, intestinal cells, epithelial and brain cells, the uptake of MnPs and their cytotoxic effects have been shown (Schirinzi et al., 2017; An et al., 2020). In an investigation, it was proved that MnPs from polystyrene could induce cardiac fibrosis in rats by inducing cardiomyocyte death and activating the Wnt/β-catenin signaling pathway. Similarly, polystyrene MnPs were also shown to disrupt the blood–testis barrier in rat models under the control of the MAPK-Nrf2 signaling pathway (Li et al., 2020; Li et al., 2021). It has been demonstrated that certain MnPs can disrupt the transport and metabolism of lipids in a variety of organisms (Lu et al., 2018; Guo et al., 2020). MnPs have recently been shown to alter the secondary structures of proteins and cause protein misfolding, both of which have been shown to alter the structural integrity of proteins (Hollóczki & Gehrke, 2019). In addition, it has been demonstrated that proteins and nanoparticle interactions produce coronal protein rings that have been found to have a major impact on the endocytosis process of nanoparticles in cells. By serving as environmental pollutant vectors, MnPs may also adversely affect the health of humans (Amobonye et al., 2021). Similar to the parent material of plastic polymer, MnPs' nonpolar surfaces allow them to absorb and subsequently transport other hydrophobic substances, including POPs (Mattsson et al., 2018).

4.3 On Flora and Fauna of the Water Bodies

Numerous studies have examined the unfavorable effects of MnPs on aquatic life, including zooplankton, microorganisms, as well as seabirds, fish, and amphibians (Sana et al., 2020). Certain microorganisms like bacteria, yeast, and algae have been found to be inhibited by MnPs, which affects their crucial essential roles in many environments (Nomura et al., 2016; Sun et al., 2018; Mammo et al., 2020). In addition, it has been noted that MnPs can impede the digestive systems of zooplankton and benthic creatures like mussels and oysters, which frequently results in a lack of appetite, malnutrition, and even mortality (Lee et al., 2013).

The primary characteristics that are tested for MP toxicity are tissue distribution, growth rate, reproduction, biological activity of enzymes, and oxidative damage (Sana et al., 2020). Numerous species, including plankton, vertebrates, and invertebrates, are destroyed by the presence of MPs, causing a variety of unforeseen impacts (Sana et al., 2020). Additionally, MPs are hazardous to some microbes including fungi and bacteria. For instance, yeast cells may be adversely affected by polystyrene nanoparticles (Nomura et al., 2016). For *Halomona salkaliphila*, marine bacterium, MPs can stop their development and damage their function in the environment (Lee et al., 2013). Furthermore, MPs have a multitude of negative effects on aquatic life, including fish, seabirds, amphibians, and invertebrates like zooplankton and some benthic creatures. According to a study by Lee et al. (2013), a lack of appetite, digestive system obstruction, eating impact, sluggish growth, malnutrition, and occasionally even mortality are the main causes of the effects on zooplankton. According to Van et al.'s study (2013), mussels transport MPs into the gastrointestinal system through endocytosis, which has an impact on benthic organisms like oysters and mussels. In the end, this causes inflammation and a decrease in the stability of lysosome membranes. MPs have the potential to raise oyster mortality, limit growth, affect how well energy is absorbed, and impede reproduction and the development of offspring. Additionally, MPs already have a negative effect on some microbes like fungi and bacteria. According to a study by Mattsson et al. (2017), NP particles affected finfish behavior (measured as time of feeding, activity, and distance swam searching prey) and decreased zooplankton survival.

5 Strategies for Mitigating Micro-Nanoplastics Pollution in Water Bodies from Stormwater Runoff

In wastewater treatment plants (WWTPs), flocculation and primary settling, bioreactor system, and membrane filtration are used for the removal of MPs. Salts of iron and aluminum are used as flocculants in WWTPs. This technology removes the settled parts of the suspended MPs; whereas the nonsinkable floating MPs are removed as scum. The bioreactor system helped in the removal of MPs through microbe ingestion and sludge aggregate formation. The most popular treatment for MPs in WWTPs is membrane bioreactor technology, which shows 99.99% efficiency for MPs removal due to high concentration of mixed-liquor solids (Liu et al., 2021).

5.1 Preventive Measures

Plastic appears to be an essential commodity due to its greater sales and market share on the global market and its widespread use in daily life. Due to plastic's high durability and persistence, only a very small portion of the entire amount generated gets collected for recycling, causing plastic trash to incessantly accumulate in the environment. Additionally, there have been restrictions on the manufacture and usage of plastics, even if globally the community has focused on a variety of regulatory actions in past years to control the pollution of plastic and its harmful consequences on the environment (Peng et al., 2022). Preventive efforts concentrate on ways to stop the production of debris or stop it from reaching sea. These include reducing trash at the source, recycling, reusing, and composting garbage, converting waste to energy, keeping debris out of receiving waterways' entrance points, and implementing various waste management strategies on land. An effective method of preventing marine plastic contamination is source identification. According to Ogunola's study (2017), there are three main places where MPs come from: land, rivers, and the ocean. According to a report by Pettipas et al. (2016), it is highly challenging to identify garbage origins in multi-industry areas and residential plastic disposal is difficult to manage due to insufficient resources for audits. They contend that when typical objects are found in the ocean, on beaches, or along coastlines, it helps to identify specific targets and allow for further action. One of the strategies to reduce plastic waste is to completely ban or restrict the usage of plastic bags. According to the study by Bergmann et al. (2015), the first country to restrict bags made of polythene was Bangladesh in 2002, followed by Myanmar, China, and other African nations. Except for personal care goods, it is illegal to produce and trade items that contain MPs in Canada (Ogunola, 2017). Although all signatories to the International Convention for Prevention of Marine Pollution for Ships (MARPOL), United Nations Convention on the Law of the Sea (UNCLOS), and Honolulu Conventions reiterated a complete ban on the plastic waste disposal at sea by shipping vessels, these regulations are lax because of ineffective monitoring, enforcement, and surveillance (Province of Nova-Scotia, 2011, for example).

5.2 *Remediation/Management Strategies*

The most common methods for degrading MnPs, namely, biodegradation and chemical degradation, are discussed here. Microbes like fungi, bacteria, and algae break down contaminants (including plastics) through a process called biodegradation. Due to its ease and environmental neutrality, plastic biodegradation has achieved considerable acceptance. The range of bacteria, the form of polymers, the composition of plastics, and the choice of enzymes are all significant candidates that influence plastic biodegradation, in addition to ecological factors. Research is being done on biodegradation by bacteria, enzymes, fungi, and collective biodegradation. There are numerous forms of biodegradation, each with a distinct mechanism of action (Peng et al., 2022).

The majority of all living things, bacteria, make up a large group of microorganisms. They predominantly inhabit water, soil, and the atmosphere and have the capacity to digest contaminants. Numerous studies have looked into using microbes to degrade MP as of late. Pure bacterial cultures have been the focus of studies on the breakdown of MPs mediated by microbes in lab settings. Although some were obtained from culture collections, enrichment culturing was predominantly used to separate these cultures from sediment, sludge, and wastewater. It is practical to examine metabolic pathways or assess the impact of various environmental variables on the degradation of MP when employing pure strains to conduct investigations on the degradation of MPs. Additionally, it is possible to keep a careful eye on the degradation of MPs by functional bacteria and the changes caused by it in MPs (Yuan et al., 2020). Pure bacterial cultures were isolated from mangrove sediment and used in a study (Auta et al., 2018) to degrade polypropylene (PP) MPs. Strains of both *Rhodococcus* and *Bacillus* caused PP MP to lose weight after 40 days of incubation, albeit at different rates—6.4% and 4.0%, respectively.

Fungi have the capacity to cling to and use MP. The hydrophobicity of MPs can be reduced by fungi by promoting the synthesis of various chemical bonds (functional groups like carbonyl, ester, and carboxyl) in MPs. Fungi are a diverse group of microorganisms that have a broad distribution and powerful reproduction capacities (Yuan et al., 2020). *Zalerion maritimum* (a fungus of marine origin) used polyethylene as a growth substrate and drastically decreased the fraction of polyethylene in 28 days, according to a laboratory study by Paço et al. (2017). Polyester yarn mineral media demonstrated filamentous fungal tip development in the presence of *F. solani* and *F. oxysporum*, according to Taniguchi et al.'s (2019) study.

The biological degradation of polymeric material can be successfully carried out using microalgae, their enzymes, and their toxins. The fundamental benefit is that, in contrast to the bacterial system, they can grow with relatively low supply of carbon and can adapt to a wide range of habitats, where the majority of the MPs are found. In wastewater streams, microalgae colonize plastic surfaces, and this adherence causes the ligninolytic and exopolysaccharide enzymes to start breaking down the plastic. These polymers primarily act as a source of carbon, increasing cellular proteins and carbohydrates while also speeding up growth (Anand et al., 2023). *Chlamydomonas reinhardtii*, a green microalga, was genetically altered to produce polyethylene terephthalate hydrolase, which can break down terephthalic acid and polyethylene terephthalate films (Kim et al., 2020).

5.3 Policy and Governance Measures

The environment and public health are becoming more and more concerned with MPs and NPs. To lessen the effects of these contaminants, policy and governance actions are required. With a focus on single-use plastics, governments all over the world have created regulatory and legislative measures to regulate, restrict, and reduce the use of plastics. More thorough regulations, however, are required to regulate all plastics, including MPs and NPs. Table 1 represents policies undertaken to mitigate micro-nanoplastic pollution.

The extended producer responsibility (EPR) is based on the "management of the whole-life-cycle of products" approach in combination with "circular economy framework" and the "polluter-pays-principle." The German Packaging Ordinance, which was passed in 1991, introduced the EPR by requiring producers to set up distinct management and recycling of all kinds of packaging to meet required recycling quotas. Following Germany, the EPR system was adopted by a large number of developed nations, including France, Austria, and Belgium. The industry established a nonprofit organization in Norway, Belgium, Italy, and France that is in charge of finance, coordination, and recycling. In contrast, stores retain monetary deposits that are returned to customers in the UK, Denmark, Finland, Germany, and Sweden after recycling. Japan, the first country in Asia to implement the EPR strategy, passed the recycling of containers and packaging law in 1995. In 2011, India used EPR to treat plastic trash and e-waste (Johannes et al., 2021; Pani & Pathak, 2021).

Table 1 Policies around the world to mitigate micro-nanoplastic pollution

Sr. No	Country	Policies
1	United States	California banned shopping bags made of plastic
2	Rwanda	Ban on all nonbiodegradable plastic
3	United States	San Francisco banned sale of plastic water bottles
4	United States	Passed a bill to ban beauty/skin-care products containing plastic microbeads
5	India	Restrained the use of plastic bags <50 microns
6	France	"Plastic Ban" law in 2016 banning all plastic by 2020
7	Ireland	Charges for plastic bags (10 cents) and recycling program (4 cents)
8	Canada	Introduction of environmental protection act to ban the usage of microbeads in beauty products
9	China	Passed a law in 2008, wherein the consumer pays for plasticbag use
10	India	Ban by National Green Tribunal in Delhi and NCR on disposable plastics

Source: Laskar and Kumar (2019)

5.4 Socioeconomic Measures

Several governmental and nonprofit organizations must set up educational, outreach, and public awareness campaigns to promote change to lessen the reckless plastic waste disposal in the environment. The idea that "change begins with you" at the grassroots level should be the main consideration while managing the environment. The marine environment should be viewed by people as both common and private property that has to be adequately safeguarded and managed. They ought to quit leaving plastic rubbish on the beaches when they are there. Our educational system must introduce required ocean and environmental courses at all levels (primary, high, and postsecondary universities). For schools and the public, projects and workshops on conservation, ocean management, and protection must be organized (Ogunola, 2017). According to a study by Wiener et al. (2015), the Hawaiian natives expressed a great desire to pursue traditional ocean conservation practices. Furthermore, Rapa Nui (Chile) residents are the ones who care the most about waste management and coastal litter, according to a study by Kiessling et al. (2017). Due to their cultural heritage, biodiversity, geographic location, and primarily tourist-dependent economy, the residents of this country have a high awareness of the need for protection from outside pressure. The usage of social media for advertising, images, tales, and videos can assist in spreading messages fast to a big audience in their communities. The promotion of good behavioral change and the pursuit of marine-related jobs is also thought to be successful when youth are the target (Ogunola, 2017).

6 Conclusion and Future Directions

MnP buildup in the environment due to improper handling of plastic trash has become an indisputable issue globally. The sustainability of the environment and human health may suffer significantly. MnP pollution can be reduced by using less plastic, incorporating plastics into the circular economy, and managing waste properly, among other strategies. Biodegradable polymers are seen as an appropriate substitute to lower plastic consumption. However, depending on the circumstances, biodegradability can change. Waste plastics are receiving a lot of attention when it comes to recycling, reusing, and value addition. Promising methods for the valorization of plastic waste include the conversion of waste into chemicals, fuels, road building and construction materials, and carbon materials (adsorbents and solid catalysts). However, new technological developments for extensive industrial applications continue to constrain them. Additionally, not all plastic products may be recycled. Therefore, further research is needed to create affordable plastic substitutes and develop environmentally and financially sound plastic waste valorization techniques. The risks and fate of primary MnPs must also be assessed, and suitable regulations must be put in place for environmentally safe limits. It is difficult to promote these strategies for all sorts of plastics, even if some nations have passed legislation to decrease plastic consumption and increase plastic reuse. In addition, the ability to create these rules is influenced by socioeconomic level and geographic location. Therefore, to lessen the threat of MnPs in the surroundings, strict environmental regulations, the construction of suitable infrastructure, and plastic waste management systems that are environmentally sustainable, economically sound, and socially acceptable are essential.

Acknowledgments The authors are thankful to Nida Sarfaraz for copyediting and proofreading the manuscript.

References

Allen, S., Allen, D., Karbalaei, S., Maselli, V., & Walker, T. R. (2022). Micro (nano) plastics sources, fate, and effects: What we know after ten years of research. *Journal of Hazardous Materials Advances, 6*, 100057.

Allouzi, M. M. A., Tang, D. Y. Y., Chew, K. W., Rinklebe, J., Bolan, N., Allouzi, S. M. A., & Show, P. L. (2021). Micro (nano) plastic pollution: The ecological influence on soil-plant system and human health. *Science of the Total Environment, 788*, 147815.

Amobonye, A., Bhagwat, P., Raveendran, S., Singh, S., & Pillai, S. (2021). Environmental impacts of microplastics and nanoplastics: A current overview. *Frontiers in Microbiology, 3728*.

An, L., Liu, Q., Deng, Y., Wu, W., Gao, Y., & Ling, W. (2020). Sources of microplastic in the environment. In *Microplastics in terrestrial environments: Emerging contaminants and major challenges* Springer. (pp. 143–159).

Anand, U., Dey, S., Bontempi, E., Ducoli, S., Vethaak, A. D., Dey, A., & Federici, S. (2023). Biotechnological methods to remove microplastics: A review. *Environmental Chemistry Letters, 21*, 1–24.

Anderson, J. C., Park, B. J., & Palace, V. P. (2016). Microplastics in aquatic environments: Implications for Canadian ecosystems. *Environmental Pollution, 218*, 269–280.
Auta, H. S., Emenike, C. U., Jayanthi, B., & Fauziah, S. H. (2018). Growth kinetics and biodeterioration of polypropylene microplastics by Bacillus sp. and Rhodococcus sp. isolated from mangrove sediment. *Marine Pollution Bulletin, 127*, 15–21.
Benson, N. U., Agboola, O. D., Fred-Ahmadu, O. H., De-la-Torre, G. E., Oluwalana, A., & Williams, A. B. (2022). Micro (nano) plastics prevalence, food web interactions, and toxicity assessment in aquatic organisms: A review. *Frontiers in Marine Science, 9*, 291.
Bergmann, M., Gutow, L., & Klages, M. (2015). *Marine anthropogenic litter* (p. 447). Springer Nature.
Björklund, K. (2010). Substance flow analyses of phthalates and nonylphenols in stormwater. *Water Science and Technology, 62*(5), 1154–1160.
Boyle, K., & Örmeci, B. (2020). Microplastics and nanoplastics in the freshwater and terrestrial environment: A review. *Water, 12*(9), 2633.
Brahney, J., Hallerud, M., Heim, E., Hahnenberger, M., & Sukumaran, S. (2020). Plastic rain in protected areas of the United States. *Science, 368*(6496), 1257–1260.
Burkhardt, M., Zuleeg, S., Vonbank, R., Schmid, P., Hean, S., Lamani, X., et al. (2011). Leaching of additives from construction materials to urban storm water runoff. *Water Science and Technology, 63*(9), 1974–1982.
Chae, Y., & An, Y. J. (2017). Effects of micro-and nanoplastics on aquatic ecosystems: Current research trends and perspectives. *Marine Pollution Bulletin, 124*(2), 624–632.
Chai, B., Wei, Q., She, Y., Lu, G., Dang, Z., & Yin, H. (2020). Soil microplastic pollution in an e-waste dismantling zone of China. *Waste Management, 118*, 291–301.
Duong, T. T., & Lee, B. K. (2011). Determining contamination level of heavy metals in road dust from busy traffic areas with different characteristics. *Journal of Environmental Management, 92*(3), 554–562.
Feist, B. E., Buhle, E. R., Baldwin, D. H., Spromberg, J. A., Damm, S. E., Davis, J. W., & Scholz, N. L. (2017). Roads to ruin: Conservation threats to a sentinel species across an urban gradient. *Ecological Applications, 27*(8), 2382–2396.
Gromaire, M. C., Garnaud, S., Saad, M., & Chebbo, G. (2001). Contribution of different sources to the pollution of wet weather flows in combined sewers. *Water Research, 35*(2), 521–533.
Guo, Y., Ma, W., Li, J., Liu, W., Qi, P., Ye, Y., et al. (2020). Effects of microplastics on growth, phenanthrene stress, and lipid accumulation in a diatom, Phaeodactylum tricornutum. *Environmental Pollution, 257*, 113628.
Hahladakis, J. N., Velis, C. A., Weber, R., Iacovidou, E., & Purnell, P. (2018). An overview of chemical additives present in plastics: Migration, release, fate and environmental impact during their use, disposal and recycling. *Journal of Hazardous Materials, 344*, 179–199.
Hobbie, S. E., Finlay, J. C., Janke, B. D., Nidzgorski, D. A., Millet, D. B., & Baker, L. A. (2017). Contrasting nitrogen and phosphorus budgets in urban watersheds and implications for managing urban water pollution. *Proceedings of the National Academy of Sciences, 114*(16), 4177–4182.
Hollóczki, O., & Gehrke, S. (2019). Nanoplastics can change the secondary structure of proteins. *Scientific Reports, 9*(1), 16013.
Horton, A. A., Svendsen, C., Williams, R. J., Spurgeon, D. J., & Lahive, E. (2017). Large microplastic particles in sediments of tributaries of the River Thames, UK–Abundance, sources and methods for effective quantification. *Marine Pollution Bulletin, 114*(1), 218–226.
Järlskog, I., Strömvall, A. M., Magnusson, K., Gustafsson, M., Polukarova, M., Galfi, H., et al. (2020). Occurrence of tire and bitumen wear microplastics on urban streets and in sweepsand and washwater. *Science of the Total Environment, 729*, 138950.
Johannes, H. P., Kojima, M., Iwasaki, F., & Edita, E. P. (2021). Applying the extended producer responsibility towards plastic waste in Asian developing countries for reducing marine plastic debris. *Waste Management & Research, 39*(5), 690–702.

Kayhanian, M., Vichare, A., Green, P. G., & Harvey, J. (2009). Leachability of dissolved chromium in asphalt and concrete surfacing materials. *Journal of Environmental Management, 90*(11), 3574–3580.

Kiessling, T., Salas, S., Mutafoglu, K., & Thiel, M. (2017). Who cares about dirty beaches? Evaluating environmental awareness and action on coastal litter in Chile. *Ocean & Coastal Management, 137*, 82–95.

Kim, J. W., Park, S. B., Tran, Q. G., Cho, D. H., Choi, D. Y., Lee, Y. J., & Kim, H. S. (2020). Functional expression of polyethylene terephthalate-degrading enzyme (PETase) in green microalgae. *Microbial Cell Factories, 19*(1), 1–9.

Koutnik, V. S., Leonard, J., Alkidim, S., DePrima, F. J., Ravi, S., Hoek, E. M., & Mohanty, S. K. (2021). Distribution of microplastics in soil and freshwater environments: Global analysis and framework for transport modeling. *Environmental Pollution, 274*, 116552.

Laskar, N., & Kumar, U. (2019). Plastics and microplastics: A threat to environment. *Environmental Technology & Innovation, 14*, 100352.

Lee, K. W., Shim, W. J., Kwon, O. Y., & Kang, J. H. (2013). Size-dependent effects of micro polystyrene particles in the marine copepod *Tigriopus japonicus*. *Environmental Science & Technology, 47*(19), 11278–11283.

Li, Z., Zhu, S., Liu, Q., Wei, J., Jin, Y., Wang, X., & Zhang, L. (2020). Polystyrene microplastics cause cardiac fibrosis by activating Wnt/β-catenin signaling pathway and promoting cardiomyocyte apoptosis in rats. *Environmental Pollution, 265*, 115025.

Li, S., Wang, Q., Yu, H., Yang, L., Sun, Y., Xu, N., et al. (2021). Polystyrene microplastics induce blood–testis barrier disruption regulated by the MAPK-Nrf2 signaling pathway in rats. *Environmental Science and Pollution Research, 28*(35), 47921–47931.

Liu, A., Hong, N., Zhu, P., & Guan, Y. (2018). Understanding benzene series (BTEX) pollutant load characteristics in the urban environment. *Science of the Total Environment, 619*, 938–945.

Liu, W., Zhang, J., Liu, H., Guo, X., Zhang, X., Yao, X., et al. (2021). A review of the removal of microplastics in global wastewater treatment plants: Characteristics and mechanisms. *Environment International, 146*, 106277.

Loganathan, P., Vigneswaran, S., & Kandasamy, J. (2013). Road-deposited sediment pollutants: A critical review of their characteristics, source apportionment, and management. *Critical Reviews in Environmental Science and Technology, 43*(13), 1315–1348.

Lu, L., Wan, Z., Luo, T., Fu, Z., & Jin, Y. (2018). Polystyrene microplastics induce gut microbiota dysbiosis and hepatic lipid metabolism disorder in mice. *Science of the Total Environment, 631*, 449–458.

Lundy, L., Ellis, J. B., & Revitt, D. M. (2012). Risk prioritisation of stormwater pollutant sources. *Water Research, 46*(20), 6589–6600.

Magnusson, K., Eliasson, K., Fråne, A., Haikonen, K., Hultén, J., Olshammar, M., Stadmark, J., & Voisin, A. (2016). *Swedish sources and pathways for microplastics to the marine environment*. (No. C 183). IVL Svenska Miljöinstitutet, Stockholm, Sverige.

Mammo, F. K., Amoah, I. D., Gani, K. M., Pillay, L., Ratha, S. K., Bux, F., & Kumari, S. (2020). Microplastics in the environment: Interactions with microbes and chemical contaminants. *Science of the Total Environment, 743*, 140518.

Mason, S. A., Garneau, D., Sutton, R., Chu, Y., Ehmann, K., Barnes, J., et al. (2016). Microplastic pollution is widely detected in US municipal wastewater treatment plant effluent. *Environmental Pollution, 218*, 1045–1054.

Mattsson, K., Johnson, E. V., Malmendal, A., Linse, S., Hansson, L. A., & Cedervall, T. (2017). Brain damage and behavioural disorders in fish induced by plastic nanoparticles delivered through the food chain. *Scientific Reports, 7*(1), 11452.

Mattsson, K., Jocic, S., Doverbratt, I., & Hansson, L. A. (2018). Nanoplastics in the aquatic environment. In *Microplastic contamination in aquatic environments* Elsevier. (pp. 379–399).

McCarthy, S. G., Incardona, J. P., & Scholz, N. L. (2008, Aug). Coastal storms, toxic runoff, and the sustainable conservation of fish and fisheries. In *American Fisheries Society symposium* (Vol. 64, No. 12, pp. 7–27).

Müller, A., Österlund, H., Marsalek, J., & Viklander, M. (2020). The pollution conveyed by urban runoff: A review of sources. *Science of the Total Environment, 709*, 136125.
Napper, I. E., Bakir, A., Rowland, S. J., & Thompson, R. C. (2015). Characterisation, quantity and sorptive properties of microplastics extracted from cosmetics. *Marine Pollution Bulletin, 99*(1–2), 178–185.
National Research Council. (2008). *Urban stormwater management in the United States*. National Academy Press.
Nomura, T., Tani, S., Yamamoto, M., Nakagawa, T., Toyoda, S., Fujisawa, E., et al. (2016). Cytotoxicity and colloidal behavior of polystyrene latex nanoparticles toward filamentous fungi in isotonic solutions. *Chemosphere, 149*, 84–90.
Ogunola, O. S. (2017). Current advances in strategies to mitigate the impacts of micro/nano plastics: A review. *Journal of Environmental & Analytical Toxicology, 7*(447), 2161–0525.
Oliveira, M., & Almeida, M. (2019). The why and how of micro (nano) plastic research. *TrAC Trends in Analytical Chemistry, 114*, 196–201.
Paço, A., Duarte, K., da Costa, J. P., Santos, P. S., Pereira, R., Pereira, M. E., et al. (2017). Biodegradation of polyethylene microplastics by the marine fungus *Zalerion maritimum*. *Science of the Total Environment, 586*, 10–15.
Pani, S. K., & Pathak, A. A. (2021). Managing plastic packaging waste in emerging economies: The case of EPR in India. *Journal of Environmental Management, 288*, 112405.
Peng, L., Mehmood, T., Bao, R., Wang, Z., & Fu, D. (2022). An overview of micro (nano) plastics in the environment: Sampling, identification, risk assessment and control. *Sustainability, 14*(21), 14338.
Pesiakova, A. A., Gusakova, E. V., Trofimova, A. N., & Sorokina, T. Y. (2018). Migratory birds are the source of highly toxic organic pollutants for indigenous people in the Russian Arctic. In *IOP conference series: Earth and environmental science* (Vol. 107, No. 1, p. 012088). IOP Publishing.
Peter, K. T., Tian, Z., Wu, C., Lin, P., White, S., Du, B., et al. (2018). Using high-resolution mass spectrometry to identify organic contaminants linked to urban stormwater mortality syndrome in Coho salmon. *Environmental Science & Technology, 52*(18), 10317–10327.
Pettipas, S., Bernier, M., & Walker, T. R. (2016). A Canadian policy framework to mitigate plastic marine pollution. *Marine Policy, 68*, 117–122.
Pitt, R., Field, R., Lalor, M., & Brown, M. (1995). Urban stormwater toxic pollutants: Assessment, sources, and treatability. *Water Environment Research, 67*(3), 260–275.
Province of Nova-Scotia. (2011). *Our path forward: Building on the success of Nova Scotia's solid waste resource management strategy*.
Saaristo, M., Brodin, T., Balshine, S., Bertram, M. G., Brooks, B. W., Ehlman, S. M., et al. (2018). Direct and indirect effects of chemical contaminants on the behaviour, ecology and evolution of wildlife. *Proceedings of the Royal Society B, 285*(1885), 20181297.
Sana, S. S., Dogiparthi, L. K., Gangadhar, L., Chakravorty, A., & Abhishek, N. (2020). Effects of microplastics and nanoplastics on marine environment and human health. *Environmental Science and Pollution Research, 27*, 44743–44756.
Sarkar, B., Dissanayake, P. D., Bolan, N. S., Dar, J. Y., Kumar, M., Haque, M. N., et al. (2022). Challenges and opportunities in sustainable management of microplastics and nanoplastics in the environment. *Environmental Research, 207*, 112179.
Schirinzi, G. F., Pérez-Pomeda, I., Sanchís, J., Rossini, C., Farré, M., & Barceló, D. (2017). Cytotoxic effects of commonly used nanomaterials and microplastics on cerebral and epithelial human cells. *Environmental Research, 159*, 579–587.
Shen, M., Zhu, Y., Zhang, Y., Zeng, G., Wen, X., Yi, H., et al. (2019). Micro (nano) plastics: Unignorable vectors for organisms. *Marine Pollution Bulletin, 139*, 328–331.
Spromberg, J. A., Baldwin, D. H., Damm, S. E., McIntyre, J. K., Huff, M., Sloan, C. A., et al. (2016). Coho salmon spawner mortality in western US urban watersheds: Bioinfiltration prevents lethal storm water impacts. *Journal of Applied Ecology, 53*(2), 398–407.

Strungaru, S. A., Jijie, R., Nicoara, M., Plavan, G., & Faggio, C. (2019). Micro-(nano) plastics in freshwater ecosystems: Abundance, toxicological impact and quantification methodology. *TrAC Trends in Analytical Chemistry, 110*, 116–128.

Sun, X., Chen, B., Li, Q., Liu, N., Xia, B., Zhu, L., & Qu, K. (2018). Toxicities of polystyrene nano- and microplastics toward marine bacterium *Halomonas alkaliphila*. *Science of the Total Environment, 642*, 1378–1385.

Talbot, R., & Chang, H. (2022). Microplastics in freshwater: A global review of factors affecting spatial and temporal variations. *Environmental Pollution, 292*, 118393.

Taniguchi, I., Yoshida, S., Hiraga, K., Miyamoto, K., Kimura, Y., & Oda, K. (2019). Biodegradation of PET: Current status and application aspects. *ACS Catalysis, 9*(5), 4089–4105.

U.S. Environmental Protection Agency (EPA). (2005). *Stormwater phase II final rule: Construction site runoff control minimum control measure*. (Fact Sheet 2.6). U.S. Environmental Protection Agency, Office of Water.

Van Cauwenberghe, L., Vanreusel, A., Mees, J., & Janssen, C. R. (2013). Microplastic pollution in deep-sea sediments. *Environmental Pollution, 182*, 495–499.

Wiener, C. S., Manset, G., & Lemus, J. D. (2015). Ocean use in Hawaii as a predictor of marine conservation interests, beliefs, and willingness to participate: An exploratory study. *Journal of Environmental Studies and Sciences, 6*, 712–723.

Xiao, F., Simcik, M. F., & Gulliver, J. S. (2012). Perfluoroalkyl acids in urban stormwater runoff: Influence of land use. *Water Research, 46*(20), 6601–6608.

Yuan, J., Ma, J., Sun, Y., Zhou, T., Zhao, Y., & Yu, F. (2020). Microbial degradation and other environmental aspects of microplastics/plastics. *Science of the Total Environment, 715*, 136968.

Zhang, J., Hua, P., & Krebs, P. (2017). The influence of surface pavement on the distribution of polycyclic aromatic hydrocarbons (PAHs) in urban watershed. *Water, Air, & Soil Pollution, 228*, 1–12.

Zhou, Y., Kumar, M., Sarsaiya, S., Sirohi, R., Awasthi, S. K., Sindhu, R., et al. (2022). Challenges and opportunities in bioremediation of micro-nano plastics: A review. *Science of the Total Environment, 802*, 149823.

Ziajahromi, S., Drapper, D., Hornbuckle, A., Rintoul, L., & Leusch, F. D. (2020). Microplastic pollution in a stormwater floating treatment wetland: Detection of tyre particles in sediment. *Science of the Total Environment, 713*, 136356.

Micro-nanoplastics in the Environment: Current Research and Trends

Prodipto Bishnu Angon, Shitosri Mondal, Arpan Das, Md. Shakil Uddin, and Afsana Ahamed Eva

1 Introduction

Researchers and policymakers are increasingly challenged by the growing environmental implications of micro-nanoplastics (MNPs), which are being generated, dispersed, and accumulated worldwide (Kumar, 2020; Xu et al., 2020). Recently, this problem has received a lot of attention from scientific communities as well as governing bodies. Microplastics (MPs) (less than 5 mm) and nanoplastics (NPs) (typically deemed less than 1000 nm) are also a source of worry, in addition to the more obvious macroplastics (Kukkola et al., 2021). Over the last 50 years, the world has created an estimated 9200 million metric tons of plastic, marking a 20-fold increase in plastic manufacturing (Walker & Fequet, 2023). More than 6900 metric tons were dumped in landfills or, worse, added to pollutants in the environment (Brooks et al., 2018; Geyer et al., 2017). About 368 million metric tons of plastic were manufactured globally in 2019, but in the upcoming 20 years, that number has been predicted to quadruple (Walker & Fequet, 2023). The fact that most consumer plastics are made for single use and have little recyclability is partly to blame for the rise in global production and consumption that has led to record-breaking accumulations of plastic rubbish and widespread plastic pollution (Borrelle et al., 2020). Only around 9% of all plastic trash in the world has ever been recycled; another 12% has been burned, and the other 79% has accumulated in ecosystems (Geyer et al., 2017).

A significant source of MNP contamination is the usage of agricultural plastic items, which is why agricultural areas are regarded as a primary sink of MNPs (Huang et al., 2020). An estimated 12.5 million tons of plastic are utilized in agricultural production globally, of which 310,000 tons are used exclusively for crop cultivation (Zantis et al., 2023). The agriculture and horticultural industries

P. B. Angon (✉) · S. Mondal · A. Das · M. S. Uddin · A. A. Eva
Faculty of Agriculture, Bangladesh Agricultural University, Mymensingh, Bangladesh

S. A. Bhat et al. (eds.), *Management of Micro and Nano-plastics in Soil and Biosolids*,
https://doi.org/10.1007/978-3-031-51967-3_5

annually discharge between 5400 and 23,500 tons of MPs into the environment (Zantis et al., 2023). Another significant MNP pollution source is the spread of biosolids, or treated sludge, from wastewater treatment facilities on agricultural fields (Kallenbach et al., 2022). The number of MPs discharged from sewage sludge onto agricultural areas varies between 44,000 and 300,000 tons in North America and 63,000 and 430,000 tons in Europe (Nizzetto et al., 2016). Nonfarming activities, such as the use of landfills, littering (e.g., discarded tires and other materials), and air deposition, all contribute to an increase in soil contamination (Yadav et al., 2020).

MNPs may alter soil structure in agricultural fields, which may then have an impact on many creatures, including plants. MNPs are capable of being absorbed by plant roots and leaves (Sun et al., 2021). The trophic transfer along the food chain, from soil to organisms and people, has consequently caused worry (Beriot et al., 2021). Additionally, toxicological consequences have been noted in plants exposed to MNPs, including metabolic reactions and effects on growth performance (Colzi et al., 2022). However, reactions differ significantly between research. For instance, MNPs had a detrimental impact on the germination rate of cress but had no discernible impact on wheat (*Triticum aestivum* L.) or maize (*Zea mays* L.) (Gong et al., 2021). In addition to species-specific reactions, factors like the number, type, shape, and size of MNPs might affect the outcomes. For instance, different MNP types and shapes will have different impacts on plant development, and different MNP sizes will have different effects on enzyme activities (Zantis et al., 2023).

Recently, a number of reviews and studies that explore the incidence and destiny of MNPs in natural environments have been published (Kumar, 2020; Wang et al., 2019). Researchers looked at marine MPs, and it has been well established that they exist in coastal areas (Aslam et al., 2020; Kane & Clare, 2019). Another possible sink for MPs is the freshwater system (Strungaru et al., 2019). Recent agricultural intensification and plastic landfilling methods have drawn MPs' attention to the pollution of terrestrial ecosystems, which includes agricultural soil (Wang et al., 2019). Recently, it was estimated and observed that MPs occur in the terrestrial environment 4–23 times more frequently than in the marine environment (Horton, Walton, et al., 2017). A wide variety of concentrations of MPs were examined in industrial soils (L. Li et al., 2020), home garden soil, agricultural soil (Liu et al., 2018), floodplain soil (Scheurer & Bigalke, 2018), and coastal soil. The presence of MPs in terrestrial soils demonstrates a range of environmental consequences for the soil's flora and fauna, as well as toxicity (de Souza Machado et al., 2018; Rillig et al., 2019). MPs' contaminated soil had a detrimental effect on how native flora and the soil microbial ecosystem functioned. A further significant finding is that MPs are taken up by plants (crops), reaching a higher trophic level and having a greater negative effect (Sun et al., 2020).

NPs are produced from impregnated or manufactured nanoparticles or when larger plastic waste materials break down (Ferreira et al., 2019). The smaller-sized unit of MPs known as NPs has recently attracted more interest (Wagner & Reemtsma, 2019). NPs pose more danger than MPs due to their ease of entry into cells and tissues. The complexity of NP separation and identification, as well as their abundance in the environment, has mainly gone unnoticed until now. Therefore, it is possible that NPs' physical threat to human health is underestimated. A recent field of study in the environmental sciences is NPs. Clarifying the toxicological effects and mechanism of action of the neoplastic will require more study in this area.

Diverse sources of MNPs provide significant hazards to ecosystems, human health, plants, and aquatic life. MNPs enter the environment through improper waste management, contributing to pollution in the air, water bodies, and soils. In aquatic ecosystems, MNPs pose risks to marine life by causing physical harm, affecting feeding behavior, and facilitating the transfer of toxins through the food chain. Additionally, MNPs can disturb aquatic plants' growth and photosynthetic processes. Human exposure occurs through ingestion, inhalation, and skin contact, raising concerns about their impacts on health, including inflammation, DNA damage, and endocrine disruption, with potential links to cancer and neurological disorders. MNPs can impact plant traits such as growth, chlorophyll content, and root development. Research gaps regarding MNPs might include the study of nonspherical MNPs, the influence of aged particles, terrestrial pollution implications, and comprehensive ecosystem interactions. Urgent global efforts are needed to address MNP contamination, as plastic production continues to outpace recycling and waste reduction attempts. Though efforts to reduce plastic wastes are underway, they lag behind production rates. Addressing MNP contamination necessitates an overhaul of the global plastics value chain, demanding the collaboration of governments, industries, and individuals. Collective action is essential to mitigate pollution, curb plastic use, and ensure a sustainable future. The authors jointly refer to MPs and NPs in this chapter as MNPs. This chapter highlights the sources of MNPs in the environment. It also discusses their effects on the environment and ecosystem, soil organisms, human health, plants, marine fish, and aquatic organisms. Finally, the status of research in this field is focused. By analyzing this recent research, the authors find out some research gaps and indicate future research areas for MNPs.

2 Sources of MNPs

The amount of plastic pollution has been consistently rising over the last several decades, and researchers are working very hard to understand how microplastic pollution works and how it enters the natural ecosystem from urban areas. Most of the research is concerned with marine systems. Little attention is given to land-based micro-nanoplastics sources, despite having a great impact on the level of micro-nanoplastics pollution as a whole (Golwala et al., 2021). Though the manufacturing and use of plastics are mainly limited on land (Horton & Dixon, 2018), more attention is given to the marine ecosystem than the terrestrial environment when considering studies on microplastic pollution. Microplastics are widely distributed and found in the air, rivers and lakes, sediments, estuaries, oceans, and other environmental areas as a consequence of not having appropriate waste management and improper human disposal of plastics (Fig. 1).

There are mainly two types of sources for micro- and nanoplastics in the environment: primary sources and secondary sources. Secondary microplastics are created from the destruction of macro and meso-plastics via chemical, biological, and mechanical interactions, while primary microplastics are purposefully created in micro sizes for many different objectives (He et al., 2019). Primary microplastic sources include commonly used personal care products containing paint, microbeads, glitter, wastewater from washing, rubber from cities, vehicle tires, sewage sludge, and artificial turf. Because the number of vehicles is rising quickly due to urbanization, one of the most significant sources of microplastic is thought to be car tire wear in this modern era. Municipal wastes such as bottles and plastic bags, farming films, mulching material, microfibers produced from synthetic clothing, fishing wastes, and other relatively large plastic wastes are included in the secondary sources of microplastic (An et al., 2020).

The abrasion of tires from vehicles and traffic (Evangeliou et al., 2020), household and domestic products like cleaning supplies and cosmetics, fibers of synthetic origin from washing cloths and textiles (Boucher & Friot, 2017), coatings, paint, and activities leading up to painting are included as sources as inputs of micro-sized plastic to the terrestrial environment (Fig. 1). Improper waste management and direct littering, such as industrial spills and landfill site discharges, also contribute a major portion to the release of microplastics into the environment.

Plastics can enter the terrestrial environment as a result of agricultural operations, either through inappropriate bale twine and wrapping disposal or by the application of sewage sludge to the cultivated field (Corradini et al., 2019). For instance, in Europe alone, sewage sludge applications annually add between 125 and 850 tons of microplastic per million people to agricultural soils (Nizzetto et al., 2016).

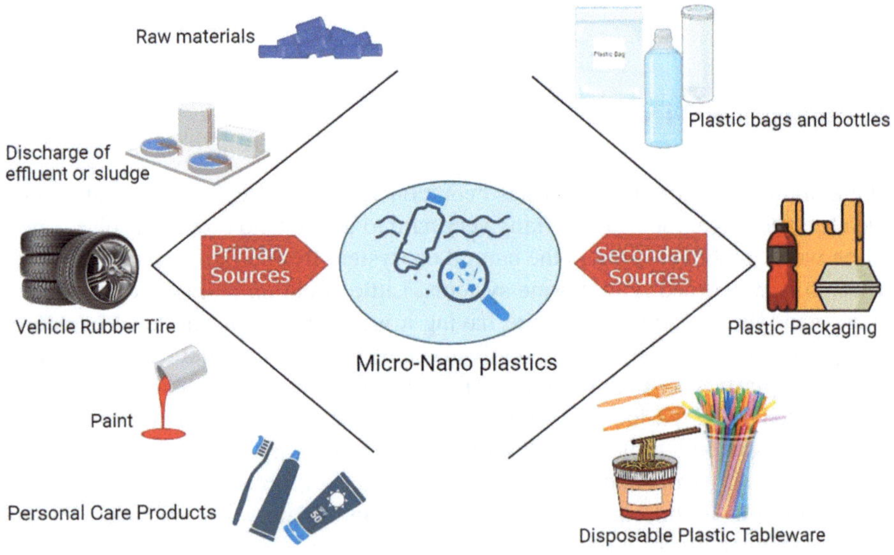

Fig. 1 Sources of MNPs in environment

Through wind transmission, air deposition, and surface runoff from neighboring areas, poor waste management or insufficient trash disposal act as sources of plastic in the aquatic environment (Xia et al., 2020).

2.1 Primary Sources

The rapid increase in the use of synthetic fibers in textile industries, microbeads from items for personal care, and cosmetics are significant primary microplastic sources worldwide (Fu & Wang, 2019).

2.1.1 Raw Materials

Granular-sized plastics or pellets with diameters of 2–5 mm and a uniform shape are used as raw materials in plastic production (Karlsson et al., 2018), which are frequently used in making different plastic products. These particles of plastic are commonly used in many aspects of day-to-day life, including household appliances, building materials, clothing materials, chemicals, and even agriculture. Besides, they are extensively utilized in the production of automotive, electrical, telecommunication, and medical instruments. Plastic pellets are persistent materials, which means that after entering the environment they degrade very slowly. They mostly affect human health because of their small particle size, easy consumption by fish and birds, and easy transmission down the environmental food chain (An et al., 2020).

There has been a rapid rise in the manufacturing of plastic particles in recent years due to their high demand. Unexpected consequences of this manufacturing include the potential release of microplastic particles into the surroundings during manufacturing, transportation, usage, storage, and recycling of plastic particles.

2.1.2 Wastewater

With the discharge of effluent or sludge, microfibers of plastic origin are released during clothing washing and go into the environment. According to certain research, the bulk of microplastics in the environment is plastic microfibers that may be discovered in rivers, oceans, and soil (Horton et al., 2017a, b).

2.1.3 Vehicle Rubber Tire

Among the main sources of micro- or nanoplastics in road dust, microplastics emitted by tires have the most effect. Air, soil, sewers, soil, and surface water are only a few of the locations where rubber particles from roadside motor vehicle tires can end up. With the increase in friction with the road, the rate of rubber particle emission from tires also increases (Vogelsang et al., 2019).

2.1.4 Paint

Typically, pigments, solvents, fillers, and trace amounts of useful ingredients make up paint. Numerous studies show that applying a layer of paint to a surface may cause the development of tiny particles of plastic, which can eventually be discharged into nature due to aging, abrasion, and erosion. So, paint is considered a major source of microplastics. Architectural coatings, such as paint that comes off of painted buildings, maritime coatings, automobile coatings, and road marking paints, are microplastics from all environmental sources (Wang et al., 2019).

2.1.5 Products for Personal Care

Microbeads are little pieces of plastic that have undergone processing to become microscopic. In order to produce cosmetic effects including exfoliating, whitening, washing, and the removal of dead skin, microbeads may largely replace the synthetic pigments currently found in personal care products. Toothpaste, UV protection lotion, shower gel, and hair dye are just a few products that contain microplastic beads. Because of their tiny dimensions, slow fragmentation, and insolubility in water, when washing wastewater enters the sewer system, microbeads do enter. The sewage treatment technology struggles to properly remove plastic microbeads. Due to the fact that sewage sludge is often used in agriculture, these microbeads end up in the environment.

2.2 *Secondary Sources*

Weathering and degradation of large microplastic materials under various environmental conditions are the main sources of secondary microplastics (Peng et al., 2020).

2.2.1 Plastic Bags and Bottles

Plastic bottles are a type of container made of materials like polyethylene terephthalate (PET), polyethylene (PE), and polypropylene (PP) and bags composed of various plastic raw materials are referred to as plastic bags. Plastic bottles are often used as disposable containers for storing liquids or solids such as drinks, agricultural veterinary drugs, culinary oils, sauces, dried fruits, and honey. Particularly in some remote areas, clean drinking water is occasionally provided in plastic bottles and bags. Currently, almost one million plastic bottles are sold per minute all over the world. Around the world, more than 5 trillion plastic bags are used annually. About 100 million plastic bags are used by customers each year in the USA (An et al., 2020).

2.2.2 Plastic Packaging

Plastic packaging is used to wrap an object to protect its original quality as well as its value during transit, storage, and distribution. Plastic packaging materials account for around 25% of the packaging materials globally, making up about 39.7% of the total amount of plastic manufactured.

2.2.3 Tableware Made of Disposable Plastic

Utensils composed of resin or similar thermoplastic substances that are used for dining or other comparable situations are referred to as disposable plastic utensils in general. Disposable plastic dinnerware includes things like lunchboxes, knives, dishes, saucers, drinking straws, containers, spoons, forks, cups, and cans; however, long-term food packaging and other similar products are not included. Polystyrene is used in the production of food service goods and packaging materials, including foam cups, instant noodle packets, and fast food cartons. Plastic tableware can enter soil, sewers, oceans, and other areas if it is not disposed of properly. Over time, it partially degrades and becomes an environmental pollutant.

3 Effects of MNPs

3.1 Effects on Environment and Ecosystem

Environmental contamination by micro-nano-plastics is being considered a new danger to ecosystem health and biodiversity. Small particles of plastic are the consequence of breaking down plastic waste that has been left in the environment, such as on land or in water bodies. Due to their capacity to accumulate in the food chain and their potential to have harmful impacts on human health, microplastics are now considered an important environmental issue. The majority of microplastic particles are produced and used on land, but their accumulation affects both land and water (Mai et al., 2018). Different types of MNPs have different effect on environment. Table 1 summarizes the size, polymer types, and shape of MNPs.

Evidence indicate that microplastics may have detrimental effects on marine ecology, especially in isolated and vulnerable regions because they have spread across the sea (Avio et al., 2015). About 4- to 23-fold greater microplastic contamination was found in land ecosystems than ocean ecosystems (Horton et al., 2017a, b). Threats that microplastics pose to aquatic systems are often linked to the chance that aquatic organisms will eat them, that they will act as solid surfaces for the transfer of contaminants or that they could cause physical damage.

Scientists, decision-makers, managers involving natural resources, the media, and the general public have recently shown a great deal of interest in the issue of

Table 1 MNPs properties based on polymer type, shape, and size

Polymer type	Size	Polymer shape
Low density polyethylene (LDPE), Polyethylene terephthalate (PET), polypropylene (PP), Polystyrene (PS)	0.5 mm – 250 nm	Pellet, fiber, line, grainy, foam
High density polyethylene (HDPE), Polyethylene terephthalate (PET), polypropylene (PP), Polyurethane (PU), Poly Vinyl Chloride (PVC)	<2 mm – 480 nm	Fragment, film, globule, line
Polycarbonate (PC), polypropylene (PP), Poly Vinyl Chloride (PVC), Polystyrene (PS)	<5 mm – 10 nm	Fragment, foam, film, fiber,
Polyethersulfone (PES), Polyethylene terephthalate (PET)	0.1 mm – 76 μm	Fragment, filament, globule
Polyethylene terephthalate (PET), polypropylene (PP), Polystyrene (PS)	0.3 mm – 50 μm	Fragment, film, filament, globule
Poly Vinyl Chloride (PVC), Polystyrene (PS), Polyethylene terephthalate (PET)	0.5 mm – 1 μm	Fragment sheet, foam, fiber,

plastic particles, which is a ubiquitous environmental concern and may also be a substantial human health issue (Prata et al., 2021). Either directly as primary microplastics or indirectly as secondary microplastics, microplastics can invade terrestrial ecosystems (Waldman & Rillig, 2020).

Plastic pollution is complex and follows a critical environmental cycle (Bank & Hansson, 2022). The numerous polymers and additives that make up microplastics can affect their ability to influence living things. The observed effect includes impacts on ingestion, population dynamics, feeding behavior, reproduction, metabolic function, and other contaminant interactions (Kögel et al., 2020).

Micro-nanoplastics have an impact on carbon dioxide (CO_2) emissions as well as in the process of carbon conversion and carbon sources, largely via altering microbial community makeup, gene abundance, and enzyme performance. Microplastics affect nitrogen fixation, nitrogen sources, ammonification, nitrification, and denitrification by changing the number of genes, the activity of enzymes, and the makeup of microbial communities. Microplastics can also influence phosphorus content and phosphorus conversion processes via uplifting enzyme production and altering the genetic makeup of microbial populations (Wang, 2022).

There exist clear research gaps on the impacts of both micro-nanoplastics in particular taxonomical groups, despite the fact that a variety of ecotoxicological effects brought on by exposure to plastic particles have been observed in a range of organisms. Ecotoxicological studies on microalgae showed that microplastics caused a wide range of biochemical and cellular or sub-cellular effects, like changing gene expression, which is involved in metabolic pathways, slowing photosynthesis, and stopping growth (Mao et al., 2018). Changes in microbial activity may also affect critical ecosystem functions such as the carbon cycle, nitrogen cycle and nutrient cycling (Green et al., 2017). Microplastic contamination can have an effect on an organization's higher level and affect the performance of an organism (such as reproduction and feeding). These changes may have an effect on the total population and have an impact on the ecosystem. It is highly challenging for any environmental contaminant to establish a direct connection between impacts at the base of the biological hierarchy and consequences on ecosystems (Galloway et al., 2017).

The combined effects of microplastics can change the soil's structure and texture physio-chemically, which has an impact on the water cycle, ecosystem function in land systems, and a number of feedbacks in plant-soil interactions (Zheng et al., 2016). Changes in the hydrology of soils caused by microplastics may affect a variety of soil organisms. This could affect important symbiotic relationships in terrestrial ecosystems, such as N-fixing and mycorrhizal associations (de Souza Machado et al., 2018).

Nanoplastics might accumulate in land and food webs, though no conclusive evidence has been found yet. The most established way for particles to enter the body is through the intestinal lymphatic cells, which can be found in small rodents, humans, and dogs. Ingestion of microplastics by accident could be very harmful to the well-being of individuals. The majority of big plastic particles are not poisonous enough to kill. However, exposure to, consumption of, and uptake of microscopic microplastics may be hazardous and serve as a new source of very long-term environmental stress.

Microplastics may not be the most dangerous (lethal or sublethal) hazard in the environment on their own. However, there are persistent tendencies in the past, present, and future that indicate an increase in nearly permanent plastic contamination of the world's natural habitats (Geyer et al., 2017). Contamination from microplastics may have significant effects on the biodiversity of continental systems, which is almost entirely unconsidered (de Souza Machado et al., 2018).

Despite the increasing recognition of the environmental risks posed by micro-nanoplastics, there are clear research gaps in understanding their specific impacts on particular taxonomical groups. The consequences of microplastic exposure on microalgae and soil organisms have not been thoroughly studied, despite the fact that ecotoxicological effects have been seen in a variety of organisms. Furthermore, nothing is known about the possible long-term effects of microplastic exposure on continental ecosystems and biodiversity. To fully comprehend and reduce the effects of the pollution of MNPs on the environment and human health, it is essential to eliminate these discrepancies in knowledge.

To sum up, the health of ecosystems and biodiversity are significantly threatened by the presence of micro-nanoplastics in the environment. Their vast accumulation in ecosystems on land and in the sea, as well as their capacity to transport toxins up the food chain, requires immediate attention. The detrimental impacts on marine life, soil structure, and other physiological and environmental systems highlight the urgent need for in-depth research and proactive steps to reduce the growing plastic pollution and protect the ecological balance of the world.

3.2 Effects on Human Health

Microplastics, originating from various sources, including industrial waste, runoff, biomedical products, and everyday items such as bottle caps, sugar, plastic tea bags, and cosmetics, have raised alarming concerns about their impact on human health (Blackburn & Green, 2022). MNPs have a slower degradation rate due to photo and thermo-oxidative processes, posing a long-term threat to the environment and human health (Sangkham et al., 2022). Individuals are exposed to MNPs through inhalation, dietary consumption, and skin exposure (Fig. 2).

Microplastics that enter the body through ingestion can also cause various health issues, including apoptosis, necrosis, DNA damage, inflammation, and cancer (Dai et al., 2022). Food is often stored in plastic containers, susceptible to degradation from external factors like temperature. Plastics can break down under favorable temperatures and mix with food. Ingesting microplastics-contaminated food can lead to plastic particles storage in our digestive system, causing inflammation or physical harm to the tissues. A study has shown that plastic tea bags release a substantial amount of MNPs when brewed in hot water, which humans can directly ingest, contributing to chromosome alteration, weaker immunity, infertility, and obesity (Zhou et al., 2019). After infiltrating into human bodies, microplastics can generate reactive oxygen species (ROS) and disrupt cellular redox balance, leading to cellular damage and dysfunction.

Consumption of seafood, which is a good source of nutrients such as omega-3 fatty acids and minerals such as zinc, can also lead to microplastic ingestion,

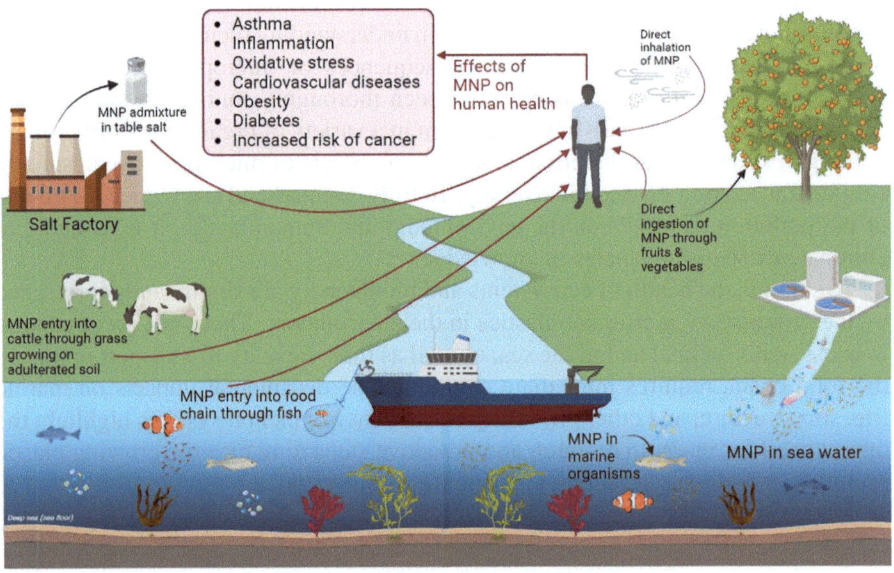

Fig. 2 Entry points for MNPs into the human body and their effects

resulting in neurological diseases. Acetylcholinesterase is an enzyme that breaks down acetylcholine, an important neurotransmitter for memory. Microplastics can inhibit the activity of acetylcholinesterase, leading to learning and memory problems. The microscopic size of the microplastics allows them to enter the bloodstream by passing through intentional walls. By entering the bloodstream, microplastics can disrupt the natural antioxidant defense of the body, which results in oxidative stress. Hence, microplastic particles can release chemicals that are detrimental to human bodies which eventually lead to inflammation. Inflammation, if persisting for a long time, can contribute to diabetes and cardiovascular diseases. Additionally, aquatic organisms consume microplastics by ingesting contaminated food materials, which can be transferred to higher levels of the food web, even to humans (Walkinshaw et al., 2020). Due to this trophic transfer, even organisms at higher levels can be affected by growth inhibition and morality from ingesting organisms that live at lower levels.

Inhalation of household and industrial dust is considered one of the prominent pathways of consuming microplastics. According to human exposure estimates, inhalation mainly exposes adults to microplastics as they migrate frequently (Bank & Hansson, 2022). After accumulating in the respiratory system through inhalation, microplastics can alter the immune system, leading to an inflammatory reaction. Plasticizers that are used in manufacturing cosmetics, such as bisphenol A (BPA) and phthalates, can lead to endocrine disruption, impact the reproductivity of humans and animals, and have potential carcinogenic properties that cause cancer and mutagenicity (Blackburn & Green, 2022). Microplastics can also absorb toxins such as epoxy resin, polyvinyl chloride (PVC), persistent organic pollutants (POP), and heavy metals from their surroundings. Hence, infection and disease-causing microorganisms and poisons can spread by attaching to nanoparticles. Upon inhaling, microplastics containing toxins can react with other compounds inside human bodies and lead to chronic health issues. Microplastic inhalation has also been linked to respiratory illnesses and increased cancer risk despite the prevailing belief that most microplastic particles can be removed from the respiratory system.

Although research on the specific health impacts of microplastics through skin exposure is still evolving, it is evident that microplastics can cause allergic reactions and skin irritation. Particularly, microplastics pose a severe threat to people living in urban landscapes, such as hospitals, garments, manufacturing industries, and markets. Microplastics have also been found in personal care products, such as cosmetics, which indicates women are more prone to be affected by microplastics compared to men. Microplastics that enter through skin can accumulate in the liver, kidney, and intestinal tract, affecting intestinal flora balance, intestinal function, and liver–lipid metabolism.

The health consequences of microplastics are a rising concern. Microplastics have the potential to disturb the endocrine system, compromise respiratory and gastrointestinal health, influence immune function, and trigger inflammation. Microplastic exposure can also lead to bioaccumulation, transfer of toxins, and food chain contamination. Limiting plastic product use and fostering environmentally responsible habits are crucial to addressing these risks. Further research is imperative to fully comprehend microplastics' long-term impact on human health and devise viable solutions to this environmental threat.

3.3 Effects on Marine Fishes and Aquatic Organisms

Microplastics have been recognized worldwide in marine environments and are becoming more common as pollutants in water ecosystems such as rivers, lakes, and aquatic systems (Yuan et al., 2022). Plastic particles can harm aquatic organisms depending on their size, concentration, and length of exposure (Goncalves & Bebianno, 2021). The most common toxicity of microplastics is physical damage, specifically the hindrance of the digestive tracts of organisms, which in turn minimizes the amount of nutrients that can be absorbed by the body (Ma et al., 2019). Hence, microplastics can accumulate in the tissues and cells of organisms, interfering with their normal functioning, and can also weaken the defense mechanisms of marine animals.

The consumption of nondegradable microplastics by marine organisms can lead to their bioaccumulation within the food chain and result in both physiological and toxicological consequences, including oxidative stress, tissue damage, immune toxicity, and genotoxicity (Porcino et al., 2022). Humans dump plastic items into water bodies, including rivers, ponds, lakes, canals, etc. Plastic items thrown into water bodies undergo decomposition by microbial communities and are stored in the stomach of fish, which are consumed by humans.

Microplastics can have long-lasting effects on fish, including variations in gene expression related to their immune systems, tissue damage, changes in the quantities of antioxidants in their bodies, histological abnormalities in the intestines, intestinal blockage, behavioral changes, and lipid metabolism changes, alteration in swimming behavior, immune response, and capacity to avoid prey (Bhuyan, 2022). A study has indicated that fish containing microplastics and nanoplastics displayed elevated levels of AChE activity in their brain and amplified Lipid peroxidation (LPO) levels in their brain and muscle tissues (Barboza et al., 2020). LPO levels in the brain are a marker for oxidative stress, characterized by the disproportionate generation of free radicals relative to the body's capacity to counteract them.

Ships and other water vehicles can discharge oil even when equipped with wastewater treatment. Altered oil properties can create larger and more persistent oil slicks, affecting light penetration, nutrient availability, and dissolved oxygen levels in the water body (Yang et al., 2021). Hypoxia, which refers to low oxygen levels, arises due to fish mortality and has detrimental impacts on the ecosystem by causing oxygen depletion. At the same time, less mobile animals cannot move to areas with more oxygen and are often killed, resulting in a significant reduction in life in affected zones and a disturbed aquatic ecosystem.

Nanoplastic can also affect fish's ability to navigate, communicate, and locate predators by interfering with their senses of smell, eyesight, and hearing. Nanoplastic can transport harmful substances such as heavy metals, endocrine disruptors (EDCs) and persistent organic pollutants (POPs), which can potentially permeate the tissues of marine organisms, particularly fish, resulting in adverse consequences. Nanoplastics can accumulate in fish tissues, decrease locomotion, change hormone levels, and interfere with physiological activities.

Small crustacean consumption of microplastics can cause population declines, while biomarkers in fish and bivalves have shown abnormal behavior, oxidative stress, and neurotoxic effects (Dioses-Salinas et al., 2019). Microplastic exposure has been observed to cause impaired growth and development in crustaceans such as copepods, amphipods, and shrimps (Zhang et al., 2023). Epibenthic copepod reproduction could be impacted by the presence of microplastics in ocean sediments (Nobre et al., 2022).

Additionally, Microplastics also have devastating impacts on aquatic plants, and they can affect plants directly or indirectly. Microplastics in aquatic environments can interfere with the photosynthetic rate and biomass production of aquatic plants by blocking light and disrupting the balance of chlorophyll-a and chlorophyll-b ratios (Li et al., 2020). The existence of microplastics in the soil or sediment in which aquatic plants thrive can alter their physical and chemical characteristics, thus impacting their capacity to absorb water and essential nutrients, as well as hindering their proper root establishment. Elevated levels of microplastic in aquatic plants can result in cell death and DNA damage that eventually can lead to impairment of development, growth, and reproduction (Mateos-Cárdenas et al., 2021).

A multifaceted approach is necessary to reduce the impact of microplastics on the marine ecosystem. One practical strategy could be reducing the amount of plastic waste entering the environment by promoting recycling and improving waste management practices. Another alternative could be building more wastewater treatment plants that are capable of filtering out microplastics from sewage and stormwater before they are discharged into water bodies.

3.4 Effects on Plants

The omnipresence of MNPs affects almost all life forms (Pantos, 2022) and plants are no exception (Rillig et al., 2019). MNPs can affect various aspects of plants in both positive and negative ways (Table 1). For example, photosynthetic pigment or chlorophyll content, seedling growth, nitrogen content of leaves, α-amylase activity, and root and shoot biomass are found to be affirmatively related in some studies. There are some adverse effects of plastic pollutants on the plant body, like altered root growth, oxidative damage, lowered photosynthetic activity, an inactive enzyme system, inhibited gene expression, and diminished biomass (Fig. 3).

MNPs can enter the plant vascular system simply by absorption. Plant root surfaces can absorb microplastic particles that range from 5 to 50 nm in size (Yu et al., 2022) and nanoplastics have a chance of being internalized into plant tissue (Yin et al., 2021). The permeability of nanoplastics is greater than that of microplastics as they are tinier, and for that reason, they can penetrate biological membranes more

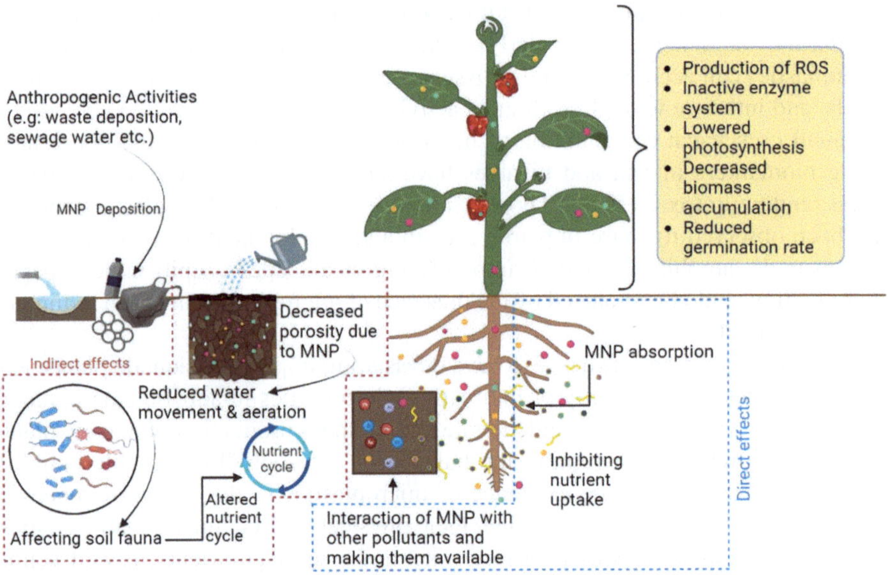

Fig. 3 Interaction of MNPs with the properties of soil and its impact on plant

easily (Bellingeri et al., 2020). Microplastics with both sharp and smooth edges directly affect plant root growth. Shorter root length is a visible effect of microplastic treatment in *Lemna minor* (Kalčíková et al., 2017). Sharp-edged microplastics mechanically damage plant roots with their rough surfaces. Nanoplastics, upon absorption, can cause oxidative damage and alter cell membrane structure (Rillig et al., 2019).

MPNs significantly reduce germination percentages in many plant species like *Lepidium sativum, Cicer arietinum, Trifolium repens, Lolium perenne, Impatiens balsamina.* Before germination, imbibition is a must for seed. During imbibition, plastic particles clog the seed surface, inhibiting germination, but upon plant growth, these plastic particles get moved from their previous position and the clogging effect is alleviated (Islam et al., 2023; Mondal et al., 2022).

Reactive oxygen species (ROS) are produced in excess compared to antioxidants, which causes oxidative damage. ROS production is a natural response of a cell to metabolism. But, in certain conditions, such as extreme temperature, excess light, drought, salinity, and heavy metal accumulation, ROS genesis is increased in response to stress. MNPs also have an impact on this ROS generation (Fig. 3). In nature, plants have a defense system to protect them from ROS, which is an antioxidant system. Many enzymes like glutathione (GSH), superoxide dismutase (SOD), catalase (CAT), and some nonenzymatic compounds like tocopherols, ascorbate (AsA), carotenoids, and phenolics serve as antioxidants. Studies show that MNP exposure can lead to the manufacture of ROS (Ekner-Grzyb et al., 2022). In onion, PS-NP significantly raises the H_2O_2 level when treated with 1000 mg/L concentration for 72 h (Giorgetti et al., 2020). In *Vicia faba,* MNPs decline CAT enzyme activity, which eventually leads to oxidative damage (Jiang et al., 2019). A study conducted on wheat shows that MNPs combined with Cd impose a negative effect on the plant antioxidant system as they reduce Superoxide dismutase (SOD), Malondialdehyde (MDA) enzyme activity (Azeem et al., 2021). MNPs have a variety of effects on both terrestrial and aquatic plants, but additional research is needed to fully understand how different plastic kinds affect plants.

3.5 *Effect of MNPs on Soil Organisms*

MNPs are the major environmental and soil health hazards. MNPs affect a number of soil characteristics, including fertility, nutrient availability, and porosity, directly and indirectly (Bouaicha et al., 2022; Maddela et al., 2023), which determine the overall health of the soil.

MNPs change the soil's bulk density, porosity, and water-holding capacity, among other physical properties (Fig. 3) (Bouaicha et al., 2022). The types and concentrations of MNPs affect how they affect soil. Soil bulk density decreases as a result of microplastics such as PES (polyester) and PA (polyamide) fibers (Salam et al., 2023). The decrease in bulk density can be attributed to MNPs because they are not as dense as the soil matrix (Salam et al., 2023). Since bulk density makes it easier for water and other substances to move through the soil, improves soil aeration, and prevents compaction, it is one of the most crucial indicators of soil health. The bulk density of soil has a direct impact on its hydrological and agroecosystem processes (Y. Lu et al., 2019). Moreover, MNPs are hydrophobic in nature (colloidal MNPs). It raises difficulties in water movement through the soil column, which reduces the amount of available water to both microbes and plants (Wang, 2022). This may lead to imbalanced soil moisture levels, plant growth inhibition, and drought stress (Lu et al., 2021). MNPs can decrease soil porosity by occupying pore spaces (Rong et al., 2022). Decrease in porosity affects soil respiration rates and microbial activity by limiting water entry and airflow in the soil (Ben-Noah & Friedman, 2018). But sometimes microplastic pollution encourages soil cracking, which could lead to an increase in soil porosity. In addition, MNPs may disrupt soil aggregation processes by adhering to soil particles or changing microbial activity (Salam et al., 2023). Due to this disruption, soil stability is weakened, and erosion susceptibility is raised.

MNPs have a significant impact on soil fertility, nutrient absorption, and the toxicological composition of pollutants (Maddela et al., 2023). MNPs can influence the availability of nutrients by interacting with nutrient ions as they have adsorption ability. This can lead to an uneven intake of nutrients by plants as well as has an effect on the toxicological composition of contaminants of soil (H. Li et al., 2022). Microplastic contamination reduces soil pH by influencing soil bacteria that release hydrogen cations and also affects nutrient cycling as it declines soil microbial population (Lu et al., 2021). Poor soil quality, change in organic matter levels, and nutrient availability affect soil fertility (Medyńska-Juraszek & Jadhav, 2022).

MNPs influence soil microbial population. It disrupts microbial breakdown processes and declines microbial diversity (Elbasiouny et al., 2022; Maddela et al., 2023). The disruption of microbial diversity has the potential to have an impact on soil health and ecological balance. Substrate-induced respiration (SIR) rate reduces drastically with the addition of MNPs, indicating changes in the function of the soil microbial community (Salam et al., 2023). Plastic pollution might potentially reduce enzyme activity (Yu et al., 2020) which may contribute to the reduction in soil respiration. It is a quite alarming aspect as enzymatic processes used by microorganisms can break down MNPs, preventing their buildup in soil. The properties of MNPs, microbial species, soil factors, and ambient circumstances impact the biodegradation process.

Altogether, MNPs constitute a serious danger to soil health as well as the ecosystem due to their effects on soil characteristics, fertility, nutrient availability, and microbial activity. To protect soil ecosystems and ensure long-term sustainability, it is crucial to understand the effects of MNPs. It is essential to develop effective mitigation strategies which will ensure the long-term health of soil ecosystems.

4 Present Status of MNPs-Related Research

Using the latest instruments of analytical chemistry, researchers are trying to discover the right analytical procedures to describe and quantify the total amount of MNPs in water samples. Based on forensic investigation of microplastics from sediments, the proportions of polyester and acrylic fibers used in everyday apparel approximated those observed in locations that receive sewage discharges and sewage effluent. Based on research using wastewater from residential washing machines, one garment can yield close to 1900 fibers per wash. Up to 360 million tons of plastic products were produced annually in 2018, and it is anticipated that 500 million tons will be produced in 2025, according to statistics (Huang et al., 2021). Nearly 75% of all plastic manufactured ends up as waste, and 87% of solid waste from uncontrolled landfills that were improperly managed, leaks into the environment, causing plastic pollution.

Oceans are estimated to receive 4.8 to 12.7 million tons of the 2.5 billion tons of trash of solid origin created each year, of which coastal countries contribute up to 275 billion tons of plastic. (Jambeck et al., 2015). The North Pacific Ocean's Great Pacific Garbage Patch, which covers over 600 km^2, is recognized as the biggest landfill for plastic in the world (L. Lebreton et al., 2018).

The majority of studies on microplastics or nanoplastics focus on problems related to their presence, characterization, absorption by organisms, and impacts on fate and transport. Though assessing MNP's quantity, density, and distribution in the aquatic environment is very difficult because of their spatial as well as temporal fluctuation brought on by ocean currents and seasonal cycles, microplastics are considered to make up a significant portion of marine trash (Wang et al., 2021).

The research on microplastic pollution in sea surfaces (Anderson et al., 2017), surface waters (La Daana et al., 2017), and sediments (Lots et al., 2017) has increased sharply. These studies mainly focus on examining the geographical distribution and abundance of microplastics as well as classifying these artificial substances according to their many forms, dimensions, colors, and shapes.

The research field's focus has been on identifying the composition of microplastics to determine their probable sources (Lo & Chan, 2018). In the samples obtained from the bottom water of the Irish Continental Shelf, fibers were identified as the primary source of microplastic contamination (85%) (Martin et al., 2017). Every year, over two million tons of microfibers are discharged into the ocean (Boucher & Friot, 2017). Due to the toxicological concerns connected with these tiny particles, microplastic uptake by marine creatures is a developing concern. Numerous studies have shown microplastic consumption by finding various particles of polymers in the intestines and stomachs of many creatures, including zooplankton, shellfish, corals (Reichert et al., 2018), fish (Jabeen et al., 2017), and marine mammals (Nelms et al., 2018).

The Ganga River Catchment comes in second with an annual release of 0.12 million tons of plastics, next to the Chinese Yangtze River's impact of roughly 0.33 million tons of plastics, based on an empirical global model, that estimated that the ocean receives between 1.15 and 2.41 million tons of plastic through streams and that twenty Asian rivers contribute 67% of the world's marine plastic pollution (Lebreton et al., 2017). The sediments of the Ganga were incredibly rich in microplastics, with a mass ranging from 11.48 to 63.79 mg/kg and a numerical quantity of 99.27 to 409.86 items/kg, in accordance with the ICAR-CIFRI study at seven sites along the river, including Buxar, Nabadwip, Patna, Bhagalpur, Barrackpore, and Frazergunj (Sarkar et al., 2019).

Even while research on MNPs in the aquatic environment is expanding, there is still a significant research gap regarding the precise toxicological effects of MNPs on marine animals. While research has shown that MNPs are present and abundant in a variety of marine habitats, little is known about the potential trophic transfer and health effects of MNPs within marine food chains. In order to properly understand the possible hazards and long-term environmental effects of MNP contamination, more research must be done on the effect of MNPs on human health, particularly in terms of MNP consumption through food sources and water.

According to extensive research on the subject, there is growing concern about the prevalence and effects of MNPs in aquatic habitats. Plastic contamination poses a severe threat to marine ecosystems due to increasing plastic production and insufficient waste management. In order to address this rising contaminant and protect the ecosystem, it is essential to understand the composition, distribution, and toxicological impacts of microplastics.

5 Research Gaps and Future Works

Although MNPs are ubiquitous in nature, the research field related to MNPs is considerably novel. There are some research gaps in this field. The following aspects should be taken under consideration to fill the remaining voids:

- MNPs used in the experiments are spherical in shape, but microplastics derived from fragmentation are more like fibers (McIlwraith et al., 2019). These fiber-like or non-spherical nanoplastics are more difficult to come by. This has a significant impact on studies that measure the phytotoxicity that plastic particles impose.
- The absorption capacity of aged plastic particles is greater than that of primary MNPs (Mao et al., 2022). Aged plastic particles affect different things, especially plants and soil. As microplastics age, they undergo fragmentation and produce nanoplastics.
- MNP studies are more confined to aquatic organisms. Terrestrial plants play a key role as producers. Recently, this field has been getting attention, but an extensive area of terrestrial plastic pollution is yet to be explored.
- Current researches are more focused on individual ecosystem and mostly toxic effects of MNPs on organisms and environmental chemistry of MNPs. But a more holistic approach should be taken, ensuring inter disciplinary measures as MNPs have complex interactions among ecosystems.

6 Conclusion

Among other pollutants, MNP pollution is recognized as a significant contaminant since it lasts longer in the environment. The sustainability of environmental, global, social, and economic systems is threatened by MNPs contamination. In the past 50 years, the manufacture, usage, and disposal of plastic, particularly single-use plastic packaging, have multiplied 20-fold. Many single-use plastic products are difficult to recycle or can only be recycled so many times before ending up in the trash.

This chapter mainly discusses the significant effects of MNPs on the ecosystem, environment, human health, aquatic organisms, plants, and soil. Environmental contamination by MNPs poses a new threat to ecosystem health and biodiversity due to their accumulation in the food chain and potential harm to human health. MNPs' harmful impacts on marine life, soil structure, and physiological systems highlight the urgency of research and proactive measures to reduce plastic pollution. MNPs also affect human health through ingestion, inhalation, and skin exposure, leading to various health issues including inflammation, endocrine disruption and cancer. The presence of MNPs in water bodies affects aquatic organisms, plants, and soil, altering nutrient cycles, microbial populations, and ecosystem functions. This chapter provided an overall summary of the status of MNP-related research.

Besides, this chapter also found out some major research gaps regarding MNPs, including the influence of aged particles, the study of nonspherical MNPs, terrestrial pollution implications, and comprehensive ecosystem interactions. To solve the problems associated with MNPs, focus must be on a holistic approach to minimize those gaps.

Governments, businesses, and consumers have all made ambitious attempts to reduce plastic wastes, but these efforts have fallen behind the unsustainable rate at which plastic is being produced and used. Extreme efforts are needed to overhaul the global plastics value chain and achieve zero plastic waste in order to address the complex global problem of MNP contamination. Beyond laws and changes generated by a policy, particular measures are needed to accomplish this. All the people around the world should come forward from their respective positions to reduce pollution and the use of plastics.

References

An, L., Liu, Q., Deng, Y., Wu, W., Gao, Y., & Ling, W. (2020). *Sources of microplastic in the environment* (pp. 143–159). Microplastics in Terrestrial Environments: Emerging Contaminants and Major Challenges.

Anderson, P. J., Warrack, S., Langen, V., Challis, J. K., Hanson, M. L., & Rennie, M. D. (2017). Microplastic contamination in Lake Winnipeg, Canada. *Environmental Pollution, 225*, 223–231.

Aslam, H., Ali, T., Mortula, M. M., & Attaelmanan, A. G. (2020). Evaluation of microplastics in beach sediments along the coast of Dubai, UAE. *Marine Pollution Bulletin, 150*, 110739.

Avio, C. G., Gorbi, S., & Regoli, F. (2015). Experimental development of a new protocol for extraction and characterization of microplastics in fish tissues: First observations in commercial species from Adriatic Sea. *Marine Environmental Research, 111*, 18–26.

Azeem, I., Adeel, M., Ahmad, M. A., Shakoor, N., Jiangcuo, G. D., Azeem, K., Ishfaq, M., Shakoor, A., Ayaz, M., & Xu, M. (2021). Uptake and accumulation of nano/microplastics in plants: A critical review. *Nanomaterials, 11*(11), 2935.

Bank, M. S., & Hansson, S. V. (2022). *The microplastic cycle: An introduction to a complex issue* (pp. 1–16). Microplastic in the Environment: Pattern and Process.

Barboza, L. G. A., Lopes, C., Oliveira, P., Bessa, F., Otero, V., Henriques, B., Raimundo, J., Caetano, M., Vale, C., & Guilhermino, L. (2020). Microplastics in wild fish from north East Atlantic Ocean and its potential for causing neurotoxic effects, lipid oxidative damage, and human health risks associated with ingestion exposure. *Science of the Total Environment, 717*, 134625.

Bellingeri, A., Casabianca, S., Capellacci, S., Faleri, C., Paccagnini, E., Lupetti, P., Koelmans, A. A., Penna, A., & Corsi, I. (2020). Impact of polystyrene nanoparticles on marine diatom *Skeletonema marinoi* chain assemblages and consequences on their ecological role in marine ecosystems. *Environmental Pollution, 262*, 114268.

Ben-Noah, I., & Friedman, S. P. (2018). Review and evaluation of root respiration and of natural and agricultural processes of soil aeration. *Vadose Zone Journal, 17*(1), 1–47.

Beriot, N., Peek, J., Zornoza, R., Geissen, V., & Lwanga, E. H. (2021). Low density-microplastics detected in sheep faeces and soil: A case study from the intensive vegetable farming in Southeast Spain. *Science of the Total Environment, 755*, 142653.

Bhuyan, M. S. (2022). Effects of microplastics on fish and in human health. *Frontiers in Environmental Science, 10*, 250.

Blackburn, K., & Green, D. (2022). The potential effects of microplastics on human health: What is known and what is unknown. *Ambio, 51*(3), 518–530.
Borrelle, S. B., Ringma, J., Law, K. L., Monnahan, C. C., Lebreton, L., McGivern, A., Murphy, E., Jambeck, J., Leonard, G. H., & Hilleary, M. A. (2020). Predicted growth in plastic waste exceeds efforts to mitigate plastic pollution. *Science, 369*(6510), 1515–1518.
Bouaicha, O., Mimmo, T., Tiziani, R., Praeg, N., Polidori, C., Lucini, L., Vigani, G., Terzano, R., Sanchez-Hernandez, J. C., & Illmer, P. (2022). Microplastics make their way into the soil and rhizosphere: A review of the ecological consequences. *Rhizosphere, 22*, 100542.
Boucher, J., & Friot, D. (2017). *Primary microplastics in the oceans: A global evaluation of sources* (Vol. 10). Iucn Gland.
Brooks, A. L., Wang, S., & Jambeck, J. R. (2018). The Chinese import ban and its impact on global plastic waste trade. *Science Advances, 4*(6), eaat0131.
Colzi, I., Renna, L., Bianchi, E., Castellani, M. B., Coppi, A., Pignattelli, S., Loppi, S., & Gonnelli, C. (2022). Impact of microplastics on growth, photosynthesis and essential elements in Cucurbita pepo L. *Journal of Hazardous Materials, 423*, 127238.
Corradini, F., Meza, P., Eguiluz, R., Casado, F., Huerta-Lwanga, E., & Geissen, V. (2019). Evidence of microplastic accumulation in agricultural soils from sewage sludge disposal. *Science of the Total Environment, 671*, 411–420.
Dai, Y., Shi, J., Zhang, N., Pan, Z., Xing, C., & Chen, X. (2022). Current research trends on microplastics pollution and impacts on agro-ecosystems: A short review. *Separation Science and Technology, 57*(4), 656–669.
de Souza Machado, A. A., Kloas, W., Zarfl, C., Hempel, S., & Rillig, M. C. (2018). Microplastics as an emerging threat to terrestrial ecosystems. *Global Change Biology, 24*(4), 1405–1416.
Dioses-Salinas, D. C., Pérez-Baca, B. L., & De-la-Torre, G. E. (2019). Ecotoxicological effects of microplastics and adsorbed contaminants on aquatic organisms. *Man, 16*(2), 173–182.
Ekner-Grzyb, A., Duka, A., Grzyb, T., Lopes, I., & Chmielowska-Bąk, J. (2022). Plants oxidative response to nanoplastic. *Frontiers in Plant Science, 13*, 1027608.
Elbasiouny, H., Mustafa, A. A., Zedan, A., Amer, S. M., Albeialy, N. O., Alkharsawey, D. S., Aeash, N. R., Abuomar, A. O., Hamd, R. E., & Elbltagy, H. (2022). The effect of microplastic pollution on soil, plants and soil microbes and its remediation. *Egyptian Journal of Soil Science, 62*(4), 325–334.
Evangeliou, N., Grythe, H., Klimont, Z., Heyes, C., Eckhardt, S., Lopez-Aparicio, S., & Stohl, A. (2020). Atmospheric transport is a major pathway of microplastics to remote regions. *Nature Communications, 11*(1), 3381.
Ferreira, I., Venâncio, C., Lopes, I., & Oliveira, M. (2019). Nanoplastics and marine organisms: What has been studied? *Environmental Toxicology and Pharmacology, 67*, 1–7.
Fu, Z., & Wang, J. (2019). Current practices and future perspectives of microplastic pollution in freshwater ecosystems in China. *Science of the Total Environment, 691*, 697–712.
Galloway, T. S., Cole, M., & Lewis, C. (2017). Interactions of microplastic debris throughout the marine ecosystem. *Nature Ecology & Evolution, 1*(5), 0116.
Geyer, R., Jambeck, J. R., & Law, K. L. (2017). Production, use, and fate of all plastics ever made. *Science Advances, 3*(7), e1700782.
Giorgetti, L., Spanò, C., Muccifora, S., Bottega, S., Barbieri, F., Bellani, L., & Castiglione, M. R. (2020). Exploring the interaction between polystyrene nanoplastics and *Allium cepa* during germination: Internalization in root cells, induction of toxicity and oxidative stress. *Plant Physiology and Biochemistry, 149*, 170–177.
Golwala, H., Zhang, X., Iskander, S. M., & Smith, A. L. (2021). Solid waste: An overlooked source of microplastics to the environment. *Science of the Total Environment, 769*, 144581.
Goncalves, J. M., & Bebianno, M. J. (2021). Nanoplastics impact on marine biota: A review. *Environmental Pollution, 273*, 116426.
Gong, W., Zhang, W., Jiang, M., Li, S., Liang, G., Bu, Q., Xu, L., Zhu, H., & Lu, A. (2021). Species-dependent response of food crops to polystyrene nanoplastics and microplastics. *Science of the Total Environment, 796*, 148750.

Green, D. S., Boots, B., O'Connor, N. E., & Thompson, R. (2017). Microplastics affect the ecological functioning of an important biogenic habitat. *Environmental Science & Technology, 51*(1), 68–77.

He, H., Liu, L., Munir, S., Bashir, N. H., Yi, W., Jing, Y., & LI, C. (2019). Crop diversity and pest management in sustainable agriculture. *Journal of Integrative Agriculture, 18*(9), 1945–1952.

Horton, A. A., & Dixon, S. J. (2018). Microplastics: An introduction to environmental transport processes. *Wiley Interdisciplinary Reviews: Water, 5*(2), e1268.

Horton, A. A., Svendsen, C., Williams, R. J., Spurgeon, D. J., & Lahive, E. (2017a). Large microplastic particles in sediments of tributaries of the River Thames, UK – Abundance, sources and methods for effective quantification. *Marine Pollution Bulletin, 114*(1), 218–226.

Horton, A. A., Walton, A., Spurgeon, D. J., Lahive, E., & Svendsen, C. (2017b). Microplastics in freshwater and terrestrial environments: Evaluating the current understanding to identify the knowledge gaps and future research priorities. *Science of the Total Environment, 586*, 127–141.

Huang, Y., Liu, Q., Jia, W., Yan, C., & Wang, J. (2020). Agricultural plastic mulching as a source of microplastics in the terrestrial environment. *Environmental Pollution, 260*, 114096.

Huang, D., Tao, J., Cheng, M., Deng, R., Chen, S., Yin, L., & Li, R. (2021). Microplastics and nanoplastics in the environment: Macroscopic transport and effects on creatures. *Journal of Hazardous Materials, 407*, 124399.

Islam, M. S., Mondal, S., Angon, P. B., & Jalil, M. A. (2023). Accumulation of microplastics in agroecosystems and its effects on terrestrial plants: A short review. *Current Applied Science and Technology*, 10–55003.

Jabeen, K., Su, L., Li, J., Yang, D., Tong, C., Mu, J., & Shi, H. (2017). Microplastics and mesoplastics in fish from coastal and fresh waters of China. *Environmental Pollution, 221*, 141–149.

Jambeck, J. R., Geyer, R., Wilcox, C., Siegler, T. R., Perryman, M., Andrady, A., Narayan, R., & Law, K. L. (2015). Plastic waste inputs from land into the ocean. *Science, 347*(6223), 768–771.

Jiang, X., Chen, H., Liao, Y., Ye, Z., Li, M., & Klobučar, G. (2019). Ecotoxicity and genotoxicity of polystyrene microplastics on higher plant *Vicia faba*. *Environmental Pollution, 250*, 831–838.

Kalčíková, G., Gotvajn, A. Ž., Kladnik, A., & Jemec, A. (2017). Impact of polyethylene microbeads on the floating freshwater plant duckweed *Lemna minor*. *Environmental Pollution, 230*, 1108–1115.

Kallenbach, E. M., Rødland, E. S., Buenaventura, N. T., & Hurley, R. (2022). Microplastics in terrestrial and freshwater environments. *Microplastic in the Environment: Pattern and Process, 87*.

Kane, I. A., & Clare, M. A. (2019). Dispersion, accumulation, and the ultimate fate of microplastics in deep-marine environments: A review and future directions. *Frontiers in Earth Science, 7*, 80.

Karlsson, T. M., Arneborg, L., Broström, G., Almroth, B. C., Gipperth, L., & Hassellöv, M. (2018). The unaccountability case of plastic pellet pollution. *Marine Pollution Bulletin, 129*(1), 52–60.

Kögel, T., Bjorøy, Ø., Toto, B., Bienfait, A. M., & Sanden, M. (2020). Micro-and nanoplastic toxicity on aquatic life: Determining factors. *Science of the Total Environment, 709*, 136050.

Kukkola, A., Krause, S., Lynch, I., Smith, G. H. S., & Nel, H. (2021). Nano and microplastic interactions with freshwater biota – Current knowledge, challenges and future solutions. *Environment International, 152*, 106504.

Kumar, S. (2020). Abiotic stresses and their effects on plant growth, yield and nutritional quality of agricultural produce. *International Journal of Agriculture and Food Science, 4*, 367–378.

La Daana, K. K., Officer, R., Lyashevska, O., Thompson, R. C., & O'Connor, I. (2017). Microplastic abundance, distribution and composition along a latitudinal gradient in the Atlantic Ocean. *Marine Pollution Bulletin, 115*(1–2), 307–314.

Lebreton, L. C., Van Der Zwet, J., Damsteeg, J.-W., Slat, B., Andrady, A., & Reisser, J. (2017). River plastic emissions to the world's oceans. *Nature Communications, 8*(1), 15611.

Lebreton, L., Slat, B., Ferrari, F., Sainte-Rose, B., Aitken, J., Marthouse, R., Hajbane, S., Cunsolo, S., Schwarz, A., & Levivier, A. (2018). Evidence that the Great Pacific Garbage Patch is rapidly accumulating plastic. *Scientific Reports, 8*(1), 1–15.

Li, L., Luo, Y., Li, R., Zhou, Q., Peijnenburg, W. J., Yin, N., Yang, J., Tu, C., & Zhang, Y. (2020). Effective uptake of submicrometre plastics by crop plants via a crack-entry mode. *Nature Sustainability, 3*(11), 929–937.

Li, H., Liu, L., Xu, Y., & Zhang, J. (2022). Microplastic effects on soil system parameters: A meta-analysis study. *Environmental Science and Pollution Research*, 1–12.

Liu, L., Li, W., Song, W., & Guo, M. (2018). Remediation techniques for heavy metal-contaminated soils: Principles and applicability. *Science of the Total Environment, 633*, 206–219.

Lo, H. K. A., & Chan, K. Y. K. (2018). Negative effects of microplastic exposure on growth and development of Crepidula onyx. *Environmental Pollution, 233*, 588–595.

Lots, F. A., Behrens, P., Vijver, M. G., Horton, A. A., & Bosker, T. (2017). A large-scale investigation of microplastic contamination: Abundance and characteristics of microplastics in European beach sediment. *Marine Pollution Bulletin, 123*(1–2), 219–226.

Lu, Y., Si, B., Li, H., & Biswas, A. (2019). Elucidating controls of the variability of deep soil bulk density. *Geoderma, 348*, 146–157.

Lu, C., Ke, L., Li, J., Zhao, H., Lu, T., Mentis, A. F. A., Wang, Y., Wang, Z., Polissiou, M. G., & Tang, L. (2021). Saffron (Crocus sativus L.) and health outcomes: A meta-research review of meta-analyses and an evidence mapping study. *Phytomedicine, 91*, 153699.

Ma, P., Wei Wang, M., Liu, H., Feng Chen, Y., & Xia, J. (2019). Research on ecotoxicology of microplastics on freshwater aquatic organisms. *Environmental Pollutants and Bioavailability, 31*(1), 131–137.

Maddela, N. R., Ramakrishnan, B., Kadiyala, T., Venkateswarlu, K., & Megharaj, M. (2023). Do microplastics and Nanoplastics pose risks to biota in agricultural ecosystems? *Soil Systems, 7*(1), 19.

Mai, L., Bao, L.-J., Wong, C. S., & Zeng, E. Y. (2018). Microplastics in the terrestrial environment. In *Microplastic contamination in aquatic environments* (pp. 365–378). Elsevier.

Mao, Y., Ai, H., Chen, Y., Zhang, Z., Zeng, P., Kang, L., Li, W., Gu, W., He, Q., & Li, H. (2018). Phytoplankton response to polystyrene microplastics: Perspective from an entire growth period. *Chemosphere, 208*, 59–68.

Mao, X., Xu, Y., Cheng, Z., Yang, Y., Guan, Z., Jiang, L., & Tang, K. (2022). The impact of microplastic pollution on ecological environment: A review. *Frontiers in Bioscience-Landmark, 27*(2), 46.

Martin, J., Lusher, A., Thompson, R. C., & Morley, A. (2017). The deposition and accumulation of microplastics in marine sediments and bottom water from the Irish continental shelf. *Scientific Reports, 7*(1), 10772.

Mateos-Cárdenas, A., van Pelt, F. N., O'Halloran, J., & Jansen, M. A. (2021). Adsorption, uptake and toxicity of micro-and nanoplastics: Effects on terrestrial plants and aquatic macrophytes. *Environmental Pollution, 284*, 117183.

McIlwraith, H. K., Lin, J., Erdle, L. M., Mallos, N., Diamond, M. L., & Rochman, C. M. (2019). Capturing microfibers–marketed technologies reduce microfiber emissions from washing machines. *Marine Pollution Bulletin, 139*, 40–45.

Medyńska-Juraszek, A., & Jadhav, B. (2022). Influence of different microplastic forms on pH and mobility of Cu2+ and Pb2+ in soil. *Molecules, 27*(5), 1744.

Mondal, N. K., Kundu, S., Debnath, P., Mondal, A., & Sen, K. (2022). Effects of polyethylene terephthalate microplastic on germination, biochemistry and phytotoxicity of Cicer arietinum L. and cytotoxicity study on Allium cepa L. *Environmental Toxicology and Pharmacology, 94*, 103908.

Nelms, S. E., Galloway, T. S., Godley, B. J., Jarvis, D. S., & Lindeque, P. K. (2018). Investigating microplastic trophic transfer in marine top predators. *Environmental Pollution, 238*, 999–1007.

Nizzetto, L., Futter, M., & Langaas, S. (2016). Are agricultural soils dumps for microplastics of urban origin? *Environmental Science & Technology, 50*, 10777–10779.

Nobre, C. R., Moreno, B. B. M. B., Alves, A. V., de Souza Abessa, D. M., Cesar, A., Choueri, R. B., Choueri, P. K. G., & Pereira, C. D. S. (2022). Can leachates of environmentally relevant concentrations of microplastics in marine sediments affect the reproduction of an epibenthic copepod? *Ecotoxicology and Environmental Contamination, 17*(1), 106–113.

Pantos, O. (2022). Microplastics: Impacts on corals and other reef organisms. *Emerging Topics in Life Sciences, 6*(1), 81–93.

Peng, L., Fu, D., Qi, H., Lan, C. Q., Yu, H., & Ge, C. (2020). Micro-and nano-plastics in marine environment: Source, distribution and threats—A review. *Science of the Total Environment, 698*, 134254.

Porcino, N., Bottari, T., & Mancuso, M. (2022). Is wild marine biota affected by microplastics? *Animals, 13*(1), 147.

Prata, J. C., Reis, V., da Costa, J. P., Mouneyrac, C., Duarte, A. C., & Rocha-Santos, T. (2021). Contamination issues as a challenge in quality control and quality assurance in microplastics analytics. *Journal of Hazardous Materials, 403*, 123660.

Reichert, J., Schellenberg, J., Schubert, P., & Wilke, T. (2018). Responses of reef building corals to microplastic exposure. *Environmental Pollution, 237*, 955–960.

Rillig, M. C., Lehmann, A., de Souza Machado, A. A., & Yang, G. (2019). Microplastic effects on plants. *New Phytologist, 223*(3), 1066–1070.

Rong, H., Li, M., He, L., Zhang, M., Hsieh, L., Wang, S., Han, P., & Tong, M. (2022). Transport and deposition behaviors of microplastics in porous media: Co-impacts of N fertilizers and humic acid. *Journal of Hazardous Materials, 426*, 127787.

Salam, M., Zheng, H., Liu, Y., Zaib, A., Rehman, S. A. U., Riaz, N., Eliw, M., Hayat, F., Li, H., & Wang, F. (2023). Effects of micro (nano) plastics on soil nutrient cycling: State of the knowledge. *Journal of Environmental Management, 344*, 118437.

Sangkham, S., Faikhaw, O., Munkong, N., Sakunkoo, P., Arunlertaree, C., Chavali, M., Mousazadeh, M., & Tiwari, A. (2022). A review on microplastics and nanoplastics in the environment: Their occurrence, exposure routes, toxic studies, and potential effects on human health. *Marine Pollution Bulletin, 181*, 113832.

Sarkar, D. J., Sarkar, S. D., Das, B. K., Manna, R. K., Behera, B. K., & Samanta, S. (2019). Spatial distribution of meso and microplastics in the sediments of river Ganga at eastern India. *Science of the Total Environment, 694*, 133712.

Scheurer, M., & Bigalke, M. (2018). Microplastics in Swiss floodplain soils. *Environmental Science & Technology, 52*(6), 3591–3598.

Strungaru, S.-A., Jijie, R., Nicoara, M., Plavan, G., & Faggio, C. (2019). Micro-(nano) plastics in freshwater ecosystems: Abundance, toxicological impact and quantification methodology. *TrAC Trends in Analytical Chemistry, 110*, 116–128.

Sun, X.-D., Yuan, X.-Z., Jia, Y., Feng, L.-J., Zhu, F.-P., Dong, S.-S., Liu, J., Kong, X., Tian, H., & Duan, J.-L. (2020). Differentially charged nanoplastics demonstrate distinct accumulation in Arabidopsis thaliana. *Nature Nanotechnology, 15*(9), 755–760.

Sun, H., Lei, C., Xu, J., & Li, R. (2021). Foliar uptake and leaf-to-root translocation of nanoplastics with different coating charge in maize plants. *Journal of Hazardous Materials, 416*, 125854.

Vogelsang, C., Lusher, A., Dadkhah, M. E., Sundvor, I., Umar, M., Ranneklev, S. B., Eidsvoll, D., & Meland, S. (2019). *Microplastics in road dust–characteristics, pathways and measures*. NIVA-Rapport.

Wagner, S., & Reemtsma, T. (2019). Things we know and don't know about nanoplastic in the environment. *Nature Nanotechnology, 14*(4), 300–301.

Waldman, W. R., & Rillig, M. C. (2020). Microplastic research should embrace the complexity of secondary particles. *Environmental Science & Technology, 54*, 7751–7753.

Walker, T. R., & Fequet, L. (2023). Current trends of unsustainable plastic production and micro (nano) plastic pollution. *TrAC Trends in Analytical Chemistry, 116984*.

Walkinshaw, C., Lindeque, P. K., Thompson, R., Tolhurst, T., & Cole, M. (2020). Microplastics and seafood: Lower trophic organisms at highest risk of contamination. *Ecotoxicology and Environmental Safety, 190*, 110066.

Wang, X. (2022). Managing land carrying capacity: Key to achieving sustainable production systems for food security. *Land, 11*(4), 484.

Wang, X.-M., Wang, X.-K., Su, Y.-B., & Zhang, H.-X. (2019). Land pavement depresses photosynthesis in urban trees especially under drought stress. *Science of the Total Environment, 653*, 120–130.

Wang, L., Wu, W.-M., Bolan, N. S., Tsang, D. C., Li, Y., Qin, M., & Hou, D. (2021). Environmental fate, toxicity and risk management strategies of nanoplastics in the environment: Current status and future perspectives. *Journal of Hazardous Materials, 401*, 123415.

Xia, W., Rao, Q., Deng, X., Chen, J., & Xie, P. (2020). Rainfall is a significant environmental factor of microplastic pollution in inland waters. *Science of the Total Environment, 732*, 139065.

Xu, B., Liu, F., Cryder, Z., Huang, D., Lu, Z., He, Y., Wang, H., Lu, Z., Brookes, P. C., & Tang, C. (2020). Microplastics in the soil environment: Occurrence, risks, interactions and fate – A review. *Critical Reviews in Environmental Science and Technology, 50*(21), 2175–2222.

Yadav, V., Sherly, M., Ranjan, P., Tinoco, R. O., Boldrin, A., Damgaard, A., & Laurent, A. (2020). Framework for quantifying environmental losses of plastics from landfills. *Resources, Conservation and Recycling, 161*, 104914.

Yang, B., Liu, J., Zhao, X., & Zheng, S. (2021). Evaporation and cracked soda soil improved by fly ash from recycled materials. *Land Degradation & Development, 32*(9), 2823–2832.

Yin, L., Wen, X., Huang, D., Du, C., Deng, R., Zhou, Z., Tao, J., Li, R., Zhou, W., & Wang, Z. (2021). Interactions between microplastics/nanoplastics and vascular plants. *Environmental Pollution, 290*, 117999.

Yu, H., Fan, P., Hou, J., Dang, Q., Cui, D., Xi, B., & Tan, W. (2020). Inhibitory effect of microplastics on soil extracellular enzymatic activities by changing soil properties and direct adsorption: An investigation at the aggregate-fraction level. *Environmental Pollution, 267*, 115544.

Yu, J., Adingo, S., Liu, X., Li, X., Sun, J., & Zhang, X. (2022). Micro plastics in soil ecosystem – A review of sources, fate, and ecological impact. *Plant, Soil and Environment, 68*(1), 1–17.

Yuan, Z., Nag, R., & Cummins, E. (2022). Ranking of potential hazards from microplastics polymers in the marine environment. *Journal of Hazardous Materials, 429*, 128399.

Zantis, L. J., Borchi, C., Vijver, M. G., Peijnenburg, W., Di Lonardo, S., & Bosker, T. (2023). Nano-and microplastics commonly cause adverse impacts on plants at environmentally relevant levels: A systematic review. *Science of the Total Environment, 867*, 161211.

Zhang, S., Wu, H., & Hou, J. (2023). Progress on the effects of microplastics on aquatic crustaceans: A review. *International Journal of Molecular Sciences, 24*(6), 5523.

Zheng, W., Morris, E. K., Lehmann, A., & Rillig, M. C. (2016). Interplay of soil water repellency, soil aggregation and organic carbon. A meta-analysis. *Geoderma, 283*, 39–47.

Zhou, N., Price, L., Yande, D., Creyts, J., Khanna, N., Fridley, D., Lu, H., Feng, W., Liu, X., & Hasanbeigi, A. (2019). A roadmap for China to peak carbon dioxide emissions and achieve a 20% share of non-fossil fuels in primary energy by 2030. *Applied Energy, 239*, 793–819.

Part II
Micro-Nano-plastics in the Soil Systems

Beneath the Surface: Unraveling the Impact of Micro and Nanoplastics on Plant Performance

Shiamita Kusuma Dewi, Sartaj Ahmad Bhat, Yongfen Wei, and Fusheng Li

1 Introduction

Plastic has become an essential material and inseparable in our daily lives. Plastic production reached 311 million tons (MT) in 2014, a 20-fold increase from 1964 (Agenda, 2016). By 2015, 6300 million tons of plastic waste had been produced. Only 9% was recycled, 12% was incinerated, and 79% ended up in landfills or the environment (Geyer et al., 2017). Furthermore, during the coronavirus disease 2019 (COVID-19) pandemic, the production of plastic waste has significantly risen. The increase in plastic waste is attributed to the widespread use of personal protective equipment like gloves and masks, as well as the resurgence of plastic shopping bags to avoid potential contamination. Moreover, the lockdown has caused a spike in e-commerce, leading to a substantial demand for materials used in packaging and containment (Gorrasi et al., 2021). Medical waste has increased by 370%, and there has been up to a 40% increase in demand for packaging plastic (Patrício Silva et al., 2021).

Plastic waste is discharged across diverse environments, encompassing both marine and terrestrial ecosystems. Since 1950, nearly half of the planet's plastic waste has been disposed of in landfills or released into the environment, while only 9% of utilized plastic undergoes appropriate recycling (Woldemar D'ambrières, 2019). Annually, a range of 4–12 million metric tons of plastic waste is discharged into the oceans (Jambeck et al., 2015). These plastic wastes can undergo fragmentation due to various weather-related factors, such as sunlight (UV radiation), temperature fluctuations, wind, and moisture, which can contribute to the degradation

S. K. Dewi (✉)
United Graduate School of Agricultural Science, Gifu University, Gifu, Japan

S. A. Bhat · Y. Wei · F. Li
River Basin Research Center, Gifu University, Gifu, Japan

S. A. Bhat et al. (eds.), *Management of Micro and Nano-plastics in Soil and Biosolids*,
https://doi.org/10.1007/978-3-031-51967-3_6

and breakdown of plastic waste into smaller particles called microplastics (MPs) and nanoplastics (NPs), in the environment. Microplastics are generally defined as plastic particles with a size of less than 5 mm (Barnes et al., 2009), while nanoplastics are even smaller, typically ranging from 1 nm to 1000 nm in size (Gigault et al., 2018). Due to their small size and the toxic additives and chemicals they absorb, the ingestion of MPs can cause harm to organisms by reducing fecundity, affecting growth rate, and posing a threat to the ecosystem (Lu et al., 2018; Prata et al., 2020). These plastics have saturated the environment. They have invaded plants and animals, and now they are even finding their way into the human body.

MNPs have an impact on various ecosystems. Due to the accumulation of human-generated waste, rising amounts of microplastics are present in water bodies such as streams, reservoirs, seas, and oceans. They are not only harming the marine environment but also affecting marine life. Aquatic organisms, including sea cucumbers, fish, copepods, turtles, amphipods, and birds, ingest these microplastics (Guzzetti et al., 2018; Ivleva et al., 2017; Xu et al., 2020), leading to various adverse effects such as infertility, growth retardation, injuries, and blockage of body tracts (Cole et al., 2015; Ogonowski et al., 2016; Sussarellu et al., 2016). Additionally, microplastics can act as vectors for carrying pollutants, such as hydrocarbons, dyes, and heavy metals, further increasing ocean toxicity (Gola et al., 2021). Microplastics in the marine environment pose a significant threat to the health of aquatic organisms and the overall ecosystem.

Microplastic pollution in terrestrial ecosystems is a growing concern. Studies have indicated that the presence of microplastics can alter soil properties, including structure, water retention, bulk density, and nutrient content (De Souza Machado et al., 2018b, 2019; Liu et al., 2017). These alterations have the potential to directly impact plant performance through changes in root characteristics, growth performance, and nutrient absorption (de Souza Machado et al., 2019; Qi et al., 2018; Rillig et al., 2019). In contrast to research focused on marine ecosystems, there is a limited number of studies examining the influence of microplastics on soil ecosystems, particularly regarding their effects on plant performance. This chapter will highlight the specific impact of this emerging pollutant on plant life.

2 Plastic in the Soil: A Silent Intruder

In general, MPs found in the environment originate from two main sources. Primary sources are released directly into nature as small particles, including plastic pellets, paint, personal care products containing microbeads, sewage sludge, plastic running tracks in schools, washing wastewater, artificial turf, rubber roads in cities, and vehicle tire wear. Meanwhile, secondary sources originate from larger pieces of plastic such as plastic bags, bottles, agricultural film, fishing gear, and other large plastic debris, which are fragmented into smaller pieces and form MPs and NPs in nature.

However, the main source of micro-nanoplastics (MNPs) contamination in agricultural farmland is believed to be the use of microplastic-containing fertilizers, such as sewage sludge (Nizzetto et al., 2016) and bio-waste (Weithmann et al., 2018) processed from households or commercial bindings. This sewage sludge can introduce significant amounts of microplastics to the soil, with an estimated annual addition of 63,000 to 430,000 tons in European farmlands (Nizzetto et al., 2016). Polyethylene (PE) is the most common type of plastic found on agricultural farmland in southeast Germany, accounting for 67.90% of all microplastic pieces and 62.50% of microplastic particles (Piehl et al., 2018). In the suburbs of Shanghai and Shandong, China, the most common type of plastic identified is polypropylene (PP), accounting for approximately 50.51% and 45.0% (Liu et al., 2018) (Yu et al., 2021).

The plastic polymer can be degraded under biotic and abiotic conditions (Shah et al., 2008). Abiotic degradation of plastics refers to changes in the physical or chemical properties caused by abiotic factors such as light, water, temperature, air, and mechanical forces. In general, abiotic degradation is expected to precede biodegradation due to the low bioavailability of plastics (Andrady et al., 2003). While biotic degradation of plastics refers to the degradation of plastics caused by living organisms. Organisms can break down plastics physically by biting, chewing, or digestive fragmentation or biologically by biochemical processes (Cau et al., 2020; Danso et al., 2019; Dawson et al., 2018 2018; Jang et al., 2018; Mateos-Cárdenas et al., 2020; Porter et al., 2019). In the natural environment, plastics undergo degradation at varying rates, generally adhering to this sequence: Polyethylene (PE) (approximately 1000 years) > Polystyrene (PS) (approximately 500 years) > Polyethylene terephthalate (PET) (approximately 450 years) > Polyvinyl chloride (PVC) (approximately 100 years) > Polypropylene (PP) (20–30 years) (Zhang et al., 2021b). Due to their chemical structure, lack of biodegradability, high physical durability, continuous production and usage, and slow natural degradation, plastics tend to accumulate over time in various environments, including soil, water bodies, and ecosystems.

3 Uptake and Transport: Plastic's Covert Journey into Plants

The impact of MNPs on terrestrial plants has not been well explored yet. Several pathways for plant–plastic–particle interaction are proposed. First, MP and NP particles larger than 200 nm may adhere to plant tissues externally (Mateos-Cárdenas et al., 2021). In plants, the cell wall is the primary physical barrier defense against foreign substances, including MNPs (Milewska-Hendel et al., 2019), which blocks particles larger than the cell wall pores from entering. When MNPs adhere to a surface, they can be either adsorbed or entrapped in its structure. However, a study showed that particles larger than cell wall pores can penetrate through cracks in the wall (Li et al., 2020).

The potential pathways for NPs to enter plant cells include plasmodesmata channels, aquaporins, ion transporters, and endocytosis (Khan et al., 2019; Zhou et al., 2021). MNPs can be exposed through two pathways: foliar foliage and roots. NPs absorbed by root hairs or tips are carried to other organs, like stems and leaves (Sun et al., 2020; Abdolahpur Monikh et al., 2022; Giorgetti et al., 2020), which are primarily influenced by the pull of transpiration (Li et al., 2020). The mechanism of NP uptake by plant roots involves the transport of NPs over the plasma membrane of root cells, followed by their loading and translocation through the xylem pathway (Lian et al., 2020).

For foliar application, it was discovered that NPs attached to leaves can be absorbed through the opening and movement of the stomata (Lian et al., 2021; Sun et al., 2021). Then, they are transported by phloem (Lian et al., 2021). The positively charged NPs are more associated with leaf surfaces than the negatively charged NPs, likely due to electrostatic attraction to the negatively charged cell wall. Once taken up by the leaves, NPs can be translocated to the roots through the plant's vascular bundle (Sun et al., 2021).

An uptake, translocation, and accumulation depend on several factors, including size, shape, and charge of NPs, the movement of aquaporin, changes in transpiration pull in plants, exposure time, presence of other substances, kind of media growth, and plant species (Ekner-Grzyb et al., 2022).

4 Beyond the Roots: Systemic Effects on Plant Health

MNPs can negatively impact plant health due to their presence in the soil and water ecosystems. While research on this topic is ongoing and evolving, here are some ways in which micro and nanoplastics can impact different aspects of plant health, including growth, nutrient uptake, and reproduction:

(1) Disturbing photosynthesis and respiration: The NPs' surface charge can also influence their physiological effects on plants. Positively charged NPs have been found to inhibit photosynthesis more and stimulate antioxidant systems stronger than negatively charged NPs (Sun et al., 2021). NPs can activate antioxidant levels in plants and reduce photosynthesis. The toxicity of NPs to plants may result in the activation of anti-stress mechanisms and an increase in the generation of reactive oxygen species (ROS) (Yang & Gao, 2022).
(2) Hindering plant growth: Negatively charged NPs had a stronger influence on root development, but positively charged NPs generated more ROS accumulation and hindered plant growth and seedling development more strongly (Sun et al., 2020). NPs also decreased above-ground fresh weights, shoot/root ratio (S/R) (Lian et al., 2020), leaf area, plant height, and dry weight (Lian et al., 2021; Yang & Gao, 2022), and inhibited seedling growth.
(3) Disrupting balance of nutrient uptake: A study shows a decrease in the content of Ca, Fe, and Mg in cucumber in 14 days of treatment with PSNPs (Li

et al., 2020 2020), while polymethyl methacrylate (PMMA) NP resulted in reduced content of P, Mn, Fe, Zn, Ca, Mo, Cu, K, and B (Yildiztugay et al., 2022).

(4) Adverse effects on genetic material: The formation of mitosis abnormalities results in an increase in micronuclei (extra nuclei that can form within a cell as a result of lagging chromosomes or chromosome fragments caused by mitotic errors or DNA damage) (Cimini et al., 2003, 2004; Fenech et al., 2011), C-metaphases (a type of chromosome preparation that is used to analyze structural chromosomal abnormalities and rearrangements), and formation of sticky chromosomes, and slowed cell division (Giorgetti et al., 2020; Jiang et al., 2019; Spanò et al., 2022). For example, a study by Giorgetti et al. (2020) found an increase in micronuclei formation in the root meristems of *Allium cepa.* Similar findings in *Vicia faba* found that 5 mm and 100 nm polystyrene nanoplastic (PSNP)s increased the frequency of micronuclei (MNs) in the cells (Jiang et al., 2019). Meanwhile, PSNPs-treated meristems of rice (*Oryza sativa* L.) showed higher percentages of total metaphases and anaphases/telophases compared to the control. This increase in metaphase and anaphase anomalies, particularly C-metaphases, suggests a blockage or slowing of the mitotic cycle, possibly due to mitotic spindle failure or disruption (Spanò et al., 2022).

(5) Oxidative response: ROS in plants can originate from endogenous or exogenous sources. Endogenous means ROS produced within the plant cell. Major organelles producing ROS include chloroplasts, peroxisomes, and mitochondria. Cell walls and plasma membranes also contain peroxidases and amine oxidases, respectively, leading to ROS generation in case of environmental stress. On the other hand, exogenous is defined as the production of ROS due to some effect caused by abiotic or biotic stresses such as heat, cold, salinity, water deficiency, insects, pathogens, heavy metals, and contaminants.

Since ROS are very harmful, cells pose an ROS scavenging system. Therefore, they scavenge very rapidly. It has two systems: antioxidant enzymes and nonenzymatic antioxidants. Typically, cells maintain a balance between ROS and antioxidant systems. However, during stress conditions, an imbalance arises between ROS and antioxidants. This imbalance signifies that the cell is generating elevated levels of ROS that cannot be effectively scavenged by the antioxidant systems. The excessive accumulation of ROS can lead to oxidative stress and damage to cellular components. Some studies showed that NP exposure induces an oxidative response in plants. For example, the exposure of PS nanoparticles at a concentration of 1000 mg/L resulted in a significant increase of H_2O_2 during onion seedling (Giorgetti et al., 2020) and in Arabidopsis (PS-NH_2) induced stronger H_2O_2 accumulation than negative charge NP (Sun et al., 2020).

One of the major targets of ROS-induced damage in plants is DNA. ROS can cause DNA strand breaks, base modifications, and cross-linking, leading to mutations and genomic instability. ROS can also damage proteins and lipids in plant cells and oxidize amino acid residues in proteins, resulting in protein misfolding and function loss. This can harm plant growth development, decreasing crop productivity (Mirza-Aghazadeh-Attari et al., 2020).

5 Plastic–Microbe–Plant Interactions: Unraveling the Complex Web

5.1 The Intricate Relationship Between Soil Microbes, Microplastics, and Plants

Soil microorganisms have an essential role in nutrient cycling, organic matter decomposition, nitrogen fixation, and soil structure improvement. Various environmental factors, like soil aggregates, pore spatial structure, and hydrological characteristics, play a role in shaping the function and structure of microbial communities. These factors can further influence the rate of microbial metabolism and the decomposition of organic matter (de Souza Machado et al., 2018a, b).

MNPs influence soil physicochemical properties because they can be integrated into soil aggregate, resulting in a change in soil aggregate (Horton et al., 2017), water holding capacity, and bulk density (De Souza MacHado et al., 2018b). These changes affect the function and structure of the microbial community, as shown in Table 1. In contrast, adding MNPs to the soil has been found to positively affect certain soil chemical properties. A study by Liu et al. (2017) observed that DOC content increased significantly in soils with high levels of microplastic addition compared to control soils, which indicates that microplastics aid the release of soil nutrients into the soil solution and the accumulation of dissolved organic matter. Furthermore, the addition of microplastics significantly increased the total dissolved nitrogen (TDN) content in soils from day 7 to 30 compared to control soils. This suggests that microplastics contribute to the availability of nitrogen in the soil. These findings highlight the potential role of microplastics in nutrient cycling and soil fertility.

Microplastics have the potential to impact the growth and development of plants through various mechanisms, including changes in soil structure, nutrient immobilization, contaminant transport/adsorption, direct toxicity, and alterations in the soil microbial community and root symbionts (Rillig et al., 2019), as shown in Table 1. Conceptually, MNPs can be seen as a soil physical contaminant (de Souza Machado et al., 2018b), which changes soil physical properties. This may lead to a direct decrease in root penetration resistance and a reduction in the availability of soil pore water as a consequence of soil drying and elevated evaporation. Ultimately, this could have an adverse effect on plant performance.

MNPs have hydrophobic surfaces that can adsorb harmful substances, which may harm plant roots or their symbionts. Alternatively, the adsorption of contaminants to the surface MNPs may reduce their availability to soil organisms and plants, resulting in a protective effect (Rillig et al., 2019). Meanwhile, for direct toxicity, nano-sized particles could enter the plant root, transfer to the aerial part, potentially cause damage to cells and molecules, and produce oxidative stress (Navarro et al., 2008). The evidence of the studies of nanoplastics taken up by plants is shown in Table 2.

Table 1 Impact of MNPs on soil properties affecting the structure and function of the microbial community

Polymer type	Size	Duration	Design experiment	Impact on soil properties	Reference	Impact on soil microbes
PE	2 mm, 5 mm, 10 mm	40 h	Water evaporation experiment	Increased the rate of evaporation Reduced the effective infiltration area	Wan et al. (2019)	Changes in soil porosity and water cycling alter oxygen flux, influencing the abundance of aerobic and anaerobic microbes in the soil (Rubol et al., 2013) Changes in soil aggregate, a critical process that governs soil structure in shaping the habitat of soil microbes (Rillig & Lehmann, 2020)
Polyacrylic fiber, polyamide beads, polyester fibers, PE	Diameter 18 μm, 15–20 μm, 8 μm, 643 μm	5 weeks	Soil-MP mixtures to pots and partially burying them in an experimental garden	Reduced soil bulk density Altered soil moisture and evapotranspiration	de Souza Machado et al. (2018b)	
PS, PE	0.47–0.53, 27–32, 250–300 μm	–	Soil-MP mixtures and observe the springtails' behavior	Filled up the space of soil pores, affecting soil porosity Changed structure and disrupted the formation and stability of soil aggregate	Kim & An (2019)	
PP	<250 μm	30 days	Incubation experiment	Decreased biodegradable dissolved organic carbon (DOC) Reduced the content of humic-like substances and fulvic acid	H. Liu et al. (2019)	Reduced biodegradable DOC reduces soil microbial carbon and nitrogen, enzyme activity, and microbial diversity (H. Liu et al., 2019) The decomposition rate of humic-like materials, rather than MP degradation, may affect soil nutrient cycling and microbial activity (Zhang et al., 2021a, b)
PP	<180 mm	30 days	Incubation experiment	Enhanced the activity of fluorescein diacetate hydrolase enzyme Increased the activity of phenol oxidase (PO) Increased DOC and TDN content	H. Liu et al. (2017)	

Table 2 Recent research for NP uptake by plant

Polymer type	Size and dose	Plant species	Location MNPs in plant	Reference
Root uptake				
PS	0.03 μm (10–100 mg/kg)	Mung bean (*Vigna radiata*)	Plant vasculature and leaf	Chae & An (2020)
PS, PVC	<1 μm, 250 nm (100 mg/kg)	Lettuce (*Lactuca sativa* L.)	Root and leaves	Abdolahpur Monikh et al. (2022)
PS-NH_2, PS-COOH	0.2 μm (10–100 mg/L and 0.3–1.0 g/kg)	Thale cress (*Arabidopsis thaliana*)	Root	Sun et al. (2020)
Foliar uptake				
PS-NH_2, PS-COOH	22.0 ± 1.5 nm, 24.0 ± 2.2 nm	Maize (*Zea mays* L.)	Leaves, stem, and root	Sun et al. (2021)
PS	93.6 nm (0.1 mg/L and 1 mg/L)	Lettuce (*Lactuca sativa* L.)	Leaves	Lian et al. (2021)

5.2 *How Microorganisms Can Break Down Plastics or Facilitate Their Accumulation in Plant Tissues*

Plastic waste has two fates in the environment: accumulation or breaking-down (Shen et al., 2020), and microorganisms have important roles in those fates. The fragmentation of plastic by microorganisms contributes to the accumulation of MNPs in soil and plants. The process of breaking down polymers by microorganisms can involve several steps, such as biodeterioration, biofragmentation, mineralization, and assimilation (Zeenat et al., 2021). First, the biodeterioration of plastic products. This process transforms the surface of plastics and alters their physical, chemical, and mechanical properties. Microbial degradation of plastic is a difficult task due to the lack of hydrolysable groups in the carbon-carbon backbone. In this process, environmental conditions also influence changes in polymer properties, including light, heat, stress, hydrolysis, and oxidation. When exposed to UV light, the carbonyl group present in polymers becomes vulnerable to attack by microbial enzymes (Novotný et al., 2018). The following step involves biofragmentation, where the plastic polymer is broken down by enzymatic action. MNPs makeup 70–80% of plastic in the environment, formed by macroplastic fragmentation due to shear forces (Auta et al., 2017), while primary MPs contribute only 15–30% (Mariano et al., 2021). These two steps are the potentially involved mechanisms for how microorganisms facilitate the accumulation of MNPs in soil and plants.

5.3 *The Potential for Microbial Mediation of Plastic-Related Plant Effects*

Microbes play important roles in fully biodegrading waste into CO_2, H_2O, and CH_4 through mineralization and assimilation mechanisms. Plastic polymers formed during biofragmentation enter microbial cells via cell membranes. The monomers in cells undergo oxidation to provide energy, which is then used for biomass synthesis (*Kale*, 2015; Lucas et al., 2008). During assimilation, atoms are incorporated into microbial cells to facilitate complete breakdown. Secondary metabolites are transported outside the cells or transferred to other microbes for further degradation and use. During the degradation of secondary metabolites (mineralization), the oxidized products, i.e., N_2, CH_4, CO_2, and H_2O, are released (Krzan et al., 2006; Lucas et al., 2008). Suppose microbial activity leads to this effective degradation of plastic materials. In that case, it might help to reduce the overall plastic burden in the environment, which could indirectly benefit plants by decreasing the negative impacts associated with plastic pollution.

6 Future Perspective: Mitigation and Solutions

Researchers have been exploring ways to minimize the effects of MNPs on plants. Here are some of the present strategies being considered in this evolving field:

1. Advance the research and development of biodegradable plastics (BPs) for effective and environmentally friendly usage.

Using biodegradable plastics that break down rapidly can reduce microplastic persistence and their impact on plants and ecosystems. According to some researchers, biodegradable plastics can degrade without having adverse effects due to their persistence (Haider et al., 2019). Nowadays, BPs have proven to be effective materials in various industrial and environmental protection programs. Yet, it is widely acknowledged that biodegradable plastics (BPs) do not presently serve as a viable replacement for conventional plastics (Rujnić-Sokele & Pilipović, 2017). There are still uncertainties about whether BPs can be a viable solution for addressing global plastic pollution. There are several scientific shreds of evidence that need to be discussed. First, the production of BPs is on the rise; it is still far from being a solution to the global plastic pollution problem. According to Rujnić-Sokele et al. (2017), cost is the main factor. Plastic production and use are too widespread worldwide due to its low cost and mature technology. According to the analysis, the current process produces BPs that are 3–10 times more costly than conventional plastics such as PP and PE (Luyt & Malik, 2019). In addition, certain BPs may not perform as well as conventional plastics (Shahlari & Lee, 2012). Therefore, the application remains limited due to its low plasticity and impact strength. Second, even though BPs are expected to fully degrade in the environment, they require specific environmental conditions that are not always reliable under natural conditions (Nazareth et al., 2019), which are controlled by the microbial community, temperature, moisture levels, and climate conditions that may vary from place to place. Third, BPs are a type of plastic that inevitably has negative environmental impacts during its application. Like conventional plastics, BPs also have the fate of accumulating or breaking up in the environment. The above studies have shown that the accumulation in the environment has an ecotoxicological effect on plants and other organisms. Colorants and additives utilized in plastic manufacturing may also be released potentially during degradation, adversely affecting the ecosystem. For example, lipophilic phthalate esters (PAEs) are additives present in plastic film that can harm soil microbial activity, diversity, structure, and metabolic activity by destroying cell membrane fluidity (Xie et al., 2010). Consequently, it indicates that BPs are not entirely safe for the environment. The development and research for effective and environmentally friendly BPs is required.

2. Application of soil management and remediation

Soil management by adding soil amendment can enhance soil's physical and chemical properties (Hansen et al., 2016; Spokas & Dc Reicosky, 2009), potentially reducing the negative impact of MNPs on plants. For instance, according to some researchers, Wang et al. (2020), biochar filters can remove and immobilize microplastics with a 10-μm diameter with a capacity above 95% in leaching column tests (Wang et al., 2020). Furthermore, adding biochar to soil eliminates PET's inhibitory effect on soil bacteria diversity (Han et al., 2022). This suggests that adding biochar can potentially mitigate the impact of the phytotoxicity of microplastics.

Plant growth-promoting rhizobacteria (PGPR) is a specific group of microbiomes in the crop rhizosphere and positively affects plant growth and biomass accumulation (Oleńska et al., 2020). Some PGPRs involved in resisting environmental stress, such as *Rhizobium radiobacter* (formerly *Agrobacterium tumefaciens*), reduce heavy metal stress's effects (Jian et al., 2019) and *Proteus* spp. strain DSP1, *Pseudomonas* spp. strain DSP17, *Ensifermelilotistrain* strain RhOL6, and *E. meliloti* strain RhOL8 were reported to alleviate heat stress and heavy metals in *Medicago sativa* (Raklami et al., 2019). For new stressors in the environment, such as MNPs, PGPR has the potential to enhance beneficial soil microbial communities that can contribute to improved soil health and nutrient cycling, which might help plants cope with microplastic stress.

3. Promote educational and regulatory initiatives

Solving the global plastic pollution problem requires a dual approach: changing human behavioral awareness and implementing effective and promising strategies (Shen et al., 2020). In regions with underdeveloped waste management infrastructure, rural areas often face a widespread problem of improperly handling plastics, encompassing activities such as littering, open burning, and open dumping. Raising awareness about the impacts of MNPs on plants and ecosystems can encourage responsible plastic use and disposal.

It is the responsibility of society to manage plastic rather than allow it to be released into the environment without any regulation. In Africa, 20 countries have implemented bag bans and taxes to address the environmental and health threats posed by plastic bags (J. Jambeck et al., 2018). While Thailand has implemented the National 3R Strategy and the National Master Plan for Waste Management (2016–2021) to ensure proper management of solid waste. The goal of the 3R strategy is to reduce, reuse, and recycle plastic waste. The strategy includes measures such as promoting eco-packaging design, developing material flow for plastic containers and packaging inventory, implementing the 3Rs strategy for plastic debris management, and promoting education for stakeholders in the field of plastics and alternative materials (Wichai-utcha & Chavalparit, 2019). Furthermore, in 2019, the Ministry of Housing and Local Government (MHLG) in Malaysia initiated a waste-to-energy facility primarily focused on plastic waste and proposed the construction of incineration plants in each state across Malaysia. To some extent, providing waste management regulation can positively restrain people from throwing plastic waste at random and piling up in the environment.

7 Conclusion: The Path Ahead

The unraveling of the impact of MNPs on plant performance has been broadly discussed. MNPs can impact plant health, including growth, nutrient uptake, and reproduction. Several studies reported that MNPs disturb photosynthesis and respiration, disrupt the balance of nutrient uptake, and cause adverse effects on genetic material and oxidative response, ultimately hindering plant growth. The physicochemical characteristics of soil, such as changes in soil aggregates, bulk density, and water-holding capacity, have been observed to be impacted by the presence of MNPs. These changes can ultimately affect the function and structure of the microbial community and soil nutrient cycling. Researchers have been investigating methods to reduce the impact of MNPs on plants, such as the development of biodegradable plastic, the application of soil management and remediation, and raising awareness about the impacts of MNPs on plants and ecosystems. Further research must be conducted to verify the assumption by providing empirical data about the potential mitigation by soil amendments and PGPR.

References

Abdolahpur Monikh, F., Holm, S., Kortet, R., Bandekar, M., Kekäläinen, J., Koistinen, A., Leskinen, J. T. T., Akkanen, J., Huuskonen, H., Valtonen, A., Dupuis, L., Peijnenburg, W., Lynch, I., Valsami-Jones, E., & Kukkonen, J. V. K. (2022). Quantifying the trophic transfer of sub-micron plastics in an assembled food chain. *Nano Today, 46*. https://doi.org/10.1016/j.nantod.2022.101611

Agenda, I. (2016). *The new plastics economy rethinking the future of plastics* (p. 36). The World Economic Forum.

Andrady, A. L., Hamid, H. S., & Torikai, A. (2003). *Effects of climate change and UV-B on materials*. https://doi.org/10.1039/b211085g

Auta, H. S., Emenike, C. U., & Fauziah, S. H. (2017). Distribution and importance of microplastics in the marine environmentA review of the sources, fate, effects, and potential solutions. In *Environment international* (Vol. 102, pp. 165–176). Elsevier Ltd.. https://doi.org/10.1016/j.envint.2017.02.013

Barnes, D. K. A., Galgani, F., Thompson, R. C., & Barlaz, M. (2009). Accumulation and fragmentation of plastic debris in global environments. *Philosophical Transactions of the Royal Society B: Biological Sciences, 364*(1526), 1985–1998. https://doi.org/10.1098/rstb.2008.0205

Cau, A., Avio, C. G., Dessì, C., Moccia, D., Pusceddu, A., Regoli, F., Cannas, R., & Follesa, M. C. (2020). Benthic crustacean digestion can modulate the environmental fate of microplastics in the Deep Sea. *Environmental Science and Technology, 54*(8), 4886–4892. https://doi.org/10.1021/acs.est.9b07705

Chae, Y., & An, Y.-J. (2020). Nanoplastic ingestion induces behavioral disorders in terrestrial snails: Trophic transfer effects via vascular plants. *Environmental Science: Nano, 7*(3), 975–983. https://doi.org/10.1039/C9EN01335K

Cimini, D., Moree, B., Canman, J. C., & Salmon, E. D. (2003). Merotelic kinetochore orientation occurs frequently during early mitosis in mammalian tissue cells and error correction is achieved by two different mechanisms. In. *Journal of Cell Science, 116*(20), 4213–4225. https://doi.org/10.1242/jcs.00716

Cimini, D., Cameron, L. A., & Salmon, E. D. (2004). Anaphase spindle mechanics prevent Missegregation of Merotelically oriented chromosomes of kinetochores and spindle poles. We identified pro-metaphase/metaphase cells possessing one or more merotelic kinetochores by taking a through-focus image. *Current Biology, 14*, 2149–2155.

Cole, M., Lindeque, P., Fileman, E., Halsband, C., & Galloway, T. S. (2015). The impact of polystyrene microplastics on feeding, function and fecundity in the marine copepod *Calanus helgolandicus*. *Environmental Science and Technology, 49*(2), 1130–1137. https://doi.org/10.1021/es504525u

Danso, D., Chow, J., & Streita, W. R. (2019). Plastics: Environmental and biotechnological perspectives on microbial degradation. *Applied and Environmental Microbiology, 85*(19). https://doi.org/10.1128/AEM.01095-19/FORMAT/EPUB

Dawson, A. L., Kawaguchi, S., King, C. K., Townsend, K. A., King, R., Huston, W. M., & Bengtson Nash, S. M. (2018). *Turning microplastics into nanoplastics through digestive fragmentation by Antarctic krill*. https://doi.org/10.1038/s41467-018-03465-9

de Souza Machado, A. A., Kloas, W., Zarfl, C., Hempel, S., & Rillig, M. C. (2018a). Microplastics as an emerging threat to terrestrial ecosystems. In Global change biology (24, 4, pp. 1405–1416). Blackwell Publishing Ltd. doi:https://doi.org/10.1111/gcb.14020

de Souza Machado, A. A., Lau, C. W., Till, J., Kloas, W., Lehmann, A., Becker, R., & Rillig, M. C. (2018b). Impacts of microplastics on the soil biophysical environment. *Environmental Science and Technology, 52*(17), 9656–9665. https://doi.org/10.1021/acs.est.8b02212

de Souza Machado, A. A., Lau, C. W., Kloas, W., Bergmann, J., Bachelier, J. B., Faltin, E., Becker, R., Görlich, A. S., & Rillig, M. C. (2019). Microplastics can change soil properties and affect plant performance. *Environmental Science and Technology, 53*(10), 6044–6052. https://doi.org/10.1021/acs.est.9b01339

Ekner-Grzyb, A., Duka, A., Grzyb, T., Lopes, I., & Chmielowska-Bąk, J. (2022). Plants oxidative response to nanoplastic. In *Frontiers in plant science* (Vol. 13). Frontiers Media S.A. https://doi.org/10.3389/fpls.2022.1027608

Fenech, M., Kirsch-Volders, M., Natarajan, A. T., Surralles, J., Crott, J. W., Parry, J., Norppa, H., Eastmond, D. A., Tucker, J. D., & Thomas, P. (2011). Molecular mechanisms of micronucleus, nucleoplasmic bridge and nuclear bud formation in mammalian and human cells. *Mutagenesis, 26*(1), 125–132. https://doi.org/10.1093/mutage/geq052

Geyer, R., Jambeck, J. R., & Law, K. L. (2017). *Production, use, and fate of all plastics ever made*. https://www.science.org

Gigault, J., Halle, A. T., Baudrimont, M., Pascal, P. Y., Gauffre, F., Phi, T. L., El Hadri, H., Grassl, B., & Reynaud, S. (2018). Current opinion: What is a nanoplastic? In *Environmental pollution* (Vol. 235, pp. 1030–1034). Elsevier Ltd.. https://doi.org/10.1016/j.envpol.2018.01.024

Giorgetti, L., Spanò, C., Muccifora, S., Bottega, S., Barbieri, F., Bellani, L., & Ruffini Castiglione, M. (2020). Exploring the interaction between polystyrene nanoplastics and Allium cepa during germination: Internalization in root cells, induction of toxicity and oxidative stress. *Plant Physiology and Biochemistry, 149*, 170–177. https://doi.org/10.1016/j.plaphy.2020.02.014

Gola, D., Kumar Tyagi, P., Arya, A., Chauhan, N., Agarwal, M., Singh, S. K., & Gola, S. (2021). The impact of microplastics on marine environment: A review. In *Environmental nanotechnology, monitoring and management* (Vol. 16). Elsevier B.V. https://doi.org/10.1016/j.enmm.2021.100552

Gorrasi, G., Sorrentino, A., & Lichtfouse, E. (2021). Back to plastic pollution in COVID times. In *Environmental chemistry letters* (Vol. 19). Springer Science and Business Media Deutschland GmbH. https://doi.org/10.1007/s10311-020-01129-z

Guzzetti, E., Sureda, A., Tejada, S., & Faggio, C. (2018). Microplastic in marine organism: Environmental and toxicological effects. In *Environmental toxicology and pharmacology* (Vol. 64, pp. 164–171). Elsevier B.V. https://doi.org/10.1016/j.etap.2018.10.009

Haider, T. P., Völker, C., Kramm, J., Landfester, K., & Wurm, F. R. (2019). Kunststoffe der Zukunft? Der Einfluss von bioabbaubaren Polymeren auf Umwelt und Gesellschaft. *Angewandte Chemie, 131*(1), 50–63. https://doi.org/10.1002/ange.201805766

Han, L., Chen, L., Li, D., Ji, Y., Feng, Y., Feng, Y., & Yang, Z. (2022). Influence of polyethylene terephthalate microplastic and biochar co-existence on paddy soil bacterial community structure and greenhouse gas emission. *Environmental Pollution, 292*. https://doi.org/10.1016/j.envpol.2021.118386

Hansen, V., Müller-Stöver, D., Munkholm, L. J., Peltre, C., Hauggaard-Nielsen, H., & Stoumann Jensen, L. (2016). The effect of straw and wood gasification biochar on carbon sequestration, selected soil fertility indicators and functional groups in soil: An incubation study. https://doi.org/10.1016/j.geoderma.2016.01.033

Horton, A. A., Walton, A., Spurgeon, D. J., Lahive, E., & Svendsen, C. (2017). Microplastics in freshwater and terrestrial environments: Evaluating the current understanding to identify the knowledge gaps and future research priorities. In *Science of the Total environment* (Vol. 586, pp. 127–141). Elsevier B.V. https://doi.org/10.1016/j.scitotenv.2017.01.190

Ivleva, N. P., Wiesheu, A. C., & Niessner, R. (2017). Mikroplastik in aquatischen Ökosystemen. *Angewandte Chemie, 129*(7), 1744–1764. https://doi.org/10.1002/ange.201606957

Jambeck, J. R., Geyer, R., Wilcox, C., Siegler, T. R., Perryman, M., Andrady, A., Narayan, R., & Law, K. L. (2015). Plastic waste inputs from land into the ocean. *Science, 347*(6223), 768–771. https://doi.org/10.1126/science.1260352

Jambeck, J., Hardesty, B. D., Brooks, A. L., Friend, T., Teleki, K., Fabres, J., Beaudoin, Y., Bamba, A., Francis, J., Ribbink, A. J., Baleta, T., Bouwman, H., Knox, J., & Wilcox, C. (2018). Challenges and emerging solutions to the land-based plastic waste issue in Africa. *Marine Policy, 96*, 256–263. https://doi.org/10.1016/j.marpol.2017.10.041

Jang, M., Shim, W. J., Han, G. M., Song, Y. K., & Hong, S. H. (2018). Formation of microplastics by polychaetes (Marphysa sanguinea) inhabiting expanded polystyrene marine debris. *Marine Pollution Bulletin, 131*, 365–369. https://doi.org/10.1016/j.marpolbul.2018.04.017

Jian, L., Bai, X., Zhang, H., Song, X., & Li, Z. (2019). Promotion of growth and metal accumulation of alfalfa by coinoculation with Sinorhizobium and agrobacterium under copper and zinc stress. *PeerJ, 2019*(5). https://doi.org/10.7717/peerj.6875

Jiang, X., Chen, H., Liao, Y., Ye, Z., Li, M., & Klobučar, G. (2019). Ecotoxicity and genotoxicity of polystyrene microplastics on higher plant Vicia faba. *Environmental Pollution, 250*, 831–838. https://doi.org/10.1016/j.envpol.2019.04.055

Kale, S. K., Deshmukh, A. G., Dudhare, M. S., & Patil, V. B. (2015). Microbial degradation of plastic: a review. *Journal of Biochemical Technology, 6*(2), 952–961.

Khan, M. R., Adam, V., Rizvi, T. F., Zhang, B., Ahamad, F., Jośko, I., Zhu, Y., Yang, M., & Mao, C. (2019). Nanoparticle–Plant Interactions: Two-Way Traffic. In Small (15, Issue 37). Wiley-VCH Verlag. doi:https://doi.org/10.1002/smll.201901794

Kim, S. W., & An, Y. J. (2019). Soil microplastics inhibit the movement of springtail species. *Environment International, 126*, 699–706. https://doi.org/10.1016/j.envint.2019.02.067

Krzan, A., Hemjinda, S., Miertus, S., Corti, A., & Chiellini, E. (2006). Standardization and certification in the area of environmentally degradable plastics. *Polymer Degradation and Stability, 91*(12), 2819–2833. https://doi.org/10.1016/j.polymdegradstab.2006.04.034

Li, L., Luo, Y., Li, R., Zhou, Q., Peijnenburg, W. J. G. M., Yin, N., Yang, J., Tu, C., & Zhang, Y. (2020). Effective uptake of submicrometre plastics by crop plants via a crack-entry mode. *Nature Sustainability, 3*(11), 929–937. https://doi.org/10.1038/s41893-020-0567-9

Li, Z., Li, Q., Li, R., Zhou, J., & Wang, G. (2020). The distribution and impact of polystyrene nanoplastics on cucumber plants. https://doi.org/10.1007/s11356-020-11702-2/Published

Lian, J., Wu, J., Xiong, H., Zeb, A., Yang, T., Su, X., Su, L., & Liu, W. (2020). Impact of polystyrene nanoplastics (PSNPs) on seed germination and seedling growth of wheat (Triticum aestivum L.). *Journal of Hazardous Materials, 385*. https://doi.org/10.1016/j.jhazmat.2019.121620

Lian, J., Liu, W., Meng, L., Wu, J., Chao, L., Zeb, A., & Sun, Y. (2021). Foliar-applied polystyrene nanoplastics (PSNPs) reduce the growth and nutritional quality of lettuce (Lactuca sativa L.). *Environmental Pollution, 280*. https://doi.org/10.1016/j.envpol.2021.116978

Liu, H., Yang, X., Liu, G., Liang, C., Xue, S., Chen, H., Ritsema, C. J., & Geissen, V. (2017). Response of soil dissolved organic matter to microplastic addition in Chinese loess soil. *Chemosphere, 185*, 907–917. https://doi.org/10.1016/j.chemosphere.2017.07.064

Liu, M., Lu, S., Song, Y., Lei, L., Hu, J., Lv, W., Zhou, W., Cao, C., Shi, H., Yang, X., & He, D. (2018). Microplastic and mesoplastic pollution in farmland soils in suburbs of Shanghai, China. *Environmental Pollution, 242*, 855–862. https://doi.org/10.1016/j.envpol.2018.07.051

Liu, H., Yang, X., Liang, C., Li, Y., Qiao, L., Ai, Z., Xue, S., & Liu, G. (2019). Interactive effects of microplastics and glyphosate on the dynamics of soil dissolved organic matter in a Chinese loess soil. *Catena, 182*. https://doi.org/10.1016/j.catena.2019.104177

Lu, K., Qiao, R., An, H., & Zhang, Y. (2018). Influence of microplastics on the accumulation and chronic toxic effects of cadmium in zebrafish (Danio rerio). *Chemosphere, 202*, 514–520. https://doi.org/10.1016/j.chemosphere.2018.03.145

Lucas, N., Bienaime, C., Belloy, C., Queneudec, M., Silvestre, F., & Nava-Saucedo, J. E. (2008). Polymer biodegradation: Mechanisms and estimation techniques – A review. *Chemosphere, 73*(4), 429–442. https://doi.org/10.1016/j.chemosphere.2008.06.064

Luyt, A. S., & Malik, S. S. (2019). Can biodegradable plastics solve plastic solid waste accumulation? In *Plastics to energy* (pp. 403–423). Elsevier. https://doi.org/10.1016/B978-0-12-813140-4.00016-9

Mariano, S., Tacconi, S., Fidaleo, M., Rossi, M., & Dini, L. (2021). Micro and nanoplastics identification: Classic methods and innovative detection techniques. In *Frontiers in toxicology* (Vol. 3). Frontiers Media S.A. https://doi.org/10.3389/ftox.2021.636640

Mateos-Cárdenas, A., O'Halloran, J., van Pelt, F. N. A. M., & Jansen, M. A. K. (2020). Rapid fragmentation of microplastics by the freshwater amphipod Gammarus duebeni (Lillj.). *Scientific Reports, 10*(1). https://doi.org/10.1038/s41598-020-69635-2

Mateos-Cárdenas, A., van Pelt, F. N. A. M., O'Halloran, J., & Jansen, M. A. K. (2021). Adsorption, uptake and toxicity of micro- and nanoplastics: Effects on terrestrial plants and aquatic macrophytes. In *Environmental Pollution* (Vol. 284). Elsevier Ltd.. https://doi.org/10.1016/j.envpol.2021.117183

Milewska-Hendel, A., Zubko, M., Stróż, D., & Kurczyńska, E. U. (2019). Effect of nanoparticles surface charge on the Arabidopsis thaliana (L.) roots development and their movement into the root cells and protoplasts. *International Journal of Molecular Sciences, 20*(7). https://doi.org/10.3390/ijms20071650

Mirza-Aghazadeh-Attari, M., Ekrami, E. M., Aghdas, S. A. M., Mihanfar, A., Hallaj, S., Yousefi, B., Safa, A., & Majidinia, M. (2020). Targeting PI3K/Akt/mTOR signaling pathway by polyphenols: Implication for cancer therapy. In *Life Sciences* (Vol. 255). Elsevier. https://doi.org/10.1016/j.lfs.2020.117481

Navarro, E., Baun, A., Behra, R., Hartmann, N. B., Filser, J., Miao, A. J., Quigg, A., Santschi, P. H., & Sigg, L. (2008). Environmental behavior and ecotoxicity of engineered nanoparticles to algae, plants, and fungi. *Ecotoxicology, 17*(5), 372–386. https://doi.org/10.1007/s10646-008-0214-0

Nazareth, M., Marques, M. R. C., Leite, M. C. A., & Castro, Í. B. (2019). Commercial plastics claiming biodegradable status: Is this also accurate for marine environments? *Journal of Hazardous Materials, 366*, 714–722. https://doi.org/10.1016/j.jhazmat.2018.12.052

Nizzetto, L., Futter, M., & Langaas, S. (2016). Are agricultural soils dumps for microplastics of urban origin? In *Environmental science and technology* (Vol. 50, Issue 20, pp. 10777–10779). American Chemical Society. https://doi.org/10.1021/acs.est.6b04140

Novotný, Č., Malachová, K., Adamus, G., Kwiecień, M., Lotti, N., Soccio, M., Verney, V., & Fava, F. (2018). Deterioration of irradiation/high-temperature pretreated, linear low-density polyethylene (LLDPE) by *Bacillus amyloliquefaciens*. *International Biodeterioration and Biodegradation, 132*, 259–267. https://doi.org/10.1016/j.ibiod.2018.04.014

Ogonowski, M., Schür, C., Jarsén, Å., & Gorokhova, E. (2016). The effects of natural and anthropogenic microparticles on individual fitness in daphnia magna. *PLoS One, 11*(5). https://doi.org/10.1371/journal.pone.0155063

Oleńska, E., Małek, W., Wójcik, M., Swiecicka, I., Thijs, S., & Vangronsveld, J. (2020). Beneficial features of plant growth-promoting rhizobacteria for improving plant growth and health in challenging conditions: A methodical review. In *Science of the Total Environment* (Vol. 743). Elsevier B.V. https://doi.org/10.1016/j.scitotenv.2020.140682

Patrício Silva, A. L., Prata, J. C., Walker, T. R., Duarte, A. C., Ouyang, W., Barcelò, D., & Rocha-Santos, T. (2021). Increased plastic pollution due to COVID-19 pandemic: Challenges and recommendations. In *Chemical engineering journal* (Vol. 405). Elsevier B.V. https://doi.org/10.1016/j.cej.2020.126683

Piehl, S., Leibner, A., Löder, M. G. J., Dris, R., Bogner, C., & Laforsch, C. (2018). Identification and quantification of macro- and microplastics on an agricultural farmland. *Scientific Reports, 8*(1). https://doi.org/10.1038/s41598-018-36172-y

Porter, A., Smith, K. E., & Lewis, C. (2019). The sea urchin *Paracentrotus lividus* as a bioeroder of plastic. *Science of the Total Environment, 693*. https://doi.org/10.1016/J.SCITOTENV.2019.133621

Prata, J. C., da Costa, J. P., Lopes, I., Duarte, A. C., & Rocha-Santos, T. (2020). Environmental exposure to microplastics: An overview on possible human health effects. In *Science of the total environment* (Vol. 702). Elsevier B.V. https://doi.org/10.1016/j.scitotenv.2019.134455

Qi, Y., Yang, X., Pelaez, A. M., Huerta Lwanga, E., Beriot, N., Gertsen, H., Garbeva, P., & Geissen, V. (2018). Macro- and micro- plastics in soil-plant system: Effects of plastic mulch film residues on wheat (Triticum aestivum) growth. *Science of the Total Environment, 645*, 1048–1056. https://doi.org/10.1016/J.SCITOTENV.2018.07.229

Raklami, A., Oufdou, K., Tahiri, A. I., Mateos-Naranjo, E., Navarro-Torre, S., Rodríguez-Llorente, I. D., Meddich, A., Redondo-Gómez, S., & Pajuelo, E. (2019). Safe cultivation of Medicago sativa in metal-polluted soils from semi-arid regions assisted by heat-and metallo-resistant PGPR. *Microorganisms, 7*(7). https://doi.org/10.3390/microorganisms7070212

Rillig, M. C., & Lehmann, A. (2020). Microplastic in terrestrial ecosystems research shifts from ecotoxicology to ecosystem effects and earth system feedbacks. In *Science* (Vol. 368, Issue 6498, pp. 1430–1431). American Association for the Advancement of Science. https://doi.org/10.1126/science.abb5979

Rillig, M. C., Lehmann, A., de Souza Machado, A. A., & Yang, G. (2019). Microplastic effects on plants. In *New phytologist* (Vol. 223, Issue 3, pp. 1066–1070). Blackwell Publishing Ltd.. https://doi.org/10.1111/nph.15794

Rubol, S., Manzoni, S., Bellin, A., & Porporato, A. (2013). *Modeling soil moisture and oxygen effects on soil biogeochemical cycles including dissimilatory nitrate reduction to ammonium (DNRA)*. https://doi.org/10.1016/j.advwatres.2013.09.016

Rujnić-Sokele, M., & Pilipović, A. (2017). Challenges and opportunities of biodegradable plastics: A mini review. In *Waste management and research* (Vol. 35, Issue 2, pp. 132–140). SAGE Publications Ltd.. https://doi.org/10.1177/0734242X16683272

Shah, A. A., Hasan, F., Hameed, A., & Ahmed, S. (2008). Biological degradation of plastics: A comprehensive review. *Biotechnology Advances, 26*(3), 246–265. https://doi.org/10.1016/j.biotechadv.2007.12.005

Shahlari, M., & Lee, S. (2012). Mechanical and morphological properties of poly(butylene adipate-co-terephthalate) and poly(lactic acid) blended with organically modified silicate layers. *Polymer Engineering and Science, 52*(7), 1420–1428. https://doi.org/10.1002/pen.23082

Shen, M., Song, B., Zeng, G., Zhang, Y., Huang, W., Wen, X., & Tang, W. (2020). Are biodegradable plastics a promising solution to solve the global plastic pollution? In *Environmental Pollution* (Vol. 263). Elsevier Ltd.. https://doi.org/10.1016/j.envpol.2020.114469

Spanò, C., Muccifora, S., Ruffini Castiglione, M., Bellani, L., Bottega, S., & Giorgetti, L. (2022). Polystyrene nanoplastics affect seed germination, cell biology and physiology of rice seedlings in-short term treatments: Evidence of their internalization and translocation. *Plant Physiology and Biochemistry, 172*, 158–166. https://doi.org/10.1016/j.plaphy.2022.01.012

Spokas, K. A., & Dc Reicosky. (2009). Impacts of sixteen different biochars on soil greenhouse gas production. *Annals of Environmental Science, 3*, 179–193. www.aes.northeastern.edu

Sun, X. D., Yuan, X. Z., Jia, Y., Feng, L. J., Zhu, F. P., Dong, S. S., Liu, J., Kong, X., Tian, H., Duan, J. L., Ding, Z., Wang, S. G., & Xing, B. (2020). Differentially charged nanoplastics demonstrate distinct accumulation in Arabidopsis thaliana. *Nature Nanotechnology, 15*(9), 755–760. https://doi.org/10.1038/s41565-020-0707-4

Sun, H., Lei, C., Xu, J., & Li, R. (2021). Foliar uptake and leaf-to-root translocation of nanoplastics with different coating charge in maize plants. *Journal of Hazardous Materials, 416*. https://doi.org/10.1016/j.jhazmat.2021.125854

Sussarellu, R., Suquet, M., Thomas, Y., Lambert, C., Fabioux, C., Pernet, M. E. J., Le Goïc, N., Quillien, V., Mingant, C., Epelboin, Y., Corporeau, C., Guyomarch, J., Robbens, J., Paul-Pont, I., Soudant, P., & Huvet, A. (2016). Oyster reproduction is affected by exposure to polystyrene microplastics. *Proceedings of the National Academy of Sciences of the United States of America, 113*(9), 2430–2435. https://doi.org/10.1073/pnas.1519019113

Wan, Y., Wu, C., Xue, Q., & Hui, X. (2019). Effects of plastic contamination on water evaporation and desiccation cracking in soil. *Science of the Total Environment, 654*, 576–582. https://doi.org/10.1016/j.scitotenv.2018.11.123

Wang, Z., Sedighi, M., & Lea-Langton, A. (2020). Filtration of microplastic spheres by biochar: Removal efficiency and immobilisation mechanisms. *Water Research, 184*. https://doi.org/10.1016/j.watres.2020.116165

Weithmann, N., Möller, J. N., Löder, M. G. J., Piehl, S., Laforsch, C., & Freitag, R. (2018). *Organic fertilizer as a vehicle for the entry of microplastic into the environment*. https://www.science.org

Wichai-utcha, N., & Chavalparit, O. (2019). 3Rs policy and plastic waste management in Thailand. *Journal of Material Cycles and Waste Management, 21*(1), 10–22). Springer Tokyo. https://doi.org/10.1007/s10163-018-0781-y

Woldemar D'ambrières. (2019). Field Actions Science Reports Plastics recycling worldwide: current overview and desirable changes plastics recycling worldwide: Current overview and desirable changes • Plastics Economy • Recycling • Regulation • Eco-Design.

Xie, H. J., Shi, Y. J., Zhang, J., Cui, Y., Teng, S. X., Wang, S. G., & Zhao, R. (2010). Degradation of phthalate esters (PAEs) in soil and the effects of PAEs on soil microcosm activity. *Journal of Chemical Technology and Biotechnology, 85*(8), 1108–1116. https://doi.org/10.1002/jctb.2406

Xu, S., Ma, J., Ji, R., Pan, K., & Miao, A. J. (2020). Microplastics in aquatic environments: Occurrence, accumulation, and biological effects. In *Science of the total environment* (Vol. 703). Elsevier B.V. https://doi.org/10.1016/j.scitotenv.2019.134699

Yang, C., & Gao, X. (2022). Impact of microplastics from polyethylene and biodegradable mulch films on rice (Oryza sativa L.). *Science of the Total Environment, 828*. https://doi.org/10.1016/J.SCITOTENV.2022.154579

Yildiztugay, E., Ozfidan-Konakci, C., Arikan, B., Alp, F. N., Elbasan, F., Zengin, G., Cavusoglu, H., & Sakalak, H. (2022). The hormetic dose-risks of polymethyl methacrylate nanoplastics on chlorophyll a fluorescence transient, lipid composition and antioxidant system in Lactuca sativa. *Environmental Pollution, 308*. https://doi.org/10.1016/j.envpol.2022.119651

Yu, L., Di Zhang, J., Liu, Y., Chen, L. Y., Tao, S., & Liu, W. X. (2021). Distribution characteristics of microplastics in agricultural soils from the largest vegetable production base in China. *Science of the Total Environment, 756*. https://doi.org/10.1016/j.scitotenv.2020.143860

Zeenat, E. A., Bukhari, D. A., Shamim, S., & Rehman, A. (2021). Plastics degradation by microbes: A sustainable approach. *Journal of King Saud University – Science, 33*(6) Elsevier B.V. https://doi.org/10.1016/j.jksus.2021.101538

Zhang, X., Li, Y., Ouyang, D., Lei, J., Tan, Q., Xie, L., Li, Z., Liu, T., Xiao, Y., Farooq, T. H., Wu, X., Chen, L., & Yan, W. (2021a). Systematical review of interactions between microplastics and microorganisms in the soil environment. *Journal of Hazardous Materials, 418*). Elsevier B.V. https://doi.org/10.1016/j.jhazmat.2021.126288

Zhang, S., Wang, J., Yan, P., Hao, X., Xu, B., Wang, W., & Aurangzeib, M. (2021b). Non-biodegradable microplastics in soils: A brief review and challenge. *Journal of Hazardous Materials, 409*. https://doi.org/10.1016/j.jhazmat.2020.124525

Zhou, C. Q., Lu, C. H., Mai, L., Bao, L. J., Liu, L. Y., & Zeng, E. Y. (2021). Response of rice (Oryza sativa L.) roots to nanoplastic treatment at seedling stage. *Journal of Hazardous Materials, 401*. https://doi.org/10.1016/j.jhazmat.2020.123412

Interactıon of Micro-Nanoplastics and Heavy Metals in Soil Systems: Mechanism and Implication

Eda Ceylan, Dilara Büşra Bartan, İrem Öztürk-Ufuk, Emel Topuz, and Derya Ayral-Çınar

1 Introduction

Worldwide production and use of plastics was 1.7 million tonnes in 1950, which went up to 348 million tonnes in 2017 (Europe, 2019; van den Berg et al., 2020; Zou et al., 2020a). More than 450 million metric tons (MMT) of plastic is produced each year in the world, which is expected to be doubled or tripled by 2060 (*Global Plastics Outlook*, 2022; Moeck et al., 2023). Plastic production, usage of plastic products in daily life, and their disposal after the end of their life lead to plastic pollution in the environment. Due to the physical/chemical/biological conditions in the environment, they end up in micro or nano size and are called as microplastic or nanoplastic (MnP). Various studies have observed that the presence of MnP in the environment can trigger a multitude of toxicological effects on organisms. Therefore, studies on the fate, effect, and risk assessment of MnPs have been increasing dramatically in the literature. One of the most interesting outcomes of these studies is about the effect of MnPs on the fate and toxicity of other pollutants. Although the fate and toxicity of MnPs are still not clear from many perspectives, the studies on the effect of MnPs on the other pollutants' fate and toxicity have emerged rapidly due to the significance of the preliminary results. Especially, the findings on the vector effect of MnPs on the other pollutants that might result in higher mobility and increased toxic effects of other pollutants take significant attention in the literature.

Soil compartment is regarded as one of the most significant sinks of MnPs because of human activities such as application of mulching, biosolid, compost, and reclaimed wastewater irrigation in the agricultural fields, as well as due to atmospheric deposition. However, studies on the soil environment are lacking, especially, compared to

E. Ceylan · D. B. Bartan · İ. Öztürk-Ufuk · E. Topuz · D. Ayral-Çınar (✉)
Environmental Engineering, Gebze Technical University, Kocaeli, Turkey
e-mail: deryacinar@gtu.edu.tr

S. A. Bhat et al. (eds.), *Management of Micro and Nano-plastics in Soil and Biosolids*,
https://doi.org/10.1007/978-3-031-51967-3_7

the aquatic environment. Another type of abundant pollutant in the soil environment is heavy metals like MnP introduced by similar pollution sources. Considering the alarming effect of MnP on the mobility and toxicity of other pollutants, the interaction of heavy metals and MnP in the soil environment is required to understand the associated environmental risks and their management. Hence, the aim of this chapter is to summarize the presence of MnP and heavy metals in the soil environment, understand the interaction mechanisms of MnP and heavy metals in the soil environment including the adsorption and complexation reactions, and evaluation of mobility and toxicity of MnPs and heavy metal when they coexist in the soil environment.

1.1 Occurrence of Micro-Nanoplastics in Soil Environment

Materials out of plastics are frequently used in many fields such as medical, industrial, and agricultural fields worldwide due to their low cost and durable properties, and with this, waste produced from these plastic materials has become a recent concern for environment (Atugoda et al., 2023; Yuan et al., 2020; L. Zou et al., 2020b). As of 2015, approximately 6300 million tons of plastic waste was produced where majority of this waste (91%) was either disposed of in landfills (79%) or incinerated (12%) and only a minor portion (9%) was recycled. Accordingly, around 12,000 million tons of plastic waste is predicted to be placed in landfills or in the natural environment by 2050 (Geyer, 2020; Geyer et al., 2017). Over time, larger plastic materials are degraded by sun, wind, water, microorganisms, or erosion physically, chemically or biologically. These plastic particles smaller than 5 mm are called microplastics (MP), while plastic particles in the 1–100 nm range are classified as nanoplastics (NP) (Atugoda et al., 2023; Yu et al., 2021). These tiny microscopic plastics are also recognized as a recent widespread threat since both terrestrial and aquatic organisms can ingest macro- or nanosized plastics, which likely interfere with food intake occupying intestinal space or causing direct toxic effects on tissues (Cole et al., 2013; Huerta Lwanga et al., 2017; Teng et al., 2021; Yu et al., 2021).

Micro-nanoplastics pass into the soil environment from many sources like agricultural practices such as biosolids or compost application, mulching, irrigation with treated or untreated wastewater, atmospheric deposition, and runoff (Moeck et al., 2023; Okeke et al., 2023; Tian et al., 2022; L. Zou et al., 2020b). In addition, around 144 mg/kg microcapsules from coated manure were found to accumulate in paddy fields of Japan (Katsumi et al., 2021). Therefore, the soil environment is thought to act as an important sink for micro-nanoplastics (MnP), and 4–23 times higher load of MnP is thought to be released to land compared to ocean (Atugoda et al., 2023; Horton et al., 2017; Moeck et al., 2023). Various types of plastic have been observed in the soil environment, such as polyethylene (PE), polypropylene (PP), polyethylene terephthalate (PET), polyvinyl chloride (PVC), Low-density polyethylene (LDPE), high-density polyethylene (HDPE), polystyrene (PS), polyamide (PA), polyester (PES), polyacrylonitrile (PAN), chlorinated polyethylene

(CPE), and acrylic. In addition, when the literature is reviewed, it is seen that the most common types of plastic in the soil environment are PP, PE, and PS (Chen et al., 2021; Ma et al., 2022).

Plastic mulching is one of the most known sources for MnP to enter the terrestrial environment because it is widely used to improve the efficiency and quality of crop production (Horton et al., 2017; Huang et al., 2020; Kumar & Sheela, 2020). There is plastic mulching on approximately 20 million hectares of agricultural land worldwide (Huang et al., 2020). After plastic mulches are used, they are deliberately or unintentionally separated into micro- and nano-sized particles by various environmental conditions such as wind and sun, entering the soil environment (Atugoda et al., 2023; Moeck et al., 2023; Wahl et al., 2021). When the MnPs that pass into the soil as a result of plastic mulching are examined, it is noteworthy that the plastic types commonly found in the soils are PE and PP, and they are smaller than 2 mm (Table 1). Y. Huang et al. (2020) found that MP concentration in soil mulched for 5 years was 80.3 MP parts/kg and increased to 1075.6 MP parts/kg after 24 years of mulching. MP was also observed in 10 years of mulching agricultural soils in India where the presence of MP decreased with increasing soil depth. Yet, 36.99% of MPs were found in samples taken even from 21 to 30 cm depth (Kumar & Sheela, 2020). In Table 1, it is seen that the MnP concentration in the mulched soils varies between 62–40,800 pieces/kg and 9.25–12,560 mg/kg in the studies carried out.

Sludge produced in Wastewater Treatment Plants (WWTPs) can also contribute significantly to MnP presence in the soil environment (Abeynayaka et al., 2022; Atugoda et al., 2023; J. Jiang et al., 2020a). Table 2 summarizes MnP concentrations in biosolids produced by WWTPs in various countries. More than 90% of MnPs in the influent can be removed by WWTPs and accumulate in the sewage sludge (Abeynayaka et al., 2022; Carr et al., 2016; J. Jiang et al., 2020a; Li et al., 2018; Q. Li et al., 2019b). Thus, the use of treated sewage sludge as fertilizer in agricultural soils can cause MnP contamination in soils (Huang et al., 2023; Moeck et al., 2023; Naderi Beni et al., 2023). The amount of biosolids produced worldwide is about 10×10^7 tons, and this value is estimated to increase to 17.5×10^7 tons per year by 2050 (Wijesekara et al., 2017). Moreover, in the European community, more than 30% of biosolids are used as fertilizers in agriculture, compared to 57% in Belgium and 60% in France (Wang et al., 2008). In the studies examined in Table 2, MnP concentrations in biosolids ranged between 1180 and 4740 substance MP/kg soil. In addition, various types of plastics are found, and it is seen that the size of the plastic particles is between 10 μm and 50 mm. In the studies summarized in Table 1, biosolid-derived MP concentrations in the applied soils were found to vary between 4.38–180 mg/kg soil and 2130–3060 substance/kg soil, and the size range was between 20 nm to 3 mm. In addition, an increase in biosolid applications was reported to increase the amount of MP in the soil. For example, in a study conducted on agricultural land in Chile, after a single application of biosolid, the amount of MP in the soil was 1.37 mg/kg, while the MP concentration increased to 4.38 mg/kg following the fifth application (Corradini et al., 2019).

Another agricultural application that is widely used is compost which serves as soil conditioner and fertilizer, but it can result in MnP transfer into soils. It was

Table 1 Potential source of microplastics and nanoplastics

Environment	Country	MP/NP type	Size	Concentration	References
Mulching					
Agricultural soils	China	PE	>5 mm	Year/Conc. 5 /80,3 pieces MP /kg 15/308 pieces MP /kg 24/1075,6 pieces MP / kg	Huang et al. (2020)
Agricultural fields	Sulur, India	PE	<5 mm	10 years application 0–10 cm soil %37.93 particles 11–20 cm soil %35.07 particles 21–30 cm soil %36.99 particles	Kumar M and Sheela A (2020)
Agricultural soils	China	PP, PE, PES	1.91 ± 0.13 mm	78.0 ± 12.9 pieces MP / kg	Liu et al. (2018)
Agricultural soils	China	PP, PE, PES	1.48 ± 0.11 mm	62.5 ± 13.0 pieces MP / kg	Liu et al. (2018)
Agricultural soils	China	Not specific	< 500 μm	900–40,800 pieces MP /kg	B. Huang et al. (2021a)
The horticultural soil	Arjantin	LDPE	25 μm	10% of the total sampled surface 0.11 (± 0.07) m^2 plastic /m^2 soil	Ramos et al. (2015)
Agricultural soils	Shaanxi, China	PS, PE, PP, PET, PVC, HDPE	0–0.49 mm	1430–3410 item/kg	Ding et al. (2020)
Agricultural soils	China	PE	100–800 μm	9.25 ± 0.78–369.55 ± 6.15 mg/kg	W. Li et al. (2020b)
Agricultural soils	China	PA, PP	<2 mm	320–12,560 items/kg	Chen et al. (2020a)
Agricultural soils	China	PE	7 μm–5 mm	1075.6 ± 346.8 pieces/ kg	Huang et al. (2020)
Atmospheric deposition					
London- England		Multi-plastics (PAN, PP, PE, etc. and tire wear)	20–1080 μm	575–1008 particles/m^2/ day	Wright et al. (2020)
U.S.A		Multi-MP	4–188 μm	132 particle /m^2·Day	Brahney et al. (2020)
Gdynia, Poland		Multi-plastics	5–5000 μm	The atmospheric MP deposition ranged from 0 m^{-2}·d^{-1} to 30 m^{-2}·d^{-1}	Szewc et al. (2021)

(continued)

Table 1 (continued)

Environment	Country	MP/NP type	Size	Concentration	References
France/Paris		–	<5 mm	110 (± 96) particles/m²/day	Dris et al. (2016)
Germany		PP, PE	>11 μm	99 ± 85/particles/m²/day	Kernchen et al. (2022)
China		Multi-MP (PE, PP, PS are dominant)	1 μm–5 mm	175 to 313 particles/m²/day	Cai et al. (2017)
France/Paris		Celluloses, PP	50–4.85 mm	1586 and 11,130 fibers/d/m²	Dris et al. (2017)
France/French Pyrenees		PP, PE, PET, PS	<750 μm <300 μm	Fragments: 249 items/m²; Films: 73 items/m²; Fibers: 44 items/m²	Allen et al. (2019)
Biosplid and compost					
Agricultural soils	Spain	HDMP LDMP	150–250 μm	LDMP = 2130 ± 950 microplastics kg-1 HDMP = 3060 ± 1680 microplastics kg-1	van den Berg et al. (2020)
Agricultural soils	Chile	Acrylic PES Nylon, LDPE PVC	<3 mm	1 times = 1.37 mg/kg 2 times =2.03 mg/kg 3 times =2.22 mg/kg 4 times = 2.88 mg/kg 5 times = 4.38 mg/kg	Corradini et al. (2019)
Arable soils	Germany	Multi-plastic	<1 μm	2.38–180 mg/kg	Bläsing and Amelung (2018)
Agricultural soils	France	PE, PS, PVC	20 to 150 nm	–	Wahl et al. (2021)
Agricultural soils	Japan	PE	2–3 mm	Average 144 mg/kg Microcapsule	Katsumi et al. (2021)
Irrigation and flooding					
Germany		PE, PP, PS, PA, SAN[a], PEST[a], PVC, PUR, PET[a], EVA[a], PVAL[a], ABS[a], PLA[a]	< 20 μm	1000–24,000 Items/m³	Mintenig et al. (2017)

(continued)

Table 1 (continued)

Environment	Country	MP/NP type	Size	Concentration	References
Tire wearing					
The Netherlands		Car tires	10 nm to 100 µm	8834 tonnes/year	Jan Kole et al. (2017)
Sweden				11,619 tonnes/year	
Norway				7884 tonnes/year	
Denmark				6721 tonnes/year	
Germany				125,188 tonnes/year	
United Kingdom				63,000 tonnes/year	
Italy				50,000 tonnes/year	

[a]Styrene acrylonitrile (SAN), attached to a netting made of polyester (PEST) with a small amount polyethylene (PE), polyurethane (PUR), Ethylene vinyl acetate (EVA), polyvinyl alcohol (PVAL), acronitrile butadiene styrene (ABS), polylactide (PLA)

reported that around 18 million tons of compost was generated in the EU in 2008, and it is estimated that compost production will increase by approximately 37% by 2020 (ARCADIS & EUNOMIA, 2010). Despite its benefits for agricultural purposes, the application of composted organic material to land can introduce MnP into the soil. This is because organic waste collected from homes for composting is often not properly separated from plastic materials. In addition, industrial waste that might end up in landfills could contain plastic (Stubenrauch & Ekardt, 2020). Consequently, in Germany, MP has been found in composted arable soils at concentrations of 2.38–180 mg/kg (Bläsing & Amelung, 2018). Also, Wahl et al. (2021) examined farmland in France that had received municipal waste amendments for remediation for 15–30 years and found MnPs with a minimum size of 20 nm to a maximum of 150 nm, including polymers of PE, PS, and PVC. Therefore, compost applications can increase the MP pollution of farmland.

In order to decrease resource consumption and enhance sustainability, raw or reclaimed wastewater is commonly used for agricultural irrigation (Bläsing & Amelung, 2018; Ofori et al., 2021; Ungureanu et al., 2020). It has also been reported that Europe reuses 963 Mm^3 of untreated wastewater per year, whereas about 400 m^3/h of wastewater is used to irrigate farmland in Latin America (Jaramillo & Restrepo, 2017; Panara, 2018). Reuse of untreated or reclaimed wastewater causes environmental and health risks due to pathogenic bacteria and toxic chemical content, in addition to being an important source of MnPs entering the soil (Bläsing & Amelung, 2018; Jaramillo & Restrepo, 2017). Effluent from a wastewater treatment plant in Germany was found to contain MPs at concentrations of 1000–24,000 MP/m^3 (Mintenig et al., 2017). When this wastewater is used in agricultural land as irrigation water, MP transfer to the soil is likely.

Plastic trash thrown into the environment could also result in MnP's migration to the soil since these macroplastics can decompose into MnP and enter the soil environment under various environmental conditions. Kim et al. (2006) determined plastic load of 0.85 to 6.6 kg/ha in runoff originated from highways.

Table 2 Source of MnP in biosolid

Source	Country	n	MnP	Size	Concentration	References
WWTP*	Canada	1	Multi-MP	<1 μm	4.4 ± 2.9 particle MPs/g	Gies et al. (2018)
WWTP	China	–	Most common; PA, PET, PE, PP and PS	0,5–5 mm	36.3 ± 5.7–46.3 ± 6.2 particles/g dry sludge	J. Jiang et al. (2020a)
WWTP	China	28	PE, PA, and acrylic fibers	75–830 μm	22.7 ± 12.1 × 10^3 particles/kg dry sludge	(Li et al. 2018)
WWTP	Australia	3	PET, PE, PP	>25 μm	48.5–56.5 particles/g Dry weight	Ziajahromi et al. (2021)
WWTP	Italy	1	Mainly PES, PA	0.01–5 mm	113 ± 57 particles/g sludge Dry weight	Magni et al. (2019)
WWTP	Netherland	7	–	10–5000 μm	510–760 particles / kg wet weight	Leslie et al. (2017)
WWTP	China	3	PE, PET, PA	1–5 mm	5553–13,460 MP items/kg	Q. Li et al. (2019b)
WWTP	Spain	1	PE, PP, PES, and acrylic fibers	25 μm - 50 mm	133 ± 59 particles/g	Edo et al. (2020)
Municipal Wastewater Treatment Lagooning Plant	Morocco	1	PP, PE, PA, PES	< 500 μm	40.5–36.0 × 10^3 particles/kg	EL Hayany et al. (2020)
WWTP	Spain	1	LDPE, PP, PS, PET, PE	150–500 μm	112 MP particles/g dry weight	Bretas Alvim et al., (2020)
WWTP	Ireland	7	Multi-MP	>212 μm >63 μm >45 μm < 45 μm	4196–15,385 particles/kg dry weight	Mahon et al. (2017)
STP**	USA	7	–	400–45 mm	10^3 particles /kg	Carr et al. (2016)
STP	Sweden	1	–	≥300 μm	16.7 ± 1.96 × 10^3 particles/g	Magnusson and Norén (2014)
STP	Finland	1	PES, PET, PE, PA, PP	0,5–1 mm	170.9 × 10^3 particles/kg	Lares et al. (2018)
Municipal WWTP	France	1	PS, PE	80–500 μm >500 μm	16 particles/g	Kazour et al. (2019)
Municipal WWTP	Thailand	3	PES, PE, PA, PP, PS	0.33–4.75 mm	103.4 pieces/L	Hongprasith et al. (2020)

*Wastewater Treatment Plant
**Sewage Treatment Plant

Tire wear and tear can also introduce MnP to soils. Wear products often form small particles in the low micrometer to nanometer range, and tire wear can potentially accumulate in the soil environment from nearby road runoff and through atmospheric deposition. For example, Jan Kole et al. (2017) estimated that 67% of tire particles released into the environment in the Netherlands end up in soil, potentially a major source of soil MnP. It is also assumed that the dust from tire wear probably does not remain on the roadside, but passes into the runoff and atmosphere, transporting the MnPs to the soil and water environment. The amount of worn tires due to the tire wear of the countries has been calculated and given in Table 1. The load of MnP in a size range of 10 nm to 100 μm was found as 6721–125,188 tonnes/year.

Atmospheric deposition has been identified as another potential source for MnP to enter the soil environment. MnPs are ubiquitous in the atmosphere and exposure to open air accelerates the degradation of plastics, resulting in higher MnP formation (Atugoda et al., 2023; Wahl et al., 2021). Looking at the research studies (Table 1), various types of plastics are found in the atmospheric area (Table 1). It is also seen that the size of the plastic particles is between 1 μm and 5 mm. For example, total atmospheric precipitation of MnP collected in urban and suburban areas in Paris has a mean precipitation rate of 110 ± 96 particles/m^2/days and 53 ± 38 particles/m^2/days, respectively (Dris et al., 2016). Furthermore, Allen et al. (2019) found average daily deposition rates of MnP 365 ± 69 particles/m^2/days even in remote locations such as the Pyrenees. As shown in Table 1, similar atmospheric accumulation has been reported in China, Germany, London, USA, and Poland.

1.2 Occurrence of Heavy Metals in Soil Environment

Agricultural soil is a valuable natural resource that is susceptible to contamination by heavy metals, including cadmium (Cd), lead (Pb), zinc (Zn), mercury (Hg), arsenic (As), chromium (Cr), nickel (Ni), and copper (Cu). These contaminants can enter the soil through various pathways, both natural and anthropogenic. Natural causes involve the weathering of metal-rich rocks due to rainfall and atmospheric deposition. Meanwhile, human activities, such as mining, textile industry, leather tanning, and petrochemical industry, improper management of metal-containing waste, and certain agricultural activities, also contribute to the presence of heavy metals in the soil (Campos et al., 2019; Kumar et al., 2019; Mungai et al., 2016; Obinnaa & Ebere, 2019). The presence of heavy metals in the grasslands of the Tibetan region is primarily due to natural sources such as soil properties and climatic conditions. On the other hand, in Brazilian biosolids from WWTPs, heavy metal accumulation is significantly influenced by human activities, especially related to land use, resulting in notable differences in metal concentrations (Campos et al., 2019; Liu et al., 2016).

There are studies monitoring heavy metal pollution in regions where land use varies (Table 3). Cu and Zn exhibit significant differences in the regions with

Table 3 Occurrence of heavy metals in different regions with concentrations

Heavy metal	Type of soil/source	Country	Concentration (mg/kg)	References
Pb	Industrial-old military bases	Poland	13.7–9460	Kopañska and Dudziak (2015)
	Urban		20.6–1330	
Cd	Industrial-old military bases		0.3–239	
	Urban		0.64–11.8	
Hg	Industrial-old military bases		0.1	
	Urban		0.345–2.11	
As	Industrial-old military bases		0.2–378	
	Urban		14.1–82	
Cr	Industrial-old military bases		28.3–53	
	Urban		18.1–38.4	
Zn	Industrial-old military bases		206–86,000	
	Urban		89–5540	
Ni	Industrial-old military bases		14.9–45.5	
	Urban		32.7–52.6	
Cu	Industrial-old military bases		8.89–57.4	
	Urban		18.3–3000	
Pb	Agricultural soil	Kenya	26.87	Mungai et al. (2016)
Cd			0.42	
Hg			8.06	
As			8.93	
Cr			59.69	
Zn			247.39	
Ni			12.56	
Cu			88.59	
Pb	Grassland	Tibet	19.43	Liu et al. (2016)
Cd			0.144	
Hg			0.078	
As			18.4	
Cr			51.89	
Zn			75.06	
Ni			26.11	
Cu			24.69	
Pb	Biosolid	Brazil	58.4	Campos et al. (2019)
Cd			1.4	
Hg			<1	
As			<1	
Cr			21.9	
Zn			531	
Ni			12.7	
Cu			162	

(continued)

Table 3 (continued)

Heavy metal	Type of soil/source	Country	Concentration (mg/kg)	References
Pb	Greenhouse soil	China	13.25	Wei et al. (2020)
Cd			0.174	
Hg			–	
As			–	
Cr			–	
Zn			112.85	
Ni			28.12	
Cu			41.91	
Pb	Agricultural soil	China	30.74	Yuan et al. (2021)
Cd			0.19	
Zn			85.86	
Cu			25.81	
Hg			0.074	
Cr			67.37	
Ni			27.77	
As			8.89	
Pb	Agricultural soil	China	31.32	Zhang et al. (2023)
Cd			0.69	
Zn			96.59	
Cu			28.37	
Cr			96.53	
Ni			21.48	
As			8.76	
Pb	Wastewater treatment effluent/sludge	Iranian	16.58	Soleimani et al. (2023)
Cd			2.64	
Zn			58.2	
Cu			21.1	
Cr			16.64	
Ni			22.26	
As			1.16	

different land uses compared to other metals. Cu is detected in metallurgical contaminated industrial soils with levels ranging from 8.89 to 57.4 mg/kg. Furthermore, it is noteworthy that urban soil Cu concentrations exhibit a wide spectrum, ranging from 18.3 to 3000 mg/kg. This broad range emphasizes the diverse nature of heavy metal pollution assessment across various land use contexts. Also, in soils influenced by industrial activities and contaminated with metallurgical wastes, Zn concentrations exhibit a considerable range, spanning from 206 to 86,000 mg/kg. In contrast, urban soils tend to have comparatively lower Zn levels, with a range of 89 to 5540 mg/kg (Kopańska & Dudziak, 2015). These findings highlight the considerable disparity in heavy metal content between urban and industrial soil environments, signifying the potential impact of human activities and industrial processes on soil pollution. On the other hand, the metal with the smallest difference between urban and industrial soils was Hg. Industrial soils recorded a relatively low concentration of 0.10 mg/kg, while urban soils exhibited a range of 0.345 to 2.11 mg/kg. The smaller difference in mercury levels could be due to its widespread distribution in the environment, including natural sources like volcanoes and human activities such as coal combustion and waste incineration (Kopañska & Dudziak, 2015). Also, observed Cr concentrations both in industrial and urban soils are at the same level, ranging from 28.3 to 53 mg/kg and from 18.1 to 38.4 mg/kg, respectively. Cr sources for both land uses could contribute to a similar degree; however, different types of industrial zones where the usage of Cr is different should be always considered. Nevertheless, except for Hg and Ni, detected heavy metal contamination was higher in industrial zones.

In addition to urban and industrial sources, agricultural land use can also lead to varying levels of heavy metal contamination. For instance, the study conducted by Yuan et al. in 2021 found that heavy metal concentrations listed in descending order were Cd > Pb > Zn > Cu > Hg > Cr > Ni > As (Yuan et al., 2021). However, in potato-producing agricultural areas of China, heavy metals were found in increasing order of concentrations: Zn, Cr, Pb, Cu, Ni, As, and Cd (Zhang et al., 2023). According to the comparison of these different agricultural regions, it appeared that the amount of heavy metal contamination might vary in agricultural lands depending on the natural processes, such as weathering of rocks, as well as anthropogenic activities like agricultural practices, pesticide usage, chemical fertilizers, biosolid application, and irrigation (Zhang et al., 2023). Many countries have regulations that limit the usage of wastewater for reclaimed irrigation and the application of biosolids in agricultural fields, depending on the potential of heavy metal contamination (Shoushtarian & Negahban-Azar, 2020). When treated wastewater effluent is applied to the soil, heavy metal pollution increases as evidenced by higher concentrations of Pb, Zn, Cu, Ni, Cr, and As compared to their levels before irrigation. Among the heavy metals, Zn exhibits the greatest increase, with a concentration of 58.2 mg/kg in the treated effluent, suggesting that it poses a significant risk of contaminating the soil and potentially affecting the environment and agricultural system (Soleimani et al., 2023). In addition to irrigation, the practice of applying biosolids to agricultural fields has the potential to contaminate the soil with heavy metals. For example, Cr showed the most significant difference, rising from 26.00 mg/kg to 56.33 mg/kg when biosolid is applied to an agricultural land. Cu concentrations also increased, from 11.25 mg/kg to 15.54 mg/kg. Similarly, a substantial increase of 36% was noticed in Cd concentration. The elevation in concentration highlights the impact of biosolid application on heavy metal levels in the soil and underscore the importance of monitoring such alterations to ensure compliance with environmental regulations and standards (Koyuncu, 2022).

It is important to consider that heavy metals can also be naturally occurring. In certain areas, elevated concentrations of heavy metals can be detected in the soil naturally, which can be attributed to the geological composition and soil formation processes. Natural factors like erosion and contamination involve processes such as the movement and spreading of soil particles and the wind-driven migration in parent rocks with high background levels (Zhao et al., 2020). Peng et al. found in 2019 that heavy metal pollution in the soil is because of the local karst limestone formation of the region. In another study, in South China, the soil background values of Cd, Pb, Cu, As, and Hg were notably higher compared to North China (Cheng et al., 2018; Guo et al., 2019). This is primarily due to the release and enrichment of heavy metals during the processes of weathering and soil formation where rocks characterized by a significant background value, like limestone, play a role (Kong et al., 2018).

Overall, various studies conducted in different regions have examined the presence and level of heavy metals in different soil samples. These findings emphasize the variability of heavy metals in different soil matrices and the importance of monitoring and managing to address the potential threats they present to both the environment and human health.

2 Interaction between Micro-Nanoplastics and Heavy Metals in Soil

The main interaction mechanism between MnP and heavy metals was regarded as adsorption through electrostatic interactions and complex formation (Cao et al., 2021). While complexation is due to bonding caused by one H atom leaving the functional group of microplastic surface, attraction between positive heavy metals and negative surface charges leads to electrical interaction. For example, it has been stated that Pb, Cd, and Cu are adsorbed by MP such as HDPE, LDPE, CPE, or PVC as a result of electro-interaction or complexation reaction (J. Zou et al., 2020a). Furthermore, it is noteworthy that MnPs, denoting polymer particles, exhibit noteworthy potential as proficient adsorbents facilitating physical and chemical adsorption processes. This propensity stems from the prevalent existence of hydroxyl, carboxyl, amide, and amine functional moieties situated on their surface structures (J. Zou et al., 2020a). Owing to such functional groups on their surfaces, they can perform physical adsorption, which is due to electrical interaction between heavy metals and MPs (Zou et al., 2020b; Guo et al., 2020; Wang et al., 2022b). For example, Cd adsorption is found mainly due to physical sorption on PVC, PS, PP, and PE (Guo et al., 2020). Another characteristic of MPs that affects adsorption of heavy metals is their zeta potential. Zeta potential of MPs was generally negative at pH 3–4.5-6, which causes electrical interaction between MPs and heavy metals. Surface charge can also be affected by organic pollutants found in the MP itself. These impurities were reported to increase the surface charge of MP (Holmes et al., 2012; Lin et al., 2021).

2.1 Effect of Micro-Nanoplastic Characteristics

The way heavy metals are adsorbed on MPs varies according to the MPs type. For instance, when the adsorption capacity of Pb on different MPs was examined, it was seen that the adsorption capacity on PET was lower than PVC, nylon, PP, or PS. Moreover, compared to PVC, Zn was adsorbed more on nylon and PET, while it was not adsorbed on PP and PS. As another example, Cu was found to be more adsorbed on PVC than PS, but the mechanism has not been elucidated (Holmes et al., 2014). In a related vein, the scholarly exploration conducted by Li et al. (2020c) encompassed an inquiry into the adsorption dynamics of environmentally significant heavy metals, namely, Chromium (Cr), Copper (Cu), Arsenic (As), Lead (Pb), Barium (Ba), and Tin (Sn). This investigation encompassed their adsorption behavior onto two distinct substrates: polyethylene (PE) film and the biodegradable polymer polybutylene adipate-co-terephthalate (PBAT). It was found that in general, biodegradable polymers (PBAT) have a better capability for adsorbing heavy metals than regular plastics because degradation modifies surface morphology of biodegradable plastics, which also increases the surface roughness of biodegradable plastics (Li et al., 2020c). Consequently, the roughness increases the specific surface area and improves adsorption on MP surface. As the results of Guo et al. (2020) demonstrated the highest Cd sorption was on PVC (nearly 140 mg Cd^{2+} /kg soil), followed by PS (nearly 130 mg Cd^{2+} /kg soil), PP (nearly 120 mg Cd^{2+} /kg soil), and PE (nearly 100 mg Cd^{2+}/kg soil) after 50 hours, respectively.

Polarization is defined as another characteristic of the MPs, which is due to surface charge and affects their adsorptive property (Liao & Yang, 2020). Polarity increase was related with increase in adsorption capacity of MPs for metals (S. Liu et al., 2021b). For example, heavy metals like Cu^{2+}, Cd^{2+}, and Pb^{2+} were determined to be easily adsorbed on polar region of MPs (Holmes et al., 2012). The surface polarity of MPs varies depending on their physical and chemical properties such as surface area, C%, H%, and Cl% content, molecular weight, crystallinity, and purity (J. Zou et al., 2020a). For instance, presence of methylene (-CH2-) bond and carbon–carbon single bond (C–C) in HPE and LPE lead to lower polarity of these MPs. Besides, polar MPs like PVC and CPE including polar element chloride exhibited different metal adsorption because they are structurally different from each other. CPE has a dispersed molecular structure while PVC has a symmetrical molecular structure. Zou et al. (2020b) stated that this explains the difference in metal adsorption capacities. Moreover, the organic substances attached to the MP surface allow heavy metals (such as Cu^{2+}, Cd^{2+}, and Pb^{2}) to easily bind to the MP (PE) surface due to increased polarity of MPs surface. The characteristic of hydrophobicity stands as a pivotal surface attribute of MPs, exerting substantial influence over the adsorption phenomena pertaining to heavy metals onto MPs. This pivotal aspect has been scrutinized by several scholars, including Abbasi et al. (2020), Holmes et al. (2014), Li et al. (2020c), Mao et al. (2020), Wang et al. (2017), and Zhou et al. (2019). Furthermore, in conjunction with the capability to adsorb heavy metals within the charged or polar domains of their surfaces, MPs demonstrate an additional capacity to adsorb neutral metal–organic complexes due to their hydrophobic surface characteristics (Holmes et al., 2012).

Surface properties of MnPs can change over time due to several environmental conditions and MnPs that are modified in nature over time are called aged MnPs. Aging of MPs and functional groups on their surfaces are important factors in the adsorption and transformation of heavy metals (Cao et al., 2021). Even, it has been stated that the adsorption of heavy metals (Cr, Co, Ni, Cu, Zn, Cd, Pb, Ag, and Hg) on MPs (PE, PP, PS, PA, PVC, and polyformaldehyde) without any modification on the surface is negligible (Holmes et al., 2012; Ren et al., 2020; Tang et al., 2020). A lot of studies indicate that with the aging of MPs (such as PE and PVC), chemical reactions (such as complexation), and heavy metal (Pb, Cr, and Zn) adsorption on their surfaces increase. Increasing bonding capacity (Godoy et al., 2019; Tang et al., 2020, 2021) and bonding of organic matter to the MP surface as a result of deformation of the MP surface (Cao et al., 2021; Turner & Holmes, 2015) are considered as possible reasons for elevated adsorption, and it is determined by using infrared spectra (IR) and Fourier transform infrared spectroscopy (FTIR).

In addition, aging through photooxidative decomposition induces formation of new adsorption bands. For example, oxygen-containing groups such as phenol and ketones with carbonyl group (C=O), carbon–oxygen (C–O), hydroxyl group (OH-) bonds are formed, which are defined to increase the polarity of the MPs (Bandow et al., 2017; Mato et al., 2001). In a study to determine aging rate of Polyphenylene ether (PPE) and HDPE plastics, FTIR analysis was used, and carbonyl group (C=O), carbon-oxygen (C–O), hydroxyl group (OH-) and methyl group (CH_3) adsorption bands are measured. The virgin PPE has aged 6 times, while virgin HDPE has aged about 5 times, based on the carbonyl index, which also indicates aging ratio (Bhagat et al., 2022). Bandow et al. (2017) applied photooxidative and thermo-oxidative aging to PE, PVC, and PS MPs for 2000 h to observe the changes in the MP surface. It has been stated that new adsorption bands such as C=O (at 1700 cm^{-1}), C–O (at 900–1400 cm^{-1}), and OH^- (3100–3600 cm^{-1}) are formed especially in the case of photooxidative aging. As a result of the studies, the amount of total organic carbon, and desorbed amount of Cl, Ca, Cu, and Zn from the MP surface after aging, was measured. The proposition that the presence of elevated concentrations of oxygen-bearing functional groups on the aged microplastic (MP) surfaces may engender an augmented availability of interaction sites conducive to adsorption onto said microplastics has been put forth by Bhagat et al. (2022). Thus, it was concluded that deformation on the MP surface has a positive effect on the heavy metal adsorption on MPs (Yu et al., 2020).

2.2 *Effect of Heavy Metal Characteristics*

Type of the heavy metal determines its adsorption on MPs, as well. To illustrate, the investigation revealed that the adsorption kinetics of Lead (Pb) onto polyethylene terephthalate (PET) exhibit a slower pace in comparison to Cadmium (Cd) and Zinc

(Zn), corroborating findings noted within the existing body of literature (Abbasi et al., 2020). This difference in adsorption of metals may be the result of the elements' collective exposure to the adsorption sites on the PET surfaces. In natural and man-made matrices with constrained surface areas, the competition effect is a frequently noticed phenomenon (Abbasi et al., 2020). The mechanism of competition between different metals should be investigated in detail. In addition to heavy metal type, heavy metal concentration in soils has an impact on interaction between heavy metals and MPs. For example, Cr, Ba, As, and Pb metals tended to be adsorbed on MPs when their concentration was lower in MP structure compared to the soil. However, metals were gradually desorbed when their concentration in MP was substantially higher than in soil. More specifically, MP desorbed Sn, which was higher in MP than soil (Li et al., 2020c). In a study examining the interaction between MPs and heavy metals in different terrain types, it was found that heavy metal pollution in soil was associated with heavy metal concentration in MPs. Cd, Cr, Pb, Ag, Cu, Sb, Hg, Fe, and Mn contents of MPs in vacant land, woodland, and vegetable plots examined in the study were measured. If the metal concentration is higher in the soil compared to the MPs, the MPs adsorb metal from the soil. If there is a higher metal concentration in the MPs, then metal desorption takes place from the MPs to the soil. In particular, the concentrations of Cd, Pb, Hg, and Mn heavy metals in soil and MPs were found to be positively correlated with each other (Holmes et al., 2014). Yet, the effect of soil type or other parameters was not examined in the study (Zhou et al., 2019).

2.3 *Effect of Environmental Factors*

In addition to MnP and heavy metal properties, environmental factors have various effects on the interaction mechanisms between MnP and heavy metals (Holmes et al., 2014; Liu et al., 2021a). Among these environmental factors, soil aggregate size is an important physical parameter that affects the interaction mechanisms between MnP and heavy metals. According to the study of Yu et al. (2020), the interaction between MP and heavy metal is delayed as the size of the soil aggregate decreases. As a result, heavy metal presence in soil takes longer which in turn causes longer-term heavy metal pollution in soils composed of small aggregates (Yu et al., 2020).

Yu et al. (2020) investigated the exchangeable fractions of heavy metals (Cu, Cr, and Ni) such as carbonate-bound fractions, Fe–Mn oxide-bound fractions, and the organic-bound fractions in the presence of PE MP with different aggregate sizes (2000–250 μm, 250–53 μm, and < 53 μm) in the soil. The study concluded that MPs reduce the exchangeable, carbonate bond, and Fe–Mn oxide bond of metals (Yu et al., 2020). At the end of the study, organic bound fractions of Cr and Ni increased in all aggregate sizes, while organic bound fractions of Cu remained almost constant.

Among environmental factors, pH is considered important because it affects the electrostatic attraction between heavy metal and MP through changing electrostatic properties on the surface and the zeta potential of MP (Guo et al., 2020; Lin et al., 2021). It was stated that the zeta potential of CPE and PVC becomes more negative, whereas zeta potential of LPE turns from positive to negative values while the pH of soil increases (Liu et al., 2021a, b, c; J. Zou et al., 2020a). These alterations in zeta potential characteristics contribute to the enhancement of the adsorption mechanisms concerning a spectrum of heavy metals, including Lead (Pb), Cadmium (Cd), Cobalt (Co), Nickel (Ni), Copper (Cu), and Zinc (Zn) onto microplastics (MPs) (Guo et al., 2020; Holmes et al., 2014; Lin et al., 2021; Purwiyanto et al., 2020; Tang et al., 2020; Wang et al., 2020b; Zou et al., 2020a).

Additionally, Cr^{6+} exhibits a decrease in adsorption to virgin MPs as pH rises. This was explained by the assumption that this metal remains as Cr^{6+} in the tests, as a result; exists in oxyanionic state (more precisely, as $HCrO_4^-$ and CrO_4^{-2}). Thus, weak coulombic effects were observed between $HCrO_4^-$ and CrO_4^{-2} species and MPs leading to lower adsorption on MP. On the contrary, in the study conducted by Holmes et al. (2014), it was observed that Cr^{6+} adsorption on MP decreased as the pH of soil got higher, whereas Cd adsorption increased at pH values above 6 (Holmes et al., 2014). Soil pH also indirectly affects heavy metal adsorption on MPs by modifying adsorption on other components of soil. For instance, reduced Ph lowers the amount of negative charge on clay surface, which simultaneously decreases the adsorption of heavy metal ions on the clay. As a result, the amount of exchangeable heavy metals in the soil increases. For example, low soil pH was found to reduce Cd, Cu, Mn, Ni, and Zn adsorption on soil particulates (Sungur et al., 2014; Yu et al., 2020). Then heavy metals that could not be adsorbed on clays have a chance to adsorb on other substances in soil like MPs. Furthermore, the adsorptive dynamics of heavy metals are subject to the influence of soil pH, which induces alterations in various aspects including adsorption sites within the soil matrix, stability of adsorption surfaces, chemical speciation, and coordination attributes of these heavy metals.

The occurrence of organic matter within the environmental milieu has been documented to amplify and facilitate the adsorption of certain metals onto polymer surfaces (Godoy et al., 2019; Refaey et al., 2014; Saeedi et al., 2018). Notably, the organic matter content present in soil plays a pivotal role in modulating the interaction mechanisms between MnPs and heavy metals. This phenomenon extends to alterations in adsorption/desorption capabilities, as well as the implications for bioaccumulation and the toxicity profiles of these heavy metals. Adsorption capacity of MP with their large surface area can increase even further as a result of the adsorption of biofilms and natural organic substances on MPs (Ateia et al., 2020; Yu et al., 2020). Organic substances that bind to the MP surface allow heavy metals to bind easily to the MP surface (Bradney et al., 2019; R. Li et al., 2020a). In a study conducted in aqueous environment, the impact of MP contamination in urban wastewater, irrigation water, and pure water on heavy metal adsorption was investigated. It has been observed that Cu adsorption of MP increases from 1.592 mg g^{-1} to 6.241 mg g^{-1} when organic matter in the aquatic environment increases. A similar effect can be expected and worth to study in the soil environment. For instance, in the study of (Guo et al., 2020) Cd^{2+} sorption on the MPs (PE, PP, PVC, and PS) increased as the amount of humus material in the environment increased. Adsorption of heavy metals such as Ag, Pb, Cr, and Cu on MPs (PE, PP, PVC, and PS) was also found to be elevated by presence of organic substances (Abdolahpur Monikh et al., 2020; Godoy et al., 2019; Qiao et al., 2019). This higher adsorption of heavy metals on MP was also supported by Liu et al. (2021a, b, c) who stated that increasing adsorption sites on the MP surface due to organic matter in the environment. The effect of organic matter in the environment can also be observed with the formation of a biofilm on the surface. In the study conducted by Richard et al. (2019), it was shown that the amount of biofilm formed on the MP surface and the amount of heavy metal adsorbed on the MP are proportional. At the end of a 28-day study, regardless of the MP type, similar amounts of Ba, Cs (cesium), Fe (iron), Ga (gallium), Ni, and Rb (rubidium) metals were adsorbed by MP thanks to the biofilms formed on the MP surface in the water environment. Biofilm formation on MPs might be observed in soil environment, too. In conclusion, owing to organic substances changing surface area and surface charge, the adsorption of heavy metals on the MP surface via complexation increases because complex formation also increases the surface roughness and hydrophobicity of the MP (Qiao et al., 2019). This mechanism may occur between heavy metals and MPs in the soil. Since there are not many studies on the effect of the presence of organic matter in the soil environment on heavy metal adsorption by MP, it should be investigated in the future.

On the other hand, the adsorption behavior of heavy metals on MPs can be affected through interactions with dissolved organic matter (DOM) (Baken et al., 2011; Xu et al., 2016), which negatively affects heavy metal adsorption on MPs. The reason is that heavy metals adsorbed on these DOMs are transferred to dissolved phase rather than being adsorbed on MPs. For example, fulvic acid (FA) is commonly found in soil and an important DOM (Duval et al., 2005). In a study by Tang et al. (2020) it was observed that Pb^{2+} adsorption by aged nylon MP was negatively affected in the presence of high concentrations of FA. As the FA concentration in the medium increased, Pb^{2+} adsorption by the aged MP tended to decrease. This is due to the complexation between active functional groups such as carbonyl, hydroxyl, and carboxyl of FA and Pb^{2+}. Due to this complexation, FA binds strongly with heavy metals, which retains heavy metals in solution and decreases heavy metal adsorption on MP. Similar results can be observed in the soil pore water due to DOM content, and it is worth to study the mobility of MPs and heavy metals between soil and pore water.

Another environmental factor affecting the interaction between MPs and heavy metals is ionic strength. Although there is no study investigating the impact of ionic strength on heavy metal-MP interaction in soil, research in aqueous media can be used to characterize potential effect of ionic strength of soil pore water. Increasing ionic strength engenders an effect contrary to that of pH, leading to a diminished propensity for heavy metal adsorption onto microplastics (MPs). Consequently, an escalation in ionic strength has been associated with a reduction in the adsorption of heavy metals, including Cadmium (Cd), Copper (Cu), Cobalt (Co), Nickel (Ni), and Lead (Pb), onto microplastics (Holmes et al., 2014; Guo et al., 2020; Lian et al., 2020; Tang et al., 2020; Lin et al., 2021).

Additionally, in their study, Zou et al. (2020) stated that the increased ionic strength was found to affect competition of metals such as Cu^{2+}, Pb^{2+}, and sodium ion (Na^{+}) for sorption sites of MPs, so they could decrease the activity of heavy metals in aqueous environment due to adsorption on MPs. Contrary to this situation was observed for Cr^{6+} where increasing ionic strength increases Cr^{6+} adsorption on PE (Holmes et al., 2014). The ionic strength of the water in the soil pores is expected to affect the heavy metal-MP interaction likewise.

Ionic strength has an additional impact on heavy metal-MP interaction through affecting complexation capacity of MPs. For example, when Cu adsorption on CPE, PVC, HPE, and LPE was investigated, it was reported that increasing ionic strength decreased the adsorption except for CPE. CPE, which is a polar MP with chlorine content, adsorbed almost 10 times more Cu^{2+} compared to other MPs. The reason for this is that chlorine, which is dispersed in CPE's molecular structure, provides CPE with the ability to form complexes with bivalent cations. Thus, higher adsorption at higher ionic strength was attributed to that characteristic of CPE. Although PVC contains chlorine similar to CPE, chlorines in its structure are symmetrical and regular, which makes PVC not being able to form as many complexes as in CPE. In addition, another situation that supports the higher Cu^{2+} adsorption of CPE is the stronger complexing ability of Cu^{2+} compared to other metals in the study (J. Zou et al., 2020a). Studies examining the effect of salinity as an environmental parameter have shown that salinity does not have a significant effect on Cu^{2+} and Pb adsorption of PE pellets in aqueous environment (Holmes et al., 2014; Purwiyanto et al., 2020).

As soil attributes exert a direct influence on the interplay between MPs and heavy metals, it is pertinent to note that the presence of MPs can engender an indirect impact on this interaction, attributable to the capacity of MPs to modify soil properties. Beyond these considerations, it should be acknowledged that MPs have the potential to detrimentally influence the bioavailability of heavy metals by eliciting alterations in the physical and chemical attributes of soils. These modifications encompass factors such as soil structure, bulk density, stability of aggregation, water retention capability, moisture availability, dissolved organic matter (DOM) content, and pH. Pertinent investigations have been conducted, as exemplified by the works of (Liu et al., 2017; De Souza MacHado et al., 2018; De Souza Machado et al., 2019).

After the addition of PE, the Dissolved Organic Carbon (DOC) content of all soil aggregates decreased by an average of 36.8%, and their pH decreased by an average of 18%, which are critical parameters in soils (Liu et al., 2017). This was explained by adsorption of organic matter in the soil by MPs. This decrease in DOC in the soil increases adsorption of metals by the soil and decreases their solubility and extractability (Yu et al., 2020). In particular, many studies have shown that MPs increase the bioavailability of heavy metals (Li et al., 2020c; Zhang et al., 2020a; Chappaz & Curtis, 2013; Karlsson et al., 2006). In addition, since MPs can also adsorb organic matter, they can become a component of soil organic matter. Consequently, they lead to an elevation in the concentration of heavy metals bound to organic matter within the soil matrix (Hüffer & Hofmann, 2016).

3 Environmental Implications of Interactions between Heavy Metals and Micro-Nanoplastics in Soil

3.1 *Mobility of Micro-Nanoplastics and Heavy Metals*

3.1.1 Mobility of Micro-Nanoplastics

MnPs have a high propensity to migrate and dispersed vertically and horizontally in the soil (Nizzetto et al., 2016; Okeke et al., 2023; Xu et al., 2020). MPs in the soil environment can reach deeper parts of the soil through rain and irrigation (Keller et al., 2020). Also, traditional tillage practices such as harvesting or plowing can cause movement of plastics from the topsoil to deeper layers, as they overturn the surface and deep soils (Moeck et al., 2023). Additionally, worms, microarthropods, and other bioturbators can transport MnPs deeper from the topsoil (Huerta Lwanga et al., 2017; Lwanga et al., 2022; Moeck et al., 2023). Rillig et al. (2017a, b) observed that worms (*Lumbricus terrestris L.*) carry 710 to 2800 μm PE beads up to 10 cm in 21 days. In addition, worms, microarthropods, and other bioturbators may create preferential flow paths, resulting in vertical transport of MnPs up to 40–50 cm depth (Huerta Lwanga et al., 2017; Ren et al., 2021; Rillig et al., 2017a, b). Weather conditions such as wet-dry cycles are associated with soil cracks and can also affect vertical MnP transport. Thus, MnPs transported downstream in the soil can mix with groundwater and migrate to different environmental media (Moeck et al., 2023; Okeke et al., 2023; Wang et al., 2022a).

In addition to physical mechanisms, chemical conditions in soil, such as pH and ionic strength, have the ability to affect the transport of MnPs (Moeck et al., 2023). Increased MnP transport was observed with increasing pH. In a study pH 5 caused replacement of PS MPs 43.1%, whereas pH 9 caused replacement of 88.7% in a quartz media (Tan et al., 2021). Also, presence of ionic compounds might affect MnP transport. A positive correlation was determined between PS MPs retention in loamy sand, silty loam, and silty clay soils and the amount of positively charged Fe or Al oxides (Wu et al., 2020). Also, MPs are more agglomerated in the presence of high valence cations ($Ca^{2+,}$ $Mg^{2+)}$ and more likely to be immobilized in soil. Therefore, due to increasing ionic strength (Na+, Ca2+, and Mg2+), MP retention increases and MP transport rate decreases as MP retention increases (Moeck et al., 2023; Zhang et al., 2022). For example, with increasing solution ionic strength, the transport of PS NPs in Desert soil (DS) and Black soil (BS) decreased significantly (Wu et al., 2020). Additionally, with increasing organic matter content, MnP transport was observed to increase (Alimi et al., 2021; Keller et al., 2020; Moeck et al., 2023). For example, it was observed that the mobility of PS NPs (28 nm) along the sand-filled column increased significantly if natural organic matter (NOM) existed. The calculated elution (C/C0) value in the presence of organic matter increased from 0.024 to 0.92 (Alimi et al., 2021).

3.1.2 Mobility of Heavy Metal in the Presence of Micro-Nanoplastics

MnP exist in the soil environment along with various pollutants, and MnPs are recognized as important vectors for the accumulation and transport of these pollutants on their surfaces (Azeem et al., 2021; Feng et al., 2022; Ren et al., 2021; J. Wang et al., 2019b). Hence, transport of MnPs together with the pollutant they adsorb can cause significant environmental problems in the soil environment due to the increased mobility of other pollutants (He et al., 2018; Hüffer et al., 2019; Ren et al., 2021; Tian et al., 2022). Heavy metals are pollutants commonly found in the soil environment and MnPs can act as a vector for metals as they can affect the transport, mobility, and bioavailability of heavy metals in the soil environment (Kinigopoulou et al., 2022; M. Li et al., 2021a; Liu et al., 2021a, b, c; Medyńska-Juraszek & Jadhav, 2022; Ren et al., 2021; S. Zhang et al., 2020b).

In the studies, it is seen that the presence of MnPs (HDPE, PE, PP, PET, PP, and PS) can affect Cu, Cr, Cd, Ni, Pb, Zn, and As heavy metal mobility and bioavailability in the soil environment (Hao et al., 2023; M. Li et al., 2021a; Ma et al., 2022; Medyńska-Juraszek & Jadhav, 2022; Yu et al., 2020; S. Zhang et al., 2020b). For instance, the extractable fractions of Pb and Zn from the soil were 10.10% and 18.29%, respectively, in pure soil, while these values increased to 12.35% and 23.96%, respectively, when the PE MP concentration was increased to 10%. In the same study, while the adsorption of Zn and Pb in soil without MP was 4.94 mg metal/g soil and 4.44 mg metal/g, respectively, the adsorption values of Zn and Pb in the presence of 10% MP decreased to 4.56 mg/g soil and 3.73 mg/g, respectively (M. Li et al., 2021a). Similarly, in the presence of MnPs (PeS, Biodegradable PE, PET, PE), extractable Cu^{2+} and Pb^{2+} in the soil increased by up to 5.5% (Medyńska-Juraszek & Jadhav, 2022), while Cd mobility increased by approximately 22% in a column experiment simulating a river sediment environment. In another study, it has been determined that PE MPs (100 μm) can reduce their bioavailability by affecting the speciation of Cu, Cr, and Ni heavy metals in the soil. As a result of the incubation experiments, on average, the exchangeable fraction (F1) of Cu, Cr, and Ni decreased by 31.7%, 36.5%, and 29.2%, respectively, while the organic bound fraction of Cu, Cr, and Ni (F4) increased by 28.4%, 197.1%, and 159.2%, respectively (Yu et al., 2020).

Moreover, the increasing concentrations of PP MPs (from 0.05% to 10%) decreased the pore volume where saturation concentration is reached (from 230 to 170) and this is used to explain the increased transport rate of Cd^{2+} (Y. Li et al., 2021b).

Addition of HDPE MP (48 μm to 2 mm) lowered Cd adsorption capacity of soil and raised Cd desorption from soil. In the study, the amount of Cd adsorbed (qe) decreased from 1469.6 mg/kg to 1460.0 mg/kg, and the amount of Cd desorbed (qde) increased from 179.90 mg /kg to 241.48 mg/kg when the MP dose increased from 0% to10%. So, it increased the mobility of Cd heavy metal (S. Zhang et al., 2020b).

Ma et al. (2022) investigated the adsorption characteristics of Cd in bulk soil and three soil solid fractions (i.e., particulate organic matter (POM), organic-mineral compounds (OMC), and mineral with or without PP (50 μm) MPs. Experiment results showed that bulk soil, POM, and OMC fractions decreased the Cd adsorption capacity with the addition of PP MP. Such that, with the addition of 10% concentration of PP MP, the POM adsorption capacity decreased significantly from 3.21 to 1.77 mg/g.

Also, Hao et al. (2023) studied the co-transport dynamics of As and PS MnP particles in saturated soil. Batch adsorption experiments found that the maximum adsorption capacity of As on PS increased from 1.64 to 2.43 mg/g as MnP size decreased from 50 μm to 0.1 μm. This reported As adsorption on PS leads to change in transport behavior of As in the column. It was reported that in column containing NP with the size of 0.1 μm saturation concentration of As was reached at 1.6 pore volume whereas As saturation was reached in 1.4 pore volume in column without NPs. Related with this observation, the transport velocity of As decreased from 0.65 to 0.52 cm/min. On the other side, in the presence of PS MP with the size of 5 μm the saturation concentration was obtained after 1.2 pore volume indicating faster movement of As. However, velocity was not found to be affected.

MPs can reduce mobility and availability by leading to the conversion of heavy metal species from existing mobile fractions (replaceable) to stable organic forms (carbonate forms) (Fuentes et al., 2008; Tessier et al., 1979). For this reason, MnPs can affect the mobility of heavy metals in the soil environment, increasing the risk of mixing with groundwater and other environmental environments.

MnPs can be transferred from the soil environment into the plants. Recent studies have demonstrated the ability of MP and NP particles to move through cell membranes and plant cells (Feng et al., 2022; He et al., 2021; Hua et al., 2023; Okeke et al., 2023). For example, one study found that both MnP PS can be taken up by rice roots and then transported to the above-ground parts (Liu et al., 2022). Similarly, PS NPs were detected in the cell walls of lettuce plants exposed to PS NPs (Hua et al., 2023; L. Li et al., 2019a). It is estimated that Mexico consumes an average of 840 plastic particles per year through food intake (Jin et al., 2022). Considering that MnPs are detected in human feces, this indicates that MnPs enter human body through food chain (Schwabl et al., 2019). Moreover, MnPs can carry toxic pollutants such as heavy metals from the soil environment to plant roots and above-ground parts of plants such as stems and leaves (Jia et al., 2022; Okeke et al., 2023; Yue et al., 2023). Thus, with the consumption of these plants, it seems likely that heavy metals will enter the food chain from the soil environment (Mamathaxim et al., 2023).

The coexistence of MnP and heavy metal in the soil environment may affect their uptake by the plant. For example, in one study, it was observed that 0.1% PE MP (293 μm) in the soil would facilitate the transfer of Pb and Cu heavy metals into rapeseed plants. The results of the study showed that the PE MP content in the soil could increase the heavy metal concentration in rapeseed plants (Jia et al., 2022). In the same study, with the presence of 0.1% PE MP, Cu^{2+}, and Pb^{2+} concentrations reached 38.9 mg/kg and 9.4 mg/kg from 20.7 mg/kg and 2.6 mg/kg, respectively.

Similarly, PS NPs (82 nm, 50 mg/L) have been observed to promote As accumulation in the tissues and leaves of rice seedlings and to increase As accumulation by 12.4–36.7% in rice leaves (Mamathaxim et al., 2023). In another study, it was found that especially 10% PE MPs (8.68–500 μm) increased Cd accumulation by increasing Cd bioavailability in lettuce grown in Cd-contaminated soil (F. Wang et al., 2021a). Cd content in lettuce was determined. It was observed that while PE MP concentration increased from 0.1% to 10%, shoot and root Cd contents increased significantly from 9.50 to 62.0% and from 9.61 to 61.4%, respectively.

In another case, the presence of MnP prevents heavy metal's passage to the plant by adsorbing heavy metal (Dong et al., 2021). In one study, the presence of PS NP (0,01–10 mg/L) was observed to reduce the accumulation of Fe, Mn, Cu, and Zn in wheat seedlings (Lian et al., 2020). In the study, the contents of Mn, Cu, Zn, and Fe heavy metals were determined. While Fe concentration after 21 days of PS NP (1 mg/L) exposure decreased from 386.2 mg/kg to 344.7 mg/kg (10.8%) in shoots ($p < 0.05$), it decreased from 2472.1 mg/kg to 1791.6 mg/kg (27.5%) in roots ($p < 0.01$). Similarly, after exposure to PS NPs, shoot Mn contents decreased between 28.3% and 53.2% ($p < 0.01$), Cu accumulation in roots decreased between 26.0% and 65.1% ($p < 0.01$), and Zn content in roots decreased around 80.9% ($p < 0.05$).

Moreover, the presence of PS MnP was found to alleviate Cu/Cd accumulation in wheat seedlings (Yue et al., 2023; Zong et al., 2021). In the study, Cu and Cd contents in wheat roots and leaves were determined. In the presence of PS MP (0.5 μm, 100 mg/L), Cu and Cd contents in wheat roots decreased from 299.0 and 155.4 μg/g dry weight to 270.5 and 138.9 μg/g dry weight, respectively. This corresponded to a decrease of 9.5% ($p < 0.05$) for Cu and 10.63% ($p < 0.05$) for Cd. Likewise, the Cu and Cd contents of wheat leaves were 29.52 and 15.68 μg/g dry weight, respectively, in the presence of PS MP (0.5 μm, 100 mg/L), which decreased to 20.72 and 11.86 μg/g dry weight. This corresponded to a decrease of 29.80% ($p < 0.05$) for Cu and 24.33% ($p < 0.05$) for Cd. In another study, the transport of heavy metals by MPs to the plant root zone and the adsorption of Pb, Cd, and Zn were investigated in a wheat field. PET particles adsorbed 7.2–8.5%, 5.3–9.8%, and 29.8–68.5% of Zn, Cd, and Pb solution, respectively. Pb is adsorbed lower than Cd and Zn. This result is compatible with literature because Pb is highly adsorbed on PVC, nylon, PP, or PS than PET (Abbasi et al., 2020).

3.2 *Toxic Impacts*

MnPs are becoming a growing concern due to their potential to harm both global ecosystems and human well-being. According to the analysis of Möhrke et al., 2022 regarding the types, sizes, and shapes of the most studied MPs and tested environments, a significant proportion of the experimental trials were conducted using natural soil as the testing medium. The focus of research has primarily been on MPs rather than NPs. In the investigation of MPs toxicity for soil organisms, the primary focus has been on PE and PS. When Web of Science was searched for "soil toxicity" and "polyethylene," there were a total of 266 studies. The other types of MPs were studied less frequently compared to PE and they were in the order of PS (186 studies), PP (49 studies), and PVC (42 studies). The number of studies highlights that investigation of the soil toxicity of MPs is in its initial stage. Moreover, studies on the effect of MPs on the soil toxicity of other pollutants such as heavy metals have been initiated recently and the number of studies is very limited (37 studies for PE, 23 studies for PS, 9 studies for PP, and 5 studies for PVC). Upon reviewing the abstracts of the articles, it was observed that among the various forms of MPs, fragmented pieces constituted the majority of the studies. On the other hand, film-type MPs were relatively less explored in the studies.

3.2.1 Toxic Impacts of MnPs

The impacts of MnPs on soil organisms have been predominantly studied in terms of endpoints, including cellular response, mortality, reproduction, behavior, and development (Lahive et al., 2022; Ma et al., 2020; Selonen et al., 2020). These studies have shed light on the complex and diverse nature of responses observed in different organisms. For instance, *Eisenia fetida* exhibited mortality when exposed to PS MP at concentrations of 0.1 μg/kg –1000 μg/kg, while its development remained unaffected (X. Jiang et al., 2020b). In another study, *E. fetida* exposed to PS MP at concentrations ranging from 0.4 to 1 milligram per milliliter resulted in inhibited growth and an increased mortality rate of 40% to 50% (Cao et al., 2017). Similarly, the exposure to PS microplastics led to a decreased survival rate of about 20% and a reduction in the body length of the *Lobella sokamensis* (Kim et al., 2019). On the contrary, the survival of *Eisenia andrei* exposed to PE MP was not affected significantly, but the *E. andrei* tissue and immune system were shown to be damaged (Rodriguez-Seijo et al., 2017). Additionally, reproductive effects were observed in *Folsomia candida* exposed to PE MP concentrations ranging from 0.5% to 1% where their reproductive ability was inhibited, and alterations in their gut microbiome were detected. The half-maximal effective concentration (EC50) value for reproduction in *F. candida* was found to be 0.29% MPs (Ju et al., 2019). In terms of behavior, *Lobella sokamensis* exposed to both PE and PS MP showed disrupted movement (Kim & An, 2019; X. Jiang et al., 2020b; Selonen et al., 2020). Additionally, when *Achatina fulica* was exposed to PET MP at a concentration of up to 0.70 grams per kilogram of dry soil weight, they showed avoidance of food intake and excretion (Song et al., 2019). Moving on to cellular response, a significant decrease (by 50% to 60%) in glutathione peroxidase and total antioxidant capacity of *Achatina fulica* was observed. Similarly, when considering LDPE MP concentrations spanning from 0.1 to 1.5 g/kg *Eisenia fetida* exposed to LDPE MP exhibited surface damage, oxidative stress, and neurotoxicity (Chen et al., 2020b; Song et al., 2019). PS microplastics of sizes 100 nanometers and 130 nanometers caused damage to the intestinal cells and DNA of the earthworms (Cao et al., 2017). Moreover, the exposure of *Caenorhabditis elegans* to polystyrene (PS) microplastics at a concentration of 14.23 milligrams per kilogram caused perturbation of metabolites involved in energy metabolism.

In addition to their impact on animals, MPs have been found to significantly affect various plants (Enyoh et al., 2019). These effects include inhibiting plant growth (Bhagat et al., 2021), lowering chlorophyll content (Z. Li et al., 2020a), altering antioxidant enzymatic activities, inducing cytogenotoxic effects (Jiang et al., 2019), and bringing about changes in gene transcription (Sun et al., 2020). MnP can exert diverse phytotoxic effects on plants through various modes of action. The phytotoxic effects of MPs stem from their ability to modify the soil's carbon-to-nitrogen ratio, influencing plant growth and biomass production. Upon entering plant cells, these particles can spread to various cellular components, including the nucleus, cytoplasm, and chloroplasts (M. Li et al., 2021a; Zhou et al., 2021).

3.2.2 Toxic Impacts of Heavy Metals in the Presence of MnPs

While MPs have been studied extensively for their potential impacts on various organisms, their interactions with other environmental stressors also took attention due to their potentially complex outcomes. MPs exhibit toxicity on their own and function as carriers, capable of adsorbing and transporting various pollutants such as heavy metals across the food web and environment (Brennecke et al., 2016). Given the interaction between MnPs and heavy metals, it is imperative to assess the combined impact of MnPs and heavy metals on terrestrial organisms (Dong et al., 2021; Hodson et al., 2017; C. Huang et al., 2021b; M. Li et al., 2021a; H. T. Wang et al., 2019a; Xu et al., 2023).

The coexistence of MP with heavy metals in the soil can lead to various physical and chemical consequences on the soil invertebrates. It has been demonstrated that the co-exposure of MnP with heavy metals can result in synergistic effects in this context (C. Huang et al., 2021b; Ozturk et al., 2023; H. Zhou et al., 2020a; Y. Zhou et al., 2020b). The effect of MPs on heavy metal toxicity in terms of basic endpoints such as mortality, growth rate, and reproduction has been identified (Gudeta et al. 2023; Holzinger et al. 2022; Mai et al. 2023; Ozturk et al. 2023). The addition of PE MP and Cd had a significant impact on the growth rate of *E. fetida*. *E. fetida* exposed to 2 mg/kg of Cd-contaminated soil containing 30% MPs lost its biomass by 17.1% compared to the control group. A further decrease in *E. fetida* biomass (24.8%) was observed at a higher concentration of Cd (10 mg/kg) (C. Huang et al., 2021b). Similarly, coexistence of PS-MP and Cd amplified heavy metal toxicity in the soil. The EC50 for Cd was 102 mg/kg dry soil without PS-MPs but significantly decreased to 38 mg/kg dry soil in the presence of PS-MPs with concentration of 0.5%. This enhanced toxicity can be attributed to the elevated exposure of Cd due to its adsorption on PS-MPs (Ozturk et al. 2023). Additionally, *E. fetida* has the ability to retain PP MPs, with quantities ranging from 4.3 to 67.2 particles per gram of earthworm biomass, consequently leading to an elevation in Cd accumulation within earthworms by a margin of 9.7% to 161.3% (Y. Zhou et al., 2020b). When MP and heavy metals were present together, oxidative stress responses were also triggered, and cellular reactions were more severe compared to when they were exposed individually. When *E. fetida* were exposed to soil contaminated with

10 mg/kg of Cd and the concentration of PP MPs rose beyond 15%, there was a significant increase in tail DNA, exceeding 5%, and the OTM value also went beyond 2. This suggests that a strong connection exists between a high amount of PE MPs and Cd contamination, leading to considerable damage to the sperm DNA of *E. fetida* (C. Huang et al. 2021b). In another study, when As is administered at a concentration of 40 mg/kg and PVC microplastics at 2000 mg/kg to *Metaphire californica*, the adverse effects of arsenic on the gut bacterial communities are increased by 0.41%. This may occur through the inhibition of As(V) reduction (H. T. Wang et al., 2019a). Microplastics can coexist not only with a single metal but also with a combination of different metals, exerting an impact on toxicity. Co-exposing earthworms to a 1% concentration of PP MPs alongside a blend of Cu, Zn, and Cr metals resulted in mortality, with a 100% within a 14-day exposure period. This outcome is likely due to the combined toxic effects of the PP MPs and the heavy metal mixture, potentially leading to a decline in the earthworms'natural self-defense capabilities (Xu et al. 2023).

Even though some studies suggest that MPs can increase the toxic effects of heavy metals in earthworms, other studies have found contradicting results (Liang et al. 2022). During an investigation involving *E. fetida*, it was found that PE MP mitigated Cd toxicity. The study examined the concentrations of Cd (ranging from 1.5 mg/kg to 48.1 mg/kg) and MPs (378 mg/kg), and an antagonistic effect was observed in the mixture treatments, where the effects on CAT (catalase) activity, GST (glutathione S-transferase) activity, and MDA (malondialdehyde) content were smaller than expected based on the respective null models (Liang et al. 2022). Moreover, PS MP applications, at a concentration of 0.5%, had a decreasing effect on Ni toxicity on *E. crypticus*, as evidenced by the increase in the EC50 value from 47 mg/kg to 208 mg/kg in the presence of 10% PS MPs (Ulutuğ and Topuz, 2022).

The toxicity of heavy metals is influenced by the type and surface properties of MP (S. Liu et al., 2021b). The toxicity of various MP, including their polymer composition, shape, type, and their mixtures, still requires reliable determination. For example, the mortality effect of PP MP with sizes smaller than 150 micrometers has been observed on *Eisenia fetida*, histopathological changes, oxidative stress, and DNA damage were assessed, indicating adverse effects on its development (X. Jiang et al. 2020b). The study also examined the effects of another type of PE MP on a smaller organism, *Enchytraeus crypticus*. Interestingly, *Enchytraeus crypticus* exhibited reproductive effects in response to PE MP, with a clear decrease of 20%, 20%, and 30% observed at long fiber concentrations of 0.17%, 0.5%, and 1.5%, respectively (Selonen et al. 2020). Due to limited data on the effect of MP properties on soil toxicity, systematic investigations into how particle properties influence their ecotoxicological behavior in soil are rare, making it difficult to determine toxicity mechanisms related to MP properties (Büks and Kaupenjohann 2020; Kim & An 2019). The influence of environmental conditions, encompassing factors such as organic matter, pH, and salinity, serves to underscore the paramount significance of soil properties in the mixture toxicity of microplastics and heavy metals within the soil (Khalid et al. 2021). In this context, the introduction of biosolids brought about a noteworthy reduction in Cd toxicity. The application of 0.6 g biosolids yielded an

EC50 value of 193 mg Cd/kg dry soil, while the 0.9 g biosolids application resulted in an EC50 of 473 mg Cd/kg dry soil. Moreover, when biosolids coexisted with PS-MPs, a reduction in reproductive toxicity similar to that of biosolids was observed for Cd, as reflected in the EC50 value of 305 mg Cd/kg dry soil. It is worth noting that the bioaccumulation of Cd exhibited a distinct pattern, positively correlating only with its initial concentration in the soil, irrespective of the presence of PS-MPs or biosolids (Ozturk et al. 2023). While the characteristics of microplastics influence toxicity, soil properties also have the potential to alter toxicity. However, studies concerning the impact of organic matter and soil properties on combined toxicity of heavy metals and MPs are inadequate.

Also, as MPs age, they become better at binding to heavy metals due to increased adsorption capabilities (Khalid et al. 2021). In a study investigating the impact of aged PE MP on the bioavailability of heavy metals in soil, it was observed that even at a low concentration of 1%, aged PE MP increased the bioavailability of heavy metals. However, the aging process had a negligible effect on the overall environmental toxicity of MP. Cd concentration is measured in the body of *E. fetida* in the context of *E. fetida* exposure to the Cd and aged PE MPs. Bioaccumulation of Cd was significantly higher in the presence of aged PE MPs (5.91 mg/kg) compared to single Cd treatment. Additionally, the presence of aged PE-MP intensified the heavy metal toxicity on soil organisms, as supported by higher levels of reactive oxygen species (ROS) and malondialdehyde (MDA) in the soil. This suggests that the combination of aged PE-MP with heavy metals led to increased oxidative stress and lipid peroxidation, potentially affecting the health and survival of soil organisms (Li et al. 2023). To sum up, aged MPs had a greater toxic effect on *E. fetida* than non-aged MPs (Cheng et al. 2020). This was attributed to the aged MPs' enhanced adsorption capacity toward heavy metals, facilitated by the development of a larger specific surface area, a rougher surface structure, and the presence of oxygen-containing functional groups, all of which resulted from the aging process (Z. Wang et al. 2021b).

Various terrestrial plant species, including *Zea mays, Triticum aestivum, Lactuca sativa,* and *Allium species* have been the subject of research to explore the adverse effects of MnP on plants (Mateos-Cárdenas et al. 2021). MP in the soil reduce the ability of soil to bind heavy metals, making them more available to plants. For PS MP the research focused on their impact on *Zea mays*, shedding light on their accumulation and distribution patterns. The accumulation of beads with sizes ranging from 0.2 to 2.0 μm in root tissues varied based on their sizes, while the larger PS beads (5.0 μm) appeared to remain unabsorbed by the maize seedling (Li et al. 2023). Additionally, it was revealed that PS MP in the range of 25–400 mg/L could induce cytogenotoxicity in *Allium cepa* (Maity et al. 2020). On the other hand, PP MP exhibited severe negative effects on *Zea mays*, adversely affecting both growth and crop quality (Zhou et al., 2023). In comparison to the control group, the introduction of MP at a concentration of 1.0 mg/mL resulted in a reduction of 12% in parameters such as fresh and dry leaf and root weights, plant height, leaf count, and root length. Similarly, exposure to PE MP resulted in significant reductions in *Lactuca sativa* growth, photosynthetic parameters, and chlorophyll content. The chlorophyll content exhibited a more pronounced decrease in treated plants as the MP exposure increased from 0.25 to 0.50–1.00 mg/mL. Specifically, the chlorophyll content decreased by approximately 10% compared to the control. (Gao et al., 2019). Interestingly, the growth and yield of *Triticum aestivum* remained unaffected even with the presence of LDPE microplastics.

The co-occurrence of Cd and 10% PS and HDPE MPs leads to adverse effects on plant growth, roots, and photosynthesis. Moreover, their combined presence exacerbates the negative impact on root biomass. In the case of Cd application, the root content decreased from 1.3 mg to 0.5 mg in the presence of HDPE. Likewise, in the soil treated with PS microplastics, the dry weight decreased from 0.9 mg to 0.6 mg. It is indicated potential synergistic interactions between Cd and microplastics in influencing plant health and development (F. Wang et al. 2020a). Different types of MP have different effects on heavy metal bioavailability and soil properties. For example, PE microfibers in soils contaminated with cadmium can interfere with nitrogen and carbohydrate metabolism pathways (Pinto-Poblete et al. 2022). MP can influence the uptake and accumulation of heavy metals like As in plants, impede detoxification mechanisms, and negatively affect plant growth. In the presence of PM MPs at a concentration of 10–20 mg/L, adding 1, 2, and 4 mg/L As solution reduced the root As content by 16.5%, 14.3%, and 14.5%, and the leaf As content by 11.3%, 14.0%, and 22.7%, respectively (Dong et al., 2021).

Expanding routine testing to include more soil species is essential for a better understanding of MP potential impacts on ecosystems. Robust toxicity thresholds based on dose-response testing, including chronic toxicity endpoints, need to be established to account for effects, and heavy metal leaching from plastic particles over time. Furthermore, verifying toxicity thresholds through realistic exposure scenarios involving multiple species, trophic interactions, and long-term exposure will enhance the accuracy of the findings and provide valuable insights into the ecological implications of MP pollution with other pollutants like heavy metals.

4 Conclusions

As a result of the literature review, it appeared that MnP and heavy metals frequently coexist in soil environment due to biosolid, compost or mulch application, irrigation with reclaimed wastewater, and atmospheric deposition. Understanding interaction mechanisms between MnP and heavy metals has become crucial to assess implications of this interaction. Main mechanisms between MnPs and heavy metals are defined as adsorption and complex formation, which highly depend on MnP characteristics such as functional groups present on MnP surface, zeta potential, C content, Cl content, and biodegradability of MnP because of their impact on surface charge, polarity, and hydrophobicity. In addition to MnP properties, heavy metal type and its relative concentration in soil and MnP determine their interaction. Moreover, environmental conditions such as ionic strength, pH, and organic matter content are found to affect MnP-heavy metal interaction mechanisms. There have not been studies about the direct impact of ionic strength but it is reported to modify competition between heavy metals and complexation capacity of MnPs in aqueous environment, which can be used as a clue to expect similar results at the soil-pore water interface. Surface characteristics of MnPs and speciation of heavy metals could change as a function of pH which makes pH a significant parameter, as well. Organic matter in the soil environment and organic matter that can be formed on the MnPs surface as biofilms are determined effective in terms of modifying adsorptive behavior of MnPs through changing polarity and hydrophobicity of MnPs.

Concerns about MnP pollution in soil are increasing as scientific studies have been progressing with striking results such as enhanced mobility and toxicity of other pollutants due to presence of MnPs. Both MnPs and heavy metals are contaminants frequently detected in the soil, and MnP is found potentially to elevate mobility of heavy metals through reducing adsorption capacity of soil, and increasing extractable fraction of heavy metals. Size of the plastic particles was found to affect transport rates of heavy metals in column studies where MP decreased and NP increased movement. In addition to mobilizing heavy metals and leading to groundwater contamination, presence of MnP also enhanced transfer of heavy metals into plants. and complexation reactions between soil, MnPs, and heavy metals.

For toxicity, MnPs exert substantial influence on soil organisms, influencing parameters such as mortality rates, growth patterns, reproductive capacities, and cellular reactions. The interaction between MPs and heavy metals exhibits synergistic tendencies, amplifying the toxicity experienced by these organisms. Additionally, aged MPs demonstrate an elevated affinity for heavy metal binding, intensifying the detrimental effects. Concurrently, terrestrial plants experience compromised vitality, with disruptions in growth, photosynthetic activities, and nutrient transport pathways. Moreover, bioaccumulation factor of heavy metals is triggered by the MnPs, and ecosystem services are threatening due to the food web chain transportation.

To gain a better understanding of this intricate situation, researchers must uncover the interplay between MPs and pollutants. This will lead to defining safe limits for MPs and creating practical plans for managing them. As this exploration progresses, a deeper comprehension of these interactions will guide efforts to protect terrestrial ecosystems and maintain the delicate balance of nature.

Especially, surface properties, size, and shape of MnPs should be understood and considered well during the early design phase of plastic products to eliminate the risk of mobilization of heavy metals due to MnPs. Moreover, restrictions on the MnPs type to be used for mulching could be applied to control the mobility and toxicity of heavy metals. MnP concentration, size, and shape can be added as a soil quality parameter to restrict the usage of biosolids, reclaimed wastewater irrigation, and mulching. Besides MnPs, heavy metal types and concentration determine the possibility and type of interaction with MnP; therefore, heavy metal restrictions on the soil applications should be revised counting on their interactions with MnPs. Soil properties dramatically can change the interaction of heavy metal and MnP; therefore, restrictions can be classified considering the pH, organic matter content, and ionic strength of the soil.

Acknowledgments This work was supported by generous grants from the Scientific and Technological Research Council of Turkey under the TUBITAK 2247-A Fellowship for Outstanding Researchers (no. 120C147). However, all scientific contributions made in this project are owned and approved solely by the authors.

References

Abbasi, S., Moore, F., Keshavarzi, B., et al. (2020). PET-microplastics as a vector for heavy metals in a simulated plant rhizosphere zone. *Science of the Total Environment, 744*, 140984.

Abdolahpur Monikh, F., Vijver, M. G., Guo, Z., Zhang, P., Darbha, G. K., & Peijnenburg, W. J. G. M. (2020). Metal sorption onto nanoscale plastic debris and trojan horse effects in Daphnia magna: Role of dissolved organic matter. *Water Research, 186*, 116410.

Abeynayaka, A., Werellagama, I., Ngoc-Bao, P., et al. (2022). Microplastics in wastewater treatment plants. In *Current developments in biotechnology and bioengineering: Advances in biological wastewater treatment systems* (pp. 311–337). Elsevier.

Alimi, O. S., Farner, J. M., & Tufenkji, N. (2021). Exposure of nanoplastics to freeze-thaw leads to aggregation and reduced transport in model groundwater environments. *Water Research, 189*, 116533.

Allen, S., Allen, D., Phoenix, V. R., et al. (2019). Atmospheric transport and deposition of microplastics in a remote mountain catchment. *Nature Geoscience, 12*, 339–344.

ARCADIS and EUNOMIA. (2010). *Final report on assessment of the options to improve the management of Bio-waste in the European Union*. Assessment.

Ateia, M., Zheng, T., Calace, S., Tharayil, N., Pilla, S., & Karanfil, T. (2020). Sorption behavior of real microplastics (MPs): Insights for organic micropollutants adsorption on a large set of well-characterized MPs. *Science of the Total Environment, 720*, 137634.

Atugoda, T., Piyumali, H., Wijesekara, H., et al. (2023). Nanoplastic occurrence, transformation and toxicity: A review. *Environmental Chemistry Letters, 21*, 363–381.

Azeem, I., Adeel, M., Ahmad, M. A., et al. (2021). Uptake and accumulation of nano/microplastics in plants: A critical review. *Nanomaterials, 11*, 2935.

Baken, S., Degryse, F., Verheyen, L., Merckx, R., & Smolders, E. (2011). Metal complexation properties of freshwater dissolved organic matter are explained by its aromaticity and by anthropogenic ligands. *Environmental Science & Technology, 45*, 2584–2590.

Bandow, N., Will, V., Wachtendorf, V., & Simon, F. G. (2017). Contaminant release from aged microplastic. *Environment and Chemistry, 14*, 394–405.

Bhagat, J., Nishimura, N., & Shimada, Y. (2021). Toxicological interactions of microplastics/nanoplastics and environmental contaminants: Current knowledge and future perspectives. *Journal of Hazardous Materials, 405*, 123913.

Bhagat, K., Barrios, A. C., Rajwade, K., Kumar, A., Oswald, J., Apul, O., & Perreault, F. (2022). Aging of microplastics increases their adsorption affinity towards organic contaminants. *Chemosphere, 298*, 134238.

Bläsing, M., & Amelung, W. (2018). Plastics in soil: Analytical methods and possible sources. *Science of the Total Environment, 612*, 422–435.

Bradney, L., Wijesekara, H., Palansooriya, K. N., Obadamudalige, N., Bolan, N. S., Ok, Y. S., Rinklebe, J., Kim, K. H., & Kirkham, M. B. (2019). Particulate plastics as a vector for toxic trace-element uptake by aquatic and terrestrial organisms and human health risk. *Environment International, 131*, 104937.

Brahney, J., Hallerud, M., Heim, E., et al. (2020). Plastic rain in protected areas of the United States. *Science, 368*, 1257–1260.

Brennecke, D., Duarte, B., Paiva, F., et al. (2016). Microplastics as vector for heavy metal contamination from the marine environment. *Estuarine, Coastal and Shelf Science, 178*, 189–195.

Bretas Alvim, C., Bes-Piá, M. A., & Mendoza-Roca, J. A. (2020). Separation and identification of microplastics from primary and secondary effluents and activated sludge from wastewater treatment plants. *Chemical Engineering Journal, 402*, 126293.

Büks, F., & Kaupenjohann, M. (2020). Global concentrations of microplastics in soils – A review. *The Soil, 6*, 649–662.

Cai, L., Wang, J., Peng, J., et al. (2017). Characteristic of microplastics in the atmospheric fallout from Dongguan city, China: Preliminary research and first evidence. *Environmental Science and Pollution Research, 24*, 24928–24935.

Campos, T., Chaer, G., dos Santos, L. P., et al. (2019). Leaching of heavy metals in soils conditioned with biosolids from sewage sludge. *Floresta e Ambiente, 26*, e20180399.

Cao, D., Wang, X., Luo, X., et al. (2017). Effects of polystyrene microplastics on the fitness of earthworms in an agricultural soil. In *IOP conference series: Earth and environmental science*. Institute of Physics Publishing.

Cao, Y., Zhao, M., Ma, X., Song, Y., Zuo, S., Li, H., & Deng, W. (2021). A critical review on the interactions of microplastics with heavy metals: Mechanism and their combined effect on organisms and humans. *Science of the Total Environment, 788*, 147620.

Carr, S. A., Liu, J., & Tesoro, A. G. (2016). Transport and fate of microplastic particles in wastewater treatment plants. *Water Research, 91*, 174–182.

Chappaz, A., & Curtis, P. J. (2013). Integrating empirically dissolved organic matter quality for WHAM VI using the DOM optical properties: A case study of cu-Al-DOM interactions. *Environmental Science & Technology, 47*, 2001–2007.

Chen, Y., Leng, Y., Liu, X., & Wang, J. (2020a). Microplastic pollution in vegetable farmlands of suburb Wuhan, Central China. *Environmental Pollution*, 257, 113449.

Chen, Y., Liu, X., Leng, Y., & Wang, J. (2020b). Defense responses in earthworms (Eisenia fetida) exposed to low-density polyethylene microplastics in soils. *Ecotoxicology and Environmental Safety, 187*, 109788.

Chen, X., Gu, X., Bao, L., et al. (2021). Comparison of adsorption and desorption of triclosan between microplastics and soil particles. *Chemosphere, 263*, 127947.

Cheng, X., Danek, T., Drozdova, J., et al. (2018). Soil heavy metal pollution and risk assessment associated with the Zn-Pb mining region in Yunnan, Southwest China. *Environmental Monitoring and Assessment, 190*, 1–16.

Cheng, Y., Zhu, L., Song, W., et al. (2020). Combined effects of mulch film-derived microplastics and atrazine on oxidative stress and gene expression in earthworm (Eisenia fetida). *Science of the Total Environment, 746*, 141280.

Cole, M., Lindeque, P., Fileman, E., et al. (2013). Microplastic ingestion by zooplankton. *Environmental Science & Technology, 47*, 6646–6655.

Corradini, F., Meza, P., Eguiluz, R., et al. (2019). Evidence of microplastic accumulation in agricultural soils from sewage sludge disposal. *Sci. Total Environ., 671*, 411–420.

De Souza MacHado, A. A., Lau, C. W., Till, J., Kloas, W., Lehmann, A., Becker, R., & Rillig, M. C. (2018). Impacts of microplastics on the soil biophysical environment. *Environmental Science & Technology, 52*, 9656–9665.

De Souza Machado, A. A., Lau, C. W., Kloas, W., Bergmann, J., Bachelier, J. B., Faltin, E., Becker, R., Görlich, A. S., & Rillig, M. C. (2019). Microplastics can change soil properties and affect plant performance. *Environmental Science & Technology, 53*, 6044–6052.

Ding, L., Zhang, S., Wang, X., et al. (2020). The occurrence and distribution characteristics of microplastics in the agricultural soils of Shaanxi Province, in North-Western China. *Sci. Total Environ., 720*, 137525.

Dong, Y., Gao, M., Qiu, W., & Song, Z. (2021). Uptake of microplastics by carrots in presence of as (III): Combined toxic effects. *Journal of Hazardous Materials, 411*, 125055.

Dris, R., Gasperi, J., Saad, M., et al. (2016). Synthetic fibers in atmospheric fallout: A source of microplastics in the environment? *Marine Pollution Bulletin, 104*, 290–293.

Dris, R., Gasperi, J., Mirande, C., et al. (2017). A first overview of textile fibers, including microplastics, in indoor and outdoor environments. *Environmental Pollution, 221*, 453–458.

Duval, J. F. L., Wilkinson, K. I., Van Leeuwen, H. P., & Buffle, J. (2005). Humic substances are soft and permeable: Evidence from their electrophoretic mobilities. *Environmental Science & Technology, 39*, 6435–6445.

Edo, C., González-Pleiter, M., Leganés, F., et al. (2020). Fate of microplastics in wastewater treatment plants and their environmental dispersion with effluent and sludge. *Environmental Pollution, 259*, 113837.

El Hayany, B., El Fels, L., Quénéa, K., et al. (2020). Microplastics from lagooning sludge to composts as revealed by fluorescent staining- image analysis, Raman spectroscopy and pyrolysis-GC/MS. *Journal of Environmental Management, 275*, 111249.

Enyoh, C. E., Verla, A. W., Verla, E. N., et al. (2019). Airborne microplastics: A review study on method for analysis, occurrence, movement and risks. *Environmental Monitoring and Assessment, 191*, 1–17.

Feng, X., Wang, Q., Sun, Y., et al. (2022). Microplastics change soil properties, heavy metal availability and bacterial community in a Pb-Zn-contaminated soil. *Journal of Hazardous Materials, 424*, 127364.

Fuentes, A., Lloréns, M., Sáez, J., et al. (2008). Comparative study of six different sludges by sequential speciation of heavy metals. *Bioresource Technology, 99*, 517–525.

Gao, M., Liu, Y., & Song, Z. (2019). Effects of polyethylene microplastic on the phytotoxicity of di-n-butyl phthalate in lettuce (Lactuca sativa L. var. ramosa Hort). *Chemosphere, 237*, 124482.

Geyer, R. (2020). *Production, use, and fate of synthetic polymers*. Elsevier.

Geyer, R., Jambeck, J. R., & Law, K. L. (2017). Production, use, and fate of all plastics ever made. *Science Advances, 3*, 25–29.

Gies, E. A., LeNoble, J. L., Noël, M., et al. (2018). Retention of microplastics in a major secondary wastewater treatment plant in Vancouver, Canada. *Marine Pollution Bulletin, 133*, 553–561.

Godoy, V., Blázquez, G., Calero, M., Quesada, L., & Martín-Lara, M. A. (2019). The potential of microplastics as carriers of metals. *Environmental Pollution*, 255, 113363.

Gudeta, K., Kumar, V., Bhagat, A., et al. (2023). Ecological adaptation of earthworms for coping with plant polyphenols, heavy metals, and microplastics in the soil: A review. *Heliyon, 9*, e14572.

Guo, G., Zhang, D., & Wang, Y. (2019). Probabilistic human health risk assessment of heavy metal intake via vegetable consumption around Pb/Zn smelters in Southwest China. *International Journal of Environmental Research and Public Health, 16*, 3267.

Guo, X., Hu, G., Fan, X., & Jia, H. (2020). Sorption properties of cadmium on microplastics: The common practice experiment and a two-dimensional correlation spectroscopic study. *Ecotoxicology and Environmental Safety, 190*, 110118.

Hao, X., Sun, H. G., Zhang, Y., et al. (2023). Co-transport of arsenic and micro/nano-plastics in saturated soil. *Environmental Research, 228*, 115871.

He, D., Luo, Y., Lu, S., et al. (2018). Microplastics in soils: Analytical methods, pollution characteristics and ecological risks. *TrAC – Trends in Analytical Chemistry, 109*, 163–172.

He, D., Zhang, Y., & Gao, W. (2021). Micro(nano)plastic contaminations from soils to plants: Human food risks. *Current Opinion in Food Science, 41*, 116–121.

Hodson, M. E., Duffus-Hodson, C. A., Clark, A., et al. (2017). Plastic bag derived-microplastics as a vector for metal exposure in terrestrial invertebrates. *Environmental Science & Technology, 51*, 4714–4721.

Holmes, L. A., Turner, A., & Thompson, R. C. (2012). Adsorption of trace metals to plastic resin pellets in the marine environment. *Environmental Pollution, 160*, 42–48.

Holmes, L. A., Turner, A., & Thompson, R. C. (2014). Interactions between trace metals and plastic production pellets under estuarine conditions. *Marine Chemistry, 167*, 25–32.

Holzinger, A., Mair, M. M., Lücker, D., et al. (2022). Comparison of fitness effects in the earthworm Eisenia fetida after exposure to single or multiple anthropogenic pollutants. *Science of the Total Environment, 838*, 156387.

Hongprasith, N., Kittimethawong, C., Lertluksanaporn, R., et al. (2020). IR microspectroscopic identification of microplastics in municipal wastewater treatment plants. *Environmental Science and Pollution Research, 27*, 18557–18564.
Horton, A. A., Walton, A., Spurgeon, D. J., et al. (2017). Microplastics in freshwater and terrestrial environments: Evaluating the current understanding to identify the knowledge gaps and future research priorities. *Science of the Total Environment, 586*, 127–141.
Hua, Z., Ma, S., Ouyang, Z., et al. (2023). The review of nanoplastics in plants: Detection, analysis, uptake, migration and risk. *TrAC – Trends in Analytical Chemistry, 158*, 116889.
Huang, Y., Liu, Q., Jia, W., et al. (2020). Agricultural plastic mulching as a source of microplastics in the terrestrial environment. *Environmental Pollution, 260*, 114096.
Huang, B., Sun, L., Liu, M., et al. (2021a). Abundance and distribution characteristics of microplastic in plateau cultivated land of Yunnan Province, China. *Environmental Science Pollution Research, 28*, 1675–1688.
Huang, C., Ge, Y., Yue, S., et al. (2021b). Microplastics aggravate the joint toxicity to earthworm Eisenia fetida with cadmium by altering its availability. *Science of the Total Environment, 753*, 142042.
Huang, H., Mohamed, B. A., & Li, L. Y. (2023). Accumulation and fate of microplastics in soils after application of biosolids on land: A review. *Environmental Chemistry Letters, 21*, 1745–1759.
Huerta Lwanga, E., Gertsen, H., Gooren, H., et al. (2017). Incorporation of microplastics from litter into burrows of Lumbricus terrestris. *Environmental Pollution, 220*, 523–531.
Hüffer, T., & Hofmann, T. (2016). Sorption of non-polar organic compounds by micro-sized plastic particles in aqueous solution. *Environmental Pollution, 214*, 194–201.
Hüffer, T., Metzelder, F., Sigmund, G., et al. (2019). Polyethylene microplastics influence the transport of organic contaminants in soil. *Science of the Total Environment, 657*, 242–247.
Jan Kole P, Löhr AJ, Van Belleghem FGAJ, Ragas AMJ (2017) Wear and tear of tyres: A stealthy source of microplastics in the environment. International Journal of Environmental Research and Public Health 14, 1265.
Jaramillo, M. F., & Restrepo, I. (2017). Wastewater reuse in agriculture: A review about its limitations and benefits. *Sustainability, 9*, 1734.
Jia, H., Wu, D., Yu, Y., et al. (2022). Impact of microplastics on bioaccumulation of heavy metals in rape (Brassica napus L.). *Chemosphere, 288*, 132576.
Jiang, X., Chen, H., Liao, Y., et al. (2019). Ecotoxicity and genotoxicity of polystyrene microplastics on higher plant Vicia faba. *Environmental Pollution, 250*, 831–838.
Jiang, J., Wang, X., Ren, H., et al. (2020a). Investigation and fate of microplastics in wastewater and sludge filter cake from a wastewater treatment plant in China. *Science of the Total Environment, 746*, 141378.
Jiang, X., Chang, Y., Zhang, T., et al. (2020b). Toxicological effects of polystyrene microplastics on earthworm (Eisenia fetida). *Environmental Pollution, 259*, 113896.
Jin, T., Tang, J., Lyu, H., et al. (2022). Activities of microplastics (MPs) in agricultural soil: A review of MPs pollution from the perspective of agricultural ecosystems. *Journal of Agricultural and Food Chemistry, 70*, 4182–4201.
Ju, H., Zhu, D., & Qiao, M. (2019). Effects of polyethylene microplastics on the gut microbial community, reproduction and avoidance behaviors of the soil springtail, Folsomia candida. *Environmental Pollution, 247*, 890–897.
Karlsson, T., Persson, P., & Skyllberg, U. (2006). Complexation of copper(II) in organic soils and in dissolved organic matter - EXAFS evidence for chelate ring structures. *Environmental Science & Technology, 40*, 2623–2628.
Katsumi, N., Kusube, T., Nagao, S., & Okochi, H. (2021). Accumulation of microcapsules derived from coated fertilizer in paddy fields. *Chemosphere, 267*. https://doi.org/10.1016/j.chemosphere.2020.129185

Kazour, M., Terki, S., Rabhi, K., et al. (2019). Sources of microplastics pollution in the marine environment: Importance of wastewater treatment plant and coastal landfill. *Marine Pollution Bulletin, 146*, 608–618.

Keller, A. S., Jimenez-Martinez, J., & Mitrano, D. M. (2020). Transport of Nano- and microplastic through unsaturated porous media from sewage sludge application. *Environmental Science & Technology, 54*, 911–920.

Kernchen, S., Löder, M. G. J., Fischer, F., et al. (2022). Airborne microplastic concentrations and deposition across the Weser River catchment. *Science of the Total Environment, 818*, 151812.

Khalid, N., Aqeel, M., Noman, A., et al. (2021). Interactions and effects of microplastics with heavy metals in aquatic and terrestrial environments. *Environmental Pollution, 290*, 118104.

Kim, S. W., & An, Y. J. (2019). Soil microplastics inhibit the movement of springtail species. *Environment International, 126*, 699–706.

Kim, L. H., Kang, J., Kayhanian, M., et al. (2006). Characteristics of litter waste in highway storm runoff. *Water Science and Technology, 53*, 225–234.

Kim, H. M., Lee, D. K., Long, N. P., et al. (2019). Uptake of nanopolystyrene particles induces distinct metabolic profiles and toxic effects in Caenorhabditis elegans. *Environmental Pollution, 246*, 578–586.

Kinigopoulou, V., Pashalidis, I., Kalderis, D., & Anastopoulos, I. (2022). Microplastics as carriers of inorganic and organic contaminants in the environment: A review of recent progress. *Journal of Molecular Liquids, 350*, 118580.

Kong, X., Liu, T., Yu, Z., et al. (2018). Heavy metal bioaccumulation in rice from a high geological background area in Guizhou Province, China. *International Journal of Environmental Research and Public Health, 15*, 2281.

Kopańska, D., & Dudziak, M. (2015). Occurrence of heavy metals in selected made grounds. *Ecological Chemistry and Engineering. A, 22*(2), 137–149.

Koyuncu, S. (2022). Occurrence of organic micropollutants and heavy metals in the soil after the application of stabilized sewage sludge. *Journal of Environmental Health Science and Engineering, 20*, 385–394.

Kumar, M. V., & Sheela, A. M. (2020). Effect of plastic film mulching on the distribution of plastic residues in agricultural fields. *Chemosphere, 273*, 128590–128590.

Kumar, V., Singh, J., & Kumar, P. (2019). Heavy metals accumulation in crop plants: Sources, response mechanisms, stress tolerance and their effects. In *Contaminants in agriculture and environment: Health risks and remediation* (pp. 38–57). Agro Environ Media – Agriculture and Ennvironmental Science Academy.

Lahive, E., Cross, R., Saarloos, A. I., et al. (2022). Earthworms ingest microplastic fibres and nanoplastics with effects on egestion rate and long-term retention. *Science of the Total Environment, 807*, 151022.

Lares, M., Ncibi, M. C., Sillanpää, M., & Sillanpää, M. (2018). Occurrence, identification and removal of microplastic particles and fibers in conventional activated sludge process and advanced MBR technology. *Water Research, 133*, 236–246.

Leslie, H. A., Brandsma, S. H., van Velzen, M. J. M., & Vethaak, A. D. (2017). Microplastics en route: Field measurements in the Dutch river delta and Amsterdam canals, wastewater treatment plants, North Sea sediments and biota. *Environment International, 101*, 133–142.

Li, X., Chen, L., Mei, Q., et al. (2018). Microplastics in sewage sludge from the wastewater treatment plants in China. *Water Research, 142*, 75–85.

Li, L., Zhou, Q., Yin, N., et al. (2019a). Uptake and accumulation of microplastics in an edible plant. *Kexue Tongbao/Chinese Science Bulletin, 64*, 928–934.

Li, Q., Wu, J., Zhao, X., et al. (2019b). Separation and identification of microplastics from soil and sewage sludge. *Environmental Pollution, 254*, 113076.

Li, R., Liu, Y., Sheng, Y., Xiang, Q., Zhou, Y., & Cizdziel, J. V. (2020a). Effect of prothioconazole on the degradation of microplastics derived from mulching plastic film: Apparent change and interaction with heavy metals in soil. *Environmental Pollution, 260*, 113988.

Li, W., Wufuer, R., Duo, J., et al. (2020b). Microplastics in agricultural soils: Extraction and characterization after different periods of polythene film mulching in an arid region. *Science of the Total Environment, 749*, 141420.

Li, Z., Li, Q., Li, R., et al. (2020c). Physiological responses of lettuce (Lactuca sativa L.) to microplastic pollution. *Environmental Science and Pollution Research, 27*, 30306–30314.

Li, M., Liu, Y., Xu, G., et al. (2021a). Impacts of polyethylene microplastics on bioavailability and toxicity of metals in soil. *Science of the Total Environment, 760*, 144037.

Li, Y., Chen, M., Gong, J., et al. (2021b). Effects of virgin microplastics on the transport of cd (II) in Xiangjiang River sediment. *Chemosphere, 283*, 131197.

Li, M., Jia, H., Gao, Q., et al. (2023). Influence of aged and pristine polyethylene microplastics on bioavailability of three heavy metals in soil: Toxic effects to earthworms (Eisenia fetida). *Chemosphere, 311*, 136833.

Lian, J., Wu, J., Xiong, H., et al. (2020). Impact of polystyrene nanoplastics (PSNPs) on seed germination and seedling growth of wheat (Triticum aestivum L.). *Journal of Hazardous Materials, 385*, 121620.

Liang, X., Zhou, D., Wang, J., et al. (2022). Evaluation of the toxicity effects of microplastics and cadmium on earthworms. *Science of the Total Environment, 836*.

Liao, Y. L., & Yang, J. Y. (2020). Microplastic serves as a potential vector for Cr in an in-vitro human digestive model. *Science of the Total Environment, 703*, 134805.

Lin, L., Tang, S., Wang, X., Sun, X., & Yu, A. (2021). Hexabromocyclododecane alters malachite green and lead(II) adsorption behaviors onto polystyrene microplastics: Interaction mechanism and competitive effect. *Chemosphere, 265*, 129079.

Liu, Y. R., He, Z. Y., Yang, Z. M., et al. (2016). Variability of heavy metal content in soils of typical Tibetan grasslands. *RSC Advances, 6*, 105398–105405.

Liu, H., Yang, X., Liu, G., Liang, C., Xue, S., Chen, H., Ritsema, C. J., & Geissen, V. (2017). Response of soil dissolved organic matter to microplastic addition in Chinese loess soil. *Chemosphere, 185*, 907–917.

Liu, M., Lu, S., Song, Y., et al. (2018). Microplastic and mesoplastic pollution in farmland soils in suburbs of Shanghai, China. *Environmental Pollution, 242*, 855–862.

Liu, G., Dave, P. H., Kwong, R. W. M., Wu, M., & Zhong, H. (2021a). Influence of microplastics on the mobility, bioavailability, and toxicity of heavy metals: A review. *Bulletin of Environmental Contamination and Toxicology, 107*, 710–721. https://doi.org/10.1007/s00128-021-03339-9

Liu, S., Shi, J., Wang, J., Dai, Y., Li, H., Li, J., Liu, X., Chen, X., Wang, Z., & Zhang, P. (2021b). Interactions between microplastics and heavy metals in aquatic environments: A review. *Frontiers in Microbiology, 12*, 1–14.

Liu, Y., Shao, H., Liu, J., et al. (2021c). Transport and transformation of microplastics and nanoplastics in the soil environment: A critical review. *Soil Use and Management, 2014*, 1–19.

Liu, Y., Guo, R., Zhang, S., et al. (2022). Uptake and translocation of nano/microplastics by rice seedlings: Evidence from a hydroponic experiment. *Journal of Hazardous Materials, 421*, 126700.

Lwanga, E. H., Beriot, N., Corradini, F., et al. (2022). Review of microplastic sources, transport pathways and correlations with other soil stressors: A journey from agricultural sites into the environment. *Chemical and Biological Technologies in Agriculture, 9*, 1–20.

Ma, J., Sheng, G. D., & O'Connor, P. (2020). Microplastics combined with tetracycline in soils facilitate the formation of antibiotic resistance in the Enchytraeus crypticus microbiome. *Environmental Pollution, 264*, 114689.

Ma, X., Zhou, X., Zhao, M., et al. (2022). Polypropylene microplastics alter the cadmium adsorption capacity on different soil solid fractions. *Frontiers of Environmental Science & Engineering, 16*, 1–12.

Magni, S., Binelli, A., Pittura, L., et al. (2019). The fate of microplastics in an Italian wastewater treatment plant. *Science of the Total Environment, 652*, 602–610.

Magnusson, K., & Norén, F. (2014). *Screening of microplastic particles in and down-stream a wastewater treatment plant.*

Mahon, A. M., O'Connell, B., Healy, M. G., et al. (2017). Microplastics in sewage sludge: Effects of treatment. *Environmental Science & Technology, 51*, 810–818.

Mai, H., Thien, N. D., Dung, N. T., & Valentin, C. (2023). Impacts of microplastics and heavy metals on the earthworm Eisenia fetida and on soil organic carbon, nitrogen, and phosphorus. *Environmental Science and Pollution Research, 30*, 64576–64588.

Maity, S., Chatterjee, A., Guchhait, R., et al. (2020). Cytogenotoxic potential of a hazardous material, polystyrene microparticles on Allium cepa L. *Journal of Hazardous Materials, 385*, 121560.

Mamathaxim, N., Song, W., Wang, Y., & Habibul, N. (2023). *Effects of microplastics on arsenic uptake and distribution in rice seedlings* (p. 862). Sci.

Mao, R., Lang, M., Yu, X., Wu, R., Yang, X., & Guo, X. (2020). Aging mechanism of microplastics with UV irradiation and its effects on the adsorption of heavy metals. *Journal of Hazardous Materials, 393*, 122515.

Mateos-Cárdenas, A., van Pelt, F. N. A. M., O'Halloran, J., & Jansen, M. A. K. (2021). Adsorption, uptake and toxicity of micro- and nanoplastics: Effects on terrestrial plants and aquatic macrophytes. *Environmental Pollution, 284*, 117183.

Mato, Y., Isobe, T., Takada, H., Kanehiro, H., Ohtake, C., & Kaminuma, T. (2001). Plastic resin pellets as a transport medium for toxic chemicals in the marine environment. *Environmental Science & Technology, 35*, 318–324.

Medyńska-Juraszek, A., & Jadhav, B. (2022). Influence of different microplastic forms on pH and mobility of Cu2+ and Pb2+ in soil. *Molecules, 27*, 1744.

Mintenig, S. M., Int-Veen, I., Löder, M. G. J., et al. (2017). Identification of microplastic in effluents of waste water treatment plants using focal plane array-based micro-Fourier-transform infrared imaging. *Water Research, 108*, 365–372.

Moeck, C., Davies, G., Krause, S., & Schneidewind, U. (2023). Microplastics and nanoplastics in agriculture—A potential source of soil and groundwater contamination? *Grundwasser, 28*, 23–35.

Möhrke, A. C. F., Haegerbaeumer, A., Traunspurger, W., & Höss, S. (2022). Underestimated and ignored? The impacts of microplastic on soil invertebrates—Current scientific knowledge and research needs. *Frontiers in Environmental Science, 10*, 1498.

Mungai, T. M., Owino, A. A., Makokha, V. A., et al. (2016). Occurrences and toxicological risk assessment of eight heavy metals in agricultural soils from Kenya, eastern Africa. *Environmental Science and Pollution Research, 23*, 18533–18541.

Naderi Beni, N., Karimifard, S., Gilley, J., et al. (2023). Higher concentrations of microplastics in runoff from biosolid-amended croplands than manure-amended croplands. *Communications Earth & Environment, 4*, 42.

Nizzetto, L., Futter, M., & Langaas, S. (2016). Are agricultural soils dumps for microplastics of urban origin? *Environmental Science & Technology, 50*, 10777–10779.

Obinnaa, I. B., & Ebere, E. C. (2019). A review: Water pollution by heavy metal and organic pollutants: Brief review of sources, effects and progress on remediation with aquatic plants. *Analytical Methods in Environmental Chemistry Journal, 2*, 5–38.

Ofori, S., Puškáčová, A., Růžičková, I., & Wanner, J. (2021). Treated wastewater reuse for irrigation: Pros and cons. *Science of the Total Environment, 760*, 144026.

Okeke, E. S., Chukwudozie, K. I., Addey, C. I., et al. (2023). Micro and nanoplastics ravaging our agroecosystem: A review of occurrence, fate, ecological impacts, detection, remediation, and prospects. *Heliyon, 9*, e13296.

Ozturk, I., Ozkul, F., & Topuz, E. (2023). The effect of polystyrene microplastic and biosolid application on the toxicity and bioaccumulation of cadmium for Enchytraeus crypticus. *Integrated Environmental Assessment and Management, 19*, 489–500.

Panara, A. M. (2018). Uncovering the super-plant: Improving domestic water reuse. In *2018 IEEE global humanitarian technology conference (GHTC)* (pp. 1–7). IEEE.

Pinto-Poblete, A., Retamal-Salgado, J., López, M. D., et al. (2022). Combined effect of microplastics and cd alters the enzymatic activity of soil and the productivity of strawberry plants. *Plants, 11*, 536.

Purwiyanto, A. I. S., Suteja, Y., Trisno Ningrum, P. S., Putri, W. A. E., Rozirwan Agustriani, F., Fauziyah Cordova, M. R., & Koropitan, A. F. (2020). Concentration and adsorption of Pb and cu in microplastics: Case study in aquatic environment. *Marine Pollution Bulletin, 158*, 111380.

Qiao, R., Lu, K., Deng, Y., Ren, H., & Zhang, Y. (2019). Combined effects of polystyrene microplastics and natural organic matter on the accumulation and toxicity of copper in zebrafish. *Science of the Total Environment, 682*, 128–137.

Ramos, L., Berenstein, G., Hughes, E. A., et al. (2015). Polyethylene film incorporation into the horticultural soil of small periurban production units in Argentina. *Science of the Total Environment, 523*, 74–81.

Refaey, Y., Jansen, B., El-Shater, A.-H., El-Haddad, A. A., & Kalbitz, K. (2014). The role of dissolved organic matter in adsorbing heavy metals in clay-rich soils. *Vadose Zone Journal, 13*, 7.

Ren, X., Tang, J., Liu, X., & Liu, Q. (2020). Effects of microplastics on greenhouse gas emissions and the microbial community in fertilized soil. *Environmental Pollution, 256*, 113347.

Ren, Z., Gui, X., Xu, X., et al. (2021). Microplastics in the soil-groundwater environment: Aging, migration, and co-transport of contaminants – A critical review. *Journal of Hazardous Materials, 419*, 126455.

Richard, H., Carpenter, E. J., Komada, T., Palmer, P. T., & Rochman, C. M. (2019). Biofilm facilitates metal accumulation onto microplastics in estuarine waters. *Science of the Total Environment, 683*, 600–608.

Rillig, M. C., Ingraffia, R., & De Souza Machado, A. A. (2017a). Microplastic incorporation into soil in agroecosystems. *Frontiers in Plant Science, 8*, 8–11.

Rillig, M. C., Ziersch, L., & Hempel, S. (2017b). Microplastic transport in soil by earthworms. *Scientific Reports, 7*, 1362.

Rodriguez-Seijo, A., Lourenço, J., Rocha-Santos, T. A. P., et al. (2017). Histopathological and molecular effects of microplastics in Eisenia andrei Bouché. *Environmental Pollution, 220*, 495–503.

Saeedi, M., Li, L. Y., & Grace, J. R. (2018). Effect of organic matter and selected heavy metals on sorption of acenaphthene, fluorene and fluoranthene onto various clays and clay minerals. *Environment and Earth Science, 77*, 1–12.

Schwabl, P., Koppel, S., Konigshofer, P., et al. (2019). Detection of various microplastics in human stool: A prospective case series. *Annals of Internal Medicine, 171*, 453–457.

Selonen, S., Dolar, A., Jemec Kokalj, A., et al. (2020). Exploring the impacts of plastics in soil – The effects of polyester textile fibers on soil invertebrates. *Science of the Total Environment, 700*, 134451.

Shoushtarian, F., & Negahban-Azar, M. (2020). World wide regulations and guidelines for agriculturalwater reuse: A critical review. *Water, 12*, 25.

Soleimani, H., Mansouri, B., Kiani, A., et al. (2023). Ecological risk assessment and heavy metals accumulation in agriculture soils irrigated with treated wastewater effluent, river water, and well water combined with chemical fertilizers. *Heliyon, 9*, 3.

Song, Y., Cao, C., Qiu, R., et al. (2019). Uptake and adverse effects of polyethylene terephthalate microplastics fibers on terrestrial snails (Achatina fulica) after soil exposure. *Environmental Pollution, 250*, 447–455.

Stubenrauch, J., & Ekardt, F. (2020). Plastic pollution in soils: Governance approaches to foster soil health and closed nutrient cycles. *Environments – MDPI, 7*, 1–18.

Sun, X. D., Yuan, X. Z., Jia, Y., et al. (2020). Differentially charged nanoplastics demonstrate distinct accumulation in Arabidopsis thaliana. *Nature Nanotechnology, 15*, 755–760.

Sungur, A., Soylak, M., & Ozcan, H. (2014). Investigation of heavy metal mobility and availability by the BCR sequential extraction procedure: Relationship between soil properties and heavy metals availability. *Chemical Speciation and Bioavailability, 26*, 219–230.

Szewc, K., Graca, B., & Dołęga, A. (2021). Atmospheric deposition of microplastics in the coastal zone: Characteristics and relationship with meteorological factors. *Science of the Total Environment*, 761, 143272.

Tan, M., Liu, L., Zhang, M., et al. (2021). Effects of solution chemistry and humic acid on the transport of polystyrene microplastics in manganese oxides coated sand. *Journal of Hazardous Materials, 413*, 125410.

Tang, S., Lin, L., Wang, X., Feng, A., & Yu, A. (2020). Pb(II) uptake onto nylon microplastics: Interaction mechanism and adsorption performance. *Journal of Hazardous Materials, 386*, 121960.

Tang, S., Lin, L., Wang, X., Yu, A., & Sun, X. (2021). Interfacial interactions between collected nylon microplastics and three divalent metal ions (cu(II), Ni(II), Zn(II)) in aqueous solutions. *Journal of Hazardous Materials, 403*, 123548.

Teng, J., Zhao, J., Zhu, X., et al. (2021). Toxic effects of exposure to microplastics with environmentally relevant shapes and concentrations: Accumulation, energy metabolism and tissue damage in oyster Crassostrea gigas. *Environmental Pollution, 269*, 116169.

Tessier, A., Campbell, P. G. C., & Bisson, M. (1979). Sequential extraction procedure for the speciation of particulate trace metals. *Analytical Chemistry, 51*, 844–851.

Tian, L., Jinjin, C., Ji, R., et al. (2022). Microplastics in agricultural soils: Sources, effects, and their fate. *Current Opinion in Environmental Science & Health, 25*, 100311.

Turner, A., & Holmes, L. A. (2015). Adsorption of trace metals by microplastic pellets in fresh water. *Environment and Chemistry, 12*, 600–610.

Ulutuğ, F. C., & Topuz, E. (2022). Mikroplastik ve Biyokatı Varlığında Toprakta Nikel Toksisitesinin Belirlenmesi. *Iğdır Üniversitesi Fen Bilimleri Enstitüsü Dergisi, 12*, 1386–1394.

Ungureanu, N., Vlăduţ, V., & Voicu, G. (2020). Water scarcity and wastewater reuse in crop irrigation. *Sustainability, 12*, 1–19.

van den Berg, P., Huerta-Lwanga, E., Corradini, F., & Geissen, V. (2020). Sewage sludge application as a vehicle for microplastics in eastern Spanish agricultural soils. *Environmental Pollution, 261*, 114198.

Wahl, A., Le Juge, C., Davranche, M., et al. (2021). Nanoplastic occurrence in a soil amended with plastic debris. *Chemosphere, 262*, 127784.

Wang, X., Chen, T., Ge, Y., & Jia, Y. (2008). Studies on land application of sewage sludge and its limiting factors. *Journal of Hazardous Materials, 160*, 554–558.

Wang, J., Peng, J., Tan, Z., Gao, Y., Zhan, Z., Chen, Q., & Cai, L. (2017). Microplastics in the surface sediments from the Beijiang River littoral zone: Composition, abundance, surface textures and interaction with heavy metals. *Chemosphere, 171*, 248–258.

Wang, H. T., Ding, J., Xiong, C., et al. (2019a). Exposure to microplastics lowers arsenic accumulation and alters gut bacterial communities of earthworm Metaphire californica. *Environmental Pollution, 251*, 110–116.

Wang, J., Liu, X., Li, Y., et al. (2019b). Microplastics as contaminants in the soil environment: A mini-review. *Science of the Total Environment, 691*, 848–857.

Wang, F., Zhang, X., Zhang, S., et al. (2020a). Effects of co-contamination of microplastics and cd on plant growth and cd accumulation. *Toxics, 8*.

Wang, T., Wang, L., Chen, Q., Kalogerakis, N., Ji, R., & Ma, Y. (2020b). Interactions between microplastics and organic pollutants: Effects on toxicity, bioaccumulation, degradation, and transport. *Science of the Total Environment, 748*, 142427.

Wang, F., Wang, X., & Song, N. (2021a). Polyethylene microplastics increase cadmium uptake in lettuce (Lactuca sativa L.) by altering the soil microenvironment. *Science of the Total Environment, 784*, 147133.

Wang, Z., Fu, D., Gao, L., et al. (2021b). Aged microplastics decrease the bioavailability of coexisting heavy metals to microalga Chlorella vulgaris. *Ecotoxicology and Environmental Safety, 217*.

Wang, F., Feng, X., Liu, Y., et al. (2022a). Micro(nano)plastics and terrestrial plants: Up-to-date knowledge on uptake, translocation, and phytotoxicity. *Resources, Conservation & Recycling Advances, 185*, 106503.

Wang, F., Wang, Q., Adams, C. A., Sun, Y., & Zhang, S. (2022b). Effects of microplastics on soil properties: Current knowledge and future perspectives. *Journal of Hazardous Materials, 424*, 127531.

Wei, B., Yu, J., Cao, Z., et al. (2020). The availability and accumulation of heavy metals in greenhouse soils associated with intensive fertilizer application. *International Journal of Environmental Research and Public Health, 17*, 1–13.

Wijesekara, H., Bolan, N. S., Thangavel, R., et al. (2017). The impact of biosolids application on organic carbon and carbon dioxide fluxes in soil. *Chemosphere, 189*, 565–573.

Wright, S. L., Ulke, J., Font, A., et al. (2020). Atmospheric microplastic deposition in an urban environment and an evaluation of transport. *Environment International, 136*, 105411.

Wu, X., Lyu, X., Li, Z., et al. (2020). Transport of polystyrene nanoplastics in natural soils: Effect of soil properties, ionic strength and cation type. *Science of the Total Environment, 707*, 136065.

Xu, J., Tan, W., Xiong, J., Wang, M., Fang, L., & Koopal, L. K. (2016). Copper binding to soil fulvic and humic acids: NICA-Donnan modeling and conditional affinity spectra. *Journal of Colloid and Interface Science, 473*, 141–151.

Xu, C., Zhang, B., Gu, C., et al. (2020). Are we underestimating the sources of microplastic pollution in terrestrial environment? *Journal of Hazardous Materials, 400*, 123228.

Xu, G., Lin, X., & Yu, Y. (2023). Different effects and mechanisms of polystyrene micro- and nano-plastics on the uptake of heavy metals (cu, Zn, Pb and cd) by lettuce (Lactuca sativa L.). *Environmental Pollution, 316*, 120656.

Yu, H., Hou, J., Dang, Q., Cui, D., Xi, B., & Tan, W. (2020). Decrease in bioavailability of soil heavy metals caused by the presence of microplastics varies across aggregate levels. *Journal of Hazardous Materials, 395*, 122690.

Yu, Y., Griffin-LaHue, D. E., Miles, C. A., et al. (2021). Are micro- and nanoplastics from soil-biodegradable plastic mulches an environmental concern? *Journal of Hazardous Materials Advances, 4*, 100024.

Yuan, J., Ma, J., Sun, Y., et al. (2020). Microbial degradation and other environmental aspects of microplastics/plastics. *Science of the Total Environment, 715*, 136968.

Yuan, X., Xue, N., & Han, Z. (2021). A meta-analysis of heavy metals pollution in farmland and urban soils in China over the past 20 years. *Journal of Environmental Sciences (China), 101*, 217–226.

Yue, Y., Li, X., Wei, Z., et al. (2023). Recent advances on multilevel effects of micro(Nano)plastics and coexisting pollutants on terrestrial soil-plants system. *Sustainability, 15*, 4504.

Zhang, H., Pap, S., Taggart, M. A., Boyd, K. G., James, N. A., & Gibb, S. W. (2020a). A review of the potential utilisation of plastic waste as adsorbent for removal of hazardous priority contaminants from aqueous environments. *Environmental Pollution, 258*, 113698.

Zhang, S., Han, B., Sun, Y., & Wang, F. (2020b). Microplastics influence the adsorption and desorption characteristics of cd in an agricultural soil. *Journal of Hazardous Materials, 388*, 121775.

Zhang, H., Huang, Y., An, S., & Zhu, Z. (2022). A review of microplastics in soil: Distribution within pedosphere compartments, environmental fate, and effects. *Water, Air, and Soil Pollution, 233*, 380.

Zhang, J., Liu, K., He, X., et al. (2023). Evaluation of heavy metal contamination of soil and the health risks in four potato-producing areas. *Frontiers in Environmental Science, 11*, 1071353.

Zhao, K., Zhang, L., Dong, J., et al. (2020). Risk assessment, spatial patterns and source apportionment of soil heavy metals in a typical Chinese hickory plantation region of southeastern China. *Geoderma, 360*, 114011.

Zhou, Y., Liu, X., & Wang, J. (2019). Characterization of microplastics and the association of heavy metals with microplastics in suburban soil of Central China. *Science of the Total Environment, 694*, 133798.

Zhou, H., Zhang, T., Zhuang, J., et al. (2020a). Study on the regulation of earthworm physiological function under cadmium stress based on a compound mathematical model. *Environmental Toxicology and Pharmacology, 80*, 103499.

Zhou, Y., Liu, X., & Wang, J. (2020b). Ecotoxicological effects of microplastics and cadmium on the earthworm Eisenia foetida. *Journal of Hazardous Materials, 392*, 122273.

Zhou, C. Q., Lu, C. H., Mai, L., et al. (2021). Response of rice (Oryza sativa L.) roots to nanoplastic treatment at seedling stage. *Journal of Hazardous Materials, 401*, 123412.

Zhou, W., Wang, Q., Wei, Z., et al. (2023). Effects of microplastic type on growth and physiology of soil crops: Implications for farmland yield and food quality. *Environmental Pollution, 326*, 121512.

Ziajahromi, S., Neale, P. A., Telles Silveira, I., et al. (2021). An audit of microplastic abundance throughout three Australian wastewater treatment plants. *Chemosphere, 263*, 128294.

Zong, X., Zhang, J., Zhu, J., et al. (2021). Effects of polystyrene microplastic on uptake and toxicity of copper and cadmium in hydroponic wheat seedlings (Triticum aestivum L.). *Ecotoxicology and Environmental Safety, 217*, 112217.

Zou, J., Liu, X., Zhang, D., & Yuan, X. (2020a). Adsorption of three bivalent metals by four chemical distinct microplastics. *Chemosphere, 248*, 126064.

Zou, L., Gusnawan, P., Zhang, G. S., et al. (2020b). Microplastics in soils: Analytical methods, pollution characteristics and ecological risks. *Environmental Pollution, 50*, 25–32.

Effects of Micro-Nanoplastics Exposure to Earthworms in the Soil System

Sartaj Ahmad Bhat, Zaw Min Han, Shiamita Kusuma Dewi, Guangyu Cui, Yongfen Wei, and Fusheng Li

1 Introduction

Emerging contaminants such as microplastics (size <5 mm) and nanoplastics (<100 nm) have drawn attention globally due to their abundant dissemination in nature, along with negative impacts on the environment (UNEP, 2016). About 300 million tons of plastics are produced annually, half of which are primarily single-use plastic (Chen et al., 2021; Allouzi et al., 2021). Huge amounts of plastics are continuously discharged into the environment as a result of poor waste management, where they are converted into micro-nanoplastics (MNPs). Various researchers have revealed that terrestrial ecosystems particularly soil systems as one of the largest reservoirs of MNPs (Rillig, 2012; de Souza Machado et al., 2018; Rillig & Lehmann, 2020). The maintenance of soil microclimates and biodiversity, the maintenance of water sources, and the provision of food are all vital functions of the soil in ecosystems (Rillig & Lehmann, 2020). Unfortunately, increasing soil MNP contamination is presently endangering these processes, as well as the related plant and animal life (Boots et al., 2019; Gong et al., 2021; Dang et al., 2022). Earthworms are well recognized as ecosystem engineers because they are crucial to ecosystem processes (Bhat et al., 2017, 2018). Earthworms have also been employed as model

S. A. Bhat (✉) · Y. Wei · F. Li
River Basin Research Center, Gifu University, Gifu, Japan

Z. M. Han
Graduate School of Engineering, Gifu University, Gifu, Japan

S. K. Dewi
United Graduate School of Agricultural Science, Gifu University, Gifu, Japan

G. Cui
School of Environment and Energy, Peking University Shenzhen Graduate School, Shenzhen, China

S. A. Bhat et al. (eds.), *Management of Micro and Nano-plastics in Soil and Biosolids*,
https://doi.org/10.1007/978-3-031-51967-3_8

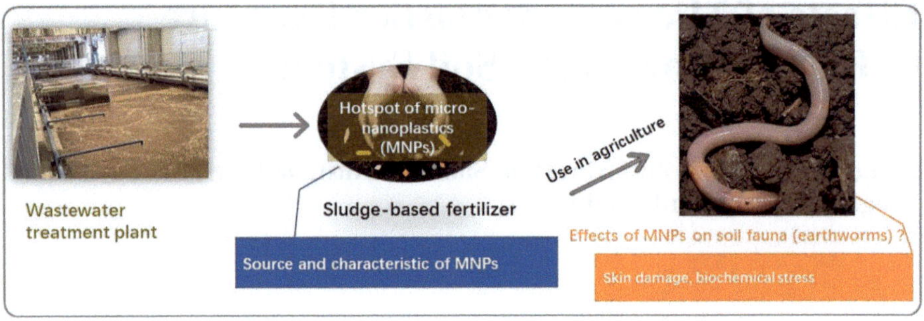

Fig. 1 Effect of MNPs exposure to earthworms in the soil environment

species in ecotoxicological investigations for the last four decades because they are reasonably simple to keep under controlled laboratory settings and have a short reproduction time (OECD, 1984, 2004). Given the current concern over the detrimental effects of MNPs on the soil systems, attention has recently focused on how toxic they are to earthworms, particularly in terms of metabolic disorders, DNA damage, and other negative effects on the earthworms' growth, reproduction, oxidative stress, and avoidance behavior as caused by MNPs (Rodriguez-Seijo et al., 2018; Lahive et al., 2019; Bhat et al., 2022; Khaldoon et al., 2022). Keeping in view the above facts, this chapter focuses on the MNPs, including their source and characteristics in the soil environment, and their toxicity to earthworms. In addition, the combined effect of MNPs with heavy metals and organic pollutants as well as the crucial challenges and future perspectives in reducing MNPs in soil environment are also considered. The effect of MNPs exposure to earthworms in the soil environment is shown in Fig. 1.

2 Source and Characteristics of MNPs in Soil Systems

MNP sources in the soil system can be categorized into three groups: use of sludge as fertilizer, the impact of runoff and deposition, and the disintegration of larger plastic materials. Using biosolids/sludge as fertilizer for crops is common. Although MNPs from wastewater can be efficiently eliminated by wastewater treatment plants (WWTPs), these eliminated MNPs can build up in biosolids (Corradini et al., 2019). The concentrations of MNPs in the sludge vary greatly depending on the country or region. MNPs are now substantially more likely to enter the terrestrial ecosystem through land-based sludge use (Zhang et al., 2020). The application of sludge or biosolid to farmland is the main source of soil MNP contamination since it will increase the amount of MNPs in the agricultural soil (Edo et al., 2020). The use of plastic mulch in agriculture may also produce MNPs. This method is frequently used to boost agricultural yields and control pests. After the use of plastic mulch, the plastic films frequently leave behind pieces (Steinmetz et al., 2016). This debris

Fig. 2 Source of MNPs in soil systems

may build up in soils and subsequently disintegrate particles as small as a nanometer (Briassoulis et al., 2015). Surrounding soils may also get contaminated with MNPs by runoff from roadways or metropolitan areas that are not contained by sewer systems. Additionally, atmospheric transport certainly contributes to a portion of the MNPs in soils and has the ability to carry plastics in the smallest size classes across large distances (Dris et al., 2016). Disintegration of large plastics results in the formation of MNPs (Horton et al., 2017). MNPs may also be generated in situ due to the breakdown of larger plastic materials. The degradation process is thought to be the main source of MNPs in the environment (da Costa et al., 2016; Lambert & Wagner, 2016). The source of MNPs in soil systems is shown in Fig. 2.

3 Effect of MNPs on Earthworms

MNPs can associate with soil fauna especially earthworms, cause physical and physiological harm, and alter the microbial community, potentially having an adverse effect on the health, reproduction, and survival of the soil fauna (Agathokleous et al., 2021). Effect of micro-nanoplastics (MNPs) on earthworms is presented in Table 1. The various harmful effects of MNPs change depending on their sizes and concentrations. MNPs have a considerable inhibitory effect on soil organism growth, intake, and reproduction at significant exposure levels (Huerta Lwanga et al., 2016). According to Liu et al. (2022a) histological harm, disruption of the gut flora, immunological response, and changes in gene expression are the main causes of the toxicity of microplastics. In contrast, nanoplastics cause some harmful effects when they enter biological membranes, cells, or tissue, including

Table 1 Effect of micro-nanoplastics (MNPs) on earthworms

S. no.	MNP type and concentration	MNP size and exposure time	Earthworm species	Effects on earthworms (histopathological damage, oxidative stress, DNA damage)	References
1	Low-density polyethylene (LDPE) 0, 62, 125, 250, 500, and 1000 mg MPs kg^{-1} $soil_{dw}$	Between 250 and 1000 μm 28 days	*Eisenia andrei*	1. Histological damages including gut inflammation were observed.	Rodriguez-Seijo et al. (2017)
2	LDPE 0.1, 0.25, 0.5, 1.0, 1.5 g/kg dry	< 400 μm 28 days	*Eisenia fetida*	1. Surface damage to earthworms. 2. Induced oxidative stress. 3. Stimulated neurotoxicity response.	Chen et al. (2020)
3	Polystyrene microplastics (PS-MPs) 100 and 1000 μg	100 nm and 1300 nm 14 days	*Eisenia fetida*	1. Intestinal cells of *E. fetida* were damaged. 2. PS-MPs changed the glutathione (GSH) level and superoxide dismutase (SOD) activity in earthworms. 3. PS-MPs also induced DNA damage.	Jiang et al. (2020)
4	PS-MPs and nanoplastics (NPs) 10 mg/kg and 100 mg/kg	MPs (10 μm and 100 μm) NPs (100 nm) 21 days	*Eisenia fetida*	1. MPs with high concentration caused serious oxidative damage to earthworms and also disturbed the earthworm immune and metabolic systems. 2. Greater genotoxicity to earthworms was detected for MPs than NPs.	Xu et al. (2021a)
5	PS-MPs 0–100 g kg^{-1} and 0–1.5 g kg^{-1}	Average size of PS-MPs was 24.65 ± 5.20 μm 28 days	*Eisenia fetida*	1. Epidermis damage and setae lacerations in *E. fetida* were observed. 2. Ps-MPs caused more severe oxidative stress to earthworms.	Liu et al. (2022b)

(continued)

Table 1 (continued)

S. no.	MNP type and concentration	MNP size and exposure time	Earthworm species	Effects on earthworms (histopathological damage, oxidative stress, DNA damage)	References
6	LDPE, polylactic acid (PLA), and polybutylene adipate terephthalate (PBAT) 1% of microplastic, dw/dw, and pure MPs in the Petri Dish	20–648 μm 49 days	*Lumbricus terrestris*	1. No mortality was observed in MPs 1% (dw/dw), whereas MPs caused lethal damage in earthworms that were fed with solely MPs. 2. Ingested MPs undergo fragmentation and depolymerization in the earthworm gut.	Meng et al. (2023)
7	Polystyrene nanoplastics (PS-NPs) 0.02% w:w	90–110 nm 14 days	*Eisenia fetida*	1. Intestinal epithelium and chloragogenous tissues were damaged by PS-NPS in earthworms. 2. In earthworms PS-NPs also cause oxidative and inflammatory stress.	Tang et al. (2023)

inflammation, alterations in membrane permeability, oxidative stress, and decreased enzyme activity (Jeong et al., 2016). Earthworms are frequently used to determine the toxicity of pollutants (Xing et al., 2018). Many researchers have reported that the soils contaminated with MNPs affect coelomocyte viability, expand the oxidative stress and mortality, trigger male reproductive organ destruction, and make DNA damage of earthworms. (Kwak & An, 2021; Huerta Lwanga et al., 2016; Jiang et al., 2020). The effect of MNPs in earthworms is discussed in Table 1.

3.1 *Histopathological Damage*

According to Chen et al. (2020), high concentrations of MNP exposure are responsible for skin damage to earthworms. The authors used 100–200 μm size and 1000 mg/kg-1500 mg/kg concentration of low-density polyethylene (LDPE) and confirmed the internal damage to the *Eisenia fetida* species of earthworms. Since MNPs are mostly made of organic carbon, earthworms may consume them and suffer deleterious effects. According to Rodriguez-Seijo et al. (2017), *Eisenia andrei* had only minor physical injury when polyethylene pellet concentrations were 62.5 mg kg^{-1} $soil_{dw}$; however, the damage significantly increased when concentrations reached 1000 mg kg^{-1}. Recently Jiang et al. (2020) studied the histopathological changes induced by polystyrene microplastics (PS-MPs) in the intestines of *E. fetida* after 14 days of contact time. The authors revealed that the intestinal cells of *E. fetida* were damaged after exposure to PS-MPs. Similarly, Liu et al. (2022b) observed physical damage including epidermis damage and setae lacerations in *E. fetida* at higher concentrations of PS-MPs. The setae play a major role for the earthworm's locomotion and are crucial to the wellbeing of earthworms. In general, the toxic impact of MNPs on earthworms may be described by the ingestion of MNPs that leads to damage, ulceration, and obstacles to the digestive tract and also attaching MNPs to the body of earthworms causing physical destruction. Recently Tang et al. (2023) studied the molecular toxicity in earthworms induced by polystyrene nanoplastics (PS-NPs). Intestinal epithelium and chloragogenous tissues were damaged by PS-NPS in *E. fetida* and also caused inflammatory stress to the earthworm species.

3.2 Oxidative Stress

Studies have shown that earthworms exposed to MNPs suffer from oxidative stress, histopathological changes, neurotoxicity, and oxidative and DNA damage (Prendergast-Miller et al., 2019; Wang et al., 2019a; Kwak & An, 2021). Earthworms subjected to microplastics may experience oxidative stress, which can change their enzyme activity (Jiang et al., 2020). The author studied that superoxide dismutase activities were lowered, and glutathione level was increased in *E. fetida* earthworm species when exposed to microplastics. The effects of LDPE microplastics on *E. fetida* were examined by Chen et al. (2020) at concentrations of 0.1, 0.25, 0.5, 1.0, and 1.5 g/kg. The authors revealed that following exposure to 1.0 g/kg of LDPE microplastics for 28 days, catalase (CAT) activity and malondialdehyde (MDA) content increased dramatically. Enzymatic activities have been regarded as reliable markers of contaminants in the environment (Liang et al., 2017). Superoxide dismutase (SOD) is a vital component of the enzymatic defense system since it is an oxygen-free radical scavenger (Liu et al., 2012). According to Jiang et al. (2020), earthworms exposed to PS-MPs with particle sizes larger than 100 nm experienced oxidative stress and DNA damage and considerably changed glutathione (GSH) level and SOD activity. Liu et al. (2022b) also observed that the antioxidant enzyme activities indicated that microplastics (MPs), especially at the concentration of 100 mg/kg, caused more severe oxidative stress to *E. fetida* than nanoplastics (NPs). The authors also observed that NPs demonstrated a more severe effect on the intestinal microflora structure of *E. fetida* than MPs.

3.3 DNA Damage

Comet assay is an efficient technique for evaluating DNA damage caused by MNPs. Jiang et al. (2020) observed that 100 and 1300 nm PS-MPs caused great levels of DNA damage on earthworm *E. fetida*. The authors believe that the ROS accumulation in *E. fetida* tissues may be liable for the DNA damage in earthworm coelomocytes. Xu et al. (2021b) also observed that the micron-size MPs (10 μm) showed greater DNA damage to earthworm coelomocytes than nano-size MPs. Sobhani et al. (2022) assessed the acute and chronic effects of polyethylene microplastics (PEMP) on *E. fetida* species of earthworms. The tail DNA percent (TD%) and olive tail movement (OTM) were used to measure the DNA damage in earthworm coelomocytes. The authors revealed that the PEMP triggered DNA damage in *E. fetida* even at 0.1% concentration. There have also been reports of DNA damage in earthworms exposed to phthalates in the past (Feng et al., 2016). Overall, the PEMP-contaminated soils can damage earthworms' DNA and have a negative impact on earthworms' reproduction. Huang et al. (2021) studied the co-exposure of polyethylene and cadmium on the *E. fetida* species of earthworm. The authors indicated that with increasing MPs concentration, the tailing and nucleolar elongation apparently occurred, indicating that the *E. fetida* sperm endured DNA damage.

3.4 Gut Microbiota

In earthworms, gut microbiota is directly associated with health, immunity, and metabolism together with soil decomposition processes. Recently, more interest has been given to the influence of MNPs on the gut microbiota of earthworms. Exposure to microplastics may modify earthworms' feeding habits, which in turn may alter the habitat of their gut and, ultimately, alter the intestinal microbiota of earthworms (Xu et al., 2021b). Cheng et al. (2021) observed that polypropylene microplastics (PP-MPs) modified the soil microbiota instead of gut microbiota of the *Metaphire guillelmi* earthworm. These findings will contribute to better know the impact of MNPs on the soil and earthworm gut microbiota. According to Wang et al. (2019b), microplastics (polyvinyl chloride) exposure decreases arsenic buildup and modifies gut bacterial communities in *Metaphire californica* species of earthworms. These studies will help to better understand the impacts of combined pollutants on soil system.

3.5 Combined Effects of MNPs and Other Pollutants on Earthworms

MNPs are very hydrophobic and have a substantial specific surface area, which permits them to adsorb various pollutants from the contaminated environment (Horton et al., 2017). Hence, the combined effects of MNPs and other pollutants adsorbed on earthworms received a lot of interest (Huang et al., 2021). Zhou et al. (2020) revealed that exposure to MPs and heavy metals collectively had additional harmful effects on earthworm growth than exposure to MPs alone. Since, MPs have the ability to boost the bioavailability of heavy metal ions, which can also stop earthworms' development (Huang et al., 2021). According to Zhang et al. (2022), the co-exposure to polyethylene microplastics (PE-MPs) and ZnO nanoparticles indicated that adding PE-MPs caused larger weight loss in earthworms. Since MNPs absorb heavy metals in the soil and then desorb them in earthworm intestines, they may be contributing to the weight loss in earthworms by promoting the accumulation of heavy metals. Recently Liu et al. (2022a) observed that the co-exposure to pyrene and MPs (10 μm) triggered greater biomass loss to earthworms than co-exposure to pyrene and NPs (100 nm). This might be a result of the more severe physical harm that larger-sized MPs inflict on earthworms, including damage to their skin and coelomocytes, which increases the buildup of other pollutants. In contrast, NPs are quickly expelled (Liu et al., 2022a, b), thereby inflicting less harm to earthworms.

4 Conclusions and Perspectives

This chapter discusses the toxic effects of MNPs on earthworms mainly in terms of histopathological damage, oxidative stress, DNA damage, gut microbiota, and the combined effect of MNPs with heavy metals and other pollutants in the soil environment. Overall, MNP exposure can produce more serious damage to earthworms and other biota in the soil ecosystem. Moreover, MNPs can trigger physical injury to earthworms and can modify the bioavailability of other pollutants thus boosting accumulation. Most of the findings were performed under lab conditions and the exposure times remained comparatively short, compared with the real condition. Future research should thus concentrate on field investigations, particularly on real MNPs exposure dosages and the cumulative effects of MNPs and other pollutants on earthworms over an extended period of time. To reduce MNPs in soils, the use of biodegradable plastics is essential and may be utilized greatly in the future.

Acknowledgments Sartaj Ahmad Bhat acknowledges the Japan Society for the Promotion of Science (JSPS) for the JSPS International Postdoctoral Fellowship.

References

Agathokleous, E., Iavicoli, I., Barceló, D., & Calabrese, E. J. (2021). Ecological risks in a 'plastic'world: A threat to biological diversity? *Journal of Hazardous Materials, 417*, 126035.

Allouzi, M. M. A., Tang, D. Y. Y., Chew, K. W., Rinklebe, J., Bolan, N., Allouzi, S. M. A., & Show, P. L. (2021). Micro (nano) plastic pollution: The ecological influence on soil-plant system and human health. *Science of the Total Environment, 788*, 147815.

Bhat, S. A., Singh, J., & Vig, A. P. (2017). Amelioration and degradation of pressmud and bagasse wastes using vermitechnology. *Bioresource Technology, 243*, 1097–1104.

Bhat, S. A., Singh, S., Singh, J., Kumar, S., & Vig, A. P. (2018). Bioremediation and detoxification of industrial wastes by earthworms: Vermicompost as powerful crop nutrient in sustainable agriculture. *Bioresource Technology, 252*, 172–179.

Bhat, S. A., Cui, G., Yaseera, N., Lei, X., Ameen, F., & Li, F. (2022). Removal potential of microplastics in organic solid wastes via biological treatment approaches. In P. Chowdhary, S. Mani, & P. Chaturvedi (Eds.), *Microbial biotechnology: Role in ecological sustainability and research*. https://doi.org/10.1002/9781119834489.ch14

Boots, B., Russell, C. W., & Green, D. S. (2019). Effects of microplastics in soil ecosystems: Above and below ground. *Environmental Science & Technology, 53*, 11496–11506.

Briassoulis, D., Babou, E., Hiskakis, M., & Kyrikou, I. (2015). Analysis of long-term degradation behaviour of polyethylene mulching films with pro-oxidants under real cultivation and soil burial conditions. *Environmental Science and Pollution Research, 22*, 2584–2598.

Chen, Y., Liu, X., Leng, Y., & Wang, J. (2020). Defense responses in earthworms (*Eisenia fetida*) exposed to low-density polyethylene microplastics in soils. *Ecotoxicology and Environmental Safety, 187*, 109788.

Chen, Y., Awasthi, A. K., Wei, F., Tan, Q., & Li, J. (2021). Single-use plastics: Production, usage, disposal, and adverse impacts. *Science of the Total Environment, 752*, 141772.

Cheng, Y., Song, W., Tian, H., Zhang, K., Li, B., Du, Z., Zhang, W., Wang, J., Wang, J., & Zhu, L. (2021). The effects of high-density polyethylene and polypropylene microplastics on the soil and earthworm Metaphire guillelmi gut microbiota. *Chemosphere, 267*, 129219.

Corradini, F., Meza, P., Eguiluz, R., Casado, F., Huerta-Lwanga, E., & Geissen, V. (2019). Evidence of microplastic accumulation in agricultural soils from sewage sludge disposal. *Science of the Total Environment, 671*, 411–420.

da Costa, J. P., Santos, P. S., Duarte, A. C., & Rocha-Santos, T. (2016). (Nano) plastics in the environment–sources, fates and effects. *Science of the Total Environment, 566*, 15–26.

Dang, F., Wang, Q., Huang, Y., Wang, Y., & Xing, B. (2022). Key knowledge gaps for One Health approach to mitigate nanoplastic risks. *Eco-Environment & Health, 1*, 11–22.

de Souza Machado, A. A., Kloas, W., Zarfl, C., Hempel, S., & Rillig, M. C. (2018). Microplastics as an emerging threat to terrestrial ecosystems. *Global Change Biology, 24*, 1405–1416.

Dris, R., Gasperi, J., Saad, M., Mirande, C., & Tassin, B. (2016). Synthetic fibers in atmospheric fallout: A source of microplastics in the environment? *Marine Pollution Bulletin, 104*(1–2), 290–293.

Edo, C., González-Pleiter, M., Leganés, F., Fernández-Piñas, F., & Rosal, R. (2020). Fate of microplastics in wastewater treatment plants and their environmental dispersion with effluent and sludge. *Environmental Pollution, 259*, 113837.

Feng, Q., Zhong, L., Xu, S., Qiu, L., & Wu, S. (2016). Biomarker response of the earthworm (*Eisenia fetida*) exposed to three phthalic acid esters. *Environmental Engineering Science, 33*(2), 105–111.

Gong, W., Zhang, W., Jiang, M., Li, S., Liang, G., Bu, Q., Xu, L., Zhu, H., & Lu, A. (2021). Species-dependent response of food crops to polystyrene nanoplastics and microplastics. *Science of the Total Environment, 796*, 148750.

Horton, A. A., Walton, A., Spurgeon, D. J., Lahive, E., & Svendsen, C. (2017). Microplastics in freshwater and terrestrial environments: Evaluating the current understanding to identify the knowledge gaps and future research priorities. *Science of the Total Environment, 586*, 127–141.

Huang, C., Ge, Y., Yue, S., Zhao, L., & Qiao, Y. (2021). Microplastics aggravate the joint toxicity to earthworm *Eisenia fetida* with cadmium by altering its availability. *Science of the Total Environment, 753*, 142042.

Huerta Lwanga, E., Gertsen, H., Gooren, H., Peters, P., Salánki, T., Van Der Ploeg, M., Besseling, E., Koelmans, A. A., & Geissen, V. (2016). Microplastics in the terrestrial ecosystem: Implications for Lumbricus terrestris (Oligochaeta, Lumbricidae). *Environmental Science & Technology, 50*(5), 2685–2691.

Jeong, C. B., Won, E. J., Kang, H. M., Lee, M. C., Hwang, D. S., Hwang, U. K., Zhou, B., Souissi, S., Lee, S. J., & Lee, J. S. (2016). Microplastic size-dependent toxicity, oxidative stress induction, and p-JNK and p-p38 activation in the monogonont rotifer (Brachionus koreanus). *Environmental Science & Technology, 50*(16), 8849–8857.

Jiang, X., Chang, Y., Zhang, T., Qiao, Y., Klobučar, G., & Li, M. (2020). Toxicological effects of polystyrene microplastics on earthworm (*Eisenia fetida*). *Environmental Pollution, 259*, 113896.

Khaldoon, S., Lalung, J., Maheer, U., Kamaruddin, M. A., Yhaya, M. F., Alsolami, E. S., Alorfi, H. S., Hussein, M. A., & Rafatullah, M. (2022). A review on the role of earthworms in plastics degradation: Issues and challenges. *Polymers, 14*, 4770.

Kwak, J. I., & An, Y. J. (2021). Microplastic digestion generates fragmented nanoplastics in soils and damages earthworm spermatogenesis and coelomocyte viability. *Journal of Hazardous Materials, 402*, 124034.

Lahive, E., Walton, A., Horton, A. A., Spurgeon, D. J., & Svendsen, C. (2019). Microplastic particles reduce reproduction in the terrestrial worm Enchytraeus crypticus in a soil exposure. *Environmental Pollution, 255*, 113174.

Lambert, S., & Wagner, M. (2016). Characterisation of nanoplastics during the degradation of polystyrene. *Chemosphere, 145*, 265–268.

Liang, J., Xia, X., Zhang, W., Zaman, W. Q., Lin, K., Hu, S., & Lin, Z. (2017). The biochemical and toxicological responses of earthworm (*Eisenia fetida*) following exposure to nanoscale zerovalent iron in a soil system. *Environmental Science and Pollution Research, 24*, 2507–2514.

Liu, S., Zhou, Q., & Chen, C. (2012). Antioxidant enzyme activities and lipid peroxidation in earthworm Eisenia fetida exposed to 1, 3, 4, 6, 7, 8-hexahydro-4, 6, 6, 7, 8, 8-hexamethyl-cyclopenta-γ-2-benzopyran. *Environmental Toxicology, 27*(8), 472–479.

Liu, Y., Xu, G., & Yu, Y. (2022a). Effects of polystyrene microplastics on accumulation of pyrene by earthworms. *Chemosphere, 296*, 134059.

Liu, J., Qin, J., Zhu, L., Zhu, K., Liu, Z., Jia, H., & Lichtfouse, E. (2022b). The protective layer formed by soil particles on plastics decreases the toxicity of polystyrene microplastics to earthworms (*Eisenia fetida*). *Environment International, 162*, 107158.

Meng, K., Lwanga, E. H., van der Zee, M., Munhoz, D. R., & Geissen, V. (2023). Fragmentation and depolymerization of microplastics in the earthworm gut: A potential for microplastic bioremediation? *Journal of Hazardous Materials*, 130765.

OECD (Organization for Economic Co-operation and Development). (1984). No. 207.; Earthworm, acute toxicity tests. In *OECD guideline for the testing of chemicals*. OECD Publishing.

OECD (Organization for Economic Co-operation and Development). (2004). Earthworm Reproduction Test (*Eisenia fetida/Eisenia andrei*) (No. 222). In *OECD guidelines for the testing of chemicals*. OECD Publishing.

Prendergast-Miller, M. T., Katsiamides, A., Abbass, M., Sturzenbaum, S. R., Thorpe, K. L., & Hodson, M. E. (2019). Polyester-derived microfibre impacts on the soil-dwelling earthworm Lumbricus terrestris. *Environmental Pollution, 251*, 453–459.

Rillig, M. C. (2012). Microplastic in terrestrial ecosystems and the soil? *Environmental Science & Technology, 46*, 6453–6454.

Rillig, M. C., & Lehmann, A. (2020). Microplastic in terrestrial ecosystems. *Science, 368*, 1430–1431.

Rodriguez-Seijo, A., Lourenço, J., Rocha-Santos, T. A. P., Da Costa, J., Duarte, A. C., Vala, H., & Pereira, R. (2017). Histopathological and molecular effects of microplastics in *Eisenia andrei* Bouché. *Environmental Pollution, 220*, 495–503.

Rodriguez-Seijo, A., da Costa, J. P., Rocha-Santos, T., Duarte, A. C., & Pereira, R. (2018). Oxidative stress, energy metabolism and molecular responses of earthworms (Eisenia fetida) exposed to low-density polyethylene microplastics. *Environmental Science and Pollution Research, 25*, 33599–33610.

Sobhani, Z., Panneerselvan, L., Fang, C., Naidu, R., & Megharaj, M. (2022). Chronic and transgenerational effects of polyethylene microplastics at environmentally relevant concentrations in earthworms. *Environmental Technology & Innovation, 25*, 102226.

Steinmetz, Z., Wollmann, C., Schaefer, M., Buchmann, C., David, J., Tröger, J., Muñoz, K., Frör, O., & Schaumann, G. E. (2016). Plastic mulching in agriculture. Trading short-term agronomic benefits for long-term soil degradation? *Science of the Total Environment, 550*, 690–705.

Tang, R., Zhu, D., Luo, Y., He, D., Zhang, H., El-Naggar, A., Palansooriya, K. N., Chen, K., Yan, Y., Lu, X., & Ying, M. (2023). Nanoplastics induce molecular toxicity in earthworm: Integrated multi-omics, morphological, and intestinal microorganism analyses. *Journal of Hazardous Materials, 442*, 130034.

UNEP. (2016). *UNEP Frontiers 2016 Report: Emerging issues of environmental concern.* UN Environment Programme.

Wang, J., Coffin, S., Sun, C., Schlenk, D., & Gan, J. (2019a). Negligible effects of microplastics on animal fitness and HOC bioaccumulation in earthworm Eisenia fetida in soil. *Environmental Pollution, 249*, 776–784.

Wang, H. T., Ding, J., Xiong, C., Zhu, D., Li, G., Jia, X. Y., Zhu, Y. G., & Xue, X. M. (2019b). Exposure to microplastics lowers arsenic accumulation and alters gut bacterial communities of earthworm *Metaphire californica. Environmental Pollution, 251*, 110–116.

Xing, Y., Meng, X., Wang, L., Zhang, J., Wu, Z., Gong, X., Wang, C., & Sun, H. (2018). Effects of benzotriazole on copper accumulation and toxicity in earthworm (Eisenia fetida). *Journal of Hazardous Materials, 351*, 330–336.

Xu, G., Yang, Y., & Yu, Y. (2021a). Size effects of polystyrene microplastics on the accumulation and toxicity of (semi-) metals in earthworms. *Environmental Pollution, 291*, 118194.

Xu, G., Liu, Y., Song, X., Li, M., & Yu, Y. (2021b). Size effects of microplastics on accumulation and elimination of phenanthrene in earthworms. *Journal of Hazardous Materials, 403*, 123966.

Zhang, B., Yang, X., Chen, L., Chao, J., Teng, J., & Wang, Q. (2020). Microplastics in soils: A review of possible sources, analytical methods and ecological impacts. *Journal of Chemical Technology & Biotechnology, 95*(8), 2052–2068.

Zhang, S., Ren, S., Pei, L., Sun, Y., & Wang, F. (2022). Ecotoxicological effects of polyethylene microplastics and ZnO nanoparticles on earthworm *Eisenia fetida. Applied Soil Ecology, 176*, 104469.

Zhou, Y., Liu, X., & Wang, J. (2020). Ecotoxicological effects of microplastics and cadmium on the earthworm *Eisenia foetida. Journal of Hazardous Materials, 392*, 122273.

Toxicological Effects of Micro and Nanoplastics on Soil Fauna: Current Research, Advances, and Future Outlook

Irem Ozturk-Ufuk, Ashna Waseem, Meryem Vasef, Lama Ramadan, Elif Pehlivanoğlu-Mantaş, and Emel Topuz

1 Introduction

Plastics find extensive application in various aspects of daily living, including construction, agriculture, fisheries, and packaging (Ivleva et al., 2017). The production of plastics is anticipated to double in 20 years, which would exceed the current capabilities for plastic management, making it one of the major environmental challenges (Nielsen et al., 2020). As a result of significant manufacturing, a considerable quantity of plastic waste is entering the natural environment. This waste breaks down into smaller particles, exhibiting diverse shapes and size distributions. The plastics with 1 μm – 5 mm particle size are termed as microplastics (MPs), while nanoplastics (NPs) have a particle size of less than 1 μm (Gigault et al., 2018; Li et al., 2022).

Many studies have recorded the presence of microplastics and nanoplastics (M/NPs) in the natural soil, while the detection of NPs is still lacking due to difficulties in the extraction process of NPs from soil, particularly from organic-rich soils (Atugoda et al., 2023). As of now, the presence of NPs in natural soil has not been documented in the existing literature. The detection frequency and characterization of MPs in various natural soils are presented in Table 1. As indicated in Table 1, polypropylene (PP) and polyethylene (PE) accounted for the majority of the recorded MPs in different natural soils with less contribution for polyvinyl chloride (PVC), and polystyrene (PS). PE and PP are thermoplastic polymers extensively

I. Ozturk-Ufuk · A. Waseem · M. Vasef · L. Ramadan · E. Topuz (✉)
Department of Environmental Engineering, Gebze Technical University, Kocaeli, Turkey
e-mail: emeltopuz@gtu.edu.tr

E. Pehlivanoğlu-Mantaş
Department of Environmental Engineering, Gebze Technical University, Kocaeli, Turkey

Department of Environmental Engineering, Istanbul Technical University, Istanbul, Turkey

S. A. Bhat et al. (eds.), *Management of Micro and Nano-plastics in Soil and Biosolids*,
https://doi.org/10.1007/978-3-031-51967-3_9

Table 1 Occurrence and characterization of MPs in natural soils

Country/ region	Soil	Type	Concentration	Color	Particle size	Shape	Refs.
UK	Landfill and urban soils	PE, PP, PS, and PET	12.3 p/g in landfill and 15–17 p/g in urban soils	–	25–178 μm	Films and foams	Billings et al. (2023)
Southwest China	Cultivated soil	75% PE and 25% PVC	50–1158 p/kg	60% transparent and 39% black	<5 mm	83% flaky and 15.8% fibrous	Zhang et al. (2022)
Iran	Urban and industrial soils	PET and nylon	100–3135 p/kg in urban soils and 80–1220 p/kg in industrial soils	White and transparent	<250 μm	Fibrous	Javad et al. (2022)
Northwest China	Sand land soils	53% PP, 24.85% PE, 15.88% PS and 5.51 PVC.	1360–4960 p/kg	93% black and 4% transparent	≤0.5 mm	Films, pellets, fragments, and fibers	Ding et al. (2021)
Korea	Agricultural soils	PE > PP	10–7630 items/kg	–	0.2–0.3 mm	Fragment > sheet > fiber	Kim et al. (2021)
China	Farmland and grassland	PE and PP	53.2 items/kg in shallow soil and 43.9 items/kg in and deep soil	Mostly transparent. Black, white, and blue	0–3 cm: 66% < 500 μm. 11% > 1 mm. 3–6 cm: 45% <100 μm	Mostly film	Feng et al. (2021)
Brazil	Beach sand	PE	2.4–30.4 p/m^2	49.3% white or transparent	–	85% fragments. 9.8% films	Maynard et al. (2021)
Santiago, Spain	Soils under different land use	Acrylates, PE, varnish, ethylene vinyl acetate and PP.	184–306 p/kg	–	<8 mm Mostly <2 mm	Fiber (dominant), film, fragment, and pellet	Corradini et al. (2021)
China	Agricultural soils	Mostly PP	4.94–252.70 p/kg	White	0.1–0.5 mm	Fragments	Cao et al. (2021)

North-Western China	Agricultural soils	PS, PE, PP, HDPE, PVC and PET	1430–3410 items/kg	–	Mostly ≤0.49 mm	Fibers	Ding et al. (2020)
Chile	Agricultural soil	Acrylic, Polyester and nylon	1.37–45.5 mg/kg	–	1.60–2.7 mm	Mostly fibers (90% in sludge, and 97% in soil)	Corradini et al. (2019)
China	Farmland	PE and PP	10.3 p/kg	White and translucent	<1 mm	Fiber	Lv et al. (2019)
China	Coastal beach soil	PE, PP, and PS	1.0–740 p/kg (dw)	White and yellow	<5 mm	Mostly foams and fragments	Zhou et al. (2018)
Switzerland	Floodplain soils	Mostly PE	55.5 mg/kg 593 p/kg	–	125–500 μm (80%)	–	Scheurer and Bigalke (2018)
Shanghai, China	Farmland soils	50.51% PP and 43.43% PE	78.00 p/kg in shallow soil and 62.50 p/kg in deep soil.	Mostly in black or transparent	20 μm to 5 mm	Fiber, fragment, and film	Liu et al. (2018)

employed in everyday activities, primarily due to their ease of processing and cost-effective production (Xu et al., 2020). PP is considered more preferable than PE as it has the lowest density among commercial plastics, in addition to the high chemical and temperature resistance, which make PP suitable for several versatile items including bottles, funnels, trays, and particularly the jars that need to be frequently sterilized for the clinical use (Kim & Kim, 2004; Maddah, 2016). PS exhibits notable traits such as a glossy surface, rigidity, and exceptional transparency. It is extensively used in handwashing soaps, toothpaste, cleansers, and biomedical products (Siddiqui et al., 2023). PVC is considered a cheap polymer with appropriate mechanical and chemical characteristics to be used as membrane and ion exchange material (Lieberzeit et al., 2022). However, the application of PVC is limited due to the various interfering deviations caused by the presence of active chlorine atoms in its structure (Endo, 2002). Polyethylene terephthalate (PET), low-density polyethylene (LDPE), high-density polyethylene (HDPE), polyamide (PA), polyester (PES), polyacrylonitrile (PAN), and CPE (chlorinated polyethylene) are the other types of M/NPs that are studied in the literature.

M/NPs in the terrestrial environment are originated from multiple sources, including landfilling, domestic waste, sewage sludge, irrigation with wastewater, fertilization, tire abrasion, greenhouse materials, urban runoff, and the transportation of plastic particles over long distances (Chae & An, 2018; Dris et al., 2016; Hurley & Nizzetto, 2018; Koutnik et al., 2021). Municipal and industrial wastewater treatment plants (WWTPs) were recognized as the major source of M/NPs, as biosolids retain around 95% of these particles, and it is used as fertilizers and compost in agricultural applications, which act as the main vectors for M/NPs into the terrestrial environment (Bläsing & Amelung, 2018; Ravi et al., 2022). M/NPs may also be derived from agriculture plastic mulching, which is applied for pest control and increasing crop yields (Hurley & Nizzetto, 2018). Moreover, polymers such as polyurethane and PS are directly added to compost in order to improve the soil properties, which leads to M/NPs dispersion and integration across a wide area of soil (Duis & Coors, 2016).

There are two main categories for the origin source of M/NPs. The main origin of particles involves their direct emission from manufacturing processes and consumer items such as personal care products, plastic molding powders, biomedical goods, medications, and synthetic fabrics (such as PES or nylon), all contributing to the direct release of MPs (Birch et al., 2020; Patil et al., 2022). While NPs could be released directly from electronics, adhesives, and paints (Kihara et al., 2021). The secondary source of M/NPs is from where the particles are indirectly emitted by the fragmentation of macroplastics (González-Pleiter et al., 2019; Kokalj et al., 2021). Macroplastics can disseminate M/NPs via abiotic and biotic transformation processes such as physicochemical reactions, sand friction, biodegradation, thermal degradation, photodegradation, hydrolysis, and thermo-oxidative degradation (Getor et al., 2020; Hale et al., 2020; Wei et al., 2022; Weinstein et al., 2016). It has been estimated that the primary source of M/NPs accounts only for 15–30% of the released plastics, while the secondary source contributes to 70–80% of the total plastics in the environment (Mariano et al., 2021).

2 Impact of Microplastics and Nanoplastics on Soil Fauna Toxicity

M/NPs negatively impact animals' growth, reproduction, and metabolism through various toxic mechanisms alongside various intermolecular forces such as hydrophobic effects, van der Waals forces partition effects, electrostatic interactions, non-covalent bonding, and microporous filling mechanisms (Chang et al., 2022). Therefore, uptake and elimination of M/NPs are critical to understand the toxic effects at cellular and organism levels.

2.1 Uptake and Elimination of Microplastics and Nanoplastics by Soil Fauna

2.1.1 Uptake of M/NPs

The uptake of M/NPs by organisms in terrestrial ecosystems, through various pathways such as ingestion, inhalation, and direct exposure can have effects on their growth, reproduction, cellular systems, etc. (Enyoh et al., 2020; Li et al., 2020; Yao et al., 2022). Ingestion of M/NPs by terrestrial animals through contaminated food or soil is one of the most significant uptake routes. When M/NPs are found in plants or soil animals that make up a significant portion of an animal's diet, terrestrial animals have the potential for direct interaction with M/NPs within their surroundings or through the incorporation of such particles into their dietary intake (Prata & Dias-Pereira, 2023). This direct contact and ingestion of M/NPs by terrestrial organisms amplify the intricate web of interactions between plastic pollution and terrestrial ecosystems. Terrestrial organisms can ingest M/NPs, in areas where airborne MPs are present, as well (Duis & Coors, 2016).

In order to comprehend the dynamics of M/NPs within soil communities and ecosystems, a crucial step involves identifying how the animals readily ingest M/NPs and assessing their passage through digestive systems. A recently developed and validated fluorescent counterstaining technique offers insight into this realm, as exemplified by the detection of PE at 488 nm and PP at 525 nm in the gut of *Zophobas morio*, and PE at 405 nm and PVC at 445 nm in the gut of *Lumbricus terrestris*, shedding light on M/NPs interactions with terrestrial invertebrate biomass and fecal matter (Maxwell et al., 2020). In the context of M/NPs ingestion within terrestrial ecosystems, *Rhinella diptycha* demonstrated a notably higher prevalence, with 47.7% of the sampled population containing MPs (both fiber and fragments) in their stomach contents, while *Hemidactylus mabouia* showed 12.03% and *Tropidurus torquatus* exhibited 6.00%. These findings were accompanied by a consistent prevalence of fibers over fragments across all species, along with distinct coloration variations observed in *R. diptycha* and *H. mabouia* (Mackenzie &

Vladimirova, 2023). In an experimental study conducted on organic farm soil from Chile to investigate microplastic ingestion by terrestrial organisms, *L. terrestris* was found to contain MPs such as PS, PP, PET, and LDPE. Notably, the foregut contained the highest concentration at 7% w/w, followed by 5% w/w in the midgut and 2.5% w/w in the hindgut. The outcomes of the ingestion bioassay further revealed that MPs were present across all segments of the earthworm, suggesting a potential risk for this species due to their apparent inability to differentiate between regular food sources and MPs (Baeza et al., 2020).

Furthermore, it is crucial to thoroughly evaluate the uptake of M/NPs through inhalation and the relevant trajectory in the body after the inhalation process. For example, PS MPs with a diameter of 0.02 mm, when subjected to inhalation exposure, have been observed to undergo translocation and migrate to the lungs of rats (Fournier et al., 2020). Also, in the study involving *Columba livia*, *Milvus migrans*, and *Hirundo rustica*, PP and PE were discovered in *Columba livia*, while ethylene vinyl acetate was identified in *Hirundo rustica*, signifying that these avian species are exposed to inhalable MPs, with particle sizes mainly ranging from 28.0 to 70.5 μm, which can reach their lungs in addition to dietary sources (Tokunaga et al., 2023). Compared to ingestion, the studies on inhalation of M/NPs by soil organisms are quite limited.

The uptake of M/NPs is influenced by various factors. The likelihood of ingestion can be determined by the feeding traits, organism size, and characteristics of the particles they are exposed to. Among these factors, the physical properties, specifically the size, of MPs play a significant role in determining their uptake and potential impacts. For instance, when the MP concentration is held constant at 0.5% and specifically focusing on the PS NPs type, smaller particles have been found to be ingested more by *Enchytraeus crypticus* compared to larger fragments (>200 μm), leading to biomass reduction (Zhu et al., 2018). Also in another study, enchytraeids showed limited consumption of long PS fibers with size of 24 μm, in contrast to their noticeable ingestion of shorter fibers that had no noticeable impact on their survival or reproductive functions (Selonen et al., 2020). On the contrary to these studies, in a study, researchers investigated the uptake, transfer, and impact of different sizes of PS MPs on *Culex pipiens* mosquitoes, and they reported no correlation of uptake with the size of MPs. They examined both aquatic larval and terrestrial adult stages of the mosquitoes. The results showed that both 2 μm and 15 μm MPs were transferred from the larval to the adult stage of the mosquitoes, and the amount of uptake was not closely linked to the size of MPs (Al-Jaibachi et al., 2019).

The ingestion of MPs is likely influenced by their shape, suggesting that the physical form of MPs plays a crucial role in the route of uptake. The investigations revealed that soil-dwelling organisms, including annelids, arthropods, and molluscs, ingested and expelled MPs of various types and shapes. Annelida species, such as *E. fetida*, *L. terrestris*, *E. andrei*, *E. crypticus*, and *M. guillelmi*, were found to ingest or egest PE, PS, PP, and MFR spheres, as well as fragments and fibers of LLDPE, nylon, PP, tires, and HDPE. The specific uptake ways of these MPs aren't completely known. However, there are several likely mechanisms at play. One is particles attaching to the outer surface of earthworms. Another mechanism involves particles being carried along as water moves down the earthworm burrows. Additionally, the casting activity of earthworms, where they bring up soil from deeper layers, could also play a role in particle movement. Lastly, particles might be transported by earthworms after passing through their digestive systems, as evidenced by their widespread presence in the earthworm droppings (Rillig et al., 2017; Zhou et al., 2020; Sheng et al., 2021; Kuehr et al., 2022). Arthropods like *F. candida* and *P. scaber* also demonstrated ingestion and/or egestion of PVC particles, PP fiber-fragment mixtures, and PE fibers (Kwak & An, 2021). It could be hypothesized that the diverse feeding strategies, digestive capabilities, and physiological traits of different soil organisms might lead to differences in their interactions with M/NPs. For instance, earthworms like *L. terrestris* and enchytraeids such as *E. crypticus, M. guillelmi, E. fetida*, and *E. andrei* may exhibit varying degrees of ingestion and egestion based on their size, movement patterns, and habitat preferences (Jager et al., 2003). Arthropods like *F. candida* and *P. scaber* might show distinct responses like food preferences due to differences in their mouthpart structures and feeding behaviors (Potapov et al., 2022). Similarly, molluscan species like *Achatina fulica* could demonstrate unique ingestion patterns, characterized by their specific feeding mechanisms that often involve specialized mouthparts due to their specific feeding mechanisms (Lima et al.).

The uptake of M/NPs is contingent upon various chemical properties, including the characterization of specific attributes such as surface chemistry, additives, products of weathering, crystallinity, and degradation state. MNPs undergo weathering and degradation processes in the environment, causing their chemical composition to evolve over time, and they are called "aged MPs." Moreover, these particles have the capacity to adsorb various organic and inorganic compounds from their surroundings, resulting in further modifications to their chemical properties (Andrady, 2017; Ivleva et al., 2017). The aging process induces MPs to fragment into tinier particles, influencing their assimilation by living organisms. Due to anatomical constraints such as limited mouth sizes, organisms like oyster larvae and nematodes find it more feasible to ingest these diminished MPs. In the context of MP ingestion, *A. parietinus*, a suction-feeding nematode species, can only ingest PS beads smaller than 0.5 μm through its buccal cavity. The ratio between buccal cavity size and bead dimensions predicts ingestion potential, with a limited range of 1.3–3.6 (Fueser et al., 2019).

The absorption of M/NPs can differ based on various biological factors. These factors encompass feeding patterns, and reproductive status have a crucial impact on the uptake process. Feeding behavior strongly influences the pathways of plastics in terrestrial environments. As demonstrated in studies, various concentrations of MPs have been identified in animal feed, including 123.9, 5500, and 600–3500 MP/in fishmeal, as well as cellulosic microfibres at concentration of 1230 MP/kg in soybean meal. These meals are commonly employed as sources of protein in animal diets or incorporated into animal feeds. Consequently, livestock can be exposed to MPs through oral consumption. Furthermore, animals may inadvertently consume plastics while foraging, often by ingesting contaminated invertebrates. This intake of plastic-contaminated food introduces pollutants into the organisms. Over time, these pollutants have the potential to move through the food chain, potentially leading to biomagnification (Prata & Dias-Pereira, 2023).

2.1.2 Elimination of M/NPs

M/NPs can ultimately alter the energy demands (metabolism) of organisms, which can result in shifts in community arrangement and ecosystem functioning (Kratina et al., 2019). Once small-sized MPs infiltrate an organism, they can persist for extended periods within the body, interacting with cellular processes and metabolism. These particles can traverse cell membranes, infiltrate surrounding tissues, and even enter the circulatory system, where they can produce cellular and molecular toxic effects, further impacting the organism's metabolism and overall health (Li et al., 2022).

Terrestrial organisms eliminate M/NPs through various routes, including excretion and egestion. In the process of tracing the fate of M/NPs within organisms, excretion involves the expulsion of ingested M/NPs through waste, such as feces. This intricate mechanism gains insight when examining the presence of PET microfibers at the lowest concentration of 0.01 g/kg did not yield a noteworthy decrease in food consumption when contrasted with *A. fulica* that had not been subjected to the fibers. Furthermore, a study indicated a dose-dependent interference of PET microfibers with excretion processes (Song et al., 2019). The primary mechanism driving the vertical redistribution of PS NPs within soil profiles is nonlocal transport through ingestion and subsequent excretion. This phenomenon was examined through microcosm experiments involving *L. terrestris*. The study utilized metal-doped spherical PS NPs to quantitatively analyze their movement in soil samples. Through experimentation, researchers tracked the vertical redistribution of these NPs over time, shedding light on their spread through different soil layers. Additionally, the impact of earthworm bioturbation on NP movement was explored using X-ray CT scans. Notably, the excretion of PS NPs seems to be more pronounced compared to PS MPs. As a result, earthworms are likely to transport approximately 8000 mg/kg NPs, potentially covering greater distances, particularly during the initial phases of dispersion (Heinze et al., 2021).

Egestion, which involves the regurgitation of MPs that an organism cannot digest, is a vital mechanism by which organisms rid themselves of indigestible substances. For example, in the treatment with the highest concentration of MPFs, there was a high variation of 88% in the amount of feces produced among different samples, although the average amount of feces aligned with the overall decreasing trend observed. Conducting a prolonged assessment spanning beyond 7 days enabled the study to investigate the kinetics of NP elimination in earthworms at a comparatively modest concentration of exposure 464.2 μg NP/g dry weight. The research revealed that the *L. terrestris* earthworms exhibited rapid elimination of NPs, displaying a notably high elimination rate of 0.432 per day. While a minor fraction of the NPs remained inert, implying incomplete expulsion of ingested NPs by the earthworms, this residual quantity became undetectable within a single day when placed in uncontaminated soil (Lahive et al., 2022).

2.2 *Toxic Effects at Molecular, Cellular, and Organ Levels*

Within terrestrial ecosystems, the presence of M/NPs exerts a pronounced impact on cellular dynamics. A particularly captivating focus is the cellular response an area where MPs wield the potential to induce a diverse array of changes in cell behavior. These changes encompass a spectrum of effects, spanning from genetic modifications to shifts in gene expression intensity. Metabolic pathways may undergo adjustments, enzyme levels and activity can experience fluctuations, and the initiation of the generation of reactive oxygen species (ROS) can result in oxidative stress (Koner et al., 2023). For instance, M/NPs in terrestrial environments can increase reactive oxygen species (ROS) in cells, leading to reproductive issues. This includes reduced quality, viability, and quantity of oocytes or sperm, as well as damage to reproductive tissues, indicated by histopathological changes in the organs (Prokic et al., 2021). In Table 2, available studies on the toxic effects of M/NPs in soil organisms at molecular and cellular levels are presented.

In various studies, it is assessed the impact of MPs on various species, including *E. crypticus*, *E. andrei*, and predominantly, *E. fetida* (Cheng et al., 2020; Xu et al., 2021; Jiang et al., 2020). The exposure to concentration of 1000 μg/kg PS resulted in the occurrence of DNA damage in both the initial generation (F0) and the subsequent generation (F1) of *E. fetida*. At the tissue level, the presence of PS spheres led to the enlargement and distortion of cells in *E. fetida*, prompting changes in the dimensions of cell nuclei and ultimately resulting in cellular breakdown through lysis. The superoxide dismutase (SOD) activities within *E. fetida* demonstrated a decline when subjected to 1000 μg/kg PS-MPs, recording values of 9.039 U/mg protein in the untreated soil and 9.195 U/mg protein in the soil treated with 1300 nm PS-MPs (Jiang et al., 2020). For *E. andrei*, it was observed that even at concentration of 62.5 mg of LLDPE fragments per kilogram of dry soil, significant histopathological alterations occurred. These changes included the notable presence of fibrosis, infiltration of inflammatory cells, and an increase in congestion within the organism (Rodriguez et al., 2017). In *E. crypticus*, the introduction of HDPE fragments at a concentration of 8% triggered oxidative responses, evidenced by a dose-dependent increase in GST (from 0.175 to 0.250 μkat/mg protein) and CAT activity (from 0.170 to 0.260 μkat/mg protein) (Pflugmacher et al., 2020). Similarly, within the same organism, exposure to PA and PVC led to a higher abundance of antibiotic-resistance genes, approximately 60%, while the presence of mobile genetic elements remained unaffected by these exposures (Ma et al., 2020).

Table 2 Toxic effects of M/NPs in soil organisms at molecular and cellular level

Organisms	Type of MP	Endpoint	Impact	References
E. crypticus	PS (1–10%)	Reproduction, EC50	Increase in reproduction toxicity more than 50% in the presence of PS MP	Ozturk et al. (2023)
Eisenia andrei	PE microplastics (180 μm, 250 μm)	Sperm density, seminal vesicles,	Histopathological changes, adverse impact on sperm production, arrangement, and density decrease in growth of seminal vesicles, heightened germ cell disruption	Kwak & An (2021)
E. fetida	PS Sphere (1000 μg/kg)	DNA damage SOD activity	Changes in the dimensions of cell nuclei, decreased SOD activity	Jiang et al. (2020)
E. fetida	PS	DNA damage, cellular size, superoxide dismutase (SOD)	Occurrence of DNA damage and cellular breakdown, enlargement and distortion of cells, decline in SOD activities.	Xu et al. (2021)
E. andrei	LLDPE fragments (62.5 mg/kg)	Histopathological	Significant alterations, including fibrosis, inflammatory cell infiltration, and congestion	Rodriguez et al. (2017)
E. crypticus	HDPE fragments (8%)	Oxidative responses, antibiotic resistance	Increase in GST and CAT activity; higher abundance of antibiotic resistance genes, 60% increase	Pflugmacher et al. (2020)
F. candida, *E. crypticus*	PA, PVC, PE	Microbiome stability, bacterial alpha diversity	Reduction in bacterial alpha diversity	Zhu et al. (2018), Ju et al. (2019) & Ma et al. (2020)
M. guillemi	HDPE and PP (0.25%)	Bacterial alpha diversity	No impact	Cheng et al. (2021)
Gryllodes Sigillatus	Nanoplastics	Digestive system	Growth rate decrease, reproductive organ inhibition in male earthworms	Fudlosid et al. (2022)

The impact of M/NPs on terrestrial ecosystems covers gut microbiota, although research in this area remains limited. Various forms of MPs, encompassing PA (1000 mg/kg), PVC (1 g/kg), and PE (25–43%), have been observed to trigger significant transformations in the functional changes and microbiome stability within soil-dwelling organisms such as *F. candida* and *E. crypticus*. This alteration notably manifests as a reduction in bacterial alpha diversity (Zhu et al., 2018; Ju et al., 2019; Ma et al., 2020). In contrast, the presence of HDPE and PP fragments at a concentration of 0.25% seems to have no significant effect on the bacterial alpha diversity or density in *M. guillemi* (Cheng et al., 2021). There is a study conducted focusing on *Gryllodes Sigillatus* and found that the ingestion of MPs primarily affects their digestive system and nutrient absorption. Prolonged exposure to PE MPs with a concentration of 2.5–10% in their diet can impact the insect's growth rate (a 20% decrease) and adult body size. Additionally, male earthworms experienced adverse effects on their reproductive organs, specifically inhibiting spermatogenesis. However, the impact on female reproductive tissues was minimal, similar to control, suggesting that the effects of MPs may vary between sexes (Fudlosid et al., 2022).

In a comprehensive study on the impact of M/NPs in terrestrial ecosystems, researchers studied two specific sizes of PE MP spheres with 180 μm and 250 μm, which were exposed to earthworms *E. andrei* for 21 days. While earthworm ovaries exhibited negligible effects from MP exposure, the repercussions on sperm production were notably adverse. Earthworms experienced a significant reduction in mature sperm packets and sperm density, accompanied by a disruption in sperm cell arrangement. Merely 17% and 5% of earthworm seminal vesicles displayed typical growth, indicating a distinct contrast from the control group. Conversely, exposed earthworms'seminal vesicles showed heightened germ cell disruption, accompanied by changes in tissue structure. Within the control group, 65% exhibited intact seminal vesicle structures, whereas the exposure groups demonstrated 12% and 6% with moderate and substantial structural abnormalities, correspondingly. Furthermore, the detrimental impact of MP exposure on spermatogenesis was underscored, with a small percentage of earthworm seminal vesicles (17% and 5%) showcasing normal growth and germ cell arrangement after encountering 180 μm and 250 μm PE MPs. These compelling findings cast light on the intricate consequences of microplastic exposure on earthworm reproductive systems, hinting at potential disruptions within terrestrial ecosystems (Kwak & An, 2021).

2.3 *Toxic Effects at Organism Level*

The potential toxic effects of M/NPs have raised significant concerns, as emerging research suggests their capacity to impact critical aspects such as growth, mortality rates, and organ functionality within various organisms. Table 3 shows the studies on the toxicity of M/NPs detected at organism level. One of the most frequently used endpoints for the detection of toxicity is mortality. The potential harm of PS NPs (50 nm) and PS MPs (300 nm, 600 nm, 4 μm) on mice was investigated through continuous oral exposure over a 4-week period, yielding significant outcomes. The recorded mortality rates were as follows: 1.10% in the 50 nm group, 2.10% in the 300 nm and 600 nm groups, and 0.10% in the 4 μm group. Disturbing modifications in parameters such as body weight, kidney weight, as well as indicators like ALB and ALP were observed. The mice experienced a decline in body weight, kidney weight, ALB, and ALP levels (Meng et al., 2022). PE MPs with a size less than

Table 3 The toxic effects of M/NPs in soil organism at organism level

Organisms	Type of MP	Endpoint	Impact	References
Enchytraeus crypticus	Polystyrene (1–10%)	Reproduction, EC50	>50% increase in reproduction toxicity with MPs (1–10%)	Ozturk et al. (2023)
Mice	PS-NPs, PS-MPs	Mortality rates	50 nm: 1.10%, 300 nm: 2.10%, 600 nm: 2.10%, 4 μm: 0.10%	Meng et al. (2022)
Lumbricus terrestris	PE MPs, low-density PE MPs	Mortality rates	*For PE:* 28% MPs: Increased mortality; 45% MPs: 8.3%; 60% MPs: 25% *For low-density PE:* 45% MPs: 8.3% ± 14.4 60% MPs: 25% ± 43.3	Huerta Lwanga et al. (2016)
F. candida	PE MPs	Springtail mortality and reproduction	1% MPs: 70.2% abundance; EC50 for reproduction: 0.29% MPs EC50 for mortality: 29%	Ju et al. (2019)
Cantareus aspersus	LDPE particles	Mortality rates	No immediate lethal effects; growth rate variations	Colpaert et al. (2021)
L. terrestris	Polyester-derived microfibers (0.1% and 1.0% w/w, corresponding to 0.3 g and 3.0 g), 40.7 ± 3.8 μm	Mortality rates	No impact	Prendergast-Miller et al. (2019)
Achatina fulica	PS NPs (100 mg/kg)	Growth rate, foraging, feeding	Growth rate decline: 1.77%; foraging decline: 4.63%	Chae and An (2020)
Aporrectodea rosea	PLA, HDPE MPs (1 g/kg)	Biomass, growth rate	Biomass decrease: 3.1% (exposed) vs 5.7% (control)	Boots et al. (2019)

(continued)

Table 3 (continued)

Organisms	Type of MP	Endpoint	Impact	References
L. terrestris	PE MPs (<150 µm) (Concentrations: 7%, 28%, 45%, 60% w/w)	Growth rate, weight loss	60% MPs: −0.5 mg/ worm/ day; Control: 10.3 mg/ worm/day	Kim et al. (2022)
Caenorhabditis elegans	LDPE, PLA/PBAT MPs	Reproduction and body length	LDPE 10 mg L-1: Offspring reduction 267 to 206; body length decrease	Schöpfer et al. (2020)
A. fulica	PET-MPs	Growth trajectory	10% PET-MPs: 0.5 g increase in final weight; cochlea size increase	De Felice et al. (2021)
F. candida	PE MPs (<500 µm)	Behavior	Avoidance rates (0.5% and 1% MPs) reproduction ceased at 0.1%	Ju et al. (2019)

150 µm revealed significant mortality rates for *Lumbricus terrestris* earthworms, with higher concentrations of MPs (28%, 45%, and 60%) in the litter causing increased mortality compared to lower concentrations (7%) and a control group (0%). Furthermore, the study explored the effects of LDPE MPs over 60 days, mortality rate increased up to 8.3% ± 14.4 and 25% ± 43.3 at 45% and 60% microplastics, respectively (Huerta Lwanga et al., 2016). This study clearly demonstrated the effect of MP type on their toxic effects on terrestrial organisms. In a controlled laboratory setting, *Folsomia candida* was subjected to synthetic soils containing PE MPs (0.005–1%) with sizes less than 500 µm during a 28-day laboratory experiment, at the highest concentration tested (1% MPs), a substantial reduction of 70.2% in springtail abundance was noted. The study further determined the semi-maximum effective concentration (EC50) values, revealing that the EC50 for reproduction in *F. candida* was 0.29% MPs, while the EC50 for springtail survival was 29%. This implies that at a concentration of 1% MPs, more than half of the adult springtails experienced mortality (Ju et al., 2019). This study showed that the sensitivity of endpoint is important for the toxicity of MPs, as well. Although the surface characteristics and size could be different, mortality rates for PE were significantly different for *F. Candida* and *Lumbricus terrestris* in the study of Huerta Lwanga et al. (2017) and (Ju et al., 2019), respectively. Consequently, the sensitivity of the terrestrial organism also led to different toxic effects for the similar MP. Contrary to these studies, no impact of MPs on the mortality of terrestrial organisms was also reported in a few studies. Ingestion of polyester-derived microfibers (0.1% and 1.0% w/w, corresponding to 0.3 g and 3.0 g) with an average diameter of 40.7 ± 3.8 µm did not result in fatalities for *L. Terrestris* within 35 days. However,

it did prompt transcriptional responses associated with general stress (Prendergast-Miller et al., 2019). In the realm of terrestrial ecosystems, the impact of M/NPs continues to unfold. A study focused on *Cantareus aspersus* sheds light on the potential toxicity effects of LDPE particles, frequently found in food packaging, reusable bags, and agricultural films. Young snails in soil (1-month-old and weight 1.2 ± 0.13 g) were exposed to various sizes (400–500 μm, 200–250 μm, and 50–100 μm) and concentrations of (0.014, 0.14, 0.71 g kg^{-1}) of LDPE particles for a duration of 28 days. Surprisingly, mortality rates remained consistent with controlled specimens, indicating no immediate lethal effects. (Colpaert et al., 2021).

Beyond mortality, researchers commonly investigate the impacts of M/NPs on parameters like growth rate and reproduction. Consequences of NPs on terrestrial organisms were exemplified through a study involving *Achatina fulica (*African giant snail*)*. The snails were placed in an environment where they were exposed to mung bean plant leaves. These leaves had undergone prior treatment with high-concentration FluoSpheres carboxylate-modified PS microspheres (0.02 μm) at a dosage of 100 mg/kg in dry soil for a period of 10 days. The results were striking: the snails' growth rate exhibited a notable decline of 1.77% after 14 days of consuming the NPs-contaminated food. Furthermore, their foraging and feeding behaviors demonstrated a decline of 4.63% and 0.54%, respectively, highlighting the intricate ways in which NPs can disrupt fundamental ecological processes (Chae & An, 2020). In another significant experiment, the *Aporrectodea rosea* experienced a 30 days exposure to PLA and HDPE MPs (0.1%w/w, 1 g/kg dry soil) from clothing fibers. The study revealed a concerning trend as the growth rate of the exposed 80 adult *Aporrectodea rosea* exhibited a discernible decline when compared to their counterparts. Specifically, the exposed earthworms displayed an average decrease of 3.1 ± 1.1% in biomass, in stark contrast to the 5.7 ± 3.1% increase in biomass observed in the control group (Boots et al., 2019). The study examined the impact of MPs, specifically PE particles with a size less than 150 μm, on *Lumbricus terrestris* earthworms in a terrestrial ecosystem. Moreover, earthworms exposed to higher MP concentrations experienced reduced growth rates and consequent weight loss, growth rate was determined by weight before and after exposure. Before exposure, the growth rate was 10.3 ± 4 (mg $worm^{-1}$ d^{-1}); after exposure to (7%), (28%), (45%) and (60%) MPs, growth rates were 5.8 ± 3.2, 0.5 ± 5.5, 1.4 ± 2.6 and − 0.5 ± −4.7 (mg $worm^{-1}$ day^{-1}), respectively (Kim et al., 2022). A comprehensive study delved into the influence of MPs on *Caenorhabditis* elegans. The investigation encompassed two types of MPs, LDPE, and PLA/PBAT, introduced into the soil at varying concentrations (1, 10, and 100 mg/L). These MPs had distinct mean diameters of around 40 ± 57 μm and 31 ± 40 μm, respectively. Over a 6-day period in agar, changes in nematode reproduction and body length were meticulously observed. The outcomes were notable, indicating significant shifts in both reproduction and body length. Notably, the LDPE MPs at a concentration of 10 mg L-1 led to a reduction in offspring numbers, with a substantial decline from 267 to 206 offspring. Simultaneously, the nematodes' body length exhibited a significant decrease, diminishing from an average of 1470 μm to 1340 μm. In the case of PLA/PBAT MPs, the most profound reduction in reproduction occurred at a concentration of 100 mg/L, manifesting as a decline from the control average of 267 to 232 offspring. Likewise, a remarkable reduction in body length was observed at a concentration of 1 mg L-1, with the average length contracting from 1256 μm (Schöpfer et al., 2020). Within terrestrial ecosystems, the repercussions of M/NPs are still unraveling. A study focused on *Cantareus aspersus* sheds light on the potential toxicity effects of LDPE particles, commonly found in food packaging, reusable bags, and agricultural films. Young snails (1-month-old and weight 1.2 ± 0.13 g) were exposed to various sizes (400–500 μm, 200–250 μm, and 50–100 μm) and (0.014, 0.14, 0.71 g kg^{-1}) in soil concentrations of LDPE particles for a duration of 28 days. However, the growth rate of the snails exhibited noteworthy changes in response to LDPE exposure. Among the exposure scenarios, a 15% increase in growth rate was observed for all size classes when exposed to 10% LDPE. For 25% LDPE exposure, there was a 10% growth increase in large-scale and mixed classes, coupled with a 5% reduction in growth for medium and small classes. Strikingly, in the 50% LDPE exposure test, a decline in growth rate ranging from 5% to 15% was evident across all size classes. Particularly, the smallest size class exhibited the most significant decrease in growth rate. Although immediate mortality was not a concern, the study underscores the nuanced impact of LDPE exposure on the growth dynamics of *Cantareus aspersus* snails, offering insights into potential long-term ecological consequences (Colpaert et al., 2021). Examining the environmental impact of MPs, a focus was placed on PET-MPs, a product derived from one of the world's most prevalent plastics commonly used in food and beverage packaging. These were chosen for study due to their extensive application, constituting as much as 7.1% of Europe's total plastic consumption. To investigate their effects, Achatina fulica, a type of snail, was exposed to various sizes of PET-MPs grouped into four distinct size classes. These classes were: <50 μm (6.4%), 50–100 μm (8.1%), 100–500 μm (68.5%), and 500–1000 μm (17.1%), each with a specific representation in the experimental setup. Remarkably, the study revealed that the consumption of PET-MPs leads to notable changes in the growth pattern of the snails. Surprisingly, treated specimens demonstrated improved growth compared to the control samples. Particularly striking was the observed treatment impact on a model related to the upper limit of the growth curve concerning body weight. Specifically, snails exposed to 10% PET-MPs exhibited a tendency toward a higher final weight, with an approximate increase of 0.5 g compared to the untreated group. Moreover, within a 40-day timeframe, distinct increments of 5 mm and 7 mm were noted in the width and length of the snail cochlea, respectively, in comparison to the control group, which showed dimensions of 18 mm and 25 mm for width and length (De Felice et al., 2021). In an exploration of MP effects on terrestrial ecosystems, the impact of rubber particles washed from car tires on *Folsomia candida* growth was examined. Rubber particles, obtained from tires of bicycles, cars, and electric scooters (with sizes spanning from 541 to 112 μm), were added to springtail habitats at levels of 1 and 10 g/kg of dry soil. This setup was maintained for a 28-day experimental period. The study assessed both adult and offspring body lengths in comparison to controlled samples. The outcomes revealed a reduction in the body length of both adult and offspring *Folsomia candida* due to exposure. Offspring reared in controlled conditions exhibited an average growth of 500 μm, while those in the exposed groups showed decreased total body lengths of 489 ± 123 μm, 433 ± 103 μm, 413 ± 86 μm, and 474 ± 131 μm for control, bicycle tire, car tire, and electric scooter tire particles, respectively. In the case of adult *Folsomia candida*, the body length in the controlled group was approximately 1800 μm. In contrast, the exposed groups demonstrated adult *F. candida* body lengths of 1514 ± 101 μm, 1503 ± 54 μm, 1433 ± 73 μm, and 1517 ± 60 μm for control, bicycle tire, car tire, and electric scooter tire particles, respectively. (Kim et al., 2022). In a controlled laboratory setting, the study investigated the impact of PE MPs smaller than 500 μm on *Folsomia candida* in synthetic soils over 28 days. Different MP concentrations (ranging from 0.005% to 1%) were used. At 0.5% and 1% concentrations, the springtails exhibited avoidance behaviors with rates of 59% and 69%, respectively. Reproduction of springtails ceased entirely at the 0.1% MP concentration (Ju et al., 2019).

3 Bioaccumulation of Microplastics and Nanoplastics in Soil Fauna

In terrestrial ecosystems, the potential for bioaccumulation of M/NPs is an unfolding concern, impacting soil organisms. There exists evidence of M/NPs ingestion by various terrestrial organisms, including insects, earthworms, plants, and potentially larger animals. These particles can accumulate within the digestive tracts of these organisms, potentially causing repercussions for both the organisms themselves and the wider ecosystem. For instance, in the case of *Lumbricus terrestris* exposed to MPs for 48 h, these MPs were observed throughout the segments of the earthworms, with a notable concentration in the hindgut region (Baeza et al., 2020). Bioaccumulation occurs via direct and indirect routes. Organisms interact directly with air, water, and soil, and also indirectly absorb M/NPs through contaminated food. Generally, bioaccumulation ensues as pollutants amass, elevating their concentration over organisms' lifespans. Ecotoxicology reveals M/NPs accumulation across diverse organisms via foodchain, underscoring the significance of bioaccumulation studies in terrestrial organisms (Vazquez & Rahman, 2021). This significance is further magnified by the heightened importance of understanding adsorption behaviors of NPs, particularly concerning the exposure of soil organisms such as earthworms, snails, mice, and more, which rely directly and indirectly.

Considering the effect of bioaccumulation over lifespan, Yu et al. (2021) studied the bioaccumulation of 100 nm PS-NPs (100 mg/L) in *Caenorhabditis elegans* and demonstrated that exposure to NPs led to their accumulation in the mother's gonads, subsequently affecting developing eggs. The NPs were transferred to offspring, impacting subsequent generations (F1-F4) by accumulating in somatic cells and embryos. The study highlighted the potential transfer and effects of NPs on egg development due to maternal NP exposure (Yu et al., 2021).

The route of MPs within the food chain has been revealed through simulations and field studies. This concept indicates that MPs can progress from lower-trophic-level prey organisms to higher-level predators in the food chain. The implications for agriculture and food production are noteworthy, as MPs possess the capability to accumulate within the soil, subsequently entering the food chain. This carries the potential to adversely affect soil organisms and, subsequently, humans consuming contaminated food (Gao et al., 2020). Huerta Lwanga et al. (2017) contribute significantly to this discussion by providing evidence of the transfer of both macro and microplastics from soil to chickens within traditional gardens. Their study sheds light on how LDPE MPs move through terrestrial food chains. The data presents a clear pattern of MP prevalence, progressing from farm soil, through earthworms, and ultimately appearing in chicken feces, illustrating its upward journey within this ecosystem. A specific case study conducted in traditional Mayan home gardens in Southeast Mexico demonstrated the presence of plastic bottles, averaging around 11 ± 5 cm in size, on the garden surface. The research further noted a relatively consistent distribution of MPs across soil layers and chicken feces within these home gardens. Interestingly, chickens were found to predominantly ingest plastics from surface residues, resulting in a weaker correlation between MP concentrations in the soil and those in chicken feces. However, *Lumbricus terrestris* exhibited a stronger tendency to bioconcentrate MPs in their casts when soil MP content was low. In situations with limited MPs in the soil (ranging from 0 to 2 particles/g), a higher concentration of MPs was observed in the earthworm casts. The study further revealed that MPs were predominantly confined to the gizzards and chicken feces. Specifically, within the gizzard, 16.45% of the LDPE MPs were smaller than 5 mm, while the remaining 83.55% were larger. However, in the chicken crops, only MPs were detected, and the plastic particles present in the chicken feces were consistently measured between 0.1 and 1 mm in size (Huerta Lwanga et al., 2017).

Although studies on bioaccumulation are very limited, some studies demonstrated the effect of M/NP size on bioaccumulation. To better understand this, scientists conducted a study with around 800 springtails. *F. candida* were exposed to polyethylene MPs of different sizes: 2-, 34-, and 66-mm. Adult springtails aged 90–120 days were chosen randomly for the study. The results indicated that 73% ± 10% of these springtails consumed the 2 mm MP particles (MPs), whereas only 22% ± 12% ingested the 34 mm particles. Intriguingly, no internal luminescence caused by MPs was observed in *Folsomia candida* exposed to the 66 mm particles. Based on these observations, researchers concluded that the size of MPs considered safe for *F. candida* is less than 66 mm (Kim & An, 2020). In another study, bioaccumulation and potential toxicity of PS NPs measuring 50 nm and MPs with sizes of 300 nm, 600 nm, and 4 μm were explored in male mice. Following a week-long acclimation, a group of 65 mice was divided into control and different treatment groups. These mice were exposed to fluorescent PS MPs and were subsequently analyzed after a 24-hour period. The study noted that the largest particle size accumulated within the mice reached up to 4 μm)Meng et al., 2022).

Therefore, it can be said that MPs have had an impact on the life and performance cycle of microscopic living organisms in the soil, and their ecological activity has endangered the life inside the soil due to the bioaccumulation of MPs in these organisms. These risks are transferred to the soil surface through plants such as rice and generally affect global food security and human health (De-la-Torre, 2020).

4 Mixture Effects of Microplastics and Nanoplastics on Soil Fauna

Apart from the direct impact of M/NPs when introduced into soil habitats, they may interact with and influence the fate of other pollutants. Through surface adsorption, void filling, and complexation, M/NPs can load numerous pollutants in diverse settings. M/NPs are hydrophobic and have a large specific surface area, making them efficient adsorbents for organic and inorganic pollutants in the environment, and therefore capable of acting as contamination vectors in the soil biota (Dissanayake et al., 2022). Furthermore, the adsorption capacities of NPs/MPs for organic pollutants and heavy metals are almost two and six orders of magnitude larger than those of benthic zones, respectively. Previous research has shown that adsorption influences combined toxicity via the two pathways. M/NPs decrease or increase the pollutant bioaccessibility by an adsorption-enrichment action, lowering or increasing pollutant bioaccumulation. Through the adsorptive carrier effect, M/NPs transported more unstable pollutants into organisms, increasing pollutant bioaccumulation.

When M/NPs are uptaken by diverse species, they transfer the other pollutants from the environment such as heavy metals, medicines, persistent organic pollutants (POCs), microbes, and viruses. At the same time, heavy metals and various plastic additives accumulating in organisms' digestive tracts might result in significant desorption due to the effect of low pH and surfactants. For example, in human intestinal simulation, the desorption of persistent organic matter can be up to 30 times that of saltwater, significantly increasing the concentration and threat of exposure to environmental contaminants. As a result of the desorption of these pollutants in the organisms, many harmful consequences, i.e., alteration of the gut microbial ecology and an upsurge in the antibiotic resistance genes (ARG) may occur. The existing combined toxicity studies on contaminated M/NPs comprise survival and growth, genotoxicity, oxidative stress, locomotor toxicity, neurotoxicity, immunotoxicity, and metabolic toxicity as primary biological endpoints (Danopoulos et al., 2022).

M/NPs modify the environmental natures (transfer and transformation), bioavailability, and toxicity of various inorganic and organic contaminants in soil as carriers and releasers (Wang et al., 2020b). For instance, M/NPs can enable organic contaminant metabolism to more reactive metabolites; M/NPs may induce cellular damage, making cells more sensitive to pollutants. M/NPs, in addition to functioning as vectors and releasers of other pollutants, can contribute to the breakdown and permanence of organic pollutants in soils due to their ability to affect soil microbiota structure and metabolic state.

It is generally known that NPs are absorbed far more readily and swiftly by organisms as compared to their microscale equivalents (Richards & Endres, 2017). Because NPs have a higher reactivity surface than MPs, it is more susceptible (up to hundred times more) to adsorb other environmental hazardous chemicals (Pathan et al., 2020). NPs can operate as vectors for a range of intrinsic and extrinsic (e.g., plastic additives and medicines, respectively) contaminants, impacting their dispersion in the environment as well as their incorporation, toxicity, and availability to biota (Wang et al., 2022). NPs act as transporters for pollutants, which may enter the organism as an ion or free molecule and/or particle-contaminant complex (Naddafi et al., 2011; Shen et al., 2012). The complexed pollutants can then be released once within the organisms, where the initial concentrations are minimal (Wang et al., 2022). Therefore, the toxicity and bioavailability of many contaminants are increased by the NPs via the "Trojan horse effect."

M/NPs' harmful effects on other pollutants have been classified into three categories: antagonistic effect, synergistic, and no interaction impact. The effect in which M/NPs and contaminants can enhance the toxicity in the organism and the combined impact is stronger than acting alone is referred to as "synergistic enhancement." In terrestrial environments, synergistic enhancement is most prevalent in plants, microbes, and animals. For instance, the combination of heavy metals (Cd), perfluorinated compounds (perfluorooctane sulfonic acid and perfluorooctanoic acid), and insecticides (duforin) with NP enhances their accumulation and toxicity (Sobhani et al., 2021; Huang et al., 2021; Sun et al., 2021).

The antagonistic effect, on the other hand, denotes the phenomena where the toxicity of contaminants to organisms declines with the combination of M/NPs. PS, for example, can reduce the deleterious effects of sulfamethoxazole on soil microbial structure and diversity (Xu et al., 2021). Some studies hypothesized that this behavior was caused by the formation of biofilms on MPs. The biofilm lifestyle protects soil microorganisms against organic pollutants in the soil's exterior environment (Zhang et al., 2019). However, research into the effects of MPs/NPs and organic contaminants on soil microorganisms is still in its early stages, and this prediction has yet to be proven.

The choice of M/NP sizes, colors, forms, and kinds is critical since it might alter the interaction with other contaminants in the environment. The vector-mediated transport of contaminants via MPs is dependent on particle size; extremely big particles are unable to penetrate the cell surface and so have lower vector-based toxicity, whereas smaller NPs have higher vector-based toxicity to other existing pollutants (Guo et al., 2020). Lee et al. (2014) claimed that the surface area of NPs is directly related to their interaction with existing organic and inorganic pollutants. The surface area of plastics is closely linked to the kind and physical structures of plastics as well as particle size (Xiang et al., 2022). However, the toxicity of MPs and their interaction with other pollutants also depend on their particle sizes. For example, Xu et al. (2018) found different absorption and toxicity capacities of tetracycline on PP, PE, and PS with particle sizes of 280 μm, as PS > PP > PE (Xu et al., 2018). Tetracycline sorption capacity on PS MPs was greater than on PE and PP microplastics, owing to the occurrence of polar and - interactions between tetracycline and PS microplastics. Hence, the studies have confirmed that the surface properties and adsorption behavior of NPs vary with their smaller surface area.

4.1 Combined Effects of Heavy Metal and Microplastics and Nanoplastics

The interactions between M/NPs and contaminants were predominantly "Synergistic," as shown in Table 4, indicating that their combined effects were more severe than individual ones. For instance, the combined presence of PS MPs and cadmium notably demonstrated a synergistic dynamic on reproduction of *E. crypticus*, where cadmium's toxicity was remarkably amplified within soil environments (Ozturk et al., 2023). On the contrary to Cd interaction with PS MPs, the intriguing discovery surfaced in a separate study that PVC's participation in the context of arsenic counteracted the latter's effect on gut microbial diversity. The underlying mechanism suggested PVC's potential to adsorb or bind arsenic, consequently constraining its availability and mitigating its impact on gut microbiota (Wang et al., 2019). These intricate insights underscore the complicated interactions between MPs and heavy metals, which collectively impact various endpoints, such as mortality, reproduction, cellular responses, and immune functions.

In terms of bioaccumulation, synergistic effects of MPs were proven in many studies. Li et al. (2021) investigated the bioavailability of PE (0.01%, 0.05%, and 0.1%) and metals (Cu; 100 mg/kg and Ni; 40 mg/kg) in soil. The study discovered that adding PE-MPs to the soil enhanced metal bioavailability for *E. fetida*. On day 21, earthworms had the greatest Cu2+ concentration (73.3 mg/kg) of 1.55 times greater than in the individual Cu treatment. Likewise, Ni^{2+} concentrations in *E. fetida* rose with MP concentrations; the highest Ni2+ concentration in earthworms on the 21st day (36.3 mg/kg) was 1.87 times higher than the individual Ni treatment. The metal concentrations in *E. fetida* increased with MP presence in the soil, which was compatible with the bioavailability evaluated by the sequential extraction approach. Similarly, an increase in toxicity and uptake of cadmium (2 and10 mg/kg) in the presence of M/NPs (Polyethylene; concentrations of 0%, 7%, 15%, 20%, and 30% w/w) have been reported in *Eisenia foetida*. The results demonstrated that 28-day co-exposure to MPs and Cd substantially improved the bioaccumulation of Cd in earthworm from 9.7% to 161.3% (Zhou et al., 2020).

In addition to the bioconcentration studies, combined toxic effects of heavy metals and MPs at cellular and organism levels were studied in a few researches (Table 4). MPs and Cd decreased the peroxidase (POD) and superoxide enzyme (SOD) activity in *E. fetida* while boosting glutathione (GSH) and malondialdehyde (MDA) activities. The lipid peroxides levels observed in combined exposure treatment were 44.1% and 191.9% higher than individual MPs, and individual Cd treatment, respectively (Zhou et al., 2020). In another study, various biomarkers including catalase (CAT), peroxidase (POD), and malondialdehyde (MDA)

Table 4 Combined effects of M/NPs with heavy metals

Organisms	Type of MP	Other contaminant	Endpoint	Type of interaction	Impact	References
E. crypticus	Polystyrene (1–10%)	Heavy metals (cadmium; 6.4, 16, 40, 100, and 250 mg/kg)	LC50, EC50	Synergistic	Increase in reproduction toxicity more than 50% in the existence of polystyrene MP	Ozturk et al. (2023)
M. californica	PVC(2000 mg/kg)	As (40 mg/kg)	Gut microbial diversity	Antagonistic	Decreased the impact of arsenic on the gut microbiota	Wang et al. (2019)
E. fetida	PP (300, 3000, 6000, and 9000 mg kg^{-1})	Cd (8 mg/kg)	Growth rate and mortality	Synergistic	Increased accumulation of Cd from 9.7–161.3%. Increased lipid peroxides levels up to 191.9% in the case of combined exposure (44.1% when MPs alone)	Zhou et al. (2020)
E. fetida	Polyethylene MPs (0.01–0.1%)	Cu (100 mg/kg)	Bioavailability	Synergistic	PE-MPs increased Cu2+ bioavailability by up to 1.55 times	Li et al. (2021)
E. fetida	Polyethylene MPs (0.01–0.1%)	Ni (40 mg/kg)	Bioavailability	Synergistic	PE-MPs increased Ni2+ bioavailability by up to 1.87 times	Li et al. (2021)
E. fetida	Polyethylene MPs (0%, 7–30% w/w)	Cd (2-10 mg/kg)	Bioaccumulation	Synergistic	MPs increased Cd bioaccumulation (9.7–161.3%)	Zhou et al. (2020)
E. fetida	PE MPs (0.01–0.1%)	Cd	Enzyme activities	Synergistic	Combined exposure reduced SOD, POD, elevated GSH, MDA, and lipid peroxides	Zhou et al. (2020)

contents, superoxide dismutase (SOD) activity, as well as the related gene expression levels were investigated in the presence of PE (0.01%, 0.05%, and 0.1%) and metals (Cu; 100 mg/kg and Ni; 40 mg/kg) in soil.

4.2 *Combined Effects of Organic Pollutants and Microplastics and Nanoplastics*

The interaction between M/NPs and organic pollutants has been subject to a thorough investigation, uncovering synergistic or antagonistic outcomes that hold considerable ramifications for diverse organisms. Considering organic pollutants, research highlights that the presence of distinct M/NPs can alter the bioaccumulation potential. Notably, PVC MPs were observed to bolster the accumulation of perfluorooctanesulfonic acid (PFOS) and perfluorooctanoic acid (PFOA) in earthworms, showing an astonishing upswing of almost 200% (Sobhani et al., 2021). Sobhani et al. (2021) claimed that exposure to PVC MPs (0, 0.1, 1, 10, 100, 500, and 1000 mg/kg −1 dry soil) in soil caused an increase in the bioaccumulation of PFOS and PFOA (5 mg/kg each) in *Eisenia fetida*, with the bioaccumulation factor (BAF) increasing up to 200%. On the contrary to these studies, Ju et al. (2023) detected the treatments with 1250 g/ha Chlorpyrifos and Bio-MPs (28%) considerably reduced Chlorpyrifos bioaccumulation in earthworm bodies (1.1%, w/w) as compared to the Chlorpyrifos alone (1.7%).

In addition to bioaccumulation, the effects of NPs/MPs on the toxicity of organic pollutants were studied for several endpoints at cellular level and most of the studies demonstrated synergistic effects (Table 5). The study made by Barreto et al. (2023) showed the ability of Polystyrene NPs (NPls, 0.015 mg/kg) to alter the impact of diphenhydramine (DPH) and Bisphenol A (BPA) on *Folsomia candida*. After 28 days of exposure to 100 mg/kg DPH + 0.015 mg/kg NPls, glutathione S-transferase activity and lipid peroxidation levels increased compared to single treatment of DPH. These findings show that NPls (at predicted environmental concentrations of 0.015 mg/kg) may act as a vector for the other environmental pollutants of different types (e.g., pharmaceutical and industrial chemical) and change their impact on soil organisms based on the concentration and endpoint of the contaminant evaluated. Similarly, Mishra et al. (2022) claimed that joint application pesticide (Monocrotophos) of PVC and PP MPs (<150 μm), at ecologically relevant concentration in soil (12.5 g/kg) significantly amplified the oxidative stress ($p < 0.05$) of *E. eugeniae*, with enhanced lipid peroxidation level (0.12 nmol/mg) and enzyme activities (0.068 ± 0.01 U/mg) as compared to the independent treatment with only pesticide (0.035 ± 0.01 U/mg). Moreover, Cheng et al. (2020) evaluated the integrated biological response of Antrazine (ATZ; 0.02 and 2.0 mg/kg) and low-density PE (1.25 g; 0.25% w/w) in agricultural soil and claimed that this combination increased the oxidative stress ($p < 0.05$) but there was no clear trend for abnormal expression of genes in *E. fetida* found. They also evaluated the same

Table 5 Combined effects of M/NPs with organic pollutants

Organisms	Type of MP	Other contaminant	Endpoint	Type of interaction	Impact	References
Eisenia fetida	PVC (0.1–1000 mg/kg)	PFOS and PFOA (5–10 mg/kg)	Bioaccumulation	Synergistic	Increased bioaccumulation of PFASs up to 200% in the presence of PVC Significant reduction in their reproduction (juvenile number decreased 50%)	Sobhani et al. (2021)
Enchytraeus crypticus	PS (1000 mg/kg)	Tetracycline (10 mg/kg)	Genotoxicity	Synergistic	Increase in antibiotics resistance genes, i.e., Xanthomonadaceae (from 1.34% to 4.20%), Moraxellaceae (from 2.38% to 18.05%), Oxalobacteraceae (from 0.33% to 2.17%), Sphingomonadaceae (from 0.51% to 9.43%) and Bradyrhizobiaceae (from 3.90% to 9.52%)	Ma et al. (2020)
E. fetida	PVC and PP microplastics (<150 μm) (12.5 g/kg)	Monocrotophos (pesticide)	Oxidative stress	Synergistic	Increased lipid peroxidation level (0.12 nmol/mg) and enzyme activities (0.068 ± 0.01 U/mg) compared to pesticide alone	Mishra et al. (2022)
E. fetida	PVC microplastics (0, 0.1, 1, 10, 100, 500 and 1000 mg/kg)	PFOS and PFOA (5 mg/kg each)	Reproduction	Synergistic	50% reduction in juvenile count compared to single treatments	Ju et al. (2023)
F. candida	Polystyrene nanoplastics (NPls, 0.015 mg/kg)	Bisphenol A (BPA) and diphenhydramine (DPH)	Cellular effects	Synergistic	Increased lipid peroxidation levels and glutathione S-transferase activity as compared to single treatment of DPH	Barreto et al. (2023)

(continued)

Table 5 (continued)

Organisms	Type of MP	Other contaminant	Endpoint	Type of interaction	Impact	References
F. candida	PVC and PP microplastics (<150 μm) (12.5 g/kg)	Monocrotophos (pesticide)	Cellular effects	Synergistic	Increased lipid peroxidation level, enzyme activities, and oxidative stress, compared to pesticide alone	Mishra et al. (2022)
E. fetida	Low-density PE (1.25 g; 0.25% w/w)	Antrazine (ATZ; 0.02 and 2.0 mg/kg)	Oxidative stress	Synergistic	Increased oxidative stress	Cheng et al. (2020)
E. fetida	Microplastics (40–50 μm)	Dufulin (pesticide)	Metabolic profile	Synergistic	Altered metabolic pathways	Sun et al. (2021)
P. scaber	Polyester fibers and tire particles (0.5% w/w)	Chlorpyrifos (pesticide)	Bioavailability	Antagonistic	Reduced chlorpyrifos	Selonen et al. (2023)
Soil microbial communities	PE microplastics (1% w/w)	Ciprofloxacin (10 mg/kg)	Microbial diversity	Synergistic	Inhibited ciprofloxacin breakdown, decreased soil microbial diversity	Wang et al. (2020a)

endpoints for aged and nonaged MPs. Aged MPs (farmland residual mulch film-derived microplastics) had higher effects than the nonaged MPs (unused mulch film-derived MPs). The authors speculated that this could be due to the fact that the aging process of MPs can lead to the formation of new functional groups on the surface of the MPs, which may increase their toxicity. Additionally, the aging process can also lead to the degradation of MPs, which can release smaller particles that are more easily taken up by organisms and can cause more damage. Synergistic effects of MPs on the metabolic profile of *E. Fetida* were demonstrated. The combination of a pesticide, Dufulin, and MPs (low concentration MPs; 40–50 m) resulted in a shift in the abundance of 21 metabolites and 3 metabolic pathways, indicating a deterioration of the interference metabolic profile. The combination of MPs and pesticide significantly affected two metabolic pathways, including tyrosine, tryptophan, and phenylalanine metabolism and production (Sun et al., 2021).

The effect of M/NPs on the toxicity of organic pollutants was also proven at organism level in a few studies. The earthworms, *E. fetida*, exposed to PVC MP contaminated soil (0, 0.1, 1, 10, 100, 500, and 1000 mg/kg) and PFOS and PFOA (5 mg/kg each) showed a 50% reduction in the reproduction (Juvenile count) compared to single treatments (Sobhani et al., 2022). Likewise, the presence of PE MP and bio-MP (biodegradable plastics) (7% and 28% w/w) increased the mortality of *L. terrestris* when they were exposed to an insecticide, Chlorpyrifos (0 g/ha, 250 g/ha, 1250 g/ha) in sand soil. The PE MPs were more toxic than bio-MP, (mortality increased by 62.5%) probably due to the varying adsorption capacities of the MPs (Ju et al., 2023). Whereas Barreto et al. (2023) claimed that the observed impacts of BPA and DPH (100 mg/kg) on reproduction and survival of *Folsomia candida* were equivalent at single and combined doses indicating that the unfavorable (or not present) effects of these chemicals were independent of the presence of NPls (0.015 mg/kg, $p < 0.05$). They also demonstrated that the effect of M/NPs can be antagonistic or synergistic depending on the endpoint. On the contrary to these studies, Selonen et al. (2023) indicated that the effects of tire particles and polyester fibers as MPs (0.5% w/w) on the toxicity of the pesticide, chlorpyrifos, on *Folsomia candida* (springtails) and *Porcellio scaber* (woodlice) were antagonistic. Hence their findings suggest that both kinds of MPs, but particularly tire particles, reduce chlorpyrifos bioavailability.

Apart from the effects on single organism, Wang et al. (2020a) performed a soil microcosm to investigate the combined effects of PE MPs (1% w/w) and ciprofloxacin (10 mg/kg) on soil microbial communities. The co-exposure was measured by comparing the Ciprofloxacin's degradation efficiency as well as the microbial community structure in soil with the individual amendment of CIP and the blank control. The research showed that the presence of PE in soil inhibited ciprofloxacin breakdown, and co-exposure dramatically decreased soil microbial diversity. Among the microbes, *Serratia* had the highest relative abundance in the combined CIP-amended soil and MPs. Furthermore, *Acidobacteria* and *Arthrobacter* subgroup six were the most abundant in the individual CIP-amended soil, accounting for 5.59% and 8.58%, respectively. This suggests that MPs can have an effect on terrestrial creatures involved in the functioning of terrestrial ecosystems, such as microbiota, plants, and soil organisms.

Only a few research on soil creatures have been published, with the majority of them focused on earthworms as target species (Cheng et al., 2020). A large number of these research discovered synergistic effects from the combined exposure to emerging contaminants and MPs, whereas other research studies found this impact to be minimal or antagonistic. For combined toxicity, depending on the endpoint, the effects could be synergistic or antagonistic even for the same species. Moreover, type of M/NPs was clearly altering the combined effects for the same pollutant and endpoints.

5 Future Outlook

The widespread usage of plastics is dramatically increasing and their consequences on the environment are taking attention as M/NP pollution since a decade. Therefore, scientific studies have focused on the fate and toxic effects of M/NPs in the environment. However, the studies on terrestrial environment evidently lacking compared to the aquatic environment. Similar to the fate of emerging organic compounds at the beginning of the twenty-first century, measurement protocols are shown as the biggest bottleneck for both environmental compartments. As the characterization of M/NPs has been advanced, complicated studies to understand the fate and toxicity of M/NPs have emerged for the terrestrial environment. In fact, toxic effects of various MP types were investigated for several terrestrial organisms including invertebrates and vertebrates. Endpoints, especially, at organism levels such as mortality, growth rate, and reproduction were tested and the most highlighted results are about the effect of M/NPs type including the size, shape, and surface properties. However, the variety in M/NP types is very high that may prevent to achieve enough studies to make predictions like quantitative structure analysis. Endpoints at molecular and cellular levels were almost tested for all commonly used ones for terrestrial organisms. Although the similar limitation is valid for these endpoints like the ones at organism level; at least, they have started to shed light on the mechanisms behind the toxic effects. Therefore, relations between properties of M/NPs and occurred effect could be resolved for research development studies. In addition to the toxic effects on single organisms, uptake/elimination rates, and bioaccumulation were tested for various organisms considering the food web interactions, as well. Uptake and bioaccumulation of M/NPs are also closely related to M/NPs properties. Understanding the uptake and bioaccumulations requires to know about the adsorption and complexation behaviors of M/NPs since the interactions of M/NPs with soil organisms mostly rely on these processes. In fact, bioaccumulation is almost the sole parameter to investigate toxic effects of M/NPs at ecosystem services level that are clearly lacking in the literature.

In addition to the direct effects of M/NPs on terrestrial organisms, the effect of M/NPs on the toxicity of other pollutants can be as significant as the direct effects.

M/NPs have high capability for adsorbing or binding the other organic and inorganic components due to their surface properties. Therefore, they can be the carriers and releasers of other pollutants like a trojan horse effect. Compared to direct effects, investigation of indirect effects is quite limited. However, again, M/NP types play a significant role to understand the indirect effects. According to the available knowledge in the literature, determination of the vector effects for specific pollutant categories may be useful to categorize the possible synergistic or antagonistic effects.

In the light of this summary, the following outlooks are suggested:

(1) Toxicity studies investigating the cellular and organism level endpoints on single organisms can be directed toward developing specific quantitative structure analysis for M/NP types to prioritize the M/NPs to be managed emergently.
(2) Considering the capability of artificial intelligence and machine learning for facilitating the prediction of the complicating issues, existing data should be gathered and new approaches should be developed to produce scientifically predicted data for the management of M/NPs pollution.
(3) Combined toxicity of M/NPs with the other pollutants should be investigated considering the existing pollution control standards and the need for revisions of the standard values should be evaluated based on the toxicity of these pollutants in the presence of M/NPs.
(4) Toxicity studies should be planned to develop knowledge on toxicity mechanisms; hence, they can be useful for the support of material designers, especially for optimizing the surface properties to decrease toxic effects or increasing the durability of plastics to decrease waste emissions and the formation of NPs in the environment that is considered more dangerous compared to macro and microplastics.
(5) Toxicity studies for M/NPs can be oriented through producing toxic data for LCA applications, not only for plastic production but also for the other production processes that can relate to M/NP emission.
(6) Considering the increasing trend of circularity in waste management, toxicity studies on M/NPs should consider M/NPs emissions in recycled/recovered/reclaimed sources such as biosolids and irrigation water. Moreover, the consequences of plastic recycling should be investigated in terms of toxicity.
(7) Effects of environmental conditions were rarely investigated for M/NPs toxicity. Moreover, climate change scenarios are pointing out the dramatic changes in temperature, rain regime, wind profile, etc. that should be investigated to understand their effect on the toxicity of M/NPs.
(8) In the context of sustainability assessment, environmental impacts of M/NPs may be needed more deeply.

Acknowledgments This work was partly supported by generous grants from the Scientific and Technological Research Council of Turkey under the TUBITAK BIDEB 2247-A Fellowship for Outstanding Researchers (Project No: 120C147) and under TUBITAK 1001-The Scientific and Technological Research Projects Funding Program (Project No: 121Y383). However, all scientific contributions made in this project are owned and approved solely by the authors.

References

Atugoda, T., Piyumali, H., Wijesekara, H., et al. (2023). Nanoplastic occurrence, transformation and toxicity: A review. *Environmental Chemistry Letters, 21*, 363–381.

Al-Jaibachi, R., Cuthbert, R. N., & Callaghan, A. (2019). Examining effects of ontogenic microplastic transference on Culex mosquito mortality and adult weight. *Science of the Total Environment, 651*, 871–876. https://doi.org/10.1016/j.scitotenv.2018.09.236

Andrady, A. L. (2017). The plastic in microplastics: A review. *In Marine Pollution Bulletin, 119*(1), 12–22. Elsevier Ltd.

Baeza, C., Cifuentes, C., González, P., Araneda, A., & Barra, R. (2020). Experimental exposure of Lumbricus terrestris to microplastics. *Water, Air, & Soil Pollution, 231*, 1–10.

Barreto, A., Santos, J., Almeida, L., Tavares, V., Pinto, E., Celeiro, M., et al. (2023). First approach to assess the effects of nanoplastics on the soil species Folsomia candida: A mixture design with bisphenol A and diphenhydramine. *NanoImpact, 29*, 100450.

Billings, A., Carter, H., Cross, R. K., et al. (2023). Co-occurrence of macroplastics, microplastics, and legacy and emerging plasticisers in UK soils. *Science of the Total Environment journal, 880*, 163258.

Birch, Q. T., Potter, P. M., Pinto, P. X., et al. (2020). *Sources, transport, measurement and impact of nano and microplastics in urban watersheds*. Springer.

Bläsing, M., & Amelung, W. (2018). Plastics in soil: Analytical methods and possible sources. *Science of the Total Environment, 612*, 422–435.

Boots, B., Russell, C. W., & Green, D. S. (2019). Effects of microplastics in soil ecosystems: Above and below ground. *Environmental Science & Technology, 53*(19), 11496–11506.

Cao, L., Wu, D., Liu, P., et al. (2021). Occurrence, distribution and affecting factors of microplastics in agricultural soils along the lower reaches of Yangtze River, China. *Science of the Total Environment, 794*.

Chae, Y., & An, Y. J. (2018). Current research trends on plastic pollution and ecological impacts on the soil ecosystem: A review. *Environmental Pollution, 240*, 387–395.

Chae, Y., & An, Y. J. (2020). Nanoplastic ingestion induces behavioral disorders in terrestrial snails: Trophic transfer effects via vascular plants. *Environmental Science: Nano, 7*(3), 975–983.

Chang, X., Fang, Y., Wang, Y., Wang, F., Shang, L., & Zhong, R. (2022). Microplastic pollution in soils, plants, and animals: A review of distributions, effects and potential mechanisms. In *Science of the Total Environment* (Vol. 850). Elsevier B.V.

Cheng, Y., Zhu, L., Song, W., Jiang, C., Li, B., Du, Z., et al. (2020). Combined effects of mulch film-derived microplastics and atrazine on oxidative stress and gene expression in earthworm (Eisenia fetida). *Science of the Total Environment, 746*, 141280.

Cheng, Y., Song, W., Tian, H., et al. (2021). The effects of high-density polyethylene and polypropylene microplastics on the soil and earthworm Metaphire guillelmi gut microbiota. *Chemosphere, 267*. https://doi.org/10.1016/j.chemosphere.2020.129219

Colpaert, R., Petit dit Grézériat, L., Louzon, M., de Vaufleury, A., & Gimbert, F. (2021). Polyethylene microplastic toxicity to the terrestrial snail Cantareus aspersus: Size matters. *Environmental Science and Pollution Research*, 1–10.

Corradini, F., Meza, P., Eguiluz, R., et al. (2019). Evidence of microplastic accumulation in agricultural soils from sewage sludge disposal. *Science of the Total Environment, 671*, 411–420.

Corradini, F., Casado, F., Leiva, V., et al. (2021). Microplastics occurrence and frequency in soils under different land uses on a regional scale. *Science of the Total Environment, 752*, 141917.

De Felice, B., Ambrosini, R., Bacchetta, R., Ortenzi, M. A., & Parolini, M. (2021). Dietary exposure to polyethylene terephthalate microplastics (PET-MPs) induces faster growth but not oxidative stress in the giant snail Achatina reticulata. *Chemosphere, 270*, 129430.

De-la-Torre, G. E. (2020). Microplastics: An emerging threat to food security and human health. *Journal of Food Science and Technology, 57*, 1601–1608.

Ding, L., Zhang, S., Wang, X., et al. (2020). The occurrence and distribution characteristics of microplastics in the agricultural soils of Shaanxi Province, in North-Western China. *Science of the Total Environment, 720*, 137525.

Ding, L., Wang, X., Ouyang, Z., et al. (2021). The occurrence of microplastic in Mu Us sand land soils in Northwest China: Different soil types, vegetation cover and restoration years. *Journal of Hazardous Materials, 403*, 123982.

Dissanayake, P. D., Kim, S., Sarkar, B., Oleszczuk, P., Sang, M. K., et al. (2022). Effects of microplastics on the terrestrial environment: A critical review. *Environmental Research, 209*, 112734.

Dris, R., Gasperi, J., Saad, M., et al. (2016). Synthetic fibers in atmospheric fallout: A source of microplastics in the environment? *Marine Pollution Bulletin, 104*, 290–293.

Duis, K., & Coors, A. (2016). Microplastics in the aquatic and terrestrial environment: Sources (with a specific focus on personal care products), fate and effects. *Environmental Sciences Europe, 28*, 1–25.

Danopoulos, E., Twiddy, M., West, R., & Rotchell, J. M. (2022). A rapid review and meta-regression analyses of the toxicological impacts of microplastic exposure in human cells. *Journal of Hazardous Materials, 427*, 127861.

Endo, K. (2002). Synthesis and structure of poly(vinyl chloride). *Progress in Polymer Science, 27*, 2021–2054.

Enyoh, C. E., Shafea, L., Verla, A. W., Verla, E. N., Qingyue, W., Chowdhury, T., & Paredes, M. (2020). Microplastics exposure routes and toxicity studies to ecosystems: An overview. In Environmental Health and Toxicology. *Korean Society of Environmental Health and Toxicology, 35*(1). https://doi.org/10.5620/eaht.e2020004

Feng, S., Lu, H., & Liu, Y. (2021). The occurrence of microplastics in farmland and grassland soils in the Qinghai-Tibet plateau: Different land use and mulching time in facility agriculture. *Environmental Pollution, 279*, 116939.

Fournier, S. B., D'Errico, J. N., Adler, D. S., Kollontzi, S., Goedken, M. J., Fabris, L., Yurkow, E. J., & Stapleton, P. A. (2020). Nanopolystyrene translocation and fetal deposition after acute lung exposure during late-stage pregnancy. *Particle and Fibre Toxicology, 17*(1). https://doi.org/10.1186/s12989-020-00385-9

Fueser, H., Mueller, M. T., Weiss, L., Höss, S., & Traunspurger, W. (2019). Ingestion of microplastics by nematodes depends on feeding strategy and buccal cavity size. *Environmental Pollution, 255*. https://doi.org/10.1016/j.envpol.2019.113227

Fudlosid, S., Ritchie, M. W., Muzzatti, M. J., Allison, J. E., Provencher, J., & MacMillan, H. A. (2022). Ingestion of microplastic fibres, but not microplastic beads, impacts growth rates in the Tropical House Cricket Gryllodes Sigillatus. *Frontiers in Physiology, 13*, 725.

Gao, D., Li, X. Y., & Liu, H. T. (2020). Source, occurrence, migration and potential environmental risk of microplastics in sewage sludge and during sludge amendment to soil. *Science of the Total Environment, 742*, 140355. https://doi.org/10.1016/j.scitotenv.2020.140355

Getor, R. Y., Mishra, N., & Ramudhin, A. (2020). The role of technological innovation in plastic production within a circular economy framework. *Resources, Conservation and Recycling, 163*, 105094.

Gigault, J., Ter, H. A., Baudrimont, M., et al. (2018). Current opinion: What is a nanoplastic? *Environmental Pollution, 235*, 1030–1034.

González-Pleiter, M., Tamayo-Belda, M., Pulido-Reyes, G., et al. (2019). Secondary nanoplastics released from a biodegradable microplastic severely impact freshwater environments. *Environmental Science. Nano, 6*, 1382–1392.

Guo, F., Huang, X., Chen, Z., Sun, H., & Chen, L. (2020). Prominent co-catalytic effect of CoP nanoparticles anchored on high-crystalline g-C3N4 nanosheets for enhanced visible-light photocatalytic degradation of tetracycline in wastewater. *Chemical Engineering Journal, 395*, 125118.

Hale, R. C., Seeley, M. E., La Guardia, M. J., et al. (2020). A global perspective on microplastics. *Journal of Geophysical Research, Oceans, 125*, 1–40.

Heinze, W. M., Mitrano, D. M., Lahive, E., Koestel, J., & Cornelis, G. (2021). Nanoplastic Transport in Soil via Bioturbation by Lumbricus terrestris. *Environmental Science and Technology, 55*(24), 16423–16433.

Huang, C., Ge, Y., Yue, S., Zhao, L., & Qiao, Y. (2021). Microplastics aggravate the joint toxicity to earthworm Eisenia fetida with cadmium by altering its availability. *Science of the Total Environment, 753*, 142042.

Huerta Lwanga, E., Gertsen, H., Gooren, H., Peters, P., Salánki, T., Van Der Ploeg, M., & Geissen, V. (2016). Microplastics in the terrestrial ecosystem: Implications for Lumbricus terrestris (Oligochaeta, Lumbricidae). *Environmental Science & Technology, 50*(5), 2685–2691.

Huerta Lwanga, E., Gertsen, H., Gooren, H., Peters, P., Salanki, T., van der Ploeg, M., Besseling, E., Koelmans, A. A., Geissen, V. (2017). Incorporation of microplastics from litter into burrows of Lumbricus terrestris. *Environmental Pollution, 220*, 523–531.

Hurley, R. R., & Nizzetto, L. (2018). Science direct fate and occurrence of micro (nano) plastics in soils: Knowledge gaps and possible risks. *Current Opinion in Environmental Science & Health, 1*, 6–11.

Ivleva, N. P., Wiesheu, A. C., & Niessner, R. (2017). Microplastic in aquatic ecosystems. *Angewandte Chemie – International Edition, 56*, 1720–1739.

Javad, M., Keshavarzi, B., Mohit, F., et al. (2022). Science of the total environment microplastic occurrence in urban and industrial soils of Ahvaz metropolis: A city with a sustained record of air pollution. *Science of the Total Environment, 819*, 152051.

Ju, H., Zhu, D., & Qiao, M. (2019). Effects of polyethylene microplastics on the gut microbial community, reproduction and avoidance behaviors of the soil springtail, Folsomia candida. *Environmental Pollution, 247*, 890–897. https://doi.org/10.1016/j.envpol.2019.01.097

Ju, H., Yang, X., Osman, R., & Geissen, V. (2023). Effects of microplastics and chlorpyrifos on earthworms (Lumbricus terrestris) and their biogenic transport in sandy soil. *Environmental Pollution, 316*, 120483.

Jager, T., Fleuren, R. H. L. J., Roelofs, W., & De Groot, A. C. (2003). Feeding activity of the earthworm Eisenia andrei in artificial soil. www.elsevier.com/locate/soilbio

Jiang, X., Chang, Y., Zhang, T., Qiao, Y., Klobučar, G., & Li, M. (2020). Toxicological effects of polystyrene microplastics on earthworm (Eisenia fetida). *Environmental Pollution, 259*, 113896.

Kihara, S., Köper, I., Mata, J. P., & McGillivray, D. J. (2021). Reviewing nanoplastic toxicology: It's an interface problem. *Advances in Colloid and Interface Science, 288*, 102337.

Kim, S. W., & An, Y. J. (2020). Edible size of polyethylene microplastics and their effects on springtail behavior. *Environmental Pollution, 266*, 115255.

Kim, S. S., & Kim, S. (2004). Pyrolysis characteristics of polystyrene and polypropylene in a stirred batch reactor. *Chemical Engineering Journal, 98*, 53–60.

Kim, S. K., Kim, J. S., Lee, H., & Lee, H. J. (2021). Abundance and characteristics of microplastics in soils with different agricultural practices: Importance of sources with internal origin and environmental fate. *Journal of Hazardous Materials, 403*, 123997.

Kim, L., Lee, T. Y., Kim, H., & An, Y. J. (2022). *Toxicity of tire particles released from personal Mobilities (bicycles, cars, and electric scooters) on soil organisms*. Available at SSRN 4082684.

Kokalj, A. J., Hartmann, N. B., Drobne, D., et al. (2021). Quality of nanoplastics and microplastics ecotoxicity studies: Refining quality criteria for nanomaterial studies. *Journal of Hazardous Materials, 415*, 125751.

Koutnik, V. S., Alkidim, S., Leonard, J., et al. (2021). Unaccounted microplastics in wastewater sludge: Where do they go? *ACS Environmental Science and Technology Water, 1*, 1086–1097.

Kwak, J. I., & An, Y. J. (2021). Microplastic digestion generates fragmented nanoplastics in soils and damages earthworm spermatogenesis and coelomocyte viability. *Journal of Hazardous Materials, 402*, 124034. https://doi.org/10.1016/j.jhazmat.2020.124034

Kuehr, S., Windisch, H., Schlechtriem, C., Leon, G., Gasparini, G., & Gimeno, S. (2022). Are Fragrance Encapsulates Taken Up by Aquatic and Terrestrial Invertebrate Species? *Environmental Toxicology and Chemistry, 41*(4), 931–943.

Kratina, P., Watts, T. J., Green, D. S., Kordas, R. L., & O'Gorman, E. J. (2019). Interactive effects of warming and microplastics on metabolism but not feeding rates of a key freshwater detritivore. *Environmental Pollution, 255*.

Koner, S., Florance, I., Mukherjee, A., & Chandrasekaran, N. (2023). Cellular response of THP-1 macrophages to polystyrene microplastics exposure. *Toxicology, 483*, 153385.

Li, L., Luo, Y., Li, R., Zhou, Q., Peijnenburg, W. J., Yin, N., & Zhang, Y. (2020). Effective uptake of submicrometre plastics by crop plants via a crack-entry mode. *Nature Sustainability, 3*(11), 929–937.

Li, M., Liu, Y., Xu, G., Wang, Y., & Yu, Y. (2021). Impacts of polyethylene microplastics on bioavailability and toxicity of metals in soil. *Science of the Total Environment, 760*, 144037.

Li, Z., Gao, Y., Wu, Q., et al. (2022). Quantifying the occurrence of polystyrene nanoplastics in environmental solid matrices via pyrolysis-gas chromatography/mass spectrometry. *Journal of Hazardous Materials, 440*, 129855.

Lian, J., Wu, J., Xiong, H., Zeb, A., Yang, T., Su, X., & Liu, W. (2020a). Impact of polystyrene nanoplastics (PSNPs) on seed germination and seedling growth of wheat (Triticum aestivum L.). *Journal of Hazardous Materials, 385*, 121620.

Lian, J., Wu, J., Zeb, A., Zheng, S., Ma, T., Peng, F., et al. (2020b). Do polystyrene nanoplastics affect the toxicity of cadmium to wheat (Triticum aestivum L.)? *Environmental Pollution, 263*, 114498.

Lieberzeit, P., Bekchanov, D., & Mukhamediev, M. (2022). Polyvinyl chloride modifications, properties, and applications: Review. *Polymers for Advanced Technologies, 33*, 1809–1820.

Liu, M., Lu, S., Song, Y., et al. (2018). Microplastic and mesoplastic pollution in farmland soils in suburbs of Shanghai, China. *Environmental Pollution, 242*, 855–862.

Lv, W., Zhou, W., Lu, S., et al. (2019). Microplastic pollution in rice-fish co-culture system: A report of three farmland stations in Shanghai, China. *Science of the Total Environment, 652*, 1209–1218.

Lwanga, E. H., Gertsen, H., Gooren, H., Peters, P., Salánki, T., van der Ploeg, M., et al. (2017). Incorporation of microplastics from litter into burrows of Lumbricus terrestris. *Environmental Pollution, 220*, 523–531.

Lahive, E., Cross, R., Saarloos, A. I., Horton, A. A., Svendsen, C., Hufenus, R., & Mitrano, D. M. (2022). Earthworms ingest microplastic fibres and nanoplastics with effects on egestion rate and long-term retention. *Science of the Total Environment, 807*, 151022.

Lee, H., Shim, W. J., & Kwon, J. H. (2014). Sorption capacity of plastic debris for hydrophobic organic chemicals. *Science of the total environment, 470*, 1545–1552.

Ma, J., Sheng, G. D., Chen, Q. L., & O'Connor, P. (2020). Do combined nanoscale polystyrene and tetracycline impact on the incidence of resistance genes and microbial community disturbance in Enchytraeus crypticus? *Journal of Hazardous Materials, 387*, 122012.

Maddah, H. A. (2016). Polypropylene as a promising plastic: A review. *American Journal of Polymer Science, 6*, –11.

Mariano, S., Tacconi, S., Fidaleo, M., et al. (2021). Micro and Nanoplastics identification: Classic methods and innovative detection techniques. *Frontiers in Toxicology, 3*, 1–17.

Maynard, I. F. N., Bortoluzzi, P. C., Nascimento, L. M., et al. (2021). Analysis of the occurrence of microplastics in beach sand on the Brazilian coast. *Science of the Total Environment, 771*.

Meng, X., Zhang, J., Wang, W., Gonzalez-Gil, G., Vrouwenvelder, J. S., & Li, Z. (2022). Effects of nano-and microplastics on kidney: Physicochemical properties, bioaccumulation, oxidative stress and immunoreaction. *Chemosphere, 288*, 132631. https://doi.org/10.1016/j.chemosphere.2021.132631

Mishra, C. S. K., Samal, S., Samal, R. R., et al. (2022). Polyvinylchloride and polypropylene as adsorbents of the pesticide monocrotophos enhance oxidative stress in Eudrillus eugeniae (Kinberg). *Chemosphere, 295*, 133837.

Maxwell, S. H., Melinda, K. F., Matthew, G. (2020). Counterstaining to separate Nile Red-stained microplastic particles from terrestrial invertebrate biomass. *Environmental Science & Technology, 54*(9): 5580–5588.

Mackenzie, C. M., & Vladimirova, V. (2023). Preliminary study and first evidence of presence of microplastics in terrestrial herpetofauna from Southwestern Paraguay. *Studies on Neotropical Fauna and Environment, 58*(1), 16–24.

Nielsen, T. D., Hasselbalch, J., Holmberg, K., & Stripple, J. (2020). Politics and the plastic crisis: A review throughout the plastic life cycle. Wiley Interdiscip rev. *Energy & Environment, 9*.

Naddafi, K., Zare, M. R., & Nazmara, S. (2011). Investigating potential toxicity of phenanthrene adsorbed to nano-ZnO using Daphnia magna. *Toxicological & Environmental Chemistry, 93*(4), 729–737.

Ozturk, I., Ozkul, F., & Topuz, E. (2023). The effect of polystyrene microplastic and biosolid application on the toxicity and bioaccumulation of cadmium for Enchytraeus crypticus. *Integrated Environmental Assessment and Management, 19*(2), 489–500.

Pathan, S. I., Arfaioli, P., Bardelli, T., et al. (2020). Soil pollution from micro-and nanoplastic debris: A hidden and unknown biohazard. *Sustainability, 12*(18), 7255.

Patil, S. M., Rane, N. R., Bankole, P. O., et al. (2022). An assessment of micro- and nanoplastics in the biosphere: A review of detection, monitoring, and remediation technology. *Chemical Engineering Journal, 430*, 132913.

Prendergast-Miller, M. T., Katsiamides, A., Abbass, M., Sturzenbaum, S. R., Thorpe, K. L., & Hodson, M. E. (2019). Polyester-derived microfibre impacts on the soil-dwelling earthworm Lumbricus terrestris. *Environmental Pollution, 251*, 453–459.

Prata, J. C., & Dias-Pereira, P. (2023). Microplastics in Terrestrial Domestic Animals and Human Health: Implications for Food Security and Food Safety and Their Role as Sentinels. *In Animals, 13*(4). MDPI.

Potapov, A. M., Beaulieu, F., Birkhofer, K., et al. (2022). Feeding habits and multifunctional classification of soil-associated consumers from protists to vertebrates. *Biological Reviews, 97*(3), 1057–1117.

Prokić, M. D., Gavrilović, B. R., Radovanović, T. B., Gavrić, J. P., Petrović, T. G., Despotović, S. G., & Faggio, C. (2021). Studying microplastics: Lessons from evaluated literature on animal model organisms and experimental approaches. *Journal of Hazardous Materials, 414*, 125476.

Pflugmacher, S., Sulek, A., Mader, H., et al. (2020). The influence of new and artificial aged microplastic and leachates on the germination of Lepidium sativum L. *Plants, 9*(3), 339.

Ravi, B., Harishankar, K., & Venkata, K. S. (2022). *Micro/nano-plastics occurrence, identification, risk analysis and mitigation: Challenges and perspectives*. Springer.

Rillig, M. C., Ziersch, L., & Hempel, S. (2017). Microplastic transport in soil by earthworms. *Scientific Reports, 7*(1), 1362.

Rodriguez-Seijo, A., Lourenço, J., Rocha-Santos, T. A. P., da Costa, J., Duarte, A. C., Vala, H., & Pereira, R. (2017). Histopathological and molecular effects of microplastics in Eisenia andrei Bouché. *Environmental Pollution, 220*, 495–503.

Richards, D. M., & Endres, R. G. (2017). How cells engulf: a review of theoretical approaches to phagocytosis. *Reports on Progress in Physics, 80*(12), 126601.

Scheurer, M., & Bigalke, M. (2018). Microplastics in Swiss floodplain soils. *Environmental Science & Technology, 52*, 3591–3598.

Schöpfer, L., Menzel, R., Schnepf, U., Ruess, L., Marhan, S., Brümmer, F., & Kandeler, E. (2020). Microplastics effects on reproduction and body length of the soil-dwelling nematode Caenorhabditis elegans. *Frontiers in Environmental Science, 8*, 41.

Selonen, S., Dolar, A., Jemec Kokalj, A., Skalar, T., Parramon Dolcet, L., Hurley, R., & van Gestel, C. A. M. (2020). Exploring the impacts of plastics in soil – The effects of polyester textile fibers on soil invertebrates. *Science of the Total Environment*, 700.

Selonen, S., Kokalj, A. J., Benguedouar, H., Petroody, S. S. A., Dolar, A., Drobne, D., & van Gestel, C. A. (2023). Modulation of chlorpyrifos toxicity to soil arthropods by simultaneous exposure to polyester microfibers or tire particle microplastics. *Applied Soil Ecology, 181*, 104657.

Shen, Z., Liao, Q., Hong, Q., & Gong, Y. (2012). An overview of research on agricultural non-point source pollution modelling in China. *Separation and Purification Technology, 84*, 104–111.

Sheng, Y., Ye, X., Zhou, Y., & Li, R. (2021). Microplastics (MPs) act as sources and vector of pollutants-impact hazards and preventive measures. *Bulletin of Environmental Contamination and Toxicology, 107*(4), 722–729.

Siddiqui, S. A., Singh, S., Bahmid, N. A., et al. (2023). Polystyrene microplastic particles in the food chain: Characteristics and toxicity. A review. *Science of the Total Environment, 892*, 164531.

Sobhani, Z., Fang, C., Naidu, R., & Megharaj, M. (2021). Microplastics as a vector of toxic chemicals in soil: Enhanced uptake of perfluorooctane sulfonate and perfluorooctanoic acid by earthworms through sorption and reproductive toxicity. *Environmental Technology & Innovation, 22*, 101476.

Sobhani, Z., Panneerselvan, L., Fang, C., Naidu, R., & Megharaj, M. (2022). Chronic and transgenerational effects of polyethylene microplastics at environmentally relevant concentrations in earthworms. *Environmental Technology & Innovation, 25*, 102226.

Song, Y., Cao, C., Qiu, R., Hu, J., et al. (2019). Uptake and adverse efects of polyethylene terephthalate microplastics fbers on terrestrial snails (Achatina fulica) after soil exposure. *Environmental Pollution, 250*, 447–455.

Sun, N., Liu, Q., Wang, J., He, F., et al. (2021). Probing the biological toxicity of pyrene to the earthworm Eisenia fetida and the toxicity pathways of oxidative damage: A systematic study at the animal and molecular levels. *Environmental Pollution, 289*, 117936.

Vazquez, O. A., & Rahman, M. S. (2021). An ecotoxicological approach to microplastics on terrestrial and aquatic organisms: A systematic review in assessment, monitoring and biological impact. *Environmental Toxicology and Pharmacology, 84*, 103615.

Wang, H. T., Ding, J., Xiong, C., Zhu, D., Li, G., et al. (2019). Exposure to microplastics lowers arsenic accumulation and alters gut bacterial communities of earthworm Metaphire californica. *Environmental Pollution, 251*, 110–116.

Wang, J., Liu, X., Dai, Y., Ren, J., Li, Y., et al. (2020a). Effects of co-loading of polyethylene microplastics and ciprofloxacin on the antibiotic degradation efficiency and microbial community structure in soil. *Science of the Total Environment, 741*, 140463.

Wang, Q., Bai, J., Ning, B., Fan, L., et al. (2020b). Effects of bisphenol A and nanoscale and microscale polystyrene plastic exposure on particle uptake and toxicity in human Caco-2 cells. *Chemosphere, 254*, 126788.

Wang, F., Wang, Q., Adams, C. A., Sun, Y., & Zhang, S. (2022). Effects of microplastics on soil properties: current knowledge and future perspectives. *Journal of Hazardous Materials, 424*, 127531.

Wei, X. F., Capezza, A. J., Cui, Y., et al. (2022). Millions of microplastics released from a biodegradable polymer during biodegradation/enzymatic hydrolysis. *Water Research, 211*, 118068.

Weinstein, J. E., Crocker, B. K., & Gray, A. D. (2016). From macroplastic to microplastic: Degradation of high-density polyethylene, polypropylene, and polystyrene in a salt marsh habitat. *Environmental Toxicology and Chemistry, 35*, 1632–1640.

Xu, B., Liu, F., Brookes, P. C., & Xu, J. (2018). Microplastics play a minor role in tetracycline sorption in the presence of dissolved organic matter. *Environmental Pollution, 240*, 87–94.

Xu, C., Zhang, B., Gu, C., et al. (2020). Are we underestimating the sources of microplastic pollution in terrestrial environment? *Journal of Hazardous Materials, 400*, 123228.

Xu, B., Huang, D., Liu, F., Alfaro, D., Lu, Z., et al. (2021). Contrasting effects of microplastics on sorption of diazepam and phenanthrene in soil. *Journal of Hazardous Materials, 406*, 124312.

Xiang, Y., Jiang, L., Zhou, Y., Luo, Z., Zhi, D., Yang, J., & Lam, S. S. (2022). Microplastics and environmental pollutants: key interaction and toxicology in aquatic and soil environments. *Journal of Hazardous Materials, 422*, 126843.

Yu, C. W., Luk, T. C., & Liao, V. H. C. (2021). Long-term nanoplastics exposure results in multi and trans-generational reproduction decline associated with germline toxicity and epigenetic regulation in Caenorhabditis elegans. *Journal of Hazardous Materials, 412*, 125173.

Yao, Y., Lili, W., Shufen, P., et al. (2022). Can microplastics mediate soil properties, plant growth and carbon/nitrogen turnover in the terrestrial ecosystem?. *Ecosystem Health and Sustainability, 8*(1), 2133638.

Zhang, M., Zhao, Y., Qin, X., Jia, W., et al. (2019). Microplastics from mulching film is a distinct habitat for bacteria in farmland soil. *Science of the Total Environment, 688*, 470–478.

Zhang, H., Jin, T., Geng, M., et al. (2022). Occurrence of microplastics from plastic fragments in cultivated soil of Sichuan Province: The key controls. *Water (Basel), 14*, 1417.

Zhou, Q., Zhang, H., Fu, C., et al. (2018). The distribution and morphology of microplastics in coastal soils adjacent to the Bohai Sea and the Yellow Sea. *Geoderma, 322*, 201–208.

Zhou, Y., Liu, X., & Wang, J. (2020). Ecotoxicological effects of microplastics and cadmium on the earthworm Eisenia foetida. *Journal of Hazardous Materials, 392*, 122273.

Zhu, D., Bi, Q. F., Xiang, Q., et al. (2018). Trophic predator-prey relationships promote transport of microplastics compared with the single Hypoaspis aculeifer and Folsomia candida. *Environmental Pollution, 235*, 150–154.

Long-Term Fate of Micro/Nanoplastics in Soil Systems and Their Impacts

Priyanka Sharma, Surbhi Sharma, and Jatinder Kaur Katnoria

1 Introduction

Soil plays a crucial role in plant growth, human health, and well-being (Perković et al., 2022). The expansion of agricultural production methods, however, has increased food consumption and population growth, putting stress on soil and raising the exposure of humans to pollutants (de Souza et al., 2018a). In agricultural soils, some contaminants, such as heavy metals (HMs) and pesticides, are well known and have long been prevalent, while micro/nanoplastics (MNPs) represent a new and emerging threat. Fragmented plastic particles called microplastics (MPs) (1 μm TO 5 mm) and nanoplastics (NPs) (≤0.1 μm) are produced by microbial activity, photodegradation, and mechanical forces (Bartkova et al., 2021; Perkovic et al., 2022). The shape of MNPs varies and includes fragments, fibers, microbeads, foams, and needles (de Souza et al., 2018b; Okoye et al., 2022). After entering the environment, plastics can break down into microscopic particles that are termed microplastics and with further reduction in size of particles as nanoplastics as a result of abiotic (thermo-oxidation, photo-oxidation, air oxidation, and mechanical degradation) or biotic processes (Pérez-Reverón et al., 2022). MNP occurrence represents a crucial area for research because of their physico-chemical characteristics, distribution, accumulation, bioavailability, and negative impacts (Perkovic et al., 2022).

Microplastics are ubiquitous in many ecosystems, including the sea surface, water column, atmosphere, soil, and fauna (Verla et al., 2019). The terrestrial environment has been identified as the primary source and sink of environmental MNPs (Xu et al., 2020; Rilling & Lehmann, 2020). It is related to different human

P. Sharma · S. Sharma · J. K. Katnoria (✉)
Department of Botanical and Environmental Sciences, Guru Nanak Dev University, Amritsar, Punjab, India
e-mail: jatinder.botenv@gndu.ac.in

S. A. Bhat et al. (eds.), *Management of Micro and Nano-plastics in Soil and Biosolids*,
https://doi.org/10.1007/978-3-031-51967-3_10

activities involving the use of plastic in terrestrial ecosystems (He et al., 2018). Furthermore, MNPs have a significant probability of entering the human food chain. The majority of research found that the greatest threat posed by MNPs to human health is the release of compounds known as endocrine disruptors or carcinogens, such as bisphenol A or phthalates (Gruber et al., 2023; Zolotova et al., 2022). Apart from this, the major concern is that these particles due to their small size can be taken up by a variety of organisms (Syberg et al., 2015). Their small size encourages consumption, which has severe consequences including physical pain, internal injuries, bowel blockage, malnutrition, and death (Okoye et al., 2022; Zolotova et al., 2022). Human consumption of MNPs that are prevalent through contaminated agricultural crops results in oxidative stress, inflammation, genotoxicity, neurotoxicity, and mitochondrial dysfunction (Ferrante et al., 2022; Pérez-Reverón et al., 2022). Other effects in animals include growth retardation, reduced immunity, and deformities in animal and human reproductive systems (Ferrante et al., 2022; Gruber et al., 2023; Zolotova et al., 2022).

MNPs in the soil ecosystem can interact with numerous components of the plant–soil–microbe system and alter the physicochemical properties of the soil, affecting plant growth and development (Ferrante et al., 2022; Peng et al., 2020). Despite the fact that the abundance of MNPs in a soil system leads to negative impacts on plants, the mechanisms of toxicities have been very poorly understood due to the absence of pertinent scientific literature regarding the terrestrial system (Allen et al., 2022; Maity et al., 2022). Although it is widely recognized that small plastic particles can be found in several environmental compartments, including marine, freshwater, sediments, soils, and air, current knowledge on the effects of these particles at levels found in the environment is still regarded as limited to produce the mechanism of toxicities (Akdogan & Guven, 2019). The complexity of these particles, which may contain various polymeric chains, residue monomers, additives, and components from the raw ingredients that were added accidentally, combined with the variety of possible sizes and shapes, makes it difficult to determine their real effects (Akdogan & Guven, 2019; Maity et al., 2022).

There is growing concern around the world about the occurrence, accumulation, and persistence of plastic as MP/NPs in aquatic and terrestrial ecosystems. The Sustainable Development Goals (SDGs) of the United Nations (UN), particularly Goal 14 – "Conserve and sustainably use the oceans, seas and marine resources for sustainable development," addressed the global issue of plastics (Walker, 2021). Public awareness about MNPs is crucial, and remedial strategies still need to be investigated and implemented. In recent years, macroplastic waste management methods have been developed, which include recycling, burning, bioremediation, and landfills (Cárdenas-Alcaide et al., 2022). Nonetheless, there is a potential opportunity in MNPs remediation methods and new ways must be developed to overcome constraints. In addition to these facts, a lack of agreement on research methodology has led to a variety of inconsistent outcomes (Patil et al., 2021). As a result, the purpose of this chapter is to present current viewpoints on MP/NPs considering the considerable issues brought about by the presence of MNPs in soils concerning (1) Properties of MNPs. (2) Source and occurrence of MNPs in soil. (3) Adverse impacts of MNPs in soil. (4) Mitigation of MNPs pollution and future perspective.

2 Types and Properties of Micro/Nanoplastics

2.1 Types of Microplastics

Microplastic are classified into two types primary and secondary, which differ in origin. The details are given in the following section.

2.1.1 Primary Microplastics

Primary microplastics (PMPs) are characterized as polymers manufactured to be minuscule in size for commercial use (Bonfanti et al., 2021). PMPs are frequently found in cosmetics such as facial cleansers and there is also evidence of their use in medicine as drug delivery systems (Patil et al., 2021). Synthetic textile laundry (textile microfibers) and vehicle abrasion are the main sources of primary microplastics released into the environment (Fig. 1). They predominate near metropolitan areas or wastewater discharge sources (Pérez-Reverón et al., 2022). The world releases nearly 1.5 million tons of primary microplastic into the ocean each year, with 98% of it coming from land-based operations and the remaining 2% from sea-based operations (Patil et al., 2021). Even without degradation, organisms can access these MPs in the consumable size range.

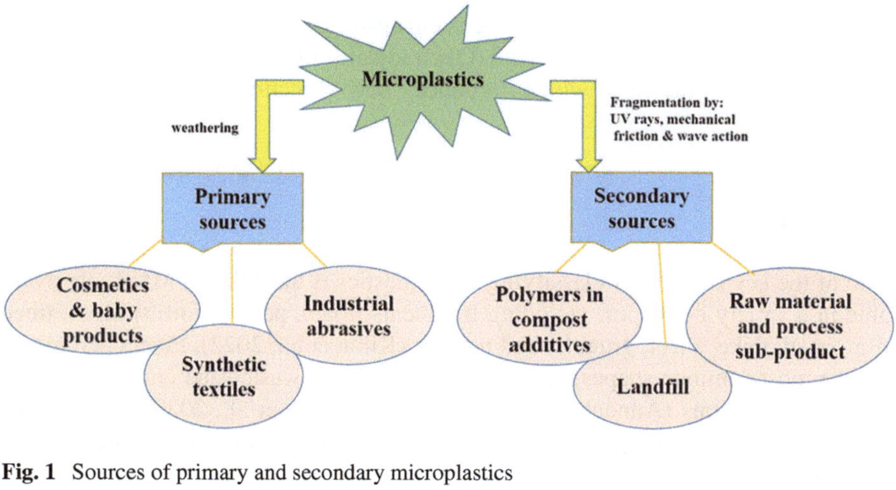

Fig. 1 Sources of primary and secondary microplastics

2.1.2 Secondary Microplastics

Secondary sources of microplastics in soil include the breakdown of plastic mulches used in agriculture and the breakdown of larger plastic debris, such as bags and packaging, which release microscopic particles into the soil via weathering processes (Fig. 1). Secondary microplastics are produced by physical, chemical, and oxidative fragmentation of large plastic debris, synthetic textile fiber, and tires (Riling, 2012). Additionally, high temperature and ultraviolet (UV) rays oxidize the plastic polymer, causing chain scission, often referred to as photo-fragmentation. They can also alter microplastics form, size, and hydrodynamics (Patil et al., 2021). Similarly, greater oxygen availability and exposure to direct sunlight at the beach promote rapid disintegration of macroplastics, resulting in yellowing brittleness and cracking (Horton et al., 2017; Patil et al., 2021).

2.2 *Properties of Micro/Nanoplastics*

Due to their distinct physical and chemical characteristics that do not exist in their larger-dimension counterparts of identical materials, micro/nanoplastics (MNPs) offer a wide range of uses. The most significant physicochemical characteristics that are altering at the nano-scale are summarized in the following section.

2.2.1 Size and Surface Area

The size of MNPs vary, but the most frequent size range for microplastics (MPs) is 1 μm to 5 mm, whereas nanoparticles (NPs) are <100 nm (Murray & Ormeci, 2020; Kiran et al., 2022). Particle size and surface area have a considerable influence on how materials interact with biological systems. MNPs surface area increases as particle size decreases. The plastic surface is changed by photo and oxidative degradation processes, which interact with OH radicals, O and N oxides, and other photo-generated radicals which produce new functional groups (Lambert et al., 2017).

2.2.2 Shape

One of the key criteria for classifying microplastics is their shape. MNPs are available in a variety of shapes, including fragments, foam, pellets, foil, spheres, fiber, film, bead, flake, sheet, granule, and nurdles (Kiran et al., 2022). Fibers and pieces are the most common shapes, accounting for 52.7% of wastewater and urban atmospheric depositions (Annenkov et al., 2021; McGivney et al., 2020). Fragmented microplastics can develop when larger plastic products are subjected to strain or UV light, while fiber particles originate as a result of mechanical abrasion and chemical weathering. (Kiran et al., 2022; Zhou et al., 2020). Pollutant transfer in the environment is impacted by the form of MNPs.

2.2.3 Color

One of the most important characteristics for identifying plastic debris and potential pollutants in sample preparation is color. Micro and nanoplastics in a variety of colors, including blue, red, orange, yellow, brown, tan, white, grey, and green, along with others have been observed (Zhang et al., 2020). Blue and red fibers are the most commonly reported colors (Kiran et al., 2022). Polypropylene is known for its transparency, while polyethylene, including low-density polyethylene (LDPE), is recognized for its opaque coloration (Kiran et al., 2022). Marine creatures frequently consume brightly colored nanoplastic particles that resemble natural meals, whereas zooplankton, euphausiids, copepods, and fish larvae consume microplastic that are mostly blue, red, green, and black (Rashid et al., 2021).

3 Sources and Occurrence of MNPs in Soil

The earth's terrestrial ecosystem has been recognized to be a significant source and sink of environmental micro/nanoplastics (MNPs) (Rillig & Lehmann, 2020; Xu et al., 2020). MNPs have been found in a variety of terrestrial habitats, such as cultivated or industrialized soils, coastal soil, urban waters, and even remote regions (Evangeliou et al., 2020; He et al., 2018; Rillig, 2020). Multiple plastic polymers, including polyethylene, polystyrene, polyurethane, polypropylene, polyvinyl chloride, and polyurethane, among others, are indicators of MNP contamination (He & Luo, 2020). Soil bound MP/NPs originate from a variety of sources, including sewage sludge/biosolids (Corradini et al., 2019; Koutnik et al., 2021; Li et al., 2018; Wang et al., 2019), organic fertilizers (Weithmann et al., 2018), domestic waste (Shi et al., 2023), wastewater irrigation (Gündoğdu et al., 2018), land fillings (He et al., 2018), plastic film mulch (Schothorst et al., 2021), atmospheric deposition (Wang et al., 2019), and tire abrasion (Kole et al., 2017).

3.1 Plastic Films

Plastic mulching, which is employed to improve crop productivity, has been recognized as a significant source of microplastics (Schothorst et al., 2021). Plastic films can be used as mulch to improve soil temperature, reduce weed development, and/or boost water efficiency. Several kinds of plastics can be used for mulching, but low-density polyethylene (LDPE) is the most common (Kasirajan & Ngouajio, 2012; Yu et al., 2022). Low-density polyethylene is a fully saturated hydrocarbon polymer that is particularly resistant to weathering (Crawford et al., 2016). As a result, LDPE mulch must be removed after harvesting, and this removal process contributes to the accumulation of LDPE debris in the environment. The slow rate of agricultural film recovery and recycling will result in a significant accumulation of waste films in the soil over time. Given its formidable resistance to decomposition in agricultural soil, it emerges as a prominent contributor to the presence of micro/nanoplastics in the soil (Yu et al., 2022).

3.2 Compost Products and Sewage Sludge

Sewage sludge and compost, crucial nutrient sources in agriculture when applied alongside other organic fertilizers, can also play a role in introducing MNPs into the environment. Compost products and sludge are frequently employed as soil amendments because they have a high concentration of plant nutrients and organic carbon. This increases the nutrient content of the soil, improves its physical and chemical qualities, and increases crop yields (Corradini et al., 2019). Microplastics can currently not be removed by the majority of sludge treatment and composting processes (Naeini & Cook, 2000; Zubris & Richards, 2005). As a result, agricultural sludge and compost play a substantial role in the contamination of agricultural soil with MNPs.

Microplastics from washed garments, personal care items, and tires/road abrasion are found in sewage sludge (Corradini et al., 2019). Urbanization is believed to have an impact on the concentration of MP in sludge. For instance, in China, improved infrastructure and industrial activity were positively connected with higher MP concentrations in sludge (Li et al., 2018). With a median of 34 items/g, the microplastic concentration in soils obtained from the dumping of sewage sludge in Chile ranged from 18 to 41 items/g (Corradini et al., 2019).

3.3 Biosolids

Biosolids can also introduce microplastics to the soil. Sludge that has been treated is referred to as biosolids and can be disposed of in landfills, burnt, or applied to land, creating pathways for microplastics (MPs) to infiltrate the ecosystem (Rolsky et al., 2020). Biosolids are used as agricultural amendments in many countries. Biosolid samples taken in Ontario, Canada, have a high microplastic concentration ranging from 8.7 × 103 items/kg to 1.4 × 104 items/kg. According to Crossman et al. (2020), yearly microplastic additions to the soils of agriculture in Ontario range between 4.1 × 1011 and 1.3 × 1012 particles.

3.4 Irrigation Water

Microplastics can accumulate in soil when crops are irrigated with polluted water (Corradini et al., 2019). This is mostly due to the fact that a significant amount of MNPs may be present in irrigational water even after it has passed through a wastewater treatment facility (Mintenig et al., 2017). In this context, Jiang et al. (2023) discovered that the abundance of microplastics in irrigation water in Taiwan ranged from 1.88 items/L to 141 items/L. According to a recent study, wastewater treatment facilities (WWTPs) may have a major impact on the release of microplastics into the environment (Browne et al., 2011). In a study conducted by Ziajahromi et al. (2021), it was found that the effluents from three different wastewater treatment plants had a significant amount of microplastics, ranging from 5880 to 11,600 million particles per day. Within these effluents, microplastics constituted a relatively small percentage, with levels ranging from 0.2% to 8%. However, a higher proportion of microplastics, ranging from 8% to 16%, was observed in the sewage sludge produced during the treatment process.

Microbeads, which are microscopic plastic particles that are frequently added to goods such as facial cleansers and toothpaste for exfoliating and abrasive reasons, can enter wastewater directly as a result of human activities (Fendall & Sewell, 2009). Furthermore, synthetic textiles such as polyester (PES) and nylon can discharge thousands of microscopic threads into wastewater when laundered. These microfibre releases lead to the prevalence of microplastics in aquatic systems and are an additional significant source of plastic pollution in the environment (Sun et al., 2019). Depending on the type of treatment units used, the WWTP may remove some of the microplastics. Nonetheless, research has demonstrated that microplastics have the capacity to bypass wastewater treatment plants (WWTPs) and end up in the environment (Browne et al., 2011).

3.5 Tire Abrasion

Another source of MNPs that are gradually released into the environment is tire abrasion, specifically soils (Baensch-Baltruschat et al., 2020). Nowadays, tires are a necessary part of every vehicle. During use, tire tread patterns become worn, resulting in progressive degradation and posing an environmental danger (Sourkova et al., 2021). Each year, as many as six million tons of microplastics produced through tire abrasion enter the environment (Khan et al., 2019). As reported by Blasing and Amelung (2018), approximately 10,000 tons of tire dust were documented in Sweden, while Germany recorded up to 110,000 tons.

3.6 Atmospheric Deposition

Another way for micro/nanoplastics to infiltrate the soil ecosystem is by atmospheric deposition. Dris et al. (2016) conducted research on the atmosphere around Paris and discovered that 29,280 pcs/m^2 of microplastics were deposited in the atmosphere on a daily basis. Furthermore, they discovered that 3–10 tons of fibrous microplastics were deposited in the region each year by atmospheric processes. Notably, the bulk of these microplastics were fibrous, accounting for 90% of the total amount, with 50% having particle sizes greater than 1000 m.

Therefore, it can be deduced that sewage sludge, waste products from wastewater treatment, wastewater irrigation, plastic mulching in agriculture, and tire abrasion are the likely primary sources of the majority of the microplastic particles (MNPs) found in a variety of soil samples.

4 Adverse Impacts of MNPs

4.1 Effects of MNPs on Soil Physicochemical Properties

MNPs have been recognized as emerging pollutants with potential detrimental effects on soil physicochemical properties. These plastic fragments can build up in soil ecosystems and come from the direct release of microbeads from personal care items or from the breaking of bigger plastic materials. Once present in the soil, MNPs can influence its physical structure and properties. It is hypothesized that the impacts of MNP pollution on soils can differ based on the type, shape, concentration, and rate of decomposition of the polymer.

The presence of MNPs in the soil can have multifaceted effects on its physicochemical properties (Table 1). MNP represents a physical pollutant in soil that alters the organization of soil, lowers resistance to root penetration, decreases soil bulk density, increases water evaporation and soil water flow, and releases (toxic) plastic leachate into soil (Maity et al., 2022). Microfiber, one of the most prevalent microplastics found in soils, improves soil water retention capacity while having a detrimental impact on bulk density. Improved soil water holding capacity enhances water accessibility to plant roots, whereas decreased soil bulk density promotes better aeration and increases soil macroporosity. Plastic mulching has a significant impact on soil C: N ratio due to the high carbon content found in the hydrocarbon structure of plastic mulch materials. Additionally, soil animals may come into contact with microplastics (Rilling et al., 2018; Boots et al., 2019). For example, earthworms have been shown to consume polystyrene beads and some studies have revealed negative impacts on earthworms.

Moreover, MNPs can introduce chemical changes to the soil environment. The additives used in the production process of the plastic particles include colors, plasticizers, and flame-retardant substances. Over time, these additives may eventually

Table 1 Effects of MNPs on the soil property

Sl. no.	Soil property	Effect of microplastics	References
1.	Physical structure	(a) Alters soil structure and reduces porosity (b) Compacts soil hindering water infiltration and drainage (c) Impedes root growth and hampers plant establishment	Wang et al. (2021)
2.	Chemical composition	(a) Releases toxic additives (flame retardants, plasticizers) (b) Alters soil chemistry and nutrient availability (c) Disrupts microbial activity and nutrient cycling	Kumar et al. (2023)
3.	Organic matter decomposition	(a) Interferes with organic matter decomposition process (b) Accumulates organic matter in the soil (c) Alters nutrient dynamics and carbon sequestration	Boots et al. (2019)
4.	Ecosystem functioning	(a) Impairs soil fertility and nutrient cycling (b) Reduces soil microbial diversity and activity (c) Disrupts ecosystem balance and overall soil health	Ding et al. (2022)

leak out of the plastic particles and contaminate the soil nearby (Wang et al., 2022; Sa'adu & Farsang, 2022). The release of these chemicals can alter soil chemistry and nutrient availability, potentially impairing plant growth and soil fertility. The presence of MNPs can also interfere with microbial communities that are essential to nutrient cycling and the breakdown of organic waste (Table 1). Disrupting these microbial processes can disrupt the balance of nutrients and carbon in the soil, further impacting its overall health and productivity (Rilling et al., 2018).

Furthermore, the build-up of MNPs in the soil can disrupt the natural decomposition process. Microorganisms responsible for breaking down organic matter may encounter difficulties in accessing and decomposing organic materials due to the presence of plastic particles. This can lead to an accumulation of organic matter in the soil, potentially altering nutrient dynamics and impacting the soil's capacity to sustain plant growth (Ding et al., 2022). Furthermore, as MNPs can linger in the environment for a long time, the presence of these particles in the soil may have lasting impacts on the ecosystem (Table 1). Comprehending the effects of MNPs on the soil ecosystem is important for sustainable land management practices and the preservation of soil fertility and ecosystem health. Efforts to address plastic pollution and devise effective plastic waste management strategies are indispensable in order to minimize the detrimental impacts on soil ecosystems (Sa'adu & Farsang, 2023; Huang et al., 2022).

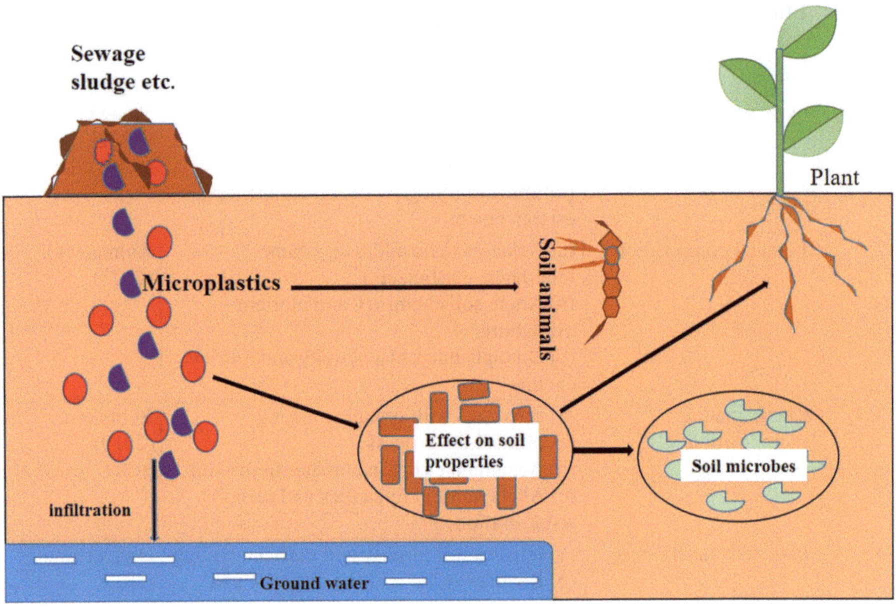

Fig. 2 Fate of MNPs in environment

4.2 MNP-Induced Toxic Effects on Soil Fauna

MNPs can undergo various transport mechanisms and interactions, potentially persisting in the environment and impacting ecosystems (Fig. 2). Nanoplastics when ingested by organisms, accumulate in their bodies and further enter the food chain (Dong et al., 2023; Pérez-Reverón et al., 2022). Microplastics have the potential to traverse blood-brain barriers and induce behavioral changes that depend on their size when they reach apex consumers at the highest level of the food chain (Pérez-Reverón et al., 2022). The particle size controls the kinetics of accumulation and the dispersion pattern inside the tissue. The most frequently researched living species of soil fauna are earthworms, snails, nematodes, collembola, rodents, and isopods (Perez-Reveron et al., 2022). In spite of the wide range of soil organisms, the majority of nanoplastic (NP) research has been conducted using nematodes, earthworms, and mice.

4.2.1 Earthworms

Earthworms are easily recognizable organism within the soil ecosystem, which make them a useful model for assessing the effects of MNPs. Earthworms can specifically consume MPs, mostly with sizes <50 μm and biodegradable components (Cui et al., 2022). Earthworms are typically more negatively impacted by microplastics of smaller sizes and higher concentrations (particularly those >0.5%, w/w) (Cui et al., 2022). Additionally, earthworm mortality rates are higher when microplastics along with additional pollutants are present in the soil (Zhang et al., 2022; He et al., 2018). The physiological status of earthworms is probably unaffected by low levels of microplastic in soil, but only a few species have been examined yet. According to Cao et al. (2017), *Eisenia fetida* was unaffected by the soil's 0.5% w/w content of polystyrene particles with a particle size of 58 μm. In a study conducted by Jiang et al. (2020), *Eisenia fetida* was treated with 100 or 1000 μg of fluorescent polystyrene microplastics (PS-MPs) at wavelengths of 100 nm and 1300 nm. The outcomes demonstrated that this exposure caused the earthworms to grow more quickly and to accumulate microplastics in their intestines along with injuries to their stomach. Concerning nanoplastics, the highest exposure resulted in large body loads and nano-sized particle accumulation in body tissue (He et al., 2018). The MNPs have reported inducing reproductive abnormalities and injury in male genitalia and impairment of spermatogenesis in *Eisenia andrei*, as well as an increase in cocoon production in *Eisenia crypticus* (Cui et al., 2022).

4.2.2 Nematodes

The presence of MNPs in the environment affects worm motor neurons, altering their movement. In particle suspensions with 0.1, 1, and 2 μm size, variations in the frequency of body flexion and the average velocity of nematode locomotion were noticed (Leonov & Tiunov, 2020). A species of nematode known as *Caenorhabditis elegans* is frequently used as a model for toxicological evaluation due to its small size, short lifespan, ease of production, and well-established lineage (Pérez-Reverón et al., 2022). Cholinergic, GABAergic, and dopaminergic neuron subtypes all have a role in controlling *C. elegans* motility. Regardless of particle size, plastic particles have been demonstrated to damage cholinergic and GABA neurons and cause oxidative stress (Leonov & Tiunov, 2020). Numerous research demonstrated reduced brood size due to germline apoptosis caused by nanoparticles translocating into reproductive organs of nematodes as well as transgenerational reproductive capacity, gonad damage, and chromosomal oocyte changes (Zaho et al., 2023). When *C. elegans* was exposed to polystyrene particles with a diameter of 0.1 μm at a concentration of 10 μg/L, it took the organism about 4.5 days to develop from a larva to an adult (Pérez-Reverón et al., 2022). Toxic effects included a reduction in brood size, the generation of reactive oxygen species in the gut, and inhibition of locomotor activity. The subsequent generation of nematodes cultured in a clean growth medium also experienced harmful effects when particle concentrations increased above 100 μg/L (Leonov & Tiunov, 2020). Given that some nanopolystyrene particles were detected in the descendants of nematodes developed in their existence, this is most likely caused by the transport of these particles to the reproductive organs and their passage on to the subsequent generation (Zhao et al., 2017).

4.2.3 Mice

Concern over the uncertain outcomes of MNPs on the well-being of individuals has recently increased as a result of the discovery of nanoplastics in samples of human blood and the placenta of pregnant women (da Silva Brito et al., 2022; Eyles et al., 2001; Jani et al., 1990). Exploring similar research investigations of microplastics and nanoplastics in mammal models, notably mice, has been undertaken. By exposing the female mice to the MNPs subsequent transgenerational impacts have recently been studied. Abdel-Zaher et al. (2023) subjected rats to various microplastic levels (6, 60, and 600 g/day) for 15 days. The results showed that exposure to 600 g/day of MPs significantly impacted RBC's normal structure, resulting in a variety of aberrant morphologies. Additionally, hematological markers were seen to decline in a concentration-dependent way and biochemical testing demonstrated that MP exposure had an effect on the liver and renal function (Abdel-Zaher et al., 2023; Li et al., 2020). MNPs not only were revealed to accumulate in reproductive organs but also crossed the pass across the placental barrier. Microplastics have been shown to translocate into distant tissues including the liver and spleen in rats after being inhaled and ingested by the animals (da Silva Brito et al., 2022). Aghaei et al. (2022), during their research, observed that the offspring of exposed mothers had declined growth and cord shrinkage, which were attributed due to a lack of nourishment through the placenta. Deng et al. (2018) observed that when mice were exposed to MNPs made of polystyrene and polyethylene via drinking water containing organic flame retardants, it resulted in increased oxidative stress within the liver. Amereh et al. (2020) observed that rodents treated with MNPs via gastrointestinal intubation had lower serum testosterone levels. Plastic exposure in mice intestines causes local inflammation, metabolic dysfunction, influenced hepatic lipid metabolism, and modified host-pathogen relationships (da Silva Brito et al., 2022). Though the results appear applicable to humans, most of the effects were caused by large MPs and NPs were at undefined endpoints that did not mimic environmental settings.

4.3 MNP-Induced Toxic Effects in Plants

Small plastic particles, including microplastics and nanoplastics, with sizes of less than 5 mm and 1 m, respectively, have been a source of concern for the oceanic ecosystem (Bartkova et al., 2021). Current research is beginning to illuminate the effects of these particles on terrestrial organisms, such as plants (Peng et al., 2020).

The development and health of plants can be adversely affected by the presence of MNPs in the soil. One toxicity mechanism involved the formation of a physical barrier that restricted root growth and hampered nutrient and water uptake by plants. Consequently, plant growth and productivity were reduced. Additionally, MNPs altered the movement and accessibility of vital nutrients in the soil by adsorbing and accumulating harmful materials such as heavy metals, organic pollutants, and other pollutants (Peng et al., 2020). This interference caused hindrance in nutrient uptake and induced imbalances in essential elements, leading to plant nutritional deficiencies. MNPs have also been demonstrated to induce oxidative stress in plants. The generation of reactive oxygen species (ROS) could harm cellular constituents and impair cellular activity, which eventually causes plant cell death (Babele & Bhatia, 2023). Another route of toxicity was shown to be hormonal disruption. MNPs could obstruct plant hormone signaling pathways, which induced an impact on growth patterns, developmental procedures, and hormonal balance (Ferrante et al., 2022). Furthermore, exposure to MNPs altered gene expression in plants, affecting stress responses, defense mechanisms, and growth regulation (Mateos-Cárdenas et al., 2021). This, in turn, could impact plant physiology and metabolism.

4.3.1 Types of Stress Induced in Plants

In recent years, the growth of MNPs in many industrial and agricultural applications has raised concerns regarding their potential adverse effects on plant ecosystems. Exploring the impacts of MNPs on plant physiology and growth is crucial to ascertain their environmental implications and develop strategies for mitigating their potential harm. Figure 3 depicts the adverse impacts of MNPs on plants.

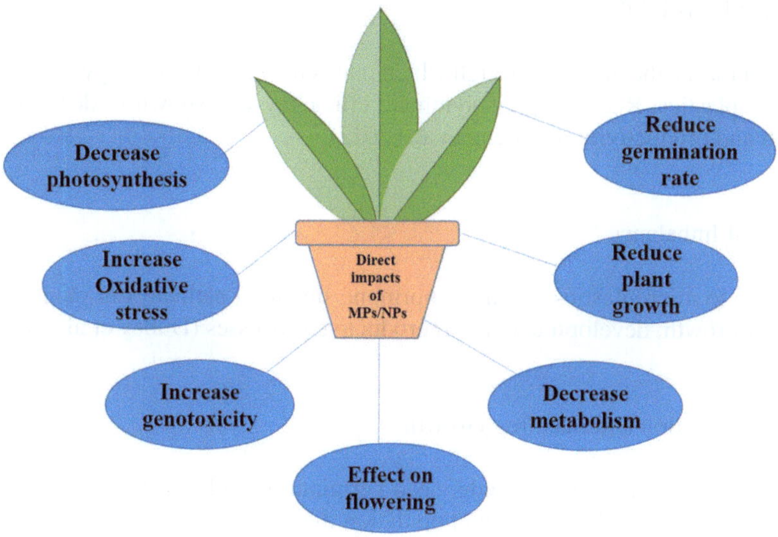

Fig. 3 Adverse impacts of MNPs on plants

Nutrient and Water Uptake

Microplastics in the soil physically blocked or reduced the movement of water and vital nutrients to plant roots. Which leads to nutrient deficiencies and water stress in plants, affecting their growth and development (Khalid et al., 2020).

Root Morphology and Function

Exposure to microplastics could lead to changes in root morphology, including stunted root growth and reduced root surface area. Weaker root systems induced negative impact on plant's ability to anchor itself and absorb nutrients efficiently (Kumar et al., 2022).

Physiological Stress

The presence of microplastics in plant tissues could induce physiological stress. Reactive oxygen species (ROS) are produced in plant cells by a variety of chemical additions found in plastics, causing oxidative stress and cellular damage (Kumar et al., 2022).

Photosynthesis Inhibition

Photosynthesis, the process through which plants turn light into energy, is hampered by microplastics. Reduced photosynthetic efficiency was shown to decrease plant growth and productivity by Zheng et al. (2022).

Hormonal Imbalance

Exposure to microplastics disrupted hormone signaling pathways in plants, affecting their growth, development, and reproductive processes (Dubey et al., 2022).

Seed Germination and Seedling Growth

Microplastics in the soil can hinder seed germination and reduce seedling growth, impacting plant establishment and population dynamics (Li et al., 2021).

Genotoxicity and Mutagenicity

Microplastics have the potential to damage genetic material and cause mutations in plant cells, leading to genetic instability and abnormal growth (Maity et al., 2023).

Impact on Mycorrhizal Associations

Microplastics can interfere with the beneficial symbiotic relationships between plants and mycorrhizal fungi, which are crucial for nutrient uptake and plant health (Liu et al., 2023).

Disruption of Plant–Microbe Interactions

Microplastics may alter the chemical communication between plants and beneficial microorganisms, affecting nutrient cycling and soil health (Khalid et al., 2020).

Translocation and Accumulation

Plants can uptake microplastics and translocate to different plant tissues. Once inside the plant, these particles may accumulate over time, potentially leading to long-term effects (Wang et al., 2022).

Reproductive Impairment

Exposure to microplastics can disrupt plant reproductive processes, affecting flower production, pollination, and seed development (Zheng et al., 2022).

Indirect Effects Through Trophic Transfer

Contaminated plants may be consumed by herbivores, leading to plastic ingestion and potential adverse effects at higher trophic levels. It is critical to evaluate the possible consequences of MPs on plants and the broader ecosystem, as plastic pollution continues to be a pressing environmental concern (Maity et al., 2022).

4.4 MNP-Induced Toxic Effects on Human Health

MNPs are widely distributed in both aquatic and terrestrial environments, facilitating their transfer up the food chain. Some studies have documented that the average MNPs were detected in food products as honey (0.10 items/g); salt (0.11 items/g); seafood (1.48 items/g); and sugar (0.44 items/g) (Kosuth et al., 2018; Yee et al., 2021). Potential sources of contamination include bottled water and plastic teabags

(Kosuth et al., 2018; Zhang et al., 2020). MPs have also been isolated from salt, ranging from 56 to 103 particles/kg of salt (Peixoto et al., 2019).

On average human beings typically consume an average intake of 39,000–52,000 MNPs; however, individuals drinking bottled water consume an additional 90,000 particles (Conti et al., 2020; Mason et al., 2018), which leads to serious health issues (Jiang et al., 2020). Barboza et al. (2020) reported that individuals in American and European countries ingest an estimated 518–3078 microplastic items per year per capita through the consumption of fish. Microplastic can potentially enter the body through the ingestion of contaminated plants (Peixoto et al., 2019). MPs infiltrate the food chain and reach top consumers, where they can cross blood-brain barriers and cause size-dependent behavioral changes or diseases (Mason et al., 2018; Mattsson et al., 2017). The three main routes for microplastic intake include ingestion, cutaneous perfusion, and inhalation. Nanoplastics, which are found in toiletries, enter the body when applied directly to the skin. Body lotion ingredients such as urea, glycerol, and hydroxyl acids enhanced the nanoparticle's ability to penetrate the epidermal barrier (Jatana et al., 2016).

The build-up of microplastics may pose health hazards such as cell death, hypersensitivity, inappropriate immunological response, and acute reactions like hemolysis (Peixoto et al., 2019; Prata et al., 2020). MNPs with a large surface area can cause oxidative stress and translocation to other tissues (Fig. 4). The persistent nature of MNPs impedes their elimination from the body, prolongs inflammation, and enhances the possibility of cancer development (Fig. 4). Conversely, microplastics when included in particulate matter may contribute to the rising prevalence of

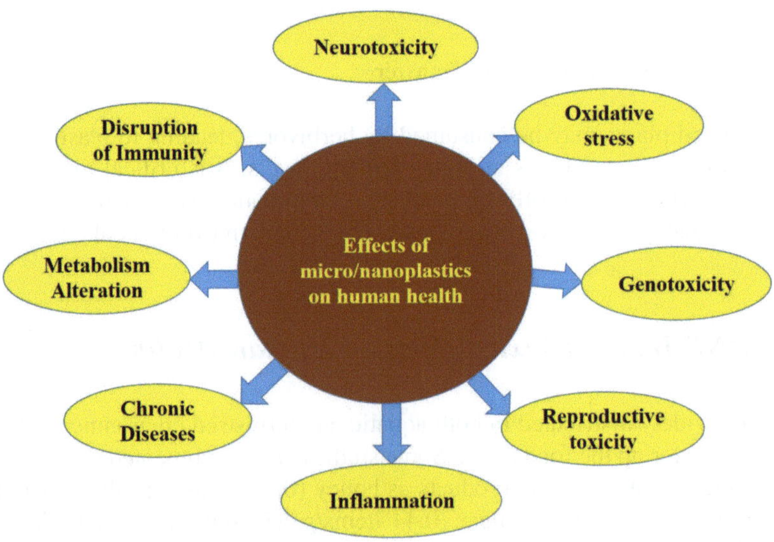

Fig. 4 Adverse impacts of MNPs on human health

immunological and neurological diseases (Hwang et al., 2019). According to Deng et al. (2017), intake of MNPs through the mouth resulted in a redox imbalance, disturbance of energy regulation, and induced neurotoxic implications. Polystyrene and polyvinyl chloride (PVC) particles (150 nm) penetrated the blood vessels and lymph nodes through the gut wall due to endocytic processes (Xu et al., 2019). Over 90% of ingested MPs were anticipated to be eliminated by the human excretory system via feces (EFSA, 2016). The most common MPs detected in human feces were polyethylene terephthalate (PET) and polypropylene, varying in size between 50 and 500 m (Schwabl et al., 2019; Shen et al., 2019).

The majority of current studies are based on in vitro cell models and only a small number of studies have concentrated on the molecular elements of MPs harmful health effects. Based on the experiments conducted on animals, researchers proposed that DNA damage, oxidative stress, and other negative consequences could result from human exposure to MNPs (Deng et al., 2017). As an illustration, mice have been documented to exhibit oxidative stress subsequent to exposure to MPs (Feng et al., 2019; Lei et al., 2018). In mice exposed to 40 nm polystyrene NPs for 35 days, a substantial decrease in testosterone and follicle-stimulating hormone was also noted (Amereh et al., 2020; Lv et al., 2019).

4.4.1 MNP-Induced Cytotoxicity and Oxidative Stress

Microplastics induced oxidative stress, disrupted energy, and lipid metabolism and resulted in neurological dysfunction (Chiu et al., 2015; Valavanidis et al., 2013). The occurrence of these effects can be attributed to MNPs, owing to their large surface area, the release of oxidizing chemicals bound to their surface, or the generation of reactive oxygen species during the inflammatory response (Deng et al., 2017). In vitro, studies were conducted using epithelial and cerebral human cells and discovered that a contributing factor to cellular toxicity was the production of oxidative stress via ROS formation (Schirinzi et al., 2017; Barboza et al., 2018). Furthermore, polystyrene (PS) (60 m) exposure generated ROS and endoplasmic reticulum stress, resulting in cell death through the process of autophagy in macrophage and lung epithelial cell cultures (Kelly & Fussel, 2012). Thus, oxidative stress and cytotoxicity were identified as key microplastic toxicity pathways.

4.4.2 MNP-Induced Diseases

MNPs can enter locally or translocate to different organs after exposure, resulting in damage to tissues located at a distance from a specific site. Inflammation can cause translocation because epithelial barriers become more permeable. According to a study by Wick et al. (2010), human placental perfusion model indicated that 240 nm polystyrene (PS) particles might cross the placental barrier. Microplastics in the bloodstream have the potential to cause pulmonary hypertension, vascular obstructions, increased blood clot formation, and cytotoxicity to blood cells (Osman et al., 2023). Following circulation, microplastics were discovered to enter the liver and kidney, which play crucial roles in xenobiotic processing and excretion.

4.4.3 MNP-Induced Disturbances in Energy Homeostasis and Metabolic Processes

Several investigations have revealed that microplastics disrupt energy homeostasis. The balance of available energy from intake and expenditure generally affects energy homeostasis. Microplastics have the ability to reduce energy consumption in marine calms, worms, and crabs by diminishing their feeding activity (Wick et al., 2010). Microplastics have the potential to affect fish foraging ability, which could be linked to neurotoxicity and disturbances in digestive capacity caused by alterations in digestive enzyme activity. This can result in increased amylase activity, decreased nutrient absorption, and trypsin activity in fish and eventually a negative influence on human consumers (Watts et al., 2015; Wen et al., 2018; Wright et al., 2013).

4.4.4 MNP-Induced Disruption of Immune Function

MNPs can trigger localized or whole-body immune defenses after exposure, based on their dispersion in the body. Environmental exposure alone, however, can affect immune function in some cases, such as those with genetic vulnerability, encouraging autoimmune diseases or immune suppression (Bernatsky et al., 2016; Wright et al., 2013). Inhaling particulate matter may contribute to the development of autoimmune illnesses by mechanisms including particle mobility, oxidative stress, and the release of immune-modulating chemicals. This eventually leads to self-antigen exposure and the development of autoantibodies (Fernandes et al., 2015). Microplastics when present as airborne particulate matter, might also play a role in disrupting immune function.

4.4.5 MNP-Induced Ovarian and Endometrial Cancer

Polycarbonate plastics, such as those utilized in food packaging and packaged water containers, are fundamentally made of bisphenols (BPA). Although it has been regarded as nontoxic as an interlinked polymer, its bonds can eventually degrade when exposed to pressure and heat or when plastic undergoes frequent washing, releasing the chemical's hazardous constituent element (Newbold et al., 2007). The capacity of BPA exposure to imitate estrogen's actions on ovarian cells (Gao et al., 2015) and chronic BPA exposure increases the risk of developing cysts in the ovaries and abnormal thickening of the endometrial lining in the uterus having the capacity to become cancerous lesions (Dekant & Volkel, 2008; Shafei et al., 2018) are two strong evidences in favor of the hypothesis that bisphenols are reproductive toxins in humans.

4.4.6 MNP-Induced Neurotoxicity and Neurodegenerative Diseases

Exposure to pollutants like MNPs can cause neurotoxicity, which is connected to neurodegenerative illness. Microplastics have been shown during in vivo studies to alter neuronal functions. In the brain of the European seabass (*Dicentrarchus labrax*), microplastics have been shown to increase the anaerobic energy production pathway, raise lipid peroxidation levels, and block acetylcholinesterase (AChE) (Gao et al., 2015).

The study concluded that microplastic could pose a variety of consequences on soil characteristics, plants, and animals. Another study reported that the impacts varied according to the types and forms of polymers in microplastics (Zhao et al., 2021).

5 Mitigation Measures for Combating MNPs Pollution

The reduction of MNP pollution is a difficult task that needs a multifaceted strategy involving several stakeholders. A number of strategies can be used to deal with this problem:

5.1 Source Reduction

The most effective way to handle MNP pollution is by preventing its release into the environment from the source of its generation (Kacprzak & Tijing, 2022; Quang et al., 2023). This process involves decreasing the overall utilization of plastics and advocating for the adaptation of alternative materials. Governments, industries, and individuals can take steps to reduce the generation of plastic waste and promote sustainable practices by adopting the measures as described below.

5.1.1 Promoting Sustainable Packaging

New policies can be introduced that promote the utilization of sustainable packaging materials and promote the development of eco-friendly alternatives to single-use plastics. Industries can adopt packaging designs that reduce the utilization of plastics and choose materials that are easily recyclable or biodegradable (Quang et al., 2023).

5.1.2 Plastic Bag Bans and Fees

Regulations such as plastic bag bans or fees to minimize the consumption of single-use plastic bags can be implemented, promoting the practice of bringing one's reusable bags when shopping, thus minimizing the waste generated (Wagner, 2017).

5.1.3 Encouraging Responsible Consumption

Awareness campaigns can be conducted to educate individuals about the environmental impact of plastic consumption. Encouraging responsible consumption habits, such as reducing the use of disposable items and choosing reusable alternatives, can significantly reduce plastic waste generation (Borg et al., 2022).

5.1.4 Extended Producer Responsibility (EPR)

The implementation of EPR programs by governments can make manufacturers accountable for the entire lifecycle of their products, including safe disposal or recycling. This encourages producers to produce goods that are easier to recycle and more environmentally friendly (Diggle & Walker, 2020).

5.2 Waste Management and Recycling

Proper waste handling is crucial to curb plastic waste from polluting the environment. Recycling programs should be improved and expanded and efforts should be made to ensure that all plastic waste is properly collected and treated. This includes implementing advanced sorting and recycling technologies to capture even the smallest plastic particles. Some of the methods to accomplish this are discussed in the following section.

5.2.1 Collection and Sorting

An effective waste management system begins with the proper collection and sorting of plastic waste. Communities and municipalities should establish comprehensive recycling programs that ensure the collection of all plastic waste from households, businesses, and public spaces. This can be done through recycling, drop-off centers, or dedicated recycling facilities (Alhazmi et al., 2021).

5.2.2 Upgrading Recycling Facilities

Recycling facilities need to be upgraded to handle the challenges posed by microplastics and nanoplastics. Advanced recycling technologies, such as fine-mesh screens, filtration systems, and centrifuges, can be employed to capture and separate microplastics during the recycling process. These technologies should ensure that even the tiniest plastic particles are intercepted and do not end up in the environment (Meegoda & Hettiarachchi, 2023; Alhazmi et al., 2021).

5.2.3 Plastic-to-Energy Conversion

In instances when recycling is not practical or economical, plastic waste can be converted into energy through processes like pyrolysis or gasification. These methods help to reduce plastic waste volumes, prevent it from entering landfills and generate energy in the process. However, it is mandatory to ensure that conversion processes are environment friendly and do not emit of detrimental gases or toxic products (Faisal et al., 2023; Tang et al., 2023).

5.3 Wastewater Treatment

MNPs often end up in water bodies through wastewater discharges (Kiran et al., 2022). The first step in putting a stop to plastic waste is manual cleaning. Furthermore, upgrading wastewater treatment plants to effectively capture and remove these particles can help prevent their release into rivers, lakes, and oceans. Advanced filtration techniques and employing activated carbon granules can be effective in removing microplastics from polluted water. Some of such methods are discussed here.

5.3.1 Granular Activated Carbon (GAC)

Granular activated carbon can be used as a filtration medium in wastewater treatment plants to adsorb and remove microplastics and nanoplastics. The high surface area of GAC provides ample sites for plastic particle adsorption, effectively reducing their presence in the treated wastewater (Wilkinson et al., 2017).

5.3.2 Coagulation and Flocculation

Chemical coagulation and flocculation processes can be employed in the treatment of wastewater to aid the elimination of microplastics and nanoplastics. The addition of coagulants to the wastewater leads to the formation of larger flocs by binding microplastics. These flocs can then be more easily separated through sedimentation or filtration processes (Wilkinson et al., 2017).

5.3.3 Biological Treatment

Biological treatment processes, such as activated sludge or biofilm reactors, can be optimized to enhance the removal of microplastics and nanoplastics. The microorganisms present in these systems can help break down and biodegrade organic matter, which may be associated with microplastics. This can contribute to the overall removal of plastic particles during the treatment process (Devi et al., 2022).

5.3.4 Sludge Management

Proper management of the sludge generated in the course of wastewater treatment process is essential to handle the MNPs' pollution. The sludge should undergo appropriate treatment and disposal procedures to prevent microplastic discharge back into the environment. Techniques such as sludge dewatering, composting, and incineration can be employed to ensure the proper handling of microplastics-containing sludge (Chen et al., 2022).

5.4 *Implementations of Regulations and Policy*

Governments can play a vital role in mitigating micro/nanoplastics pollution by implementation of regulations and policies aimed at restricting the utilization of certain types of plastics, encouraging the development of eco-friendly alternatives, and promoting sustainable manufacturing and packaging practices (Kumar et al., 2021). These measures can help create a framework for addressing plastic pollution at a systemic level. The explanation, regulations, and policies that can be effective in addressing this issue are summarized below.

5.4.1 Plastic Bans and Restrictions

Governments can introduce bans or restrictions on specific types of plastics that are major contributors to microplastic pollution, such as single-use plastics and tiny plastic particles in personal care products. By limiting the use and production of these items, governments can significantly minimize the influx of MPs into the environment (Kish, 2018).

5.4.2 Eco-Labeling and Certification

Government should establish labeling requirements that indicate the presence or absence of microplastics in products. This enables consumers to make informed choices and opt for products that are microplastic-free or have reduced microplastic content. Certification programs can also be implemented to verify and promote products that meet specific criteria for microplastics reduction (Eriksen et al., 2018).

5.5 *Public Awareness and Education*

Raising awareness about the impact of micro/nanoplastics pollution is crucial to drive behavioral change. Educational campaigns can inform the general public about the sources and impacts of plastic pollution. Promoting responsible consumption and waste management habits can significantly reduce plastic waste (Santti et al., 2020).

5.5.1 Informing About Sources and Consequences

Educational campaigns can assist people in understanding the causes of microplastic contamination, its effects on ecosystems, and how it affects human well-being. By providing clear and accessible information, people can become aware of how their actions contribute to the problem and the importance of taking preventive measures (Hammami et al., 2017).

5.5.2 Encouraging Lifestyle Changes

Awareness campaigns can inspire individuals to make conscious lifestyle changes that reduce plastic waste. This can include promoting the use of refillable water bottles, reusable shopping bags, and eco-friendly personal care products. By adopting these practices, individuals can significantly decrease their contribution to MP pollution (Heidbreder et al., 2019; Santti et al., 2020).

5.5.3 Engaging the Youth

Educating and engaging the youth is crucial for long-term behavioral change. Integrating environmental topics, including microplastic pollution, into school curricula can raise awareness among students and empower them to become advocates for sustainable practices. Youth-led initiatives and projects focused on plastic pollution can also have a significant impact on raising awareness within communities (Heidbreder et al., 2019; Santti et al., 2020).

5.5.4 Digital and Social Media Platforms

Leveraging online and digital communication platforms can enhance the reach and impact of awareness campaigns. Engaging content, infographics, videos, and interactive campaigns can effectively convey information about microplastic pollution, encouraging individuals to share and take action (Heidbreder et al., 2019; Santti et al., 2020).

5.6 Research and Innovation

Continued research and development are essential to gain deeper insights into the sources, behavior, and impacts of MNPs. Innovative technologies and materials can be developed to capture and remove microplastics from the environment. Additionally, exploring alternative materials that do not degrade into harmful particles can help prevent further pollution (Oliveira & Almeida, 2019).

5.7 International Cooperation

Since plastic pollution is a global issue, international cooperation is vital. Collaboration between countries can help establish standards and protocols for monitoring and mitigating MNP pollution. Sharing best practices, knowledge, and resources can lead to more effective solutions and coordinated global efforts. The collaboration between countries can contribute to mitigate this global issue by adopting the following measures.

5.7.1 Sharing Best Practices and Knowledge

The experiences can be shared at a global level which includes the best practices and research findings regarding monitoring, assessment, and mitigation of microplastic pollution. This collaboration can allow for the exchange of valuable information, lessons learned, and successful strategies, enabling countries to learn from one another and implement effective approaches (Williams & Rangel-Buitrago, 2022).

5.7.2 Promoting Global Advocacy and Awareness

International cooperation can amplify advocacy and awareness efforts by establishing global campaigns, events, and initiatives to raise awareness about microplastic pollution. By working together, countries can leverage their collective voice and reach to influence public opinion, engage stakeholders, and drive global action to reduce plastic pollution (Borrelle et al., 2017).

5.7.3 Strengthening Multilateral Agreements

Existing multilateral agreements, such as the Basel Convention and the Stockholm Convention, can be strengthened to address microplastic pollution comprehensively. By integrating microplastic considerations into these agreements, countries can foster greater cooperation, collaboration, and accountability in combating plastic pollution. By working together, countries can leverage their collective knowledge, resources, and influence to develop effective strategies, policies, and solutions (Simon et al., 2021). It is crucial to understand that reducing MNP pollution necessitates a thorough and long-term strategy. Combining these strategies with ongoing efforts to reduce plastic consumption and promote sustainability can help combat this environmental challenge.

6 Conclusion and Future Perspective

MNP pollution is becoming more of an issue due to its pervasiveness in the environment and its negative effects on the well-being of humans. MNPs have accumulated in the environment due to improper handling and usage of plastics, endangering the ecosystem and all living things. Recent studies revealed that both MPs and NPs were unintentionally ingested by people through their food and water. If microplastics remain in the digestive system, they may disrupt intestinal microbiota. The smaller nanoplastic accumulates in various parts of the body via penetrating into the blood and other tissues. Accumulation of MNPs in soil showed an adverse impact on soil productivity, microorganism fecundity, and hindered plant growth. As a result, controlling MNPs has become a big task in the modern world. Thus, research priority should be given to abate the MNP transfer and build-up in soil microbiota, agricultural plants, and food chain transfer. This can be achieved when plastic manufacturing, consumption, and disposal are governed by policies to reduce their uses. Essential prerequisites for promoting sustainability in plastic management include employing clean-up methods, recycling, implementing methods that minimize waste, increasing awareness and education about plastics, introducing renewable resources such as bioplastics, employing the three Rs: reuse, recycle, recover, implementing legislation and policies, and developing a long-term strategy. Additionally, providing public access to information about the health impacts of plastics is critical for raising awareness and encouraging the implementation of new policies aimed at reducing plastic pollution in the environment.

References

Abdel-Zaher, S., Mohamed, M. S., & Sayed, A. E. D. H. (2023). Hemotoxic effects of polyethylene microplastics on mice. *Frontiers in Physiology, 14*, 1072797.

Aghaei, Z., Sled, J. G., Kingdom, J. C., Baschat, A. A., Helm, P. A., Jobst, K. J., & Cahill, L. S. (2022). Maternal exposure to polystyrene micro-and nanoplastics causes fetal growth restriction in mice. *Environmental Science & Technology Letters, 9*(5), 426–430.

Akdogan, Z., & Guven, B. (2019). Microplastics in the environment: A critical review of current understanding and identification of future research needs. *Environmental Pollution, 254*, 113011.

Alhazmi, H., Almansour, F. H., & Aldhafeeri, Z. (2021). Plastic waste management: A review of existing life cycle assessment studies. *Sustainability, 13*(10), 5340.

Allen, S., Allen, D., Karbalaei, S., Maselli, V., & Walker, T. R. (2022). Micro (nano) plastics sources, fate, and effects: What we know after ten years of research. *Journal of Hazardous Materials Advances, 6*, 100057.

Amereh, F., Babaei, M., Eslami, A., Fazelipour, S., & Rafiee, M. (2020). The emerging risk of exposure to nano (micro) plastics on endocrine disturbance and reproductive toxicity: From a hypothetical scenario to a global public health challenge. *Environmental Pollution, 261*, 114158.

Annenkov, V. V., Danilovtseva, E. N., Zelinskiy, S. N., & Pal'shin, V. A. (2021). Submicro- and nanoplastics: How much can be expected in water bodies? *Environmental Pollution, 278*, 116910.

Babele, P. K., & Bhatia, R. K. (2023). Nanoplastics in agroecosystem and phytotoxicity: An evaluation of cytogenotoxicity and epigenetic regulation. *Impact of Engineered Nanomaterials in Genomics and Epigenomics*, 83–97.

Baensch-Baltruschat, B., Kocher, B., Stock, F., & Reifferscheid, G. (2020). Tyre and road wear particles (TRWP) – A review of generation, properties, emissions, human health risk, ecotoxicity, and fate in the environment. *Science of the Total Environment, 733*, 137823.

Barboza, L. G. A., Vieira, L. R., Branco, V., Figueiredo, N., Carvalho, F., Carvalho, C., & Guilhermino, L. (2018). Microplastics cause neurotoxicity, oxidative damage and energy-related changes and interact with the bioaccumulation of mercury in the European seabass, *Dicentrarchus labrax* (Linnaeus, 1758). *Aquatic Toxicology, 195*, 49–57.

Barboza, L. G. A., Lopes, C., Oliveira, P., Bessa, F., Otero, V., Henriques, B., et al. (2020). Microplastics in wild fish from north East Atlantic Ocean and its potential for causing neurotoxic effects, lipid oxidative damage, and human health risks associated with ingestion exposure. *Science of the Total Environment, 717*, 134625.

Bartkova, S., Kahru, A., Heinlaan, M., & Scheler, O. (2021). Techniques used for analyzing microplastics, antimicrobial resistance and microbial community composition: A mini-review. *Frontiers in Microbiology, 12*, 603967.

Bernatsky, S., Smargiassi, A., Barnabe, C., Svenson, L. W., Brand, A., Martin, R. V., Hudson, M., Clarke, A. E., Fortin, P. R., van Donkelaar, A., Edworthy, S., Bélisle, P., & Joseph, L. (2016). Fine particulate air pollution and systemic autoimmune rheumatic disease in two Canadian provinces. *Environmental Research, 146*, 85–91.

Bläsing, M., & Amelung, W. (2018). Plastics in soil: Analytical methods and possible sources. *Science of the Total Environment, 612*, 422–435.

Bonfanti, P., Colombo, A., Saibene, M., Motta, G., Saliu, F., Catelani, T., et al. (2021). Microplastics from miscellaneous plastic wastes: Physico-chemical characterization and impact on fish and amphibian development. *Ecotoxicology and Environmental Safety, 225*, 112775.

Boots, B., Russell, C. W., & Green, D. S. (2019). Effects of microplastics in soil ecosystems: Above and below ground. *Environmental Science & Technology, 53*(19), 11496–11506.

Borg, K., Lennox, A., Kaufman, S., Tull, F., Prime, R., Rogers, L., & Dunstan, E. (2022). Curbing plastic consumption: A review of single-use plastic behaviour change interventions. *Journal of Cleaner Production, 131077*, 131077.

Borrelle, S. B., Rochman, C. M., Liboiron, M., Bond, A. L., Lusher, A., Bradshaw, H., & Provencher, J. F. (2017). Why we need an international agreement on marine plastic pollution. *Proceedings of the National Academy of Sciences, 114*(38), 9994–9997.

Browne, M. A., Crump, P., Niven, S. J., Teuten, E., Tonkin, A., Galloway, T., & Thompson, R. (2011). Accumulation of microplastic on shorelines woldwide: Sources and sinks. *Environmental Science & Technology, 45*(21), 9175–9179.

Cao, D., Wang, X., Luo, X., Liu, G., & Zheng, H. (2017). Effects of polystyrene microplastics on the fitness of earthworms in an agricultural soil. In *IOP conference series: Earth and environmental science* (Vol. 61, No. 1, p. 12148). IOP Publishing.

Cárdenas-Alcaide, M. F., Godínez-Alemán, J. A., González-González, R. B., Iqbal, H. M., & Parra-Saldívar, R. (2022). Environmental impact and mitigation of micro (nano) plastics pollution using green catalytic tools and green analytical methods. *Green Analytical Chemistry, 3*, 100031.

Chen, Z., Wei, W., Liu, X., & Ni, B. J. (2022). Emerging electrochemical techniques for identifying and removing micro/nanoplastics in urban waters. *Water Research, 221*, 118846.

Chiu, H. W., Xia, T., Lee, Y. H., Chen, C. W., Tsai, J. C., & Wang, Y. J. (2015). Cationic polystyrene nanospheres induce autophagic cell death through the induction of endoplasmic reticulum stress. *Nanoscale, 7*(2), 736–746.

Conti, G. O., Ferrante, M., Banni, M., Favara, C., Nicolosi, I., Cristaldi, A., et al. (2020). Micro- and nano-plastics in edible fruit and vegetables. The first diet risks assessment for the general population. *Environmental Research, 187*, 109677.

Corradini, F., Meza, P., Eguiluz, R., Casado, F., Huerta-Lwanga, E., & Geissen, V. (2019). Evidence of microplastic accumulation in agricultural soils from sewage sludge disposal. *Science of the Total Environment, 671*, 411–420.

Crawford, C. B., & Quinn, B. (2016). Physiochemical properties and degradation. In *Microplastic pollutants* (pp. 57–100). Elsevier.

Crossman, J., Hurley, R. R., Futter, M., & Nizzetto, L. (2020). Transfer and transport of microplastics from biosolids to agricultural soils and the wider environment. *Science of the Total Environment, 724*, 138334.

Cui, W., Gao, P., Zhang, M., Wang, L., Sun, H., & Liu, C. (2022). Adverse effects of microplastics on earthworms: A critical review. *Science of the Total Environment, 850*, 158041.

da Silva Brito, W. A., Mutter, F., Wende, K., Cecchini, A. L., Schmidt, A., & Bekeschus, S. (2022). Consequences of nano and microplastic exposure in rodent models: The known and unknown. *Particle and Fibre Toxicology, 19*(1), 1–24.

de Souza, M. A. A., Kloas, W., Zarfl, C., Hempel, S., & Rillig, M. C. (2018a). Microplastics as an emerging threat to terrestrial ecosystems. *Global Change Biology, 24*(4), 1405–1416.

de Souza, M. A. A., Lau, C. W., Till, J., Kloas, W., Lehmann, A., Becker, R., & Rillig, M. C. (2018b). Impacts of microplastics on the soil biophysical environment. *Environmental Science & Technology, 52*(17), 9656–9665.

Dekant, W., & Völkel, W. (2008). Human exposure to bisphenol A by biomonitoring: Methods, results and assessment of environmental exposures. *Toxicology and Applied Pharmacology, 228*(1), 114–134.

Deng, Y., Zhang, Y., Lemos, B., & Ren, H. (2017). Tissue accumulation of microplastics in mice and biomarker responses suggest widespread health risks of exposure. *Scientific Reports, 7*(1), 46687.

Deng, Y., Zhang, Y., Qiao, R., Bonilla, M. M., Yang, X., Ren, H., & Lemos, B. (2018). Evidence that microplastics aggravate the toxicity of organophosphorus flame retardants in mice (Mus musculus). *Journal of Hazardous Materials, 357*, 348–354.

Devi, M. K., Karmegam, N., Manikandan, S., Subbaiya, R., Song, H., Kwon, E. E., & Govarthanan, M. (2022). Removal of nanoplastics in water treatment processes: A review. *Science of the Total Environment, 157168*.

Diggle, A., & Walker, T. R. (2020). Implementation of harmonized extended producer responsibility strategies to incentivize recovery of single-use plastic packaging waste in Canada. *Waste Management, 110*, 20–23.

Ding, L., Huang, D., Ouyang, Z., & Guo, X. (2022). The effects of microplastics on soil ecosystem: A review. *Current Opinion in Environmental Science & Health, 100344*, 100344.

Dong, X., Liu, X., Hou, Q., & Wang, Z. (2023). From natural environment to animal tissues: A review of microplastics (nanoplastics) translocation and hazards studies. *Science of the Total Environment, 855*, 158686.

Dris, R., Gasperi, J., Saad, M., Mirande, C., & Tassin, B. (2016). Synthetic fibers in atmospheric fallout: A source of microplastics in the environment? *Marine Pollution Bulletin, 104*(1–2), 290–293.

Dubey, I., Khan, S., & Kushwaha, S. (2022). Developmental and reproductive toxic effects of exposure to microplastics: A review of associated signaling pathways. *Frontiers in Toxicology, 4*, 901798.

EFSA Panel on Contaminants in the Food Chain (CONTAM). (2016). Presence of microplastics and nanoplastics in food, with particular focus on seafood. *EFSA Journal, 14*(6), e04501. https://doi.org/10.2903/j.efsa.2016.4501

Eriksen, M., Thiel, M., Prindiville, M., & Kiessling, T. (2018). Microplastic: What are the solutions. *Freshwater Microplastics*, 273.

Evangeliou, N., Grythe, H., Klimont, Z., Heyes, C., Eckhardt, S., Lopez-Aparicio, S., & Stohl, A. (2020). Atmospheric transport is a major pathway of microplastics to remote regions. *Nature Communications, 11*(1), 3381.

Eyles, J. E., Bramwell, V. W., Williamson, E. D., & Alpar, H. O. (2001). Microsphere translocation and immunopotentiation in systemic tissues following intranasal administration. *Vaccine, 19*(32), 4732–4742.

Faisal, F., Rasul, M. G., Jahirul, M. I., & Schaller, D. (2023). Pyrolytic conversion of waste plastics to energy products: A review on yields, properties, and production costs. *Science of the Total Environment, 861*, 160721.

Fendall, L. S., & Sewell, M. A. (2009). Contributing to marine pollution by washing your face: Microplastics in facial cleansers. *Marine Pollution Bulletin, 58*(8), 1225–1228.

Feng, L. J., Li, J. W., Xu, E. G., Sun, X. D., Zhu, F. P., Ding, Z., Xia, P. F., & Yuan, X. Z. (2019). Short-term exposure to positively charged polystyrene nanoparticles causes oxidative stress and membrane destruction in cyanobacteria. *Environmental Science: Nano, 6*(10), 3072–3079.

Fernandes, E. C., Silva, C. A., Braga, A. L., Sallum, A. M., Campos, L. M., & Farhat, S. C. (2015). Exposure to air pollutants and disease activity in juvenile-onset systemic lupus erythematosus patients. *Arthritis Care & Research, 67*(11), 1609–1614.

Ferrante, M. C., Monnolo, A., Del Piano, F., Mattace Raso, G., & Meli, R. (2022). The pressing issue of micro-and nanoplastic contamination: Profiling the reproductive alterations mediated by oxidative stress. *Antioxidants, 11*(2), 193.

Gao, H., Yang, B. J., Li, N., Feng, L. M., Shi, X. Y., Zhao, W. H., & Liu, S. J. (2015). Bisphenol A and hormone-associated cancers: Current progress and perspectives. *Medicine, 94*(1), e211.

Gruber, E. S., Stadlbauer, V., Pichler, V., Resch-Fauster, K., Todorovic, A., Meisel, T. C., et al. (2023). To waste or not to waste: Questioning potential health risks of micro-and nanoplastics with a focus on their ingestion and potential carcinogenicity. *Exposure and Health, 15*(1), 33–51.

Gündoğdu, S., Çevik, C., Güzel, E., & Kilercioğlu, S. (2018). Microplastics in municipal wastewater treatment plants in Turkey: A comparison of the influent and secondary effluent concentrations. *Environmental Monitoring and Assessment, 190*, 1–10.

Hammami, M. B. A., Mohammed, E. Q., Hashem, A. M., Al-Khafaji, M. A., Alqahtani, F., Alzaabi, S., & Dash, N. (2017). Survey on awareness and attitudes of secondary school students regarding plastic pollution: Implications for environmental education and public health in Sharjah city, UAE. *Environmental Science and Pollution Research, 24*, 20626–20633.

He, D., & Luo, Y. (2020). Microplastics in terrestrial environments emerging contaminants and major challenges. In *Emerging contaminants and major challenges*. Springer Nature Press.

He, D., Luo, Y., Lu, S., Liu, M., Song, Y., & Lei, L. (2018). Microplastics in soils: Analytical methods, pollution characteristics and ecological risks. *TrAC Trends in Analytical Chemistry, 109*, 163–172.

Heidbreder, L. M., Bablok, I., Drews, S., & Menzel, C. (2019). Tackling the plastic problem: A review on perceptions, behaviors, and interventions. *Science of the Total Environment, 668*, 1077–1093.

Horton, A. A., Walton, A., Spurgeon, D. J., Lahive, E., & Svendsen, C. (2017). Microplastics in freshwater and terrestrial environments: Evaluating the current understanding to identify the knowledge gaps and future research priorities. *Science of the Total Environment, 586*, 127–141.

Huang, D., Wang, X., Yin, L., Chen, S., Tao, J., Zhou, W., et al. (2022). Research progress of microplastics in soil-plant system: Ecological effects and potential risks. *Science of the Total Environment, 812*, 151487.

Hwang, J., Choi, D., Han, S., Choi, J., & Hong, J. (2019). An assessment of the toxicity of polypropylene microplastics in human derived cells. *Science of the Total Environment, 684*, 657–669.

Jani, P., Halbert, G. W., Langridge, J., & Florence, A. T. (1990). Nanoparticle uptake by the rat gastrointestinal mucosa: Quantitation and particle size dependency. *Journal of Pharmacy and Pharmacology, 42*(12), 821–826.

Jatana, S., Callahan, L. M., Pentland, A. P., & DeLouise, L. A. (2016). Impact of cosmetic lotions on nanoparticle penetration through ex vivo C57BL/6 hairless mouse and human skin: A comparison study. *Cosmetics, 3*(1), 6.

Jiang, X., Chang, Y., Zhang, T., Qiao, Y., Klobučar, G., & Li, M. (2020). Toxicological effects of polystyrene microplastics on earthworm (Eisenia fetida). *Environmental Pollution, 259*, 113896.

Jiang, J. J., Hanun, J. N., Chen, K. Y., Hassan, F., Liu, K. T., Hung, Y. H., & Chang, T. W. (2023). Current levels and composition profiles of microplastics in irrigation water. *Environmental Pollution, 318*, 120858.

Kacprzak, S., & Tijing, L. D. (2022). Microplastics in indoor environment: Sources, mitigation and fate. *Journal of Environmental Chemical Engineering, 107359*, 107359.

Kasirajan, S., & Ngouajio, M. (2012). Polyethylene and biodegradable mulches for agricultural applications: A review. *Agronomy for Sustainable Development, 32*, 501–529.

Kelly, F. J., & Fussell, J. C. (2012). Size, source and chemical composition as determinants of toxicity attributable to ambient particulate matter. *Atmospheric Environment, 60*, 504–526.

Khalid, N., Aqeel, M., & Noman, A. (2020). Microplastics could be a threat to plants in terrestrial systems directly or indirectly. *Environmental Pollution, 267*, 115653.

Khan, F. R., Halle, L. L., & Palmqvist, A. (2019). Acute and long-term toxicity of micronized car tire wear particles to Hyalella azteca. *Aquatic Toxicology, 213*, 105216.

Kiran, B. R., Kopperi, H., & Venkata Mohan, S. (2022). Micro/nano-plastics occurrence, identification, risk analysis and mitigation: Challenges and perspectives. *Reviews in Environmental Science and Bio/Technology, 21*(1), 169–203.

Kish, R. J. (2018). Using legislation to reduce one-time plastic bag usage. *Economic Affairs, 38*(2), 224–239.

Kole, P. J., Löhr, A. J., Van Belleghem, F. G., & Ragas, A. M. (2017). Wear and tear of tyres: A stealthy source of microplastics in the environment. *International Journal of Environmental Research and Public Health, 14*(10), 1265.

Kosuth, M., Mason, S. A., & Wattenberg, E. V. (2018). Anthropogenic contamination of tap water, beer, and sea salt. *PLoS One, 13*(4), e0194970.

Koutnik, V. S., Alkidim, S., Leonard, J., DePrima, F., Cao, S., Hoek, E. M., & Mohanty, S. K. (2021). Unaccounted microplastics in wastewater sludge: Where do they go? *ACS ES&T Water, 1*(5), 1086–1097.

Kumar, R., Verma, A., Shome, A., Sinha, R., Sinha, S., Jha, P. K., & Vara Prasad, P. V. (2021). Impacts of plastic pollution on ecosystem services, sustainable development goals, and need to focus on circular economy and policy interventions. *Sustainability, 13*(17), 9963.

Kumar, R., Ivy, N., Bhattacharya, S., Dey, A., & Sharma, P. (2022). Coupled effects of microplastics and heavy metals on plants: Uptake, bioaccumulation, and environmental health perspectives. *Science of the Total Environment, 836*, 155619.

Kumar, V., Singh, E., Singh, S., Pandey, A., & Bhargava, P. C. (2023). Micro-and nano-plastics (MNPs) as emerging pollutant in ground water: Environmental impact, potential risks, limitations and way forward towards sustainable management. *Chemical Engineering Journal, 459*, 141568.

Lambert, S., Scherer, C., & Wagner, M. (2017). Ecotoxicity testing of microplastics: Considering the heterogeneity of physicochemical properties. *Integrated Environmental Assessment and Management, 13*(3), 470–475.

Lei, L., Liu, M., Song, Y., Lu, S., Hu, J., Cao, C., Xie, B., Shi, H., & He, D. (2018). Polystyrene (nano) microplastics cause size-dependent neurotoxicity, oxidative damage and other adverse effects in *Caenorhabditis elegans*. *Environmental Science: Nano, 5*(8), 2009–2020.

Leonov, V. D., & Tiunov, A. V. (2020). Interaction of invertebrates and synthetic polymers in soil: A review. *Russian Journal of Ecology, 51*, 503–517.

Li, X., Chen, L., Mei, Q., Dong, B., Dai, X., Ding, G., & Zeng, E. Y. (2018). Microplastics in sewage sludge from the wastewater treatment plants in China. *Water Research, 142*, 75–85.

Li, B., Ding, Y., Cheng, X., Sheng, D., Xu, Z., Rong, Q., et al. (2020). Polyethylene microplastics affect the distribution of gut microbiota and inflammation development in mice. *Chemosphere, 244*, 125492.

Li, B., Huang, S., Wang, H., Liu, M., Xue, S., Tang, D., et al. (2021). Effects of plastic particles on germination and growth of soybean (Glycine max): A pot experiment under field condition. *Environmental Pollution, 272*, 116418.

Liu, Y., Cui, W., Li, W., Xu, S., Sun, Y., Xu, G., & Wang, F. (2023). Effects of microplastics on cadmium accumulation by rice and arbuscular mycorrhizal fungal communities in cadmium-contaminated soil. *Journal of Hazardous Materials, 442*, 130102.

Lv, W., Zhou, W., Lu, S., Huang, W., Yuan, Q., Tian, M., & He, D. (2019). Microplastic pollution in rice-fish co-culture system: A report of three farmland stations in Shanghai, China. *Science of the Total Environment, 652*, 1209–1218.

Maity, S., Guchhait, R., Sarkar, M. B., & Pramanick, K. (2022). Occurrence and distribution of micro/nanoplastics in soils and their phytotoxic effects: A review. *Plant, Cell & Environment, 45*(4), 1011–1028.

Maity, S., Guchhait, R., De, S., & Pramanick, K. (2023). High doses of nano-polystyrene aggravate the oxidative stress, DNA damage, and the cell death in onions. *Environmental Pollution, 316*, 120611.

Mason, S. A., Welch, V. G., & Neratko, J. (2018). Synthetic polymer contamination in bottled water. *Frontiers in Chemistry, 407*.

Mateos-Cárdenas, A., van Pelt, F. N., O'Halloran, J., & Jansen, M. A. (2021). Adsorption, uptake and toxicity of micro-and nanoplastics: Effects on terrestrial plants and aquatic macrophytes. *Environmental Pollution, 284*, 117183.

Mattsson, K., Johnson, E. V., Malmendal, A., Linse, S., Hansson, L. A., & Cedervall, T. (2017). Brain damage and behavioural disorders in fish induced by plastic nanoparticles delivered through the food chain. *Scientific Reports, 7*(1), 11452.

McGivney, E., Cederholm, L., Barth, A., Hakkarainen, M., Hamacher-Barth, E., Ogonowski, M., & Gorokhova, E. (2020). Rapid physicochemical changes in microplastic induced by biofilm formation. *Frontiers in Bioengineering and Biotechnology, 8*, 205.

Meegoda, J. N., & Hettiarachchi, M. C. (2023). A path to a reduction in micro and nanoplastics pollution. *International Journal of Environmental Research and Public Health, 20*(8), 5555.

Mintenig, S. M., Int-Veen, I., Löder, M. G., Primpke, S., & Gerdts, G. (2017). Identification of microplastic in effluents of waste water treatment plants using focal plane array-based micro-Fourier-transform infrared imaging. *Water Research, 108*, 365–372.

Murray, A., & Ormeci, B. (2020). Removal effectiveness of nanoplastics (<400 nm) with separation processes used for water and wastewater treatment. *Water, 12*(3), 635.

Naeini, S. M., & Cook, H. F. (2000). Influence of municipal waste compost amendment on soil water and evaporation. *Communications in Soil Science and Plant Analysis, 31*(19–20), 3147–3161.

Newbold, R. R., Jefferson, W. N., & Padilla-Banks, E. (2007). Longterm adverse effects of neonatal exposure to bisphenol A on the murine female reproductive tract. *Reproductive Toxicology, 24*(2), 253–258.

Okoye, C. O., Addey, C. I., Oderinde, O., Okoro, J. O., Uwamungu, J. Y., Ikechukwu, C. K., et al. (2022). Toxic chemicals and persistent organic pollutants associated with micro-and nanoplastics pollution. *Chemical Engineering Journal Advances, 11*, 100310.

Oliveira, M., & Almeida, M. (2019). The why and how of micro (nano) plastic research. *TrAC Trends in Analytical Chemistry, 114*, 196–201.

Osman, A. I., Hosny, M., Eltaweil, A. S., Omar, S., Elgarahy, A. M., Farghali, M., et al. (2023). Microplastic sources, formation, toxicity and remediation: A review. *Environmental Chemistry Letters, 21*(4), 2129–2169.

Patil, S., Bafana, A., Naoghare, P. K., Krishnamurthi, K., & Sivanesan, S. (2021). Environmental prevalence, fate, impacts, and mitigation of microplastics—A critical review on present understanding and future research scope. *Environmental Science and Pollution Research, 28*(5), 4951–4974.

Peixoto, D., Pinheiro, C., Amorim, J., Oliva-Teles, L., Guilhermino, L., & Vieira, M. N. (2019). Microplastic pollution in commercial salt for human consumption: A review. *Estuarine, Coastal and Shelf Science, 219*, 161–168.

Peng, L., Fu, D., Qi, H., Lan, C. Q., Yu, H., & Ge, C. (2020). Micro-and nano-plastics in marine environment: Source, distribution and threats—A review. *Science of the Total Environment, 698*, 134254.

Pérez-Reverón, R., Álvarez-Méndez, S. J., González-Sálamo, J., Socas-Hernández, C., Díaz-Peña, F. J., Hernández-Sánchez, C., & Hernández-Borges, J. (2022). Nanoplastics in the soil environment: Analytical methods, occurrence, fate and ecological implications. *Environmental Pollution, 120788*.

Perković, S., Paul, C., Vasić, F., & Helming, K. (2022). Human health and soil health risks from heavy metals, micro (nano) plastics, and antibiotic resistant bacteria in agricultural soils. *Agronomy, 12*(12), 2945.

Prata, J. C., da Costa, J. P., Lopes, I., Duarte, A. C., & Rocha-Santos, T. (2020). Environmental exposure to microplastics: An overview on possible human health effects. *Science of the Total Environment, 702*, 134455.

Quang, H. H. P., Dinh, D. A., Dutta, V., Chauhan, A., Lahiri, S. K., Gopalakrishnan, C., & Thi, L. A. P. (2023). Current approaches, and challenges on identification, remediation and potential risks of emerging plastic contaminants: A review. *Environmental Toxicology and Pharmacology, 104193*, 104193.

Rashid, C. P., Jyothibabu, R., Arunpandi, N., Abhijith, V. T., Josna, M. P., Vidhya, V., et al. (2021). Microplastics in zooplankton in the eastern Arabian Sea: The threats they pose to fish and corals favoured by coastal currents. *Marine Pollution Bulletin, 173*, 113042.

Rillig, M. C. (2012). *Microplastic in terrestrial ecosystems and the soil?*.6453–6454.

Rillig, M. C. (2020). Plastic and plants. *Nature Sustainability, 3*, 887–888.

Rillig, M. C., & Lehmann, A. (2020). Microplastic in terrestrial ecosystems. *Science, 368*(6498), 1430–1431.

Rillig, M. C., de Souza Machado, A. A., Lehmann, A., & Klümper, U. (2018). Evolutionary implications of microplastics for soil biota. *Environmental Chemistry, 16*(1), 3–7.

Rolsky, C., Kelkar, V., Driver, E., & Halden, R. U. (2020). Municipal sewage sludge as a source of microplastics in the environment. *Current Opinion in Environmental Science & Health, 14*, 16–22.

Sa'adu, I., & Farsang, A. (2023). Plastic contamination in agricultural soils: A review. *Environmental Sciences Europe, 35*(1), 13.

Sa'adu, I., & Farsang, A. (2022). Greenhouse farming as a source of macroplastic and microplastics contamination in agricultural soils: A case study from Southeast-Hungary. *Agrochemistry & Soil Science/Agrokémia és Talajtan, 71*(1), 43–57.

Santti, U., Happonen, A., & Auvinen, H. (2020). Digitalization boosted recycling: Gamification as an inspiration for young adults to do enhanced waste sorting. In *AIP conference proceedings* (Vol. 2233, No. 1, p. 050014). AIP Publishing LLC.

Schirinzi, G. F., Pérez-Pomeda, I., Sanchís, J., Rossini, C., Farré, M., & Barceló, D. (2017). Cytotoxic effects of commonly used nanomaterials and microplastics on cerebral and epithelial human cells. *Environmental Research, 159*, 579–587.

Schothorst, B. V., Beriot, N., Huerta Lwanga, E., & Geissen, V. (2021). Sources of light density microplastic related to two agricultural practices: The use of compost and plastic mulch. *Environments, 8*(4), 36.

Schwabl, P., Köppel, S., Königshofer, P., Bucsics, T., Trauner, M., Reiberger, T., & Liebmann, B. (2019). Detection of various microplastics in human stool: A prospective case series. *Annals of Internal Medicine, 171*(7), 453–457.

Shafei, A., Ramzy, M. M., Hegazy, A. I., Husseny, A. K., El-Hadary, U. G., Taha, M. M., & Mosa, A. A. (2018). The molecular mechanisms of action of the endocrine disrupting chemical bisphenol A in the development of cancer. *Gene, 647*, 235–243.

Shen, M., Zhang, Y., Zhu, Y., Song, B., Zeng, G., Hu, D., Wen, X., & Ren, X. (2019). Recent advances in toxicological research of nanoplastics in the environment: A review. *Environmental Pollution, 252*, 511–521.

Shi, Y., Chai, J., Xu, T., Ding, L., Huang, M., Gan, F., & Yang, J. (2023). Microplastics contamination associated with low-value domestic source organic solid waste: A review. *Science of the Total Environment, 857*, 159679.

Simon, N., Raubenheimer, K., Urho, N., Unger, S., Azoulay, D., Farrelly, T., & Weiand, L. (2021). A binding global agreement to address the life cycle of plastics. *Science, 373*(6550), 43–47.

Sun, J., Dai, X., Wang, Q., Van Loosdrecht, M. C., & Ni, B. J. (2019). Microplastics in wastewater treatment plants: Detection, occurrence and removal. *Water Research, 152*, 21–37.

Syberg, K., Khan, F. R., Selck, H., Palmqvist, A., Banta, G. T., Daley, J., et al. (2015). Microplastics: addressing ecological risk through lessons learned. *Environmental Toxicology and Chemistry, 34*(5), 945–953.

Šourková, M., Adamcová, D., & Vaverková, M. D. (2021). The influence of microplastics from ground tyres on the acute, subchronical toxicity and microbial respiration of soil. *Environments, 8*(11), 128.

Tang, G., Qiao, W., Wang, Z., Liu, F., He, L., Liu, M., & Liu, C. (2023). *Waste plastic to energy storage materials: A state-of-the-art review* (Vol. 25, p. 3738). Green Chemistry.

Valavanidis, A., Vlachogianni, T., Fiotakis, K., & Loridas, S. (2013). Pulmonary oxidative stress, inflammation and cancer: Respirable particulate matter, fibrous dusts and ozone as major causes of lung carcinogenesis through reactive oxygen species mechanisms. *International Journal of Environmental Research and Public Health, 10*(9), 3886–3907.

Verla, A. W., Enyoh, C. E., Verla, E. N., & Nwarnorh, K. O. (2019). Microplastic–toxic chemical interaction: A review study on quantified levels, mechanism and implication. *SN Applied Sciences, 1*(11), 1–30.

Wagner, T. P. (2017). Reducing single-use plastic shopping bags in the USA. *Waste Management, 70*, 3–12.

Walker, T. R. (2021). (micro) plastics and the UN sustainable development goals. *Current Opinion in Green and Sustainable Chemistry, 30*, 100497.

Wang, C., Zhang, Y., Deng, M., Wang, X., Tu, W., Fu, Z., & Jin, Y. (2019). Bioaccumulation in the gut and liver causes gut barrier dysfunction and hepatic metabolism disorder in mice after exposure to low doses of obs. *Environment International, 129*, 279–290.

Wang, L., Wu, W. M., Bolan, N. S., Tsang, D. C., Li, Y., Qin, M., & Hou, D. (2021). Environmental fate, toxicity and risk management strategies of nanoplastics in the environment: Current status and future perspectives. *Journal of Hazardous Materials, 401*, 123415.

Wang, F., Feng, X., Liu, Y., Adams, C. A., Sun, Y., & Zhang, S. (2022). Micro (nano) plastics and terrestrial plants: Up-to-date knowledge on uptake, translocation, and phytotoxicity. *Resources, Conservation and Recycling, 185*, 106503.

Watts, A. J., Urbina, M. A., Corr, S., Lewis, C., & Galloway, T. S. (2015). Ingestion of plastic microfibers by the crab *Carcinus maenas* and its effect on food consumption and energy balance. *Environmental Science & Technology, 49*(24), 14597–14604.

Weithmann, N., Möller, J. N., Löder, M. G., Piehl, S., Laforsch, C., & Freitag, R. (2018). Organic fertilizer as a vehicle for the entry of microplastic into the environment. *Science Advances, 4*(4), eaap8060.

Wen, B., Zhang, N., Jin, S. R., Chen, Z. Z., Gao, J. Z., Liu, Y., Liu, H. P., & Xu, Z. (2018). Microplastics have a more profound impact than elevated temperatures on the predatory performance, digestion and energy metabolism of an Amazonian cichlid. *Aquatic Toxicology, 195*, 67–76.

Wick, P., Malek, A., Manser, P., Meili, D., Maeder-Althaus, X., Diener, L., & von Mandach, U. (2010). Barrier capacity of human placenta for nanosized materials. *Environmental Health Perspectives, 118*(3), 432–436.

Wilkinson, J., Hooda, P. S., Barker, J., Barton, S., & Swinden, J. (2017). Occurrence, fate and transformation of emerging contaminants in water: An overarching review of the field. *Environmental Pollution, 231*, 954–970.

Williams, A. T., & Rangel-Buitrago, N. (2022). The past, present, and future of plastic pollution. *Marine Pollution Bulletin, 176*, 113429.

Wright, S. L., Rowe, D., Thompson, R. C., & Galloway, T. S. (2013). Microplastic ingestion decreases energy reserves in marine worms. *Current Biology, 23*(23), R1031–R1033.

Xu, M., Halimu, G., Zhang, Q., Song, Y., Fu, X., Li, Y., & Zhang, H. (2019). Internalization and toxicity: A preliminary study of effects of nanoplastic particles on human lung epithelial cell. *Science of the Total Environment, 694*, 133794.

Xu, C., Zhang, B., Gu, C., Shen, C., Yin, S., Aamir, M., & Li, F. (2020). Are we underestimating the sources of microplastic pollution in terrestrial environment? *Journal of Hazardous Materials, 400*, 123228.

Yee, M. S. L., Hii, L. W., Looi, C. K., Lim, W. M., Wong, S. F., Kok, Y. Y., & Leong, C. O. (2021). Impact of microplastics and nanoplastics on human health. *Nanomaterials, 11*(2), 496.

Yu, J., Adingo, S., Liu, X., Li, X., Sun, J., & Zhang, X. (2022). Micro plastics in soil ecosystem – A review of sources, fate, and ecological impact. *Plant, Soil and Environment, 68*(1), 1–17.

Zhang, Q., Xu, E. G., Li, J., Chen, Q., Ma, L., Zeng, E. Y., & Shi, H. (2020). A review of microplastics in table salt, drinking water, and air: Direct human exposure. *Environmental Science & Technology, 54*(7), 3740–3751.

Zhang, M., Zhang, Y., Wang, W., Cui, W., Wang, L., Sun, H., & Liu, C. (2022). Combined effects of microplastics and other contaminants on earthworms: A critical review. *Applied Soil Ecology, 180*, 104626.

Zhao, L., Qu, M., Wong, G., & Wang, D. (2017). Transgenerational toxicity of nanopolystyrene particles in the range of μg L− 1 in the nematode Caenorhabditis elegans. *Environmental Science: Nano, 4*(12), 2356–2366.

Zhao, T., Lozano, Y. M., & Rillig, M. C. (2021). Microplastics increase soil pH and decrease microbial activities as a function of microplastic shape, polymer type, and exposure time. *Frontiers in Environmental Science, 9*, 675803.

Zhao, Y., Chen, J., Wang, R., Pu, X., & Wang, D. (2023). A review of transgenerational and multigenerational toxicology in the in vivo model animal Caenorhabditis elegans. *Journal of Applied Toxicology, 43*(1), 122–145.

Zheng, X., Liu, X., Zhang, L., Wang, Z., Yuan, Y., Li, J., et al. (2022). Toxicity mechanism of nylon microplastics on Microcystis aeruginosa through three pathways: Photosynthesis, oxidative stress and energy metabolism. *Journal of Hazardous Materials, 426*, 128094.

Zhou, Y., Wang, J., Zou, M., Jia, Z., Zhou, S., & Li, Y. (2020). Microplastics in soils: A review of methods, occurrence, fate, transport, ecological and environmental risks. *Science of the Total Environment, 748*, 141368.

Ziajahromi, S., Neale, P. A., Silveira, I. T., Chua, A., & Leusch, F. D. (2021). An audit of microplastic abundance throughout three Australian wastewater treatment plants. *Chemosphere, 263*, 128294.

Zolotova, N., Kosyreva, A., Dzhalilova, D., Fokichev, N., & Makarova, O. (2022). Harmful effects of the microplastic pollution on animal health: A literature review. *PeerJ, 10*, e13503.

Zubris, K. A. V., & Richards, B. K. (2005). Synthetic fibers as an indicator of land application of sludge. *Environmental Pollution, 138*(2), 201–211.

Adsorption Behavior and Interaction of Micro-Nanoplastics in Soils and Aquatic Environment

Ajay Valiyaveettil Salimkumar, Mary Carolin Kurisingal Cleetus, Judith Osaretin Ehigie, Cyril Oziegbe Onogbosele, P. Nisha, Bindhi S. Kumar, M. P. Prabhakaran, and V. J. Rejish Kumar

1 Introduction

Micro-nanoplastics have emerged as ubiquitous pollutants in both terrestrial and aquatic ecosystems, garnering extensive research attention due to their potential to adsorb a plethora of environmental contaminants. These diminutive derivatives of larger plastic debris, either fragmented by natural processes or intentionally engineered for industrial applications, display a remarkable ability to interact with contaminants in the environment. This behavior is contingent upon a complex interplay of physicochemical properties of micro-nanoplastics and extrinsic factors or environmental conditions, making the understanding of their adsorption

A. Valiyaveettil Salimkumar
Faculty of Science and Technology, Research Centre for Experimental Marine Biology and Biotechnology (PIE-UPV/EHU), Plentzia-Biskaia, Basque Country, Spain

M. C. Kurisingal Cleetus
Faculty of Science and Technology, Research Centre for Experimental Marine Biology and Biotechnology (PIE-UPV/EHU), Plentzia-Biskaia, Basque Country, Spain

Faculty of Sciences, University of Liege, Liege, Belgium

J. O. Ehigie
Faculty of Science and Technology, Research Centre for Experimental Marine Biology and Biotechnology (PIE-UPV/EHU), Plentzia-Biskaia, Basque Country, Spain

Helmholtz Centre for Environmental Research – UFZ, Leipzig, Germany

C. O. Onogbosele
Department of Zoology, Faculty of Life Sciences, Ambrose Alli University, Ekpoma, Edo State, Nigeria

P. Nisha
Department of Biosciences, MES College, Ernakulam, Kerala, India

S. A. Bhat et al. (eds.), *Management of Micro and Nano-plastics in Soil and Biosolids*,
https://doi.org/10.1007/978-3-031-51967-3_11

mechanisms critical for predicting the environmental fate and transport of associated pollutants. As micro-nano particles offer avenues for contaminant sorption, their transport across various environments and ecosystems, can significantly alter the bioavailability and fate of associated pollutants, impacting organisms at various trophic levels and even posing potential health risks to humans. This chapter elucidates the physicochemical properties and adsorptive behavior of micro and nanoplastics, it also highlights the crucial role they play in the environmental mobilization of organic micropollutants, the challenges it poses, and the imperative for a more holistic understanding to inform mitigation strategies.

1.1 Definition and Properties of Micro and Nanoplastics

Plastics have an indefinite shelf life. However, mechanisms such as thermal breakdown, hydrolysis, physical wear and tear, thermos-oxidative degradation, light-induced degradation, and microbial breakdown (Mattsson et al., 2015) reduce and change the structural integrity, thereby degrading them into smaller pieces, fragments, or even to molecular components, making it a perpetual process and leaving a prominent presence in water, food, soil, and air. Based on the particle size, plastic debris can be categorized as macroplastics (>25 mm), mesoplastics (5–25 mm), microplastics (1 μm–5 mm), and nanoplastics (<1 μm) (Verschoor 2015).

Microplastics encompass artificial solid particles or polymer-based materials stemming from either intentional manufacturing (primary) or from the breakdown of larger plastics (secondary), possessing consistent or irregular forms and a size spanning from 1 μm to 5 mm, and remaining insoluble in water (Frias & Nash, 2019). The term "microplastics" was initially coined by Thompson et al., 2004, while the lower size limit of microplastics was fixed by the mesh size cutoff of the manta trawl net utilized for sampling microplastics in the environment (Gigault et al., 2018), while the upper size limit was defined by Arthur et al., 2009. Microplastics are found to be diverse when it comes to their elemental composition, size, color, and shape, as primary microplastics are deliberately produced in small

B. S. Kumar
Research Department of Fisheries and Aquaculture, St. Albert's College (Autonomous), Kochi, Kerala, India

M. P. Prabhakaran
Department of Aquatic Environment and Management, Kerala University of Fisheries and Ocean Studies, Kochi, Kerala, India

V. J. Rejish Kumar (✉)
Faculty of Ocean Science and Technology, Kerala University of Fisheries and Ocean Studies, Kochi, Kerala, India

Department of Aquaculture, Kerala University of Fisheries and Ocean Studies, Kochi, Kerala, India
e-mail: rejish.kumar@kufos.ac.in

sizes for uses in cosmetics, personal care items, industrial processes, textile industries, synthetic fabric production, air blasting, and as delivery agents in medicines. In contrast, secondary microplastics originate due to weathering and fragmentation of larger plastic items making them the predominant contributors to plastic pollution (GESAMP, 2015). Primary microplastics, on the other hand, account for 15–31% of global plastic pollution (Yu et al., 2021b). Over time, primary microplastics can undergo additional degradation, leading to the formation of even smaller particles known as nanoplastics. Nanoplastics are particles typically measuring between 1 and 1000 nm, which form as a result of the breakdown of industrial plastic objects. These particles can display colloidal behavior, such as remaining suspended and dispersed in various liquids or environments (Gigault et al., 2018). The proposed definition of nanoplastics considers only the degradation products, not the intentionally created or engineered materials like polystyrene latex beads. Nanoplastics exhibit significant variation in their physical properties and possess a diverse composition as they originate inadvertently from the breakdown of microplastic debris. Nanoplastic particles are recognized for their expansive surface area in relation to their volume, pronounced surface curvature, and enhanced surface reactivity (Mattsson et al., 2015), while microplastics are characterized by their extensive surface area relative to their size, high porosity levels, amorphous or noncrystalline structure, and also demonstrate hydrophobic properties (Wan et al., 2019).

1.2 Sources and Distribution of Microplastics and Nanoplastics in Soils and Aquatic Environments

The emission of microplastics and nanoplastics per individual differs greatly across regions. This disparity is shaped by factors such as population size, wealth levels, and the availability and efficiency of waste management systems (Ng et al., 2018). Micro and nanoparticles are consistently discharged into the terrestrial and aquatic environment through municipal, commercial, industrial, agricultural, construction, demolition waste, atmospheric deposition, wear and tear particles from vehicles and machinery, ship-generated litter, recreational activities, fishing fleets (gear), etc. As these particles are dispersed, they accumulate in soils, sediments, and water bodies and they can be transported to seas and oceans through wind and water runoff. The movement of microparticles differs according to the specific kind of microplastic, considering their unique basic attributes such as density, contact angle, and dielectric constant. The density of microplastic plays an important role in its diffusion and sedimentation rates, while contact angle can influence the interparticle forces between the microplastic and the indigenous particles in the environment (Koutnik et al., 2021b). These characteristics play a significant role in how these microplastics disperse and interact with their surroundings.

Nanoplastics are employed in applications such as water-based paints, surface finishes/coatings, glues, medical products, medication delivery systems, medical diagnostics, electronics, magnetic devices, optoelectronic tools, dispersible lattices, and academic research. Production methods, including the thermal slicing of plastic foam and 3D printing, are identified as sources of nanoplastics release (Hernandez et al., 2017). The spread of microplastics across land-based environments and bodies of water is influenced by several processes. These include advection (the transport by fluid flow), dispersion (the spreading of particles), diffusion (the movement from areas of high to low concentration), degradation (breakdown into smaller components), settling (sinking due to gravity), adsorption (attachment to surfaces), and aggregation (clustering of particles together) (Koutnik et al., 2021b).

1.3 Importance of Understanding the Adsorption Behavior and Interaction of MNPs

The widespread occurrence of microplastics in various ecosystems, food chains, and even in vertebrates and invertebrates is extensively documented (Wang et al., 2020b). Microplastics can affect soil properties. It affects soil enzymatic activity through direct adsorption (Yang et al., 2022). The dimensions and density of microplastics are similar to those of the minerals absorbed by the plants through mechanical transport, the microplastics can enter the cell (Smith 2018). Microplastics have been discovered within a wide array of living organisms across all levels of the food chain. This includes algae, zooplankton, lugworms, mussels, and other shellfish, whales, and other marine mammals, in various species of fish and seabirds (Eerkes-Medrano et al., 2015). Ingestion of microplastics by animals subjects them to toxicological consequences and various metabolic and reproductive disorders (Yang et al., 2022). Heavy metals, antibiotics, and organic pollutants such as polycyclic aromatic hydrocarbons, polychlorinated biphenyls, glyphosate, and amoxicillin adsorb onto microplastics, the ability to adsorb is inversely related to their hydrophobic nature (Fajardo et al., 2022). The ability of nano and microplastics to adsorb organic micropollutants can be affected by various environmental conditions (Fig. 1). These include the pH level, salinity or ionic strength, the presence of dissolved organic matter, the concentration, and the degree of crystallinity of nano and microplastics (Yu et al., 2021b). Microplastics create new habitats and act as a transport medium for microbes and as vectors for pollutants, enabling them to enter the food chain directly (Mammo et al., 2020). Microplastics can impact the migration, toxicity, bioaccumulation, and degradation of organic pollutants through their ability to adsorb and desorb these substances (Koelmans et al., 2016). The pervasive presence of nano and microplastics adsorbed with the organic micropollutants in the ecosystem leads to accidental ingestion by organisms. Micro and nanoplastics could release the adsorbed micropollutants upon entering an organism, potentially leading to localized high contaminant exposure (Wang et al., 2019). Micro and nanoplastics can potentially discharge the adsorbed micropollutants once they enter a living organism, potentially resulting in localized concentrated exposure to these contaminants. The digestive processes within aquatic organisms could facilitate the transfer of these ingested micro and nanoplastics into the circulatory systems. This could, in

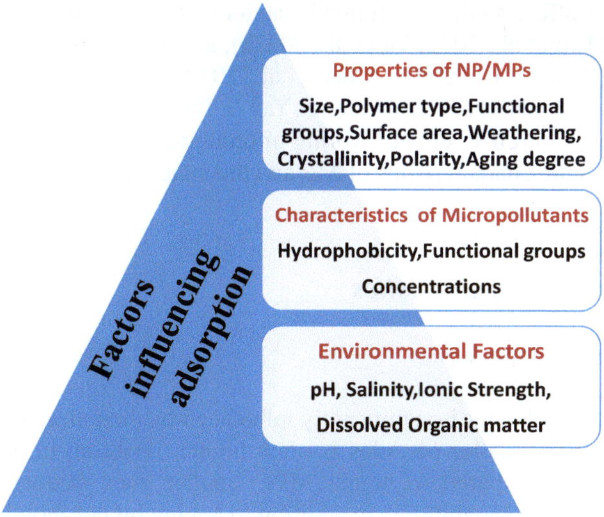

Fig. 1 Factors influencing the adsorption of organic micropollutants on MNPs

turn, cause toxic impacts on various organs within the organism. Antibiotics and heavy metals are more susceptible to adsorption by microplastics (Lu et al., 2020), while heavy metals establish an association with microplastics through two primary pathways: as pigments and stabilizers and by adsorbing onto microplastics (Zhou et al., 2019). Understanding the adsorption behavior and interaction of microplastics and nanoplastics is essential for evaluating the environmental effect, evaluating their fate and transport, understanding the ecological implications, developing remediation strategies, and facilitating risk assessment and regulations. It enables scientists, policymakers, and environmentalists to address the challenges posed by micro and nanoplastics more effectively.

2 Factors Affecting the Adsorption Behavior of MNPs in Soils and Aquatic Environments

Several factors can influence the adsorption behavior of micro and nanoplastics in soils and aquatic environments. The chemical and physical traits of polymers have an important impact on their adsorption behavior. Additionally, aging and weathering are natural processes that degrade plastics and can contribute to changes in MPs' physical and chemical characteristics (MPs). Even though there are several factors affecting the adsorption efficiency of MPs, the environmental compartment and ecosystem that exists is a dominant parameter that can determine the sorption efficiency. Additionally, environmental factors such as pH, temperature, salinity, ionic strength, and dissolved organic matter may lead to a change in the MP properties. These can act as a combination of factors that influence the adsorption. Some of these factors affecting the adsorption behavior include the dosage of plastics (Yap & Tan, 2021; Liu et al., 2022; Yu et al., 2022b), adsorption time (Liu et al., 2022), solution pH (Wang et al., 2022c; Chen et al., 2023b), solution temperature (Chen et al., 2023b), and adsorbate temperature (Yap & Tan, 2021). Besides, the effect due to these factors can reflect either favoring or reducing the adsorption efficacy. This may depend on the characteristics of MP and the attached pollutants.

2.1 Polymer Characteristics

2.1.1 Physical Properties

MPs' adsorption behavior is significantly influenced by their size. The surface area-to-volume ratio of smaller particles is often higher, resulting in more adsorption sites. Therefore, compared to bigger MPs, smaller ones frequently have more increased adsorption capabilities. In addition to having higher mobility than bigger MPs, smaller MPs are also more likely to be suspended in water. They may be more easily in contact with adsorbent due to their increased mobility, which might enhance the efficiency of their ability to absorb toxins or other compounds from the environment. The research by Ma et al. (2019) showed that the adsorption behaviors and load capacities of MPs with different particle sizes may vary for environmental contaminants. The later investigation for adsorption of triclosan on two different sizes of polyvinyl chloride (PVC) MPs concluded that smaller particles have more adsorption capability. Similarly, the adsorption capability between the different sizes of two MP polymers (polystyrene – PS and polyethylene – PE) for two kinds of triazole fungicides (flusilazole – FLU and Epoxiconazole – EPO) has been studied and found that smaller particles have stronger adsorption capability. According to Yu et al. (2020), who examined the antibiotic levofloxacin's adsorption behavior on various MPs, smaller adsorbent particles had higher adsorption efficacy. This may be because of their greater specific surface area and number of adsorption sites, which increased the likelihood that the antibiotic and MPs would come into contact. Thus, the specific surface area has a greater impact on the adsorption capacity of MPs than particle size. Therefore, MPs tend to have stronger adsorption capabilities when they have more surface area per unit mass. Additionally, in comparison to smooth surfaces, MPs with a rough or porous surface offer more adsorption-active sites.

MPs can be classified as primary and secondary based on whether they are created specifically to be micron-sized and whether they are the result of macroplastics being broken down, respectively. It can take the form of fibers, spheres, fragments, or other asymmetrical-shaped structures. The secondary MPs could form vast irregular shapes according to the fragmentation force acting upon them. Therefore, their adsorption behavior can vary depending on their shape. The type of MP may determine the interactions that have occurred on an MP surface to affect the adsorption process (Lu et al., 2023). Several variables, including the characteristics of the MP, adsorbent material, and the environment, might affect how an MP type affects adsorption. For example, the adsorption capacity differs with the types of MPs and heavy metals, as a consequence of the different physical characteristics of various MPs and heavy metals (Han et al., 2021). Due to the favorable physicochemical properties of PS-MP film, it has a higher capacity to adsorb Cu (II) than PVC, PE fishing line fibers, and bottle cap particles (Almeida et al., 2020). Small polypropylene (PP) particles have greater adsorption affinity to microcystins, which is a cyanotoxin, compared to the small polyethylene terephthalate (PET) particles due to their glassiness, polarity, and crystallinity (Moura et al., 2022). Depending on their composition, MPs can have varied densities. It can affect the adsorption behavior of MP in water bodies and has an influence on how they are distributed vertically. The denser particles may sink and can facilitate different adsorption characteristics compared to the buoyant plastics.

2.1.2 Chemical Factors

Surface chemistry differs amongst MPs. The adsorption behavior of MPs is strongly influenced by their surface chemistry, in particular the presence of functional groups and charges. The surfaces of MPs may contain hydroxyl (-OH), carbonyl (C=O), or carboxyl (-COOH) groups, among other functional compounds. Oxygen-containing functional groups (C=O, C-O, and -OH) produced by UV treatment on MPs have strong metal ion complexing properties (Cao et al., 2021). Interactions of these functional groups with other molecules by hydrogen bonds, van der Waals forces, and electrostatic interactions can influence MP adsorption behavior. External factors can induce alterations in the MP surface, and they can aid in the adsorption of pollutants. For example, the surface charge of the adsorbent may be influenced by the pH of the aqueous solution (Lin et al., 2021). The MPs can have a positive or negative charge according to their composition and surface function groups. Their contact with other charged species in the environment may be determined by their charge, and this interaction can affect the extent that they can adsorb.

Crystallinity is a property of polymer structures; the higher the crystallinity, the more organized and fixed the structure. The capacity of a contaminant and rate of absorption into the polymer matrix decrease as crystallinity increases. The amorphous region will be small in this scenario, making it potentially unfavorable for pollutant adsorption (Hartmann et al., 2017). Thus, the higher the degree of crystallinity, the lower the polymer's permeability. The adsorption of antibiotics to MP may be affected by the degree of crystallinity (Li et al., 2018). Similarly, the crystallinity of PP and PET MPs has shown an effect on microcystin adsorption (Moura et al., 2022). However, crystallinity may not necessarily be the primary factor affecting adsorption.

Polarity and electrostatic repulsion of MPs are two of the factors that affect the adsorption of specific molecules. Wang et al. (2015) explained the electrostatic repulsion between the anionic Perfluorooctanesulfonate (PFOS) molecules and the negative surface of PS particles that prevent the sorption of PFOS on PS particles. At the same time, the high polarity of PS-MP and PVC is due to the presence of one hydrogen atom in the benzene ring and a chlorine atom for each unit (Wang et al.,

2015). Therefore, the presence of benzene rings in PS plastics could result in a strong steric hindrance effect on bond rotation and a decrease in free volume between chains, which could result in lower sorption of Perfluoro sulfonamide octane FOSA on PS plastics (Wang et al., 2015). This indicates that Polar compounds' adsorption levels can affect the polarity of MPs. Li et al. (2018) further stated that polar compound antibiotics have a strong sorption capacity toward polar MPs. The polarity of MP is not sufficient to explain the differences in adsorption capacities, although distinct MPs have varied polarities and crystallinities.

2.2 Age and Weathering of Plastics in the Environment

Most of the plastic litter cannot degrade quickly. Plastic weathers and disintegrates into microscopic fragments as "microplastics" and "nanoplastics" as a result of UV radiation, mechanical abrasion, temperature fluctuations, and biodegradation (White & Turnbull, 1994). These parameters also influence how MPs absorb organic contaminants, as well as their adsorption capabilities. Weathering alters the physical and chemical factors of MPs, resulting in changes in their environmental behavior. MPs exhibit an increasing trend in improved sorption capability as they age (Hanun et al., 2023). As a result of the fragmentation, the surface area of MPs increases and creates additional adsorption sites, as well as weathering, which may result in the creation of surface imperfections, which makes MPs rougher and increases their capacity for adsorption. Following aging tests, MPs exposed to high temperatures, particularly in the air, exhibited greater adsorption affinity, which is related to the development of additional adsorption sites on the surface of MPs (Wang et al., 2022c). According to Lang et al. (2020), as MPs age, the MP surface becomes rougher, with different degrees of folding and even breaks which in turn leads to the increase in surface area that facilitates potential adsorption capacities. More adsorption sites will be added as MPs age, which will enhance the way that metal and organic contaminants adsorb (Yu et al., 2021a). At the same time, the aging of MPs can lead to the loss of properties of the polymer, and it can lead to a reduced adsorption of pollutants. For instance, a natural aging process of PP wire and PE film for 250 days has shown a visible effect in the loss of calcium carbonate, which is an additive, used in polymers, together with morphology and surface property changes, which decreased the adsorption potential of MPs toward Cu (II) (Zhou et al., 2022). This shows that the presence of additives enhances the adsorption ability of MPs, and the adsorbed pollutants may leach out in the aging process.

MPs undergo oxidation and fragmentation as they get older, which changes the characteristics of their surfaces. This transformation may result in rougher textures, increased cracks, larger open pores, increased crystallinity, greater specific surface area, and expanded pore volume, all of which would boost the MP adsorption characteristic (Chang et al., 2023).

Table 1 shows the studies published to date in ScienceDirect (https://www.sciencedirect.com/) database. The references were retrieved with the term "microplastics" in combination with "adsorption," "adsorption factors," "physical factors," "chemical factors," "vector," "soils," and "aquatic environment." In total, 26 references have been sorted and arranged with the physical, chemical, and aging properties of MPs affecting the adsorption.

Table 1 Summary of physical, chemical, and aging properties affecting the adsorption

Adsorbed pollutant	MP type	Physical/chemical/ aging factors	References
Aniline	Polystyrene (PS)	Particle size	Chen et al. (2023b)
Antibiotics – Sulfadiazine (SDZ), amoxicillin (AMX), tetracycline (TC), ciprofloxacin (CIP), and trimethoprim (TMP)	Polyethylene (PE), polystyrene (PS), polypropylene (PP), polyamide (PA), polyvinyl chloride (PVC)	Specific surface area, polarity, degree of crystallinity	Li et al. (2018)
Antibiotics – Sulfonamides (SAs), tetracyclines (TCs), fluoroquinolones (FQs), and macrolides (MLs).	Polypropylene (PP)	Aging (more developed pore structures)	Yu et al. (2022a)
Benzotriazoles, benzothiazoles	Polyvinyl chloride (PVC)	Morphology (rough surface), specific surface area, porous structure, electronegativity	Yu et al. (2022b)
Cadmium	Polystyrene (PS)	Particle size	Yu et al. (2021a)
Chlorophenylacetonitriles	Polyethylene (PE), polyethylene terephthalate (PET), polystyrene (PS)	Crystallinity, specific surface area	Yang et al. (2023)
Chlorpyrifos	Polybutylene adipate-co-terephthalate (PBAT), polylactic acid (PLA), polypropylene (PP)	Specific surface area, pore volume	Zhang et al. (2023b)
Copper	Polyvinyl chloride (PVC), polystyrene (PS)	Surface area, polarity	Brennecke et al. (2016)
Copper cu(II)	Polystyrene (PS), polyethylene terephthalate (PET)	Particle size	Wang et al. (2022c)
Cyanotoxins	Polypropylene (PP), polyethylene terephthalate (PET)	Polymer type, particle size	Moura et al. (2022)
Diclofenac (DCF), metronidazole (MNZ)	Polystyrene (PS), polyethylene terephthalate (PET), polypropylene (PP), high-density polyethylene (HDPE)	Particle size, hydrophobicity, aging	Munoz et al. (2021)
Flusilazole (FLU) and epoxiconazole (EPO)	Polystyrene (PS), polyethylene (PE)	Particle size	Liu et al. (2022)

(continued)

Table 1 (continued)

Adsorbed pollutant	MP type	Physical/chemical/ aging factors	References
Humic acid	Polystyrene (PS), polyvinyl chloride (PVC)	Particle size	Zhang et al. (2023a)
Lead Pb (II)	High-density polyethylene (HDPE)	Particle size	Nguyen et al. (2022)
Naphthalene, 2-hydroxy naphthalene, phenanthrene, pyrene	Polyvinyl chloride (PVC)	Hydrophobicity	Bao et al. (2023)
Nitro polycyclic aromatic hydrocarbons	Polyethylene (PE), polypropylene (PP), polystyrene (PS)	Crystallinity, surface area, pore volume, rough surface	Kong et al. (2023)
Nonylphenol	Polyethylene (PE), polypropylene (PP), polyvinyl chloride (PVC)	MP type, particle size	Lu et al. (2023)
Oxybenzone	Polyethylene terephthalate (PET)	Functional groups, specific surface area, crystallinity	Hanun et al. (2023)
Perfluoroalkyl substances	Polyamide (PA)	Particle size	Mejías et al. (2023)
Pesticides (malathion, diflubenzuron, difenoconazole, carbendazim)	Aged polyethylene (PE)	Specific surface area, pore volume	Lan et al. (2022)
Pesticides (α-endosulfan and chlorpyrifos) and six musk fragrances (musk xylene, musk ketone, musk moskene, galaxolide, tonalide, and celestolide)	Polyamide (PA6), polyhydroxybutyrate (PHB)	Aging	Concha-Graña et al. (2022)
Pharmaceuticals, personal care products	Polypropylene (PP)	Surface functional groups	Yao et al. (2022)
Phenanthrene, methylene blue	High-density polyethylene (HDPE), polypropylene (PP)	Aging, oxidation of MPs	Bhagat et al. (2022)
Rhodamine B	Polyvinyl chloride (PVC), polystyrene (PS), polyethylene terephthalate (PET)	Crystallinity, specific surface area, surface charge, hydrophobicity, functional groups	Du et al. (2022)
Tetracycline	Polyethylene (PE)	Specific surface area, surface charge	Wang et al. (2022a)
Thiamethoxam	Polyvinyl chloride (PVC), low-density polyethylene (LDPE), polyethylene (PE), polycaprolactone (PCL), and polybutylene succinate (PBS)	MP type, particle size	Hu et al. (2023)

3 Mechanisms of Adsorption of MNPs in Soils and Aquatic Environments

3.1 Physical Adsorption

The adsorption mechanisms and behavior of micro and nanoplastics (MNPs) in soils and aquatic environments are intricate processes. They control and determine the fate and transfer of contaminants between MNPs and these environmental compartments. In this case, the contaminant is the adsorbate, while the MNP is the adsorbent, regardless of the environmental compartment in which it occurs. Additionally, the extent to which this interaction occurs determines the adsorption characteristics (Agboola & Benson, 2021). Based on this, the adsorption mechanisms are categorized into physical adsorption (also referred to as physisorption) and chemical adsorption (also called chemisorption).

Physical adsorption deals with molecular exchanges between the adsorbent and adsorbate which are mainly by the van der Waals force, whereas chemical adsorption happens when there are chemical bonds formed between the adsorbent and adsorbate (Agboola & Benson, 2021). Other adsorption mechanisms of compounds to MNPs are pore filling, hydrophobic partitioning, surface sorption facilitated by hydrogen bonding as well as electrostatic interaction, electron–donor–acceptor (EDA) interaction, and the action of van der Waals forces (Wang et al., 2020a). In this section, we will cover physical and chemical adsorption, hydrogen bonding, electrostatic interaction, and van der Waals force (Fig. 2).

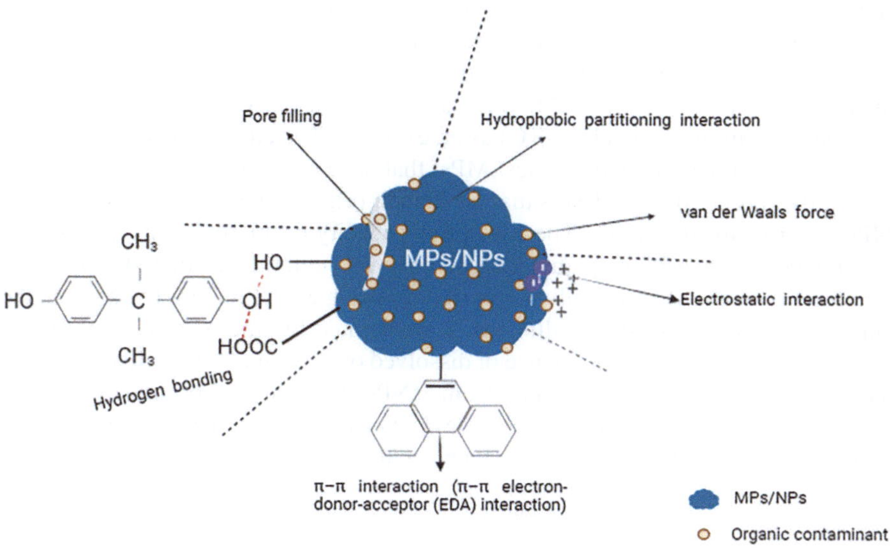

Fig. 2 The adsorption mechanisms of micro and nano plastics (MNPs)

Physical adsorption of MNPs describes the readily reversible accumulation or absorption of contaminants onto their surfaces. This type of adsorption mechanism of MNPs in soils and aquatic environments is controlled by variables including the properties of the polymer such as the type of polymer involved, the size and age of the particle, the surface area, the charge on the surface of the particle, crystal structure; properties specific to the compound; and the properties specific to the medium like ionic strength, salinity, pH, and dissolved organic matter (Wang et al., 2020a). Additionally, the surface chemistry or polarity of MNPs, which governs the adsorption of foreign molecules onto their surfaces, also influences the adsorption of contaminants onto the MPs (Harraq & Bharti, 2022).

This property of MNPs controls the extent of interaction with organic micropollutants. Since MNPs are known to be hydrophobic, their affinity for hydrophobic organic contaminants, such as PBDEs, PPCPs, POPs, and EDCs, is high (Yu et al., 2021b). This process can be further elucidated by the fact that, in order to preserve available energy when seeking to attain a state of thermodynamic balance, compounds tend to aggregate in phases that possess similar physicochemical qualities to their own, as opposed to phases that exhibit contrasting features (Heinrich et al., 2020). However, besides the high adsorption affinity for hydrophobic compounds, MNPs also have the tendency to sorb water-soluble or polar metals and organic contaminants (Lim et al., 2022).

Regarding the variables controlling adsorption mechanisms, salinity is a major specific property of the medium, whose effect on the adsorption of contaminants can differ between MNPs and the contaminants (Zhang et al., 2020). The effect of salinity is particularly evident in the marine ecosystem, where elevated saline levels in the surrounding medium, specifically water, result in a proportional augmentation of its strength and ionic valence state. This further improves their adsorption capacities, for example, in the case of phenanthrene adsorption on agricultural MPs, namely, polyurea, polyurethane, and urea-formaldehyde (Zhang et al., 2020).

The adsorption capacity of organic pollutants in soils is further improved through the exposure of microplastics (MPs) to soil environments. We will use the adsorption mechanism of tetracycline (TC) as an example. The adsorption mechanism of tetracycline (TC) onto microplastics (MPs) that have been exposed to soil (referred to as soil-exposed MPs) and MPs that are in their original state (referred to as virgin MPs) clearly elucidated this. The adsorption capacity of microplastics (MPs), specifically polylactic acid, polyethylene, and polyvinyl chloride, exhibited an increase following their exposure to soil, as reported by Wang et al. (2022a). In addition, the duration required to reach equilibrium is accelerated by microplastic exposure to soil. When considering the influence of dissolved organic matter in soil, it facilitates the desorption of organic compounds from MNPs. Again, the desorption of TC from polyethylene, polylactic acid, and polyvinyl chloride was found to be increased due to the activity of soil organic matter (Wang et al., 2022a).

3.2 *Chemical Adsorption*

Chemical adsorption (also called chemisorption) is an irreversible and specific chemical process in which covalent bonds are formed between adsorbents and adsorbate molecules, depending primarily upon surface area proportionality (Zhang et al., 2018). Here, the chemicals or otherwise, organic contaminants form covalent bonds from their interactions with the MNPs.

Chemical adsorption is characterized by strong interactions, including hydrogen bonding, as well as the creation of ionic and covalent bonds between the adsorbate and the solid surface. Chemical absorption is a process that commonly takes place even at significantly small concentrations. The species that undergo chemical absorption are often bonded to the surface in an "irreversible" manner, meaning they do not easily detach at normal room temperature. The irreversible nature of chemical adsorption poses challenges for the desorption of the chemical from the MNPs. An exception, in this case, occurs only after the covalent bonds are broken (Velzeboer et al., 2014). The process of chemical absorption is considered to be complete when all available active sites on a solid surface have been occupied by molecules that have undergone the process.

3.3 *Hydrogen Bonding*

Hydrogen bonds are noncovalent and distinctively weak interactions that involve hydrogen ions H+ (in the MNPs and organic compounds), which can influence the sorption of polymeric compounds when groups of proton acceptor and proton donor exist (Tourinho et al., 2019). Although they are weak bonds, they exhibit greater strength compared to van der Waals forces (Agboola & Benson, 2021). They are polar interactions that are involved with nonpolar (PP, PE) or weakly polar (PS) MNPs (Zhang et al., 2020). Furthermore, they can act together in one or more interactions such as electrostatic interaction/repulsion. Alternatively, they may act independently, as demonstrated by the adsorption of naphthalene on polyuria MPs (Zhang et al., 2020).

The sorption process, which involves hydrogen bonding, can be influenced by the presence of functional groups in micro and nanoplastics (MNPs) as well as the organic compounds found in soils and aquatic environments (Wang et al., 2020a). This indicates that the presence of functional groups in both the MNPs and organic compounds allows for the formation of hydrogen bonds, which may enhance the sorption affinity. For example, the amide group in polyamide (PA) has the proton donor attributes, and because of the hydrogen bond, the sorption capacity of PA for amoxicillin, tetracycline, and ciprofloxacin is higher than the polyvinyl chloride (PVC), polyethylene (PE), polystyrene (PS), and polypropylene (PP) with the same group of antibiotics (Li et al., 2018). The presence of a carbonyl group contributes to the forming of hydrogen bonding, which, in conjunction with the increased surface area of aged plastic particles, leads to an enhancement in the adsorption of contaminants (Li et al., 2018). Temperature is an additional factor that can exert an influence on hydrogen bonds since elevated or heightened temperatures have the capacity to disrupt the inherently feeble bonds (Lu et al., 2023).

3.4 Electrostatic Interactions

Electrostatic interactions arise from the attractive or repulsive forces between molecules possessing opposite or like charges, respectively (Tourinho et al., 2019). As previously mentioned in Sect. 3.1, the adsorption behavior of MNPs is influenced by many properties specific to the medium, including salinity, pH, and ionic strength. They accomplish this when they alter the surface charge and chemical speciation of MNPs. Similar to hydrogen bonding, there are also polar interactions that are involved with non-polar or weakly polar micro and nano plastics (Zhang et al., 2020). This becomes true with the involvement of electrostatic interactions since these variables affect electrostatic interactions to a large extent (Tourinho et al., 2019). Furthermore, electrostatic interactions facilitate the attraction of hydrophilic compounds toward negatively charged regions on the surface of microplastics (MPs), while hydrophobic compounds are drawn toward neutral regions through hydrophobic forces (Tourinho et al., 2019).

An electrostatic repulsion or interaction occurs depending on the charge on the surface of the adsorbent. A repulsion happens when the charge on the surface is negative and an interaction happens when the adsorbent surface is positively charged (Zaggia et al., 2016). The adsorption of the popular group, PFAS on MPs partly, involves electrostatic interactions (Gagliano et al., 2020), and this governs their adsorption onto various adsorbent materials. Meanwhile, electrostatic interactions are dominant for short-chain PFAS. Nevertheless, the adsorption of long and short-chain per- and poly-fluoroalkyl substances (PFAS) onto micro and nanoplastics (MNPs) can be influenced by the presence of organic matter (Gagliano et al., 2020). Therefore, electrostatic interactions are significant in this context, particularly in relation to the adsorption of per and poly-fluoroalkyl substances (PFAS) onto organic matter when microplastics are present. This behavior arises from the interaction between the surfaces of microplastics that have adsorbed organic matter and PFAS. Figure 3 depicts the influence of organic matter on the adsorption process of organic pollutants onto magnetic nanoparticles (MNPs).

3.5 Van Der Waals Forces

Van der Waals forces refer to intermolecular interactions that occur between molecules that lack covalent or ionic bonds. These forces are characterized by their weak nature and occur inside the molecular structure, making them intramolecular. It is

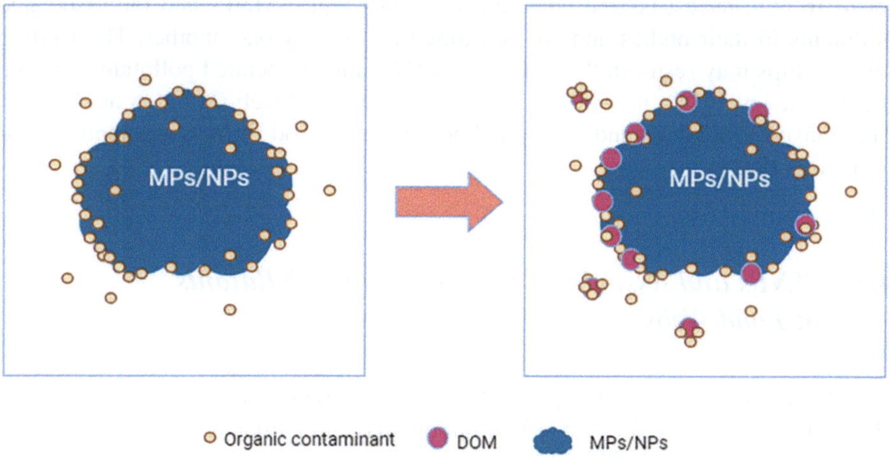

Fig. 3 The sorption behavior of organic contaminants to NPs/MPs in the presence of dissolved organic matter (DOM)

common for aliphatic polymers such as polyethylene (PE) and polyvinyl chloride (PVC) to undergo this form of interaction (Hüffer et al., 2018). Moreover, micro and nanoplastics (MNPs) possessing nonspecific functional groups, such as polypropylene (PP) and polyethylene (PE), carry out interactions with organic substances mostly driven by the van der Waals force, which arises from the characteristics of these functional groups. Such is the case of the aggregation and adsorption of ciprofloxacin onto the surface of polyethylene (PE), which were facilitated by van der Waals forces and electrostatic interactions (Agboola & Benson, 2021). Moreover, the van der Waals force might exert a dominant influence on the adsorption process occurring on microparticles (MPs), particularly for hydrophilic substances that exhibit limited dissociation in a solvent. An example is the process of adsorption involving thiacloprid (which possesses an octanol-water partition coefficient (log Kow) of 1.26), onto cellulose acetate microplastics (Li et al., 2018).

4 MNPs as Vectors for Other Environmental Pollutants in Food Webs

In addition to the primary chemical constituents (plasticizers, stabilizers, and flame retardants) of microplastics and nanoplastics (MNPs), studies have shown that MNPs are potential vectors (carriers) for other environmental pollutants in food webs (Koelmans et al., 2016; Squadrone et al., 2022). Through sorption, an association of inorganic and organic environmental pollutants with MNPs may occur and result in the transport of MNPs and the associated pollutants in environmental matrices, including biota (Rodrigues et al., 2019; Amelia et al., 2021). Different life forms in both terrestrial and aquatic ecosystems ingest MNPs and the associated pollutants in their bodies and are consumed as food by one another. The feeding relationships may result in the transfer of MNPs and associated pollutants through the trophic levels in a network of food chains, the food web (Wang et al., 2020b). The transport of MNPs and associated pollutants in food webs is a potential environmental risk.

4.1 MNPs and Associated Environmental Pollutants in Food Webs

Several organic and inorganic pollutants have been reported to interact and associate with MNPs (Table 2). Due to hydrophobicity, some organic pollutants are sorb to MNPs (Rezania et al., 2018). Nanoplastics being smaller in size than microplastics show greater interaction with other environmental pollutants than microplastics (Chen et al., 2023a). The organic pollutants reported to be associated with MNPs include persistent organic pollutants (POPs) such as polycyclic aromatic hydrocarbons (PAHs) (Wang et al., 2020b), polychlorinated biphenyls (PCBs), and organochlorine pesticides (Allen et al., 2018; Rodrigues et al., 2019). Chen et al. (2020) studied microplastics and their associated PAHs in the marine environment in Taiwan and reported a total concentration of PAHs in microplastics as 104–3595 ng/g dw with a mean value of 818 ng/g dw. Similarly, Campanale et al. (2020a) detected 16 priority PAHs associated with microplastics with concentrations ranging from 29.9 to 269.1 ng/g.

Metals as inorganic pollutants have also been reported to adsorb to MNPs (Maity et al., 2021). Xie et al. (2021) reported the detection of metals in microplastics in river estuary. The accumulation of metals (copper, zinc, lead, and cadmium) in plants (lettuce) was reported to be facilitated by their association with MNPs (Xu et al., 2023).

Table 2 Additives and other environmental pollutants associated with MNPs in food webs

Chemicals	Candidates	References
Additives	Plasticizers, stabilizers, phthalates, flame retardants, and (TBBP-A).	Campanale et al. (2020b)
Organic pollutants	PAHs (pyrene, phenanthrene, chrysene, and fluoranthene). PCBs. Organochlorine pesticides (HCH and DDT).	Rodrigues et al. (2019) Zhang et al. (2021)
Inorganic pollutants: Metals	Iron, zinc, copper, lead, cadmium, chromium, nickel, cobalt, molybdenum, manganese, and aluminum	Squadrone et al. (2022) and Xu et al. (2023)

TBBP-A tetrabromobisphenol-A, *PCBs* polychlorinated biphenyls, *PAHs* polycyclic aromatic hydrocarbons, *HCH* hexachlorocyclohexane, *DDT* dichlorodiphenyltrichloroethane

4.2 *Trophic Transfer of MNPs and Associated Environmental Pollutants*

In food webs, organisms at different trophic levels may ingest and accumulate MNPs and associated environmental pollutants (Carbery et al., 2018). However, Setälä et al. (2018) reported that in addition to ingestion, microplastics can contaminate food webs in the aquatic environment through inhalation and entanglement by the organisms. Once MNPs and associated environmental pollutants are incorporated in the bodies of organisms, they move along the network of food chains comprising various trophic levels (Fig. 4).

In terrestrial and aquatic ecosystems, autotrophic organisms (primary producers) constitute the base or the first trophic level of grazing food webs. The primary producers are usually plants in terrestrial food webs and phytoplankton (mostly algae) in freshwater, estuarine, and marine food webs. The succeeding trophic levels comprise different species of organisms with varying feeding habits classified as consumers. MNPs and associated environmental pollutants can contaminate food webs

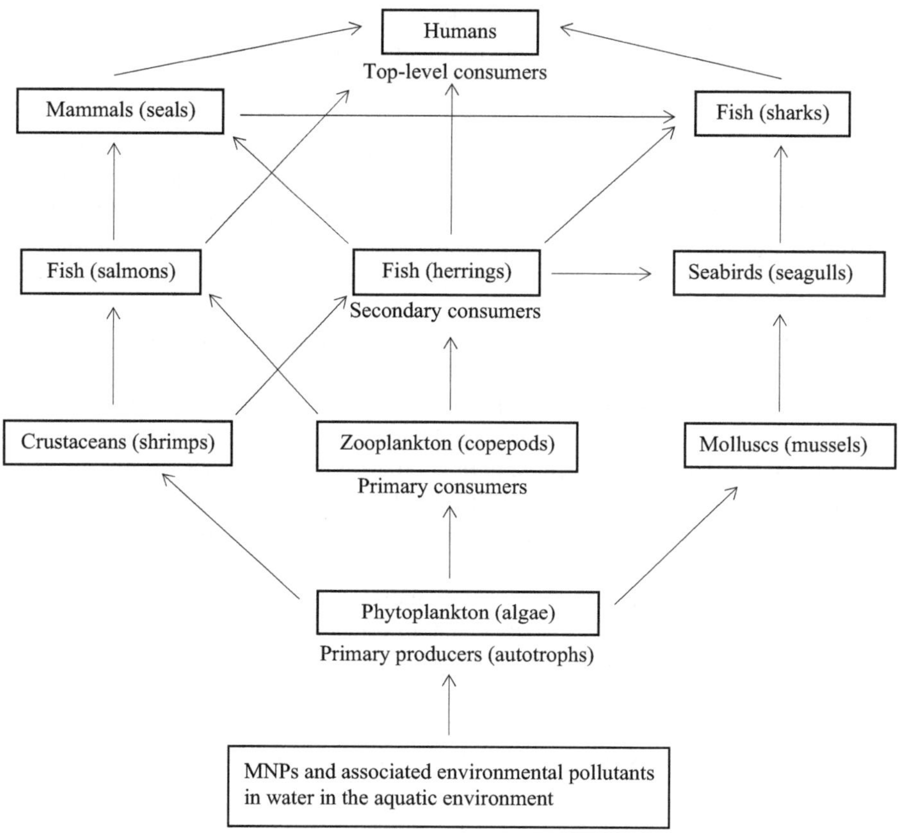

Fig. 4 Illustration of transfer of MNPs and associated environmental pollutants in the food web

through ingestion by plants and phytoplankton and their transfer along succeeding trophic levels (Huang et al., 2021). In terrestrial food webs, the uptake of MNPs and associated environmental pollutants by plants through roots and leaves, and their transport to various parts of the plant body occur (Liang et al., 2023). In addition to roots and leaves, MNPs and associated environmental pollutants have been detected in fruits, seeds, flowers, and stems of plants consumed by other organisms, making plants important sources of transfer of MNPs and associated environmental pollutants in food webs (Allouzi et al., 2021). Exposure of herbivorous animals to consumption of plants containing MNPs and associated environmental pollutants, and their transfer to other consumers (carnivores and omnivores) at higher trophic levels can occur in terrestrial food webs. Studies have shown that humans are exposed to risks of contamination with MNPs and associated pollutants due to the consumption of food (plants and animals) containing MNPs with associated environmental pollutants (Wang et al., 2022b; Dong et al., 2023).

In aquatic food webs, the incorporation of MNPs and associated environmental pollutants in the bodies of phytoplankton has been studied. Long et al. (2015) reported the interactions of microplastics with algae (diatoms) and concluded that diatoms could be a sink for microplastics in the marine environment and influence their transport in the water studies by Carbery et al. (2018) indicated trophic transfer of microplastics and associated environmental pollutants in marine food webs. MNPs and associated environmental pollutants ingested by phytoplankton occupying the first trophic level can be transported to other trophic levels comprising zooplankton (copepod crustaceans and rotifers), and other organisms including fish, amphibians, reptiles, birds, and mammals occupying higher trophic levels. Desforges et al. (2015) studied the ingestion of microplastics by zooplankton in the Northeast Pacific Ocean and reported that zooplankton and other species of primary consumers in food webs mistakenly ingest microplastics as food and consequently, expose organisms at higher trophic levels to microplastics.

In aquatic food webs, among invertebrates, the occurrence of MNPs and adsorbed pollutants in crustaceans and bivalve mollusks such as oysters, mussels, and clams has been reported. PAHs (pyrene), chlorinated pesticides (chlorofenviphos, chloropyrifos, triclosan, and chlorotoluron), and pharmaceuticals (carbamazepine) associated with MNPs have been detected in the bodies of the bivalves, *Crassostrea gigas* (oyster) and *Mytilus galloprovincialis* (mussel). The presence of MNPs resulted in the accumulation of the associated environmental pollutants in the organisms (Rios-Fuster et al., 2023).

In fish, when microplastics are ingested, they are transported to the gut and other tissues where they release the associated environmental pollutants adsorbed. In food webs, the consumption of fish by predatory animals and humans is a means of transfer of pollutants associated with microplastics to them. Birds, sharks, and mammals that feed on fish and other organisms in the aquatic food web can be exposed to contamination with MNPs and associated environmental pollutants (Mercogliano et al., 2020; Alberghini et al., 2022).

4.3 Bioavailability of MNPs and Associated Environmental Pollutants in Food Webs

It has been observed that the conditions of the environmental matrix of pollutants associated with MNPs may influence their bioavailability. Physiological conditions of the organisms contaminated with MNPs and associated environmental pollutants may facilitate desorption of the associated pollutants higher than in water in the aquatic environment (Santos et al., 2021). Pharmaceuticals associated with microplastics can be desorbed following ingestion by aquatic organisms. This influences the bioavailability, bioaccumulation, and transport of pharmaceuticals in food webs (Santos et al., 2021). However, there is a need for extensive studies on the bioavailability of several pollutants that are transported by MNPs in food webs to enhance a better understanding of the bioavailability of MNPs-associated pollutants in food webs.

5 Current Challenges and Gaps in Comprehending MNPs' Adsorption Behavior and Interaction

Various scientific studies have been conducted to assess the public perceptiveness about plastic pollution, and it demonstrates that most of the people/highest portion of the public was unaware of the micro-nanoplastics and their issues on the surroundings such as climate change, acidification of soil and its effects on living organisms (Ogunola et al., 2018). It is believed that the awareness programs on MNPs and their effects by government and nongovernment organizations will promote general responsibility to reduce the usage and disposal of plastics. In cosmetic industries and other industrial products, the less usage of microplastic beads will help to mitigate the effects of MNPs on humans, and it is mandatory to ban the overuse of microplastic beads in daily products (Anagnosti et al., 2021). Proper plastic recycling and disposal can reduce the effects of MPs, but many issues arise when the huge quantity of plastic waste materials dumped in landfills, and other natural habitats will lead to MNP contamination. Plastic recycling is not economically competitive and approaches for minimal investments, the recycling rate is higher in developed countries than in developing countries. To increase plastic recycling, actions like awareness of plastic recycling, taxes imposing on using virgin plastics and incentives by governments toward their production, etc. are needed (Calero et al., 2021).

The assessments and evaluations of the impacts of micro-nanoplastics economic scale are very challenging due to the gaps that occurred in the wide knowledge. The waste materials from plastics including nanoplastics have resulted in approximately 13.3 billion US dollars in financial loss annually (da Costa 2018). Novel biotechnological approaches are currently being explored to minimize the risks of the accumulation of MNPs in the environment. Biotechnological formulations like bioplastic made from natural resources are ecofriendly and sustainable. Some of the mitigating measures circulating to reduce plastic waste and prevent its entry into the environment, like, governments banning the use or sale of bags from plastics and covers also the use of microplastics in cosmetics and the restrictions on single use of plastic bags in around 30 countries (Ogunola et al., 2018).

5.1 Lack of Standardized Adsorption Measurement Techniques

The lack of standardized adsorption measurements for MNPs is the other issue. Commonly used measurements nowadays are not faithful in many ways. Adsorption is one of the best methods, due to being inexpensive, cost-effective, land requirements, universal existence, etc. (Badruddoza et al., 2013; Rashed, 2013). A variety of various methods are used for the removal of organic compounds such as filtration, coagulation, precipitation, flotation, ion exchange, electrodialysis, adsorption, reverse osmosis, and advanced oxidation processes (Rashed, 2013). The traditional methods, including ion exchange, membrane separation, etc., are the methods that show some of the limitations for high capital and operating costs (Manyangadze et al., 2020). The process of adherence of chemical species from the fluid phase to the solid phase like plastic debris is called sorption (Menéndez-Pedriza & Jaumot, 2020). This includes both absorption and adsorption (Fred-Ahmadu et al., 2020). The adsorption-based interactions occur between organic micropollutants and nano and microplastics. In the adsorption process, the contaminants from chemical substances are absorbed in phases between solid and fluid. Various mechanisms such as van der Wall forces, Covalent or steric bonds, and ion concentrations are involved in solid surface adsorption (Borkovec & Papastavrou, 2008). The surface polarity of micropollutants can regulate the interaction of organic micropollutants (Enfrin et al., 2019). These micropollutants have a high affinity to organic contaminants with hydrophobicity because of their hydrophobic nature. The aged or weathered MNPs affect the chemical composition, and the physicochemical properties of the surfaces of the plastics and can adsorb more hydrophobic organic contaminants (Dong et al., 2020). Similar studies explained about aging due to exposure to ozone or UV can increase the mobility of micropollutants (Liu et al., 2019) and the aged nanoplastics were shown to have more hydrophilicity and negative charge. Versatility in the pore size of MNPs from nanometers to micrometers can act as an entry route for various micropollutants (Wang et al., 2020c). Commonly, micronanoplastics adsorb organic pollutants by hydrophobic interaction, partitioning, non-covalent actions, electrostatic interactions, etc. The studies on the adsorption behavior of MPs concluded that all are focused on laboratory-scale experiments and are not sufficient to explain the mechanism of adsorption by MPs or the environmental factors that are influencing them. The microplastic existence participating in biotoxicity increases the dissolved organic matter and also influences carbon cycling (Yu et al., 2019).

5.2 Limited Knowledge of the Effect of Environmental Variables on Adsorption

Most of the public is unaware of the influence of environmental factors on the adsorption of MNPs. The aging, particle size, specific surface area, crystallinity, and functional groups of MPs can affect the organic pollutant absorbing capability. Moreover, hydrophobicity, charge of pollutants, and also pH, temperature, dissolved organic matter, and ionic strength are the major factors affecting the capacity of MNPs to adsorb or take up organic pollutants (Enfrin et al., 2019; Fred-Ahmadu et al., 2020). The crystalline, semicrystalline, and amorphous properties of polymers can also affect the adsorption behavior of MNPs (Fred-Ahmadu et al., 2020). The surface structure of MNPs and also the oxygen-containing functional groups can take part in changes by the action of environmental factors such as water, temperature, and ultraviolet radiation (Sun et al., 2021). All the plastics that are dumped in the environment can eventually go into the estuaries and then to the open seas. In the sea, the transfer of plastics can be influenced by winds, waves, rainfall, and oceanic currents (Cózar et al., 2014). On the other hand, soil erosion, agricultural runoff, and atmospheric deposits can also transfer plastics into seawater.

5.3 Inadequate Understanding of MNPs Long-Term Fate and Behavior

People all over the world are unaware of the long-term fate and behavior of MNPs. COVID-19 generated a huge amount of plastics into the environment. Face masks, gloves, sanitizer bottles, testing kits, etc., have been trashed into the environment without any proper treatment. Various studies have reported that aquatic organisms ingest the accumulated plastics in their digestive system and cause toxicological effects such as DNA damage, tissue damage, protein modulations, embryotoxicity, toxicities in all organ systems, etc. In terrestrial ecosystems, the plastic concentration is higher than in the ocean. These MP/NPs can change the soil properties, enzymatic systems, and microbes in soil and can have adverse reactions on living organisms including plants and animals. Many researchers reported the abundance of plastics in the sea is hazardous to marine life. Many studies reported the presence of MNPs in shrimps, seabirds, crustaceans, echinoderms, zooplanktons, fishes, turtles, corals, mammals, etc. (Botterell et al., 2019).

In the terrestrial ecosystem, the MNPs originated mainly from domestic wastes, irrigation, greenhouse materials, landfilling, fertilization, biosolids, sewage sludge, tire abrasion, atmospheric deposition, and agriculture organic wastes, etc. (Koutnik et al., 2021a). These MNPs are entered vertically through infiltration, tilling, and microbial actions. The presence of MNPs was reported in mangroves and mangrove sediments with negative consequences including the inhibition of growth, loss of foliage, and the effect on crustaceans that live in associated with mangroves (Ivar do Sul et al., 2014).

In human beings, the major intake of MNPs occurs through inhalation, skin perfusion, ingestion, skin penetration through cosmetics, face washes, body lotions, etc. They can cause significant health issues due to their persistent nature in the environment, ecotoxicity, and are capable to serve as a vector for chemical micropollutants (Mei et al., 2020).

6 Conclusion

MNPs in the environment are pollutants that are potential vectors for other environmental pollutants in both terrestrial and aquatic food webs. The association of inorganic and organic pollutants in the environment with MNPs may be achieved through sorption. Inorganic pollutants such as metals, and organic pollutants including PAHs, PCBs, organochlorides, and some pharmaceuticals associated with MNPs have been detected in the bodies of different species of organisms (phytoplankton, zooplankton, other invertebrates, and vertebrates, including humans) in various trophic levels in food webs. The uptake of MNPs with adsorbed environmental pollutants by plants from soils is an important route for entering the pollutants into terrestrial food webs. Trophic transfer of MNPs and associated environmental pollutants in food webs occurs due to feeding relationships involving the consumption of one organism by another. Following ingestion of MNPs and associated environmental pollutants by the organisms in food webs, the associated pollutants may desorb from the MNPs and become bioavailable or bioaccumulate. Physiological conditions of the organisms contaminated MNPs and associated environmental pollutants may facilitate the desorption of associated pollutants of MNPs higher than in water in the aquatic environment. Consumption of food composed of plant and animal products contaminated with MNPs containing other environmental pollutants is one of the exposure routes for animals and humans to MNPs and copollutants. Transport of environmental pollutants by MNPs in food webs is a risk to the environment including humans.

The adsorption mechanisms of micro- and nanoplastics (MNPs) are intricate processes. It is a vital area of MNPs research, and the elucidation of these mechanisms will not only give insights into the fate and transfer of MNPs, rather, it will pave the way for the creation of practical solutions needed to address MNPs pollution. Among the factors or variables that influence the adsorption of organic compounds to MNPs are polymeric characteristics such as particle size, particle age, type of polymer, surface area, charge, and crystal structure; compound-specific properties; and the medium-specific characteristics, such as ionic strength, pH, salinity, and dissolved organic matter. However, most studies on the adsorption mechanisms are laboratory based, which may present some limitations in the implementation. Field studies are encouraged to enable the reconciliation of laboratory findings with real-life occurrences. Lastly, more research should be directed toward demystifying these mechanisms not only in the marine environment but the freshwater and soil environments also, as all environmental compartments have their intrinsic values.

Acknowledgments The first author acknowledges the Erasmus Mundus Joint Master's Scholarship for support.

Conflict of Interest All authors declare that there is no conflict of interest.

References

Agboola, O. D., & Benson, N. U. (2021). Physisorption and chemisorption mechanisms influencing micro (nano) plastics-organic chemical contaminants interactions: A review. *Frontiers in Environmental Science, 9*, 678574.

Alberghini, L., Truant, A., Santonicola, S., Colavita, G., & Giaccone, V. (2022). Microplastics in fish and fishery products and risks for human health: A review. *International Journal of Environmental Research and Public Health, 20*(1), 789.

Allen, T., Farley, S., Draper, J., Clement, C., & Polidoro, B. (2018). Variations in sorption of organochlorine pesticides and PCBs across six different plastic polymers. *Journal of Environmental and Toxicological Studies, 2*(6).

Allouzi, M. M. A., Tang, D. Y. Y., Chew, K. W., Rinklebe, J., Bolan, N., Allouzi, S. M. A., & Show, P. L. (2021). Micro (nano) plastic pollution: The ecological influence on soil-plant system and human health. *Science of the Total Environment, 788*, 147815.

Almeida, C. M. R., Manjate, E., & Ramos, S. (2020). Adsorption of Cd and Cu to different types of microplastics in estuarine salt marsh medium. *Marine Pollution Bulletin, 151*, 110797.

Amelia, T. S., Khalik, W. M., Ong, M. C., Shao, Y. T., Pan, H., & Bhubalan, K. (2021). Marine microplastics as vectors of major ocean pollutants and its hazards to the marine ecosystem and humans. *Progress in Earth and Planetary Science, 8*(1), 1–26.

Anagnosti, L., Varvaresou, A., Pavlou, P., Protopapa, E., & Carayanni, V. (2021). Worldwide actions against plastic pollution from microbeads and microplastics in cosmetics focusing on European policies. Has the issue been handled effectively? *Marine Pollution Bulletin, 162*, 111883.

Arthur, C., Baker, J. E., & Bamford, H. A. (2009). *Proceedings of the international research workshop on the occurrence, effects, and fate of microplastic marine debris, September 9–11, 2008*. University of Washington Tacoma.

Badruddoza, A. Z. M., Shawon, Z. B. Z., Rahman, M. T., Hao, K. W., Hidajat, K., & Uddin, M. S. (2013). Ionically modified magnetic nanomaterials for arsenic and chromium removal from water. *Chemical Engineering Journal, 225*, 607–615.

Bao, Z. Z., Lu, S. Q., Wang, G., Cai, Z., & Chen, Z. F. (2023). Adsorption of 2-hydroxynaphthalene, naphthalene, phenanthrene, and pyrene by polyvinyl chloride microplastics in water and their bioaccessibility under in vitro human gastrointestinal system. *Science of the Total Environment, 871*, 162157.

Bhagat, K., Barrios, A. C., Rajwade, K., Kumar, A., Oswald, J., Apul, O., & Perreault, F. (2022). Aging of microplastics increases their adsorption affinity towards organic contaminants. *Chemosphere, 298*, 134238.

Borkovec, M., & Papastavrou, G. (2008). Interactions between solid surfaces with adsorbed polyelectrolytes of opposite charge. *Current Opinion in Colloid & Interface Science, 13*(6), 429–437.

Botterell, Z. L. R., Beaumont, N., Dorrington, T., Steinke, M., Thompson, R. C., & Lindeque, P. K. (2019). Bioavailability and effects of microplastics on marine zooplankton: A review. *Environmental Pollution (Barking, Essex : 1987), 245*, 98–110.

Brennecke, D., Duarte, B., Paiva, F., Caçador, I., & Canning-Clode, J. (2016). Microplastics as vector for heavy metal contamination from the marine environment. *Estuarine, Coastal and Shelf Science, 178*, 189–195.

Calero, M., Godoy, V., Quesada, L., & Martín-Lara, M. Á. (2021). Green strategies for microplastics reduction. *Current Opinion in Green and Sustainable Chemistry, 28*, 100442.

Campanale, C., Dierkes, G., Massarelli, C., Bagnuolo, G., & Uricchio, V. F. (2020a). A relevant screening of organic contaminants present on freshwater and pre-production microplastics. *Toxics, 8*(4), 100.

Campanale, C., Massarelli, C., Savino, I., Locaputo, V., & Uricchio, V. F. (2020b). A detailed review study on potential effects of microplastics and additives of concern on human health. *International Journal of Environmental Research and Public Health, 17*(4), 1212.

Cao, Y., Zhao, M., Ma, X., Song, Y., Zuo, S., Li, H., & Deng, W. (2021). A critical review on the interactions of microplastics with heavy metals: Mechanism and their combined effect on organisms and humans. *Science of the Total Environment, 788*, 147620.

Carbery, M., O'Connor, W., & Palanisami, T. (2018). Trophic transfer of microplastics and mixed contaminants in the marine food web and implications for human health. *Environment International, 115*, 400–409.

Chang, J., Liang, J., Fang, W., Zhang, H., Zhang, Y., Zhao, H., Zhang, R., Zhang, P., & Zhang, G. (2023). Adsorption behaviors and bioavailability of tetrabromobisphenol A in the presence of polystyrene microplastic in soil: Effect of microplastics aging. *Environmental Pollution (Barking, Essex : 1987), 334*, 122156. Advance online publication.

Chen, C. F., Ju, Y. R., Lim, Y. C., Hsu, N. H., Lu, K. T., Hsieh, S. L., Dong, C. D., & Chen, C. W. (2020). Microplastics and their affiliated PAHs in the sea surface connected to the southwest coast of Taiwan. *Chemosphere, 254*, 126818.

Chen, Z., Shi, X., Zhang, J., Wu, L., Wei, W., & Ni, B. J. (2023a). Nanoplastics are significantly different from microplastics in urban waters. *Water Research X, 19*, 100169.

Chen, Z., Yang, J., Huang, D., Wang, S., Jiang, K., Sun, W., Chen, Z., Cao, Z., Ren, Y., Wang, Q., Liu, H., Zhang, X., & Sun, X. (2023b). Adsorption behavior of aniline pollutant on polystyrene microplastics. *Chemosphere, 323*, 138187.

Concha-Graña, E., Moscoso-Pérez, C. M., López-Mahía, P., & Muniategui-Lorenzo, S. (2022). Adsorption of pesticides and personal care products on pristine and weathered microplastics in the marine environment. Comparison between bio-based and conventional plastics. *Science of the Total Environment, 848*, 157703.

Cózar, A., Echevarría, F., González-Gordillo, J. I., Irigoien, X., Ubeda, B., Hernández-León, S., Palma, A. T., Navarro, S., García-de-Lomas, J., Ruiz, A., Fernández-de-Puelles, M. L., & Duarte, C. M. (2014). Plastic debris in the open ocean. *Proceedings of the National Academy of Sciences of the United States of America, 111*(28), 10239–10244.

da Costa, J. P. (2018). Micro- and nanoplastics in the environment: Research and policymaking. *Current Opinion in Environmental Science & Health, 1*, 12–16.

Desforges, J. P., Galbraith, M., & Ross, P. S. (2015). Ingestion of microplastics by zooplankton in the Northeast Pacific Ocean. *Archives of Environmental Contamination and Toxicology, 69*(3), 320–330.

Dong, M., Zhang, Q., Xing, X., Chen, W., She, Z., & Luo, Z. (2020). Raman spectra and surface changes of microplastics weathered under natural environments. *Science of the Total Environment, 739*, 139990.

Dong, X., Liu, X., Hou, Q., & Wang, Z. (2023). From natural environment to animal tissues: A review of microplastics(nanoplastics) translocation and hazards studies. *Science of the Total Environment, 855*, 158686.

Du, H., Zhang, Y., Jiang, H., & Wang, H. (2022). Adsorption of rhodamine B on polyvinyl chloride, polystyrene, and polyethylene terephthalate microplastics in aqueous environments. *Environmental Technology & Innovation, 27*, 102495.

Eerkes-Medrano, D., Thompson, R. C., & Aldridge, D. C. (2015). Microplastics in freshwater systems: A review of the emerging threats, identification of knowledge gaps and prioritisation of research needs. *Water Research, 75*, 63–82.

Enfrin, M., Dumée, L. F., & Lee, J. (2019). Nano/microplastics in water and wastewater treatment processes – Origin, impact and potential solutions. *Water Research, 161*, 621–638.

Fajardo, C., Martín, C., Costa, G., Sánchez-Fortún, S., Rodríguez, C., de Lucas Burneo, J. J., Nande, M., Mengs, G., & Martín, M. (2022). Assessing the role of polyethylene microplastics as a vector for organic pollutants in soil: Ecotoxicological and molecular approaches. *Chemosphere, 288*, 132460.

Fred-Ahmadu, O. H., Bhagwat, G., Oluyoye, I., Benson, N. U., Ayejuyo, O. O., & Palanisami, T. (2020). Interaction of chemical contaminants with microplastics: Principles and perspectives. *Science of the Total Environment, 706*, 135978.

Frias, J. P. G. L., & Nash, R. (2019). Microplastics: Finding a consensus on the definition. *Marine Pollution Bulletin, 138*, 145–147.

Gagliano, E., Sgroi, M., Falciglia, P. P., Vagliasindi, F. G., & Roccaro, P. (2020). Removal of poly- and perfluoroalkyl substances (PFAS) from water by adsorption: Role of PFAS chain length, effect of organic matter and challenges in adsorbent regeneration. *Water Research, 171*, 115381.

GESAMP, S. (2015). *Fate and effects of microplastics in the marine environment—A global assessment (part 1)*. Joint Group of Experts on the Scientific, Aspects of Marine Environmental Protection.

Gigault, J., Halle, A. T., Baudrimont, M., Pascal, P. Y., Gauffre, F., Phi, T. L., El Hadri, H., Grassl, B., & Reynaud, S. (2018). Current opinion: What is a nanoplastic? *Environmental Pollution (Barking, Essex: 1987), 235*, 1030–1034.

Han, X., Wang, S., Yu, X., Vogt, R. D., Feng, J., Zhai, L., Ma, W., Zhu, L., & Lu, X. (2021). Kinetics and size effects on adsorption of cu(II), Cr(III), and Pb(II) onto polyethylene, polypropylene, and polyethylene terephthalate microplastic particles. *Frontiers in Marine Science, 8*, 785146.

Hanun, J. N., Hassan, F., Theresia, L., Chao, H., Bu, H. M., Rajendran, S., Kataria, N., Yeh, C., Show, P. L., Khoo, K. S., & Jiang, J. (2023). Weathering effect triggers the sorption enhancement of microplastics against oxybenzone. *Environmental Technology & Innovation, 30*, 103112.

Harraq, A. A., & Bharti, B. (2022). Microplastics through the lens of colloid science. *ACS Environmental Au, 2*(1), 3–10.

Hartmann, N. B., Rist, S., Bodin, J., Jensen, L. H., Schmidt, S. N., Mayer, P., Meibom, A., & Baun, A. (2017). Microplastics as vectors for environmental contaminants: Exploring sorption, desorption, and transfer to biota. *Integrated Environmental Assessment and Management, 13*(3), 488–493.

Heinrich, P., Hanslik, L., Kämmer, N., & Braunbeck, T. (2020). The tox is in the detail: Technical fundamentals for designing, performing, and interpreting experiments on toxicity of microplastics and associated substances. *Environmental Science and Pollution Research International, 27*(18), 22292–22318.

Hernandez, L. M., Yousefi, N., & Tufenkji, N. (2017). Are there nanoplastics in your personal care products? *Environmental Science & Technology Letters, 4*(7), 280–285.

Hu, M., Huang, L., Wang, Y., Tan, H., & Yu, X. (2023). Insight into the effect of microplastics on the adsorption and degradation behavior of thiamethoxam in agricultural soils. *Chemosphere, 337*, 139262. Advance online publication.

Huang, W., Song, B., Liang, J., Niu, Q., Zeng, G., Shen, M., Deng, J., Luo, Y., Wen, X., & Zhang, Y. (2021). Microplastics and associated contaminants in the aquatic environment: A review on their ecotoxicological effects, trophic transfer, and potential impacts to human health. *Journal of Hazardous Materials, 405*, 124187.

Hüffer, T., Weniger, A., & Hofmann, T. (2018). Sorption of organic compounds by aged polystyrene microplastic particles. *Environmental Pollution, 236*, 218–225.

Ivar do Sul, J. A., Costa, M. F., Silva-Cavalcanti, J. S., & Araújo, M. C. (2014). Plastic debris retention and exportation by a mangrove forest patch. *Marine Pollution Bulletin, 78*(1–2), 252–257.
Koelmans, A. A., Bakir, A., Burton, G. A., & Janssen, C. R. (2016). Microplastic as a vector for chemicals in the aquatic environment: Critical review and model-supported reinterpretation of empirical studies. *Environmental Science & Technology, 50*(7), 3315–3326.
Kong, X., Zhang, J., Ji, Q., Li, C., Chen, X., Cao, X., Zhu, F., Yang, S., Li, S., & He, H. (2023). Insights into adsorption mechanisms of nitro polycyclic aromatic hydrocarbons on common microplastic particles: Experimental studies and modeling. *Chemosphere, 320*, 138050.
Koutnik, V. S., Alkidim, S., Leonard, J., DePrima, F., Cao, S., Hoek, E. M., & Mohanty, S. K. (2021a). Unaccounted microplastics in wastewater sludge: Where do they go? *ACS ES&T Water, 1*(5), 1086–1097.
Koutnik, V. S., Leonard, J., Alkidim, S., DePrima, F. J., Ravi, S., Hoek, E. M. V., & Mohanty, S. K. (2021b). Distribution of microplastics in soil and freshwater environments: Global analysis and framework for transport modeling. *Environmental Pollution (Barking, Essex: 1987), 274*, 116552.
Lan, T., Cao, F., Cao, L., Wang, T., Yu, C., & Wang, F. (2022). A comparative study on the adsorption behavior and mechanism of pesticides on agricultural film microplastics and straw degradation products. *Chemosphere, 303*(Pt 2), 135058.
Lang, M., Yu, X., Liu, J., Xia, T., Wang, T., Jia, H., & Guo, X. (2020). Fenton aging significantly affects the heavy metal adsorption capacity of polystyrene microplastics. *Science of the Total Environment, 722*, 137762.
Li, J., Zhang, K., & Zhang, H. (2018). Adsorption of antibiotics on microplastics. *Environmental Pollution (Barking, Essex: 1987), 237*, 460–467.
Liang, Y., Cao, X., Mo, A., Jiang, J., Zhang, Y., Gao, W., & He, D. (2023). Micro(nano)plastics in plant-derived food: Source, contamination pathways and human exposure risks. *TrAC Trends in Analytical Chemistry, 165*, 117138.
Lim, C., Kim, N., Lee, J., & Yoon, Y. (2022). Potential of adsorption of diverse environmental contaminants onto microplastics. *Water, 14*(24), 4086.
Lin, L., Tang, S., Wang, X., Sun, X., & Yu, A. (2021). Hexabromocyclododecane alters malachite green and lead(II) adsorption behaviors onto polystyrene microplastics: Interaction mechanism and competitive effect. *Chemosphere, 265*, 129079.
Liu, J., Zhang, T., Tian, L., Liu, X., Qi, Z., Ma, Y., Ji, R., & Chen, W. (2019). Aging significantly affects mobility and contaminant-mobilizing ability of Nanoplastics in saturated loamy sand. *Environmental Science & Technology, 53*(10), 5805–5815.
Liu, X., Zhou, D., Chen, M., Cao, Y., Zhuang, L., Lu, Z., & Yang, Z. (2022). Adsorption behavior of azole fungicides on polystyrene and polyethylene microplastics. *Chemosphere, 308*, 136280.
Long, M., Moriceau, B., Gallinari, M., Lambert, C., Huvet, A., Raffray, J., & Soudant, P. (2015). Interactions between microplastics and phytoplankton aggregates: Impact on their respective fates. *Marine Chemistry, 175*, 39–46.
Lu, X. M., Lu, P. Z., & Liu, X. P. (2020). Fate and abundance of antibiotic resistance genes on microplastics in facility vegetable soil. *Science of the Total Environment, 709*, 136276.
Lu, J., Wu, J., Wu, J., & He, X. (2023). Adsorption of nonylphenol on coastal saline soil: Will microplastics play a great role? *Chemosphere, 311*(Pt 1), 137032.
Ma, J., Zhao, J., Zhu, Z., Li, L., & Yu, F. (2019). Effect of microplastic size on the adsorption behavior and mechanism of triclosan on polyvinyl chloride. *Environmental Pollution (Barking, Essex: 1987), 254*(Pt B), 113104.
Maity, S., Biswas, C., Banerjee, S., Guchhait, R., Adhikari, M., Chatterjee, A., & Pramanick, K. (2021). Interaction of plastic particles with heavy metals and the resulting toxicological impacts: A review. *Environmental Science and Pollution Research International, 28*(43), 60291–60307.
Mammo, F., Amoah, I., Gani, K., Pillay, L., Ratha, S., Bux, F., & Kumari, S. (2020). Microplastics in the environment: Interactions with microbes and chemical contaminants. *Science of the Total Environment, 743*, 140518.

Manyangadze, M., Chikuruwo, N., Narsaiah, T., Chakra, C., Radhakumari, M., & Danha, G. (2020). Enhancing adsorption capacity of nano-adsorbents via surface modification: A review. *South African Journal of Chemical Engineering, 31*, 25–32.

Mattsson, K., Hansson, L.-A., & Cedervall, T. (2015). Nano-plastics in the aquatic environment. *Environmental Science Processes and Impacts, 17*, 1712–1721.

Mei, W., Chen, G., Bao, J., Song, M., Li, Y., & Luo, C. (2020). Interactions between microplastics and organic compounds in aquatic environments: A mini review. *Science of the Total Environment, 736*, 139472.

Mejías, C., Martín, J., Santos, J. L., Aparicio, I., & Alonso, E. (2023). Adsorption of perfluoroalkyl substances on polyamide microplastics: Effect of sorbent and influence of environmental factors. *Environmental Research, 216*, 114834.

Menéndez-Pedriza, A., & Jaumot, J. (2020). Interaction of environmental pollutants with microplastics: A critical review of sorption factors, bioaccumulation and ecotoxicological effects. *Toxics, 8*(2), 40.

Mercogliano, R., Avio, C. G., Regoli, F., Anastasio, A., Colavita, G., & Santonicola, S. (2020). Occurrence of microplastics in commercial seafood under the perspective of the human food chain. A review. *Journal of Agricultural and Food Chemistry, 68*(19), 5296–5301.

Moura, D. S., Pestana, C. J., Moffat, C. F., Hui, J., Irvine, J. T. S., Edwards, C., & Lawton, L. A. (2022). Adsorption of cyanotoxins on polypropylene and polyethylene terephthalate: Microplastics as vector of eight microcystin analogues. *Environmental Pollution (Barking, Essex: 1987), 303*, 119135.

Munoz, M., Ortiz, D., Nieto-Sandoval, J., de Pedro, Z. M., & Casas, J. A. (2021). Adsorption of micropollutants onto realistic microplastics: Role of microplastic nature, size, age, and NOM fouling. *Chemosphere, 283*, 131085.

Ng, E., Huerta Lwanga, E., Eldridge, S. M., Johnston, P., Hu, H., Geissen, V., & Chen, D. (2018). An overview of microplastic and nanoplastic pollution in agroecosystems. *Science of the Total Environment, 627*, 1377–1388.

Nguyen, T., Ho, T., Huang, C., Chen, C., Chen, W., Hsieh, S., Hsieh, S., & Dong, C. (2022). Adsorption of lead(II) onto PE microplastics as a function of particle size: Influencing factors and adsorption mechanism. *Chemosphere, 304*, 135276.

Ogunola, O. S., Onada, O. A., & Falaye, A. E. (2018). Mitigation measures to avert the impacts of plastics and microplastics in the marine environment (a review). *Environmental Science and Pollution Research International, 25*(10), 9293–9310.

Rashed, M. N. (2013). Adsorption technique for the removal of organic pollutants from water and wastewater. *Organic Pollutants-Monitoring, Risk and Treatment, 7*, 167–194.

Rezania, S., Park, J., Md Din, M. F., Mat Taib, S., Talaiekhozani, A., Kumar Yadav, K., & Kamyab, H. (2018). Microplastics pollution in different aquatic environments and biota: A review of recent studies. *Marine Pollution Bulletin, 133*, 191–208.

Rios-Fuster, B., Alomar, C., & Deudero, S. (2023). Elucidating the consequences of the co-exposure of microplastics jointly to other pollutants in bivalves: A review. *Environmental Research, 216*(Pt 2), 114560.

Rodrigues, J. P., Duarte, A. C., Santos-Echeandía, J., & Rocha-Santos, T. (2019). Significance of interactions between microplastics and POPs in the marine environment: A critical overview. *TrAC Trends in Analytical Chemistry, 111*, 252–260.

Santos, L. H., Rodríguez-Mozaz, S., & Barceló, D. (2021). Microplastics as vectors of pharmaceuticals in aquatic organisms – An overview of their environmental implications. *Case Studies in Chemical and Environmental Engineering, 3*, 100079.

Setälä, O., Lehtiniemi, M., Coppock, R., & Cole, M. (2018). Microplastics in marine food webs. In *Microplastic contamination in aquatic environments* (pp. 339–363). Elsevier.

Smith, M. (2018). *Do microplastic residuals in municipal compost bioaccumulate in plant tissue?* (Doctoral dissertation, Royal Roads University (Canada)).

Squadrone, S., Pederiva, S., Bezzo, T., Sartor, R. M., Battuello, M., Nurra, N., Griglione, A., Brizio, P., & Abete, M. C. (2022). Microplastics as vectors of metals contamination in Mediterranean Sea. *Environmental Science and Pollution Research International, 29*(20), 29529–29534.

Sun, J., Zhu, Z., Li, W., Yan, X., Wang, L., Zhang, L., Jin, J., Dai, X., & Ni, B. (2021). Revisiting microplastics in landfill leachate: Unnoticed tiny microplastics and their fate in treatment works. *Water Research, 190*, 116784.

Thompson, R. C., Olsen, Y., Mitchell, R. P., Davis, A., Rowland, S. J., John, A. W., McGonigle, D., & Russell, A. E. (2004). Lost at sea: Where is all the plastic? *Science (New York, N.Y.), 304*(5672), 838.

Tourinho, P. S., Kočí, V., Loureiro, S., & van Gestel, C. A. (2019). Partitioning of chemical contaminants to microplastics: Sorption mechanisms, environmental distribution and effects on toxicity and bioaccumulation. *Environmental Pollution, 252*, 1246–1256.

Velzeboer, I., Kwadijk, C. J., & Koelmans, A. A. (2014). Strong sorption of PCBs to nanoplastics, microplastics, carbon nanotubes, and fullerenes. *Environmental Science & Technology, 48*(9), 4869–4876.

Verschoor, A. J. (2015). *Towards a definition of microplastics.* Considerations for the specification of physicochemical properties. National Institute for Public Health and the Environment. Ministry of Health, Welfare and Sport.

Wan, T., Lu, S., Cheng, W., Ren, J., Wang, M., Hu, B., Jia, Z., Li, Y., & Sun, Y. (2019). A spectroscopic and theoretical investigation of interaction mechanisms of tetracycline and polystyrene nanospheres under different conditions. *Environmental Pollution, 249*, 398–405.

Wang, F., Shih, K. M., & Li, X. Y. (2015). The partition behavior of perfluorooctanesulfonate (PFOS) and perfluorooctanesulfonamide (FOSA) on microplastics. *Chemosphere, 119*, 841–847.

Wang, J., Liu, X., Liu, G., Zhang, Z., Wu, H., Cui, B., Bai, J., & Zhang, W. (2019). Size effect of polystyrene microplastics on sorption of phenanthrene and nitrobenzene. *Ecotoxicology and Environmental Safety, 173*, 331–338.

Wang, F., Zhang, M., Sha, W., Wang, Y., Hao, H., Dou, Y., & Li, Y. (2020a). Sorption behavior and mechanisms of organic contaminants to nano and microplastics. *Molecules, 25*(8), 1827.

Wang, T., Wang, L., Chen, Q., Kalogerakis, N., Ji, R., & Ma, Y. (2020b). Interactions between microplastics and organic pollutants: Effects on toxicity, bioaccumulation, degradation, and transport. *Science of the Total Environment, 748*, 142427.

Wang, Y. L., Lee, Y. H., Chiu, I. J., Lin, Y. F., & Chiu, H. W. (2020c). Potent impact of plastic nanomaterials and micromaterials on the food chain and human health. *International Journal of Molecular Sciences, 21*(5), 1727.

Wang, K., Han, T., Chen, X., Rushimisha, I. E., Liu, Y., Yang, S., Miao, X., Li, X., Weng, L., & Li, Y. (2022a). Insights into behavior and mechanism of tetracycline adsorption on virgin and soil-exposed microplastics. *Journal of Hazardous Materials, 440*, 129770.

Wang, W., Do, A. T. N., & Kwon, J. (2022b). Ecotoxicological effects of micro- and nanoplastics on terrestrial food web from plants to human beings. *Science of the Total Environment, 834*, 155333.

Wang, X., Zhang, R., Li, Z., & Yan, B. (2022c). Adsorption properties and influencing factors of Cu(II) on polystyrene and polyethylene terephthalate microplastics in seawater. *Science of the Total Environment, 812*, 152573.

White, J. R., & Turnbull, A. (1994). Weathering of polymers: Mechanisms of degradation and stabilization, testing strategies and modelling. *Journal of Materials Science, 29*, 584–613.

Xie, Q., Li, H. X., Lin, L., Li, Z. L., Huang, J. S., & Xu, X. R. (2021). Characteristics of expanded polystyrene microplastics on island beaches in the Pearl River Estuary: Abundance, size, surface texture and their metals-carrying capacity. *Ecotoxicology, 30*, 1–12.

Xu, G., Lin, X., & Yu, Y. (2023). Different effects and mechanisms of polystyrene micro- and nano-plastics on the uptake of heavy metals (Cu, Zn, Pb and Cd) by lettuce (Lactuca sativa L.). *Environmental Pollution (Barking, Essex: 1987), 316*(Pt 2), 120656.

Yang, H., Yumeng, Y., Yu, Y., Yinglin, H., Fu, B., & Wang, J. (2022). Distribution, sources, migration, influence and analytical methods of microplastics in soil ecosystems. *Ecotoxicology and Environmental Safety, 243*, 114009.

Yang, M., Zhang, D., & Chu, W. (2023). Adsorption of highly toxic chlorophenylacetonitriles on typical microplastics in aqueous solutions: Kinetics, isotherm, impact factors and mechanism. *Science of the Total Environment, 880*, 163261.

Yao, J., Wen, J., Li, H., & Yang, Y. (2022). Surface functional groups determine adsorption of pharmaceuticals and personal care products on polypropylene microplastics. *Journal of Hazardous Materials, 423*(Pt B), 127131.

Yap, K. Y., & Tan, M. C. (2021). Oil adsorption onto different types of microplastic in synthetic seawater. *Environmental Technology & Innovation, 24*, 101994.

Yu, F., Yang, C., Zhu, Z., Bai, X., & Ma, J. (2019). Adsorption behavior of organic pollutants and metals on micro/nanoplastics in the aquatic environment. *Science of the Total Environment, 694*, 133643.

Yu, F., Li, Y., Huang, G., Yang, C., Chen, C., Zhou, T., Zhao, Y., & Ma, J. (2020). Adsorption behavior of the antibiotic levofloxacin on microplastics in the presence of different heavy metals in an aqueous solution. *Chemosphere, 260*, 127650.

Yu, A., Sun, X., Tang, S., Zhang, Y., Li, M., & Wang, X. (2021a). Adsorption mechanism of cadmium on polystyrene microplastics containing hexabromocyclododecane. *Environmental Technology & Innovation, 24*, 102036.

Yu, Y., Mo, W. Y., & Luukkonen, T. (2021b). Adsorption behaviour and interaction of organic micropollutants with nano and microplastics - A review. *Science of the Total Environment, 797*, 149140.

Yu, X., Du, H., Huang, Y., Yin, X., Liu, Y., Li, Y., Liu, H., & Wang, X. (2022a). Selective adsorption of antibiotics on aged microplastics originating from mariculture benefits the colonization of opportunistic pathogenic bacteria. *Environmental Pollution, 313*, 120157.

Yu, Y., Li, H., Chen, J., Wang, F., Chen, X., Huang, B., He, Y., & Cai, Z. (2022b). Exploring the adsorption behavior of benzotriazoles and benzothiazoles on polyvinyl chloride microplastics in the water environment. *Science of the Total Environment, 821*, 153471.

Zaggia, A., Conte, L., Falletti, L., Fant, M., & Chiorboli, A. (2016). Use of strong anion exchange resins for the removal of perfluoroalkylated substances from contaminated drinking water in batch and continuous pilot plants. *Water Research, 91*, 137–146.

Zhang, C., Yuan, X., Luo, Y., & Yu, G. (2018). Prediction of species concentration distribution using a rigorous turbulent mass diffusivity model for bubble column reactor simulation part I: Application to chemisorption process of CO2 into NaOH solution. *Chemical Engineering Science, 184*, 161–171.

Zhang, P., Huang, P., Sun, H., Ma, J., & Li, B. (2020). The structure of agricultural microplastics (PT, PU and UF) and their sorption capacities for PAHs and PHE derivates under various salinity and oxidation treatments. *Environmental Pollution (Barking, Essex: 1987), 257*, 113525.

Zhang, C., Lei, Y., Qian, J., Qiao, Y., Liu, J., Li, S., Dai, L., Sun, K., Guo, H., Sui, G., & Jing, W. (2021). Sorption of organochlorine pesticides on polyethylene microplastics in soil suspension. *Ecotoxicology and Environmental Safety, 223*, 112591.

Zhang, J., Zhan, S., Zhong, L., Wang, X., Qiu, Z., & Zheng, Y. (2023a). Adsorption of typical natural organic matter on microplastics in aqueous solution: Kinetics, isotherm, influence factors and mechanism. *Journal of Hazardous Materials, 443*, 130130.

Zhang, M., Liu, N., Hou, L., Li, C., & Li, C. (2023b). Adsorption behaviors of chlorpyrifos on UV aged microplastics. *Marine Pollution Bulletin, 190*, 114852.

Zhou, Y., Liu, X., & Wang, J. (2019). Characterization of microplastics and the association of heavy metals with microplastics in suburban soil of Central China. *Science of the Total Environment, 694*, 133798.

Zhou, J., Chen, H., Guo, Y., Chen, Q., Ren, H., & Tao, Y. (2022). Changes in metal adsorption ability of microplastics upon loss of calcium carbonate filler masterbatch through natural aging. *Science of the Total Environment, 832*, 155142.

[illegible]

Part III
Micro-Nano-plastics in the Biosolids

Dynamics of Biodegradable Plastics in the Process of Food Waste Biotreatment and Environmental Risks of Residual Plastic Fragments

Guangyu Cui, Xiaoyi Wu, Sartaj Ahmad Bhat, Fusheng Li, Pinjing He, and Qiyong Xu

1 Introduction

Plastic packaging materials, plastic containers, and agricultural plastic films have been widely used due to their advantages of convenient use, durability, and low cost. However, due to the lack of timely and appropriate management, more than half of waste plastics worldwide are still buried or abandoned in the natural environment. Waste plastic is constantly fragmented under external environmental forces such as wind, high temperature, and ultraviolet radiation, resulting in microplastics (size <5 mm). The presence of microplastics has been detected in land environments such as agricultural soil, domestic sludge, landfill leachate, and incineration plant slag (He et al., 2019; Zhang et al., 2020; Yang et al., 2021; Bhat et al., 2022). Residual agricultural mulch and sludge-derived organic fertilizers are considered important sources of microplastic pollution in soil (Weithmann et al., 2018; Zhang et al., 2020; Sanchez-Hernandez et al., 2020). There are various shapes and materials of microplastics, which are easy to adsorb heavy metals, antibiotics, and pathogenic bacteria (Zettler et al., 2013; Rillig et al., 2019; Parthasarathy et al., 2019; Pham et al., 2021). Long-term existence of microplastics not only affects the soil microenvironment and normal growth of crops (de Souza Machado et al., 2019; Boots et al., 2019) but also poses risks to animal and human health through the "amplification effect" of

G. Cui · X. Wu · Q. Xu (✉)
School of Environment and Energy, Peking University Shenzhen Graduate School, Shenzhen, China
e-mail: cuigy@pku.edu.cn; qiyongxu@pkusz.edu.cn

S. A. Bhat · F. Li
River Basin Research Center, Gifu University, Gifu, Japan

P. He
Institute of Waste Treatment and Reclamation, Tongji University, Shanghai, China

S. A. Bhat et al. (eds.), *Management of Micro and Nano-plastics in Soil and Biosolids*,
https://doi.org/10.1007/978-3-031-51967-3_12

the food chain (Duis & Coors, 2016). Therefore, it is urgent to strengthen research on soil (microplastic) pollution control.

In response to the concept of low-carbon development and the in-depth implementation of waste-free city construction actions, as a substitute for plastics, degradable plastics have low environmental impacts, and their products have been widely applied in production and daily life. Taking China as an example, in the production process of biodegradable plastics, the proportion of polylactic acid (PLA) and poly-adipic acid/butylene terephthalate (PBAT) is as high as 81% (Wen et al., 2022). The products are mainly plastic bags and lunch boxes, and most of them are disposed of in incineration plants for incineration and landfill disposal. The life cycle assessment results showed that biodegradable plastics exhibit carbon emissions comparable to conventional plastics (Wen et al., 2022). Although aerobic composting and anaerobic fermentation are not mainstream plastic waste treatment methods compared to incineration, they have good environmental benefits due to relatively low carbon emissions (the products can be used as soil amendments or agricultural organic fertilizers). Taking aerobic composting as an example, the standard (ISO 14855-1) stipulates that biodegradable plastics should be degraded within 6 months, but the actual engineering composting duration (about 30 days) is much lower than this time requirement. So, the practical significance of this standard is not significant. Therefore, the composting products of kitchen waste must also contain plastic fragments or even microplastics that have not been completely degraded (Fig. 1), and the occurrence characteristics of these plastic fragments or microplastics largely determine their microecological effects (such as soil properties and flora and fauna) after entering the soil and other receiving environments, which may affect the relevant food chain.

This chapter mainly summarizes the current status of typical biodegradable plastics in aerobic composting and anaerobic digestion of food waste; focuses on their

Fig. 1 Biopolymer degradation via thermophilic composting of food waste in a lab-scale experiment

degradation fates; analyzes the environmental effects of residual (micro)plastics in the products and proposes possible prevention and control measures in the perspectives.

2 Research Trends of (Micro)Plastic Derived from Treatment Products of Biodegradable Plastic

According to statistics from the Web of Science database (Fig. 2), the number of published studies on the theme of microplastics-biodegradable plastics-composting has shown an increasing trend year by year. The main subject areas are environmental sciences and ecology, environmental health, indicating that people are more concerned about the environmental and health risks of microplastics. In addition, countries, including China, account for the largest proportion of published papers, which may be related to the implementation of waste sorting policies in recent years and the increasing awareness of environmental protection.

3 Behavior of Biodegradable Plastic During Aerobic Composting and Anaerobic Digestion

According to the metabolic data of biodegradable plastic products in China (Wen et al., 2022), the total proportion of biodegradable plastics entering industrial composting and anaerobic fermentation is less than 0.01%, mainly composed of PBAT and PLA polymers, with plastic shopping bags and lunch boxes being the majority. Therefore, the biological treatment process of biodegradable plastics entering organic waste (such as kitchen waste) includes aerobic composting and anaerobic digestion. Table 1 summarizes the decomposition characteristics of typical biological treatment processes on biodegradable plastics. Among them, aerobic composting achieves the goal of decomposing organic matter through microbial and thermophysical interactions. In terms of biodegradable plastics, the decomposition efficiency of biodegradable (compostable) plastics under industrial composting conditions is higher than that of household composting. Research has shown that increasing temperature is beneficial for improving the decomposition rate of biodegradable plastics. The degradation performance ranking of various polymers is: PHB > PCL > PLA > PBSA>PBS > PBAT. In addition, through laboratory simulations, it was confirmed that 80% of plastic was decomposed after 120 days of treatment in an aerobic composting environment for kitchen waste using PLA/PBAT. And the degradation mechanism is closely related to the dynamic changes in the microbial phase (Chen et al., 2023). It is obvious that high-temperature composting has a good degradation effect on biodegradable plastics, but there are still plastic fragments and even microplastics in its products. Furthermore, Salomez et al. (2019)

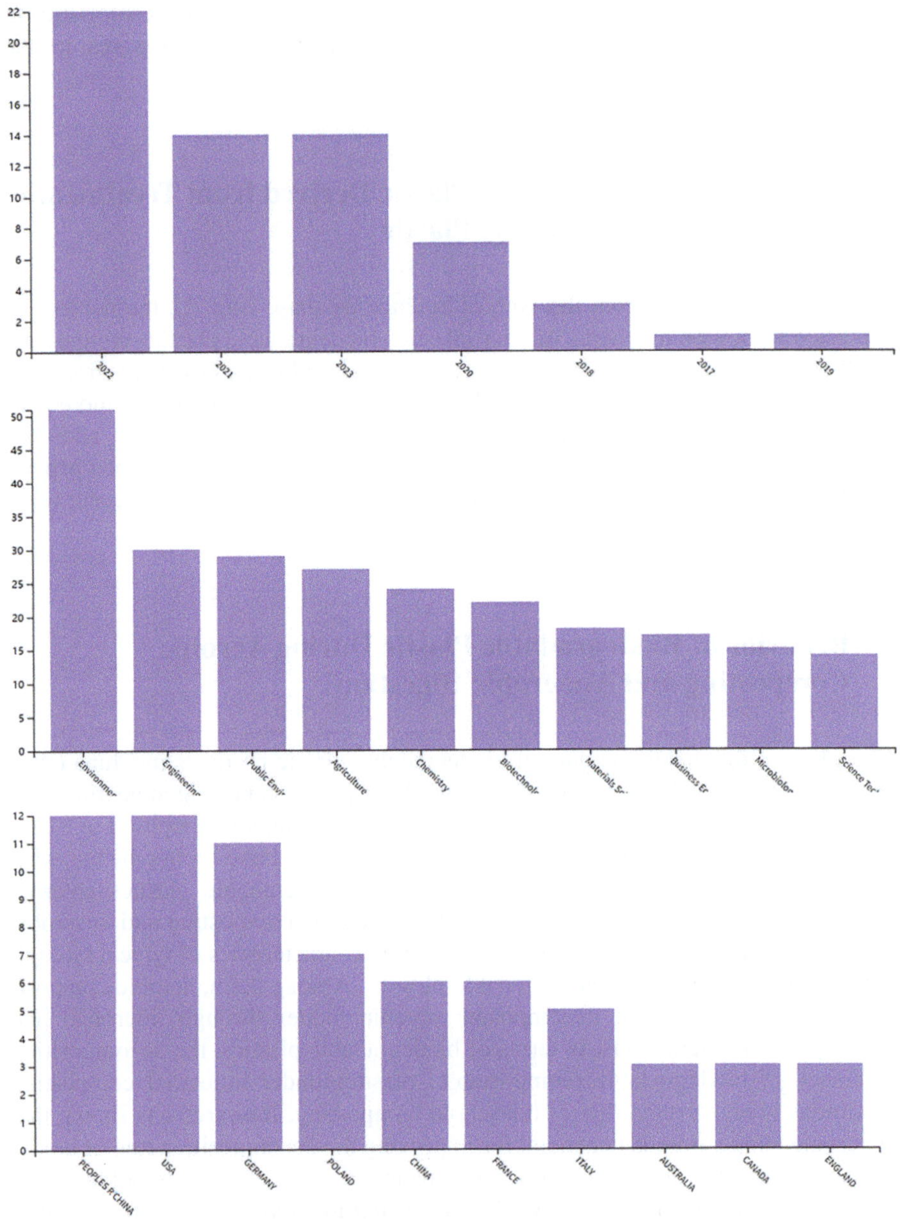

Fig. 2 Publication analysis regarding biodegradable microplastics and composting based on Web of Science (https://www.webofscience.com/wos/alldb/basic-search). (**a**) publishing year; (**b**) research area; (**c**) country/region

Table 1 Fates of biopolymers in the biological treatment (composting, anaerobic digestion) processes of organic wastes

Number	Treatment conditions	Bioplastic types	Key points	References
1	Composting; 120 days	PLA + PBAT	80% PLA/PBAT were degraded after 120 days; PLA + PBAT bacterial network had higher complexity	Chen et al. (2023)
2	Laboratory-scale composting; 58 ± 1 °C; 120 days	PHBV (Poly(3-hydroxybutyrate-co-3-hydroxyvalerate)) and PBSA (Poly(butylene succinate-co-adipate))	Differences in degradation rates for PHBV and PBSA polymers are closely related to significant differences in crystals and spatial structures for both plastics	Salomez et al. (2019)
3	Pilot and lab-scale composting; 12 weeks (ISO 16929); 60 days (ISO 14855-2)	PHBV (Poly(3-hydroxybutyrate-co-3-hydroxyvalerate))	The polymers degradation occurred from the surface to the interior	Weng et al. (2010)
4	Earthworm (*Eisenia fetida*) digestion	PLA	Decreased abundance and smaller size of plastic particles; longer half-life (45 hrs)	Wang et al. (2022)
5	Earthworm (*Lumbricus terrestris* (Oligochaeta)) digestion	PLA and PBAT	Depolymerization of PLA/PBAT microplastics did not occur in the soil environment, but only in the gut of earthworms	Meng et al. (2023)
6	Anaerobic digestion	PBAT+PLA	Lower degradation rate; these polymers after 55 °C AD presented higher biodegradability in aerobic post-treatment	Peng et al. (2022)
7	Anaerobic digestion under mesophilic and thermophilic conditions	PBSA, PLA, PCL, PPC	Degradation of PBSA and PLA presented different erosion characteristics	Jin et al. (2022)
8	Anaerobic digestion under mesophilic and thermophilic conditions	PHB, TPS (Thermoplastic starch), PCL(poly(Ɛ-caprolactone)), PLA PBS(poly(butylene succinate))	PHB degraders and lactate-utilizing bacteria had a key role in the PHB and PLA degradation process	Cazaudehore et al. (2023)

found through experiments that the difference in degradation efficiency between PHBVA and PBSA is related to the different properties of the polymer itself, such as crystallinity and molecular structure of the material. Moreover, this degradation mainly extends from the surface of the plastic to its interior (Weng et al., 2010). It is worth noting that some scholars have found that PLA and PBAT polymers can undergo depolymerization reactions through intestinal digestion of earthworms (Meng et al., 2023), which means that earthworm compost has the potential to degrade microplastics. This technology can achieve high-value conversion (producing organic fertilizer and high protein) of organic waste (such as animal manure and kitchen waste).

In terms of anaerobic digestion, under anaerobic conditions, biodegradable polymers are decomposed through the enzymatic hydrolysis of microorganisms, and their decomposition efficiency is often evaluated by the biochemical methane potential (BMP) test. Recent studies (Peng et al., 2022) confirmed that anaerobic digestion has limited degradation of PLA + PBAT polymers, with only slight changes in plastic surface morphology. But after aerobic composting, the decomposition effect is obvious. This means that the coupling of anaerobic digestion and aerobic composting can reduce the occurrence of residual plastic fragments or microplastics in the product. Based on the results of BMP testing (Cazaudehore et al., 2022), it can be seen that compared to anaerobic digestion at medium temperature (35–38 °C), anaerobic digestion at high temperature (55–58 °C) has a higher degradation rate for biodegradable plastics. Representative polymers include PHB and PLA, with a degradation efficiency of over 90%.

4 Environmental Risks

As mentioned above, organic waste products after biological treatment still contain plastic fragments and even microplastics, including large-sized ones. Like traditional microplastics, these materials are considered to be a new type of pollutant in the soil environment. As shown in Fig. 3, Due to their special surface properties (hydrophobicity, rough surface, and high specific surface area), they are prone to biofilm formation and are also seen as vectors of environmental pollution. Microplastics can adsorb/carry various pollutants such as pathogens, heavy metals, and antibiotics, which can be released back into the environment through desorption, or directly affect the health of soil microorganisms (such as earthworms, nematodes) through the food chain. A recent study (Zhou et al., 2021) also found that PHBA biodegradable plastics could promote the growth and biomass of microorganisms in the soil, increase the activities of glucose oxidase and leucine aminopeptidase, thereby changing the microbial community structure and abundance. Therefore, the residual biodegradable plastic fragments (microplastics) in compost or biogas digestate may alter the turnover of carbon and nutrients in the soil environment, which in turn will affect plant growth.

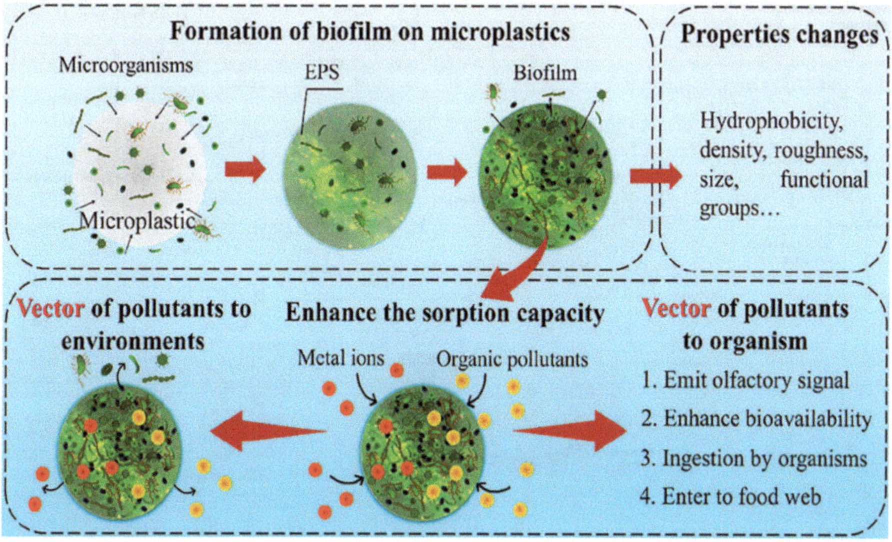

Fig. 3 The formation of microplastic biofilm and its environmental hazards as pollutants. (Reprinted (adapted) with permission from (Wang et al., 2022). Copyright (2023) American Chemical Society)

5 Conclusion and Perspectives

Currently, aerobic composting or anaerobic digestion cannot effectively degrade the carried biodegradable plastics in food waste, and there are still a large number of microplastics or plastic fragments in the products. The low degradation efficiency is mainly related to the relatively short actual composting and anaerobic digestion duration. Therefore, after these products are applied to environmental media such as soil, there are problems of microplastic pollution, but they may be lighter than traditional microplastic pollution.

In view of the current situation, where biodegradable plastics are being promoted on a large scale to enter food waste treatment facilities, the cycles in the current evaluation standards for aerobic composting and anaerobic digestion of biodegradable plastics should be adjusted (Wen et al., 2022), such as changing the standard for aerobic composting to 30 days and anaerobic digestion to 40 days (Wen et al., 2022). In addition, high plastic degradation efficiency can be achieved by controlling process parameters (high temperature for anaerobic digestion, addition of microbial agents, etc.). High-performance biodegradable polymers suitable for specific scenarios (considering food waste composting and anaerobic digestion at different temperature systems) can also be developed to reduce their residual levels in the final products (Peng et al., 2022). Additionally, further study on developing plastic degradation evaluation methods (13C labeling tracing technology) (Zumstein et al., 2018) to analyze the degradation mechanism and improve the degradation

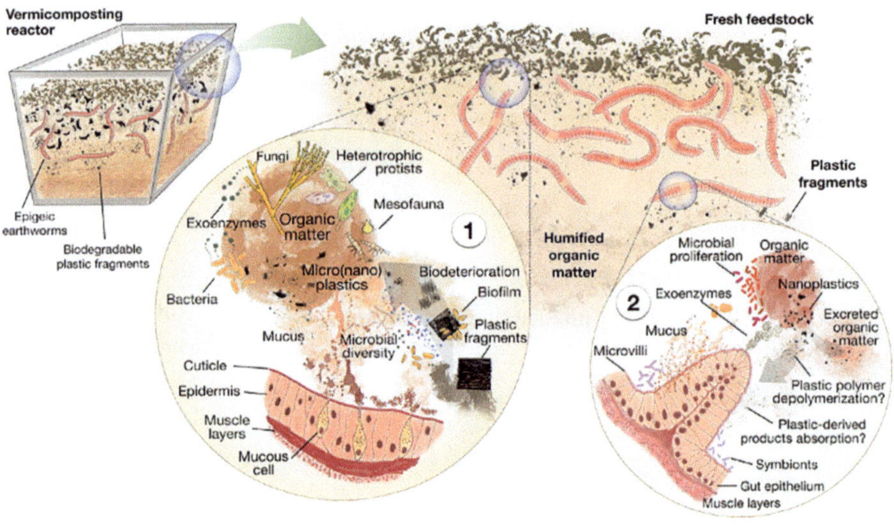

Fig. 4 Potential mechanism of biodegradable plastic residues by vermicomposting. (Reprinted (adapted) with permission from (Sanchez-Hernandez et al., 2020). Copyright (2023) American Chemical Society)

efficiency of plastic fragments regardless in soil or practical engineering environments is needed. For the possible residual microplastics in compost or biogas digestate, it is suggested to develop relevant degradation technologies. For example, earthworm bioremediation is considered to have potential in degrading microplastics, which may be related to the gut digestion process (Fig. 4). In addition, adding biochar to soil can increase the carbon transformation of soil microplastics (Kumar et al., 2023), thereby alleviating soil microplastic pollution.

References

Bhat, S. A., Cui, G., Yaseera, N., Lei, X., Ameen, F., & Li, F. (2022). Removal potential of microplastics in organic solid wastes via biological treatment approaches. In *Microbial biotechnology: Role in ecological sustainability and research* (pp. 255–263).

Boots, B., Russell, C. W., & Green, D. S. (2019). Effects of microplastics in soil ecosystems: Above and below ground. *Environmental Science & Technology, 53*(19), 11496–11506.

Cazaudehore, G., Guyoneaud, R., Evon, P., Martin-Closas, L., Pelacho, A. M., Raynaud, C., & Monlau, F. (2022). Can anaerobic digestion be a suitable end-of-life scenario for biodegradable plastics? A critical review of the current situation, hurdles, and challenges. *Biotechnology Advances, 56*, 107916.

Cazaudehore, G., Monlau, F., Gassie, C., Lallement, A., & Guyoneaud, R. (2023). Active microbial communities during biodegradation of biodegradable plastics by mesophilic and thermophilic anaerobic digestion. *Journal of Hazardous Materials, 443*, 130208.

Chen, W., Feng, Z., Chang, Y., Xu, S., Zhou, K., Shi, X., Wang, Z., Zhang, L., Wei, Y., & Li, J. (2023). Comparing the bacterial composition, succession and assembly patterns in

plastisphere and kitchen waste composting with PLA/PBAT blends. *Journal of Hazardous Materials, 454*, 131405.
de Souza Machado, A. A., Lau, C. W., Kloas, W., Bergmann, J., Bachelier, J. B., Faltin, E., Becker, R., Görlich, A. S., & Rillig, M. C. (2019). Microplastics can change soil properties and affect plant performance. *Environmental Science & Technology, 53*(10), 6044–6052.
He, P., Chen, L., Shao, L., Zhang, H., & Lü, F. (2019). Municipal solid waste (MSW) landfill: A source of microplastics?-Evidence of microplastics in landfill leachate. *Water Research, 159*, 38–34.
Jin, Y., Cai, F., Song, C., Liu, G., & Chen, C. (2022). Degradation of biodegradable plastics by anaerobic digestion: Morphological, micro-structural changes and microbial community dynamics. *Science of the Total Environment, 834*, 155167.
Kumar, R., Verma, A., Rakib, M. R. J., Gupta, P. K., Sharma, P., Garg, A., Girard, P., & Aminabhavi, T. M. (2023). Adsorptive behavior of micro (nano) plastics through biochar: Co-existence, consequences, and challenges in contaminated ecosystems. *Science of the Total Environment, 856*, 159097.
Meng, K., Lwanga, E. H., van der Zee, M., Munhoz, D. R., & Geissen, V. (2023). Fragmentation and depolymerization of microplastics in the earthworm gut: A potential for microplastic bioremediation? *Journal of Hazardous Materials, 447*, 130765.
Parthasarathy, A., Tyler, A. C., Hoffman, M. J., Savka, M. A., & Hudson, A. O. (2019). Is plastic pollution in aquatic and terrestrial environments a driver for the transmission of pathogens and the evolution of antibiotic resistance? *Environmental Science & Technology, 53*, 1744–1745.
Peng, W., Wang, Z., Shu, Y., Lü, F., Zhang, H., Shao, L., & He, P. (2022). Fate of a biobased polymer via high-solid anaerobic co-digestion with food waste and following aerobic treatment: Insights on changes of polymer physicochemical properties and the role of microbial and fungal communities. *Bioresource Technology, 343*, 126079.
Pham, D. N., Clark, L., & Li, M. (2021). Microplastics as hubs enriching antibiotic-resistant bacteria and pathogens in municipal activated sludge. *Journal of Hazardous Materials Letters, 2*, 100014.
Rillig, M. C., de Souza Machado, A. A., Lehmann, A., & Klümper, U. (2019). Evolutionary implications of microplastics for soil biota. *Environmental Chemistry, 16*(1), 3–7.
Salomez, M., George, M., Fabre, P., Touchaleaume, F., Cesar, G., Lajarrige, A., & Gastaldi, E. (2019). A comparative study of degradation mechanisms of PHBV and PBSA under laboratory-scale composting conditions. *Polymer Degradation and Stability, 167*, 102–113.
Sanchez-Hernandez, J. C., Capowiez, Y., & Ro, K. S. (2020). Potential use of earthworms to enhance decaying of biodegradable plastics. *ACS Sustainable Chemistry & Engineering, 8*(11), 4292–4316.
Wang, L., Peng, Y., Xu, Y., Zhang, J., Liu, C., Tang, X., Lu, Y., & Sun, H. (2022). Earthworms' degradable bioplastic diet of polylactic acid: Easy to break down and slow to excrete. *Environmental Science & Technology, 56*(8), 5020–5028.
Weithmann, N., Möller, J. N., Löder, M. G., Piehl, S., Laforsch, C., & Freitag, R. (2018). Organic fertilizer as a vehicle for the entry of microplastic into the environment. *Science Advances, 4*(4), eaap8060.
Wen, Z., Hu, P., & Li, H. (2022). *Environmental impact assessment of degradable plastics and policy support research project report*. http://www.ccin.com.cn/detail/154561bd68537a30a51e20508a3190d0
Weng, Y. X., Wang, Y., Wang, X. L., & Wang, Y. Z. (2010). Biodegradation behavior of PHBV films in a pilot-scale composting condition. *Polymer Testing, 29*(5), 579–587.
Yang, Z., Lü, F., Zhang, H., Wang, W., Shao, L., Ye, J., & He, P. (2021). Is incineration the terminator of plastics and microplastics. *Journal of Hazardous Materials, 401*, 123429.
Zettler, E. R., Mincer, T. J., & Amaral-Zettler, L. A. (2013). Life in the "plastisphere": Microbial communities on plastic marine debris. *Environmental Science & Technology, 47*(13), 7137–7146.

Zhang, L., Xie, Y., Liu, J., Zhong, S., Qian, Y., & Gao, P. (2020). An overlooked entry pathway of microplastics into agricultural soils from application of sludge-based fertilizers. *Environmental Science & Technology, 54*(7), 4248–4255.

Zhou, J., Gui, H., Banfield, C. C., Wen, Y., Zang, H., Dippold, M. A., Charlton, A., & Jones, D. L. (2021). The microplastisphere: Biodegradable microplastics addition alters soil microbial community structure and function. *Soil Biology and Biochemistry, 156*, 108211.

Zumstein, M. T., Schintlmeister, A., Nelson, T. F., Baumgartner, R., Woebken, D., Wagner, M., Kohler, H. E., Mcneill, K., & Sander, M. (2018). Biodegradation of synthetic polymers in soils: Tracking carbon into CO2 and microbial biomass. *Science Advances, 4*(7), eaas9024.

Occurrence and Fate of Microplastics in Anaerobic Digestion of Dewatered Sludge

Kuok Ho Daniel Tang

1 Introduction

Water treatment has been perceived as an interceptor of MPs in the environment. Water treatment processes remove MPs particularly during primary treatment, or coagulation and sedimentation depending on the types of treatment processes (Liong et al., 2021; Tang & Hadibarata, 2022). For wastewater treatment, an average of 72% of MPs was reportedly removed during primary treatment. The removal efficiency increased to 88% when the primary effluents were subjected to secondary treatment. An additional tertiary treatment could push the overall MP removal efficiency to 94% on average (Iyare et al., 2020). Most of the MPs in wastewater treatment plants (WWTPs) are removed through primary treatment, which typically comprises the screening of large floatables and grits, as well as primary sedimentation to permit settling of suspended solids by gravity after the addition of coagulants and flocculants (Lv et al., 2019). Secondary treatment commonly employs the activated sludge process where an aeration tank is paired with a secondary clarifier to activate the microorganisms in activated sludge for removal of dissolved organics followed by the settling of activated sludge in secondary sedimentation (Tang, 2022). Both primary and secondary sedimentation processes trap MPs in the sludge formed and/or settled. MPs in sludge constitute the bulk of MPs in the wastewater streams (Tang, 2022).

During drinking water treatment, coagulation and sedimentation have been reported to remove MPs in the efficiency range of 1.8% to 54.5%, with the types and doses of flocculants as important contributing factors to the efficiency (Tang & Hadibarata, 2021). The overall MPs removal in drinking water treatments (DWTPs) varies depending on the technologies employed. A study on MP removal from

K. H. D. Tang (✉)
Department of Environmental Science, The University of Arizona, Tucson, AZ, USA

S. A. Bhat et al. (eds.), *Management of Micro and Nano-plastics in Soil and Biosolids*,
https://doi.org/10.1007/978-3-031-51967-3_13

DWTPs employing a combination of coagulation/flocculation, sedimentation, sand filtration, and activated carbon revealed a removal efficiency of 70 to 80% (Pivokonsky et al., 2018). Another similar study on DWTPs with a similar water treatment train except with an addition of ozonation showed a close MPs removal range of 84.4–86.7% (Wang et al., 2020). While DWTPs generally do not remove as many MPs as WWTPs during sedimentation, it is, nonetheless, a major process where substantial MPs are trapped.

Sludge from both WWTPs and DWTPs is, therefore, an important sink of MPs, though the contents of MPs are likely to vary due to the difference in MPs removal efficiencies through sedimentation in WWTPs and DWTPs. Sludge is a semi-solid substance produced during water treatment processes (Cesaro et al., 2022). WWTP sludge typically undergoes thickening and dewatering before being digested. In some instances, dewatering ensues anaerobic sludge digestion. Sludge thickening involves the addition of clarifying agents followed by stirring, which results in the formation of larger aggregates (Tang, 2022). The aggregates subsequently settle, yielding sludge with different contents of solids. Solids usually constitute 8 to 10% of primary sludge and approximately 4% of secondary sludge (Kelessidis & Stasinakis, 2012). Thickened sludge goes through centrifugation, filtration, or evaporation to remove the water therein, and the water produced is recycled to the wastewater treatment streams (Tang, 2022). Dewatered sludge contains about 50 to 75% water. Following dewatering, the sludge is digested aerobically, anaerobically, or composted to further reduce its volume while recovering the energy and valuable resources it contains (Kelessidis & Stasinakis, 2012).

Anaerobic digestion of sludge occurs in the absence of oxygen under mesophilic or thermophilic conditions. This is further illustrated in Sect. 2. Aerobic sludge digestion is essentially the extension of activated sludge during which bacteria degrade the organic matter in sludge into carbon dioxide through endogenous respiration (Zupančič & Roš, 2008). Aerobic sludge digestion is faster, hence more cost--effective than anaerobic digestion, but requires higher costs to operate an aeration system (Zupančič & Roš, 2008). During composting, sludge is usually mixed with additives such as sawdust and wood chips which provide carbon to microorganisms involved in the process. An aerobic condition needs to be maintained by adding bulking agents such as corn cobs, bark chips, and tree cuttings, and in some instances, by an aeration system consisting of air ducts and an air diffusing layer (Tang, 2023a). Composting allows bacteria to break down the sludge and its organic additives, with heat generated concurrently. The heat kills pathogens in the sludge. Composted sludge could be used for soil amendment (Kebibeche et al., 2019). Unlike WWTP sludge, DWTP sludge contains less organic material and is typically thickened and dewatered without digestion. The dewatered DWTP sludge is disposed of to designated sites (Wang et al., 2020).

MPs have been detected in sludge at widely varying concentrations. For WWTPs, MPs up to 1,401,000 items/kg sludge have been reported, and sludge MP concentrations in the range of thousand to hundred thousand items/kg are common (Lares et al., 2018; Rasmussen et al., 2021; Zhang et al., 2021; Lee & Kim, 2018). These MPs are then bound for digestion together with the sludge after dewatering. The presence of MPs in sludge could affect anaerobic digestion. For instance, a study revealed polystyrene (PS) MPs to negatively impact anaerobic digestion at concentrations larger than 0.2 g/L, leading to a 17.9 to 19.3% decline in methane production (Zhang et al., 2020). Contrarily, thickened sludge containing 10 polyvinyl chloride (PVC) MPs/g DW was found to enhance methane production by 6%, but increasing the concentrations of PVC MPs beyond 20 particles/g DW of thickened sludge hampered methanogenesis and reduced methane yield significantly (Wei et al., 2019b). This chapter, therefore, aims to review the occurrence and fate of MPs in anaerobic digestion of wastewater treatment sludge to permit a better characterization of MPs' impacts on anaerobic digestion. Understanding MPs' fate during anaerobic sludge digestion may also shed light on its potential in the biodegradation of MPs.

2 Anaerobic Sludge Digestion

As explained earlier, water treatment sludge, whether dewatered or not, is anaerobically digested to permit the recovery of energy and resources, particularly the nutrients in sludge. This is usually performed on WWTP sludge, which is rich in organic and inorganic matters (Edo et al., 2020). During anaerobic digestion, microorganisms convert the biodegradable organics in the sludge into methane, carbon dioxide, and water. The residue of anaerobic digestion is a wet organic mixture (Meegoda et al., 2018). Anaerobic digestion can be performed on sludge with high moisture content (>85%) though, in most instances, dewatered sludge with 50–75% water content is used (Nielsen et al., 2019). Anaerobic digestion is less energy-intensive than aerobic digestion and composting, which typically require mechanical aeration. Anaerobic digestion produces less heat than aerobic digestion and may require external heating to create an optimal environment for biogas production (Shao et al., 2013). However, the heating energy requirement is low, and anaerobic digestion usually produces net positive energy due to biogas generation. Biogas is a combination of methane and carbon dioxide, used as fuel for heat and power generation (Govani et al., 2021).

Wet anaerobic digestion can be conducted on sludge with high moisture content (>85%) during which frequent mixing of the sludge with microbes through mechanical means or recycling of biogas and sludge is carried out (Nagao et al., 2012). Wet anaerobic digestion does not have strict temperature requirements and is also called mesophilic with temperature falling between 30 and 39 °C (Nagao et al., 2012). Dewatered sludge with low moisture content is channeled through dry anaerobic digestion in a plug flow reactor. Dry anaerobic digestion can withstand sludge with higher levels of contaminants and is usually thermophilic (49 to 57 °C), producing more heat (Kothari et al., 2014). Anaerobic digestion consists of four stages, namely hydrolysis, acidogenesis, acetogenesis, and methanogenesis, further explained as follows:

- Hydrolysis in anaerobic digestion is a biological process that breaks down macromolecules into simpler molecules serving as the nutrients for acidogenic bacteria. This stage could be rate limiting and is optimal at 30–50 °C and pH 5–7. Microbes in the family of *Burkholderiaceae* are typically involved in this stage (Tang, 2023b).
- During acidogenesis, volatile fatty acids, particularly acetate, propionate, and butyrate, are produced at different ratios. This stage usually occurs faster than other stages and could result in the accumulation of volatile fatty acids, which cause acidification and digester failure. An example of acidogens is those of the genus *Butyribacterium* (Huang et al., 2015; Wainaina et al., 2019).
- Acetogenesis converts volatile fatty acids as well as other monomeric and oligomeric substrates into predominantly acetate with some hydrogen. Hydrogen and carbon dioxide are first formed from volatile fatty acids, and these substrates are subsequently consumed by acetogens to produce acetate. The microbes involved in this stage comprise those of the family *Acidaminococcaceae*, *Clostridiaceae*, *Spirochaetaceae*, and *Petrotogaceae* (Kothari et al., 2014).
- Methanogenesis involves the production of methane by methanogenic microorganisms through the conversion of hydrogen, carbon dioxide, and acetate. Methanogens are obligate anaerobic archaea. There are two main groups of methanogens involved in anaerobic digestion, namely acetoclastic methanogens capable of converting acetate into methane and carbon dioxide, as well as hydrogenotrophic methanogens which convert hydrogen gas into methane. The former accounts for approx. 2/3 of the methane produced. Some methanogens are very sensitive to the contaminants present in sludge. The methanogens include those in the family of *Chitinophagaceae*, *Lachnospiraceae*, and *Rhizobiaceae* (Meegoda et al., 2018).

3 Occurrence of Microplastics in Anaerobic Sludge Digestion

Most of the existing studies on anaerobic sludge digestion focus on the concentrations of MPs in digested sludge (Li et al., 2020; Hatinoglu & Sanin, 2022). The concentrations of MPs reported in the digested sludge vary considerably from as low as 0.02 ± 0.01 MPs/g DW to 169,000 MPs/g DW sludge (Table 1). The presence of MPs in the digested sludge implies the retention of MPs during anaerobic digestion. The common anaerobic digestion methods employed for sludge treatment in the studies reviewed are thermophilic (approx. 60 °C), and mesophilic with temperatures typically ranging from 32 °C to 40 °C. An advanced anaerobic digestion combining a thermophilic process and heat treatment was differentiated from the conventional anaerobic digestion, which is typically mesophilic, indicating thermophilic anaerobic digestion is a relatively new method (Horton et al., 2021). The digested sludge from both methods contains MPs but the variability of MPs is often

Table 1 Concentrations of MPs in anaerobically digested sludge

Region	WWTP Population Equivalent	Treatment	MPs in Digested Sludge (MPs/g DW)	Reference
Spain	55,000	Thermophilic	16.45 ± 5.43	Franco et al. (2023)
	280,274	Thermophilic	57.18 ± 20.69	
	NA	Thermophilic	0.02 ± 0.01	
UK	157,946	Mesophilic (32–40 °C); 15 days retention time	180.7 (>50% particle)	Harley-Nyang et al. (2022)
Sweden	790,000	Mesophilic (35 °C); 20 days retention time	6.36 (± 0.80) × 10^3 (the digester received an average 1.2 × 10^{14} MPs/y and discharged 0.85 × 10^{14} MPs/y)	Chand et al. (2021)
UK	213–321 L/population equivalent/day	Advanced (thermophilic with heat treatment)	7652	Horton et al. (2021)
			500	
			2600	
		Conventional	2003	
Australia	2 WWTPs (234,000 – 700,000)	Not specified (NS)	56.5	Ziajahromi et al. (2021)
		NS	48.5	
Spain	NS	NS	133 ± 59 (101 ± 19 after subsequent heat-drying)	Edo et al. (2020)
Finland	NS	NS	170 ± 28.7 (161 ± 25.5 fiber)	Lares et al. (2018)
Norway	330,000	Thermophilic; 14 days retention time	553,478 – 3,394,274 MPs/m^3	Lusher et al. (2018)
	615,000	Thermophilic	492,150–662,481 MPs/m^3	
	60,000	Mesophilic; 24 days retention time	470,270 MPs/m^3	
	330,000	Mesophilic (38–40 °C); 35 days retention time	762,246 MPs/m^3	
	80,000	Thermal (60 °C); 12–15 days retention time	2,173,985 MPs/m^3	
Ireland	284,696	NS	2.743 (2.065 fiber)	Mahon et al. (2017)
	130,000	NS	5.156 (4.007 fiber)	
Denmark	NS	NS	169,000	Vollertsen and Hansen (2017)

associated with the influent characteristics of the WWTPs, particularly the volume and quality, including the contents of MPs (Lv et al., 2019). As such, a clear association between the population equivalent served by a WWTP and the MPs in the sludge cannot be observed.

A study confirms the presence of MPs in anaerobic digesters by tracking MPs input and output of a digester in a WWTP in Sweden (Chand et al., 2021). It was found that the digester received about 1.2×10^{14} MPs/y and released 0.85×10^{14} MPs/y, thus demonstrating a loss (approx. 29%) of MPs. The same study revealed that the mass of MPs reduced by 33.3% after passing through the digester. Though not statistically significant, the decline in MPs abundance and mass implied potential degradation of MPs in the digester (Chand et al., 2021). The same was observed by Mahon et al. (2017) when comparing the abundance of MPs in sludge subjected to lime stabilization, thermal drying or anaerobic digestion. Anaerobically digested sludge was found to contain reduced MPs, indicating the loss of MPs after anaerobic digestion (Mahon et al., 2017). The abundance of MPs in anaerobically digested sludge, which subsequently underwent heat drying, declined marginally in a study. Despite the loss, the study did not deem the change was significant (Edo et al., 2020). Mahon et al. (2017) revealed the morphological change of thermally treated MPs such as blistering and melting but did not notice a significant reduction in MPs abundance. Nonetheless, in parallel to these findings, the combination of thermophilic anaerobic digestion and heat treatment in advanced sludge digestion could offer the benefit of better MPs degradation, albeit to a small extent. Different analytical methods employed in determining the abundance of MPs could contribute to the variability in MPs abundance reported. Chand et al. (2021) attributed the high abundance of MPs detected in their study to their use of Focal Plane Array (FPA) – Micro Fourier Transform Infrared (μFTIR) spectroscopy hyperspectral imaging coupled with digital identification of MPs sized 10 to 500 μm, as well as Attenuated Total Reflectance (ATR)-FTIR for those with sizes between 500 and 5000 μm, instead of manual sorting and characterization of individual particles. A similar automated MPs identification method was also adopted in the study of Horton et al. (2021) which reported higher MPs abundance in digested sludge than that of Harley-Nyang et al. (2022).

Fibers are the most common MPs retrieved from anaerobically digested sludge samples, and in most instances, they constituted more than 75% of the total MPs (Franco et al., 2023; Ziajahromi et al., 2021; Lares et al., 2018; Mahon et al., 2017). Fibers and fragments frequently make up the bulk of MPs retrieved from digested samples, with fragments as the second most common MPs (Mahon et al., 2017; Ziajahromi et al., 2021; Franco et al., 2023). Filaments and flakes are present in significantly lower abundance (Franco et al., 2023). In some instances, the shapes are related to the types of MPs. Lares et al. (2018) found polyester (PES) and polyethylene terephthalate (PET) MPs in their sludge samples were predominantly fibrous, while polyethylene (PE) was mostly in the form of particles or fragments. However, Harley-Nyang et al. (2022) revealed particles (>50%) as the main MPs observed in their sludge samples. In their study, the proportion of fibrous MPs, which were predominant in the reception tank, declined along the sludge treatment train to 28.6% in the final treatment (Harley-Nyang et al., 2022). Another exception was also noted in the study of Lusher et al. (2018), who reported that microbeads were more abundant (37.6%) than fragments (31.8%) and fibers (28.9%) in their sludge samples. It is likely for MPs shape distributions to differ in sludge sampled from different WWTPs (Lusher et al., 2018).

In terms of size, most of the MPs retrieved from the sludge samples were < 1 mm, whereas those >1 mm were significantly less (Mahon et al., 2017; Lusher et al., 2018; Harley-Nyang et al., 2022). Attempts were made to characterize the size distribution of MPs < 1 mm. Franco et al. (2023) found MPs sized 100–355 μm to be most abundant in their digested sludge samples, while MPs > 212 μm were the largest fraction in the study of Mahon et al. (2017). In parallel to this, MPs recovered from digested sludge by Edo et al. (2020) were mostly in the size range of 25–104 μm (54%), followed by 104–375 μm (34%). The most common types of MPs reported are acrylic, polyamide (PA), PE, PES, PET, and polypropylene (PP) (Edo et al., 2020; Horton et al., 2021; Franco et al., 2023). The three most abundant MPs reported by Lusher et al. (2018) were PE, PET, and PP. The same was also reported by Horton et al. (2021), with PE being the most prevalent MPs in three of the five anaerobic sludge digesters studied while PET was the predominating MPs in one of the digesters. Substantial polyurethane (PU) MPs were encountered, though in relatively lower abundance than the three common MP types (Lusher et al., 2018).

White MPs were prevalent in the digested sludge samples of Franco et al. (2023), making up 33% of the total MPs. Other major colors of MPs were brown, blue, and black (Franco et al., 2023). However, the prevailing colors of MPs seem to differ in different studies. Lusher et al. (2018) found clear MPs to be the most abundant, followed by black and blue MPs with equal abundance. Green, blue, and red MPs were the three most abundant MPs in the study of Harley-Nyang et al. (2022), which revealed that the identification of clear and white MPs was often limited by the ability to differentiate these MPs from the white and clear fibers of the filters used, thus potentially resulting in their underestimation. In the study of Edo et al. (2020), white and clear MPs were referred to as clear MPs, and their findings concurred with those of Franco et al. (2023) and Lusher et al. (2018) that clear (white + clear) MPs were more abundant than the colored ones. Ziajahromi et al. (2021) observed black, blue, and transparent MPs to be most common during sewage sludge treatment. It can generally be concluded that transparent, white, black, and blue MPs are very common in anaerobically digested sludge.

4 Fates of MPs During Anaerobic Digestion

4.1 *Interactions with Organic Matters and Other Pollutants*

MPs have been reported to affect biogas production in various ways. Most of the studies on the impacts of MPs on anaerobic digestion have been conducted at lab scale under mesophilic conditions with defined quantities of MPs added (Wei et al., 2019b; Chen et al., 2023). Li et al. (2020) reported that adding PES MPs to activated sludge bound for anaerobic digestion at concentrations ranging from 1 MPs/kg to

200,000 MPs/kg resulted in a logarithmic decrease in methane production by 7.1% to 11.5%. The largest reduction (11.5%) occurred at 1000 MPs/kg. The study revealed that the addition of MPs did not significantly change the microbial profile of the anaerobic digester. Instead, they led to the retention of organic matter and nutrients, probably because they impeded digestion (Li et al., 2020). A study examining the effects of PVC on anaerobic digestion reported that PVC MPs actually increased methane production at 10 particles/g total solids, but further increase in the MPs decreased methane production by up to 24% with 60 particles/g (Wei et al., 2019b). While MPs facilitated the release of soluble chemical oxygen demand to increase the yield of volatile fatty acids favoring methane production at low concentrations, higher MP contents reduced volatile fatty acids formation, hence acidification (Fig. 1) (Wei et al., 2019b).

MPs could also interact with other pollutants in an anaerobic digester. The increasing presence of emerging pollutants, such as perfluorochemicals, and their potential interactions with MPs complicate their impacts on anaerobic digestion. Chen et al. (2020) observed perfluorooctane sulfonate (PFOS) and PS nanoparticles to have an antagonistic effect, where the former alleviated the negative impact of PS nanoparticles on hydrogen yield and the damage they inflicted on the extracellular polymers of thermophilic bacteria. PS nanoparticles and PFOS were observed to adversely affect hydrogen production from anaerobic sludge digestion, respectively. A PFOS dose of 1 mg/L caused an approximately 10 mL reduction in hydrogen yield from 100 mL at 0.1 mg/L PFOS. The presence of PS nanoparticles resulted in more than a 50% decline in cumulative hydrogen yield. When both PS nanoparticles

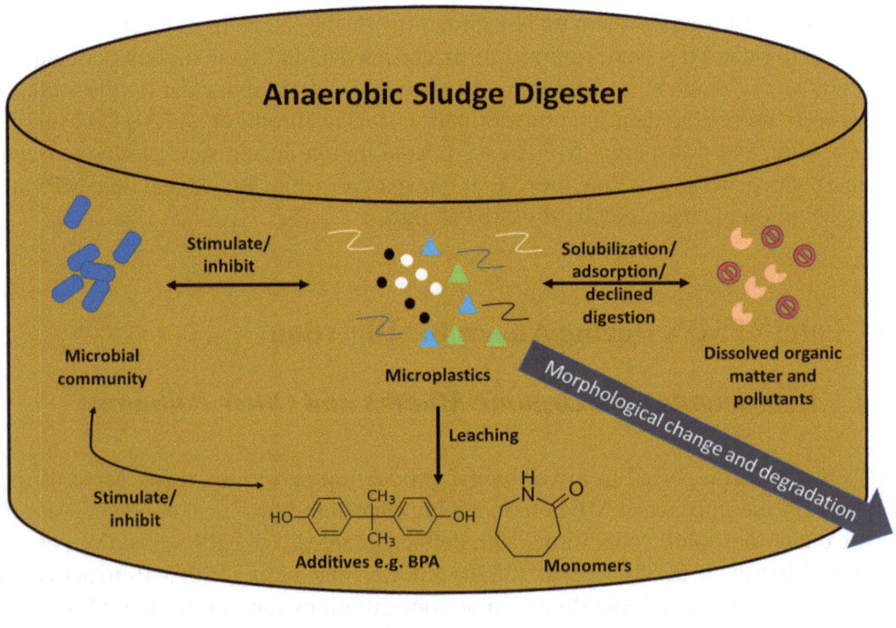

Fig. 1 Fate of MPs in anaerobic sludge digester

and PFOS were present, the inhibitory effect on hydrogen production was relieved (Chen et al., 2020). In a separate study, PS nanoparticles were also reported to result in a decrease in hydrogen production by 30%, owing probably to the inhibition of hydrolysis (Wang et al., 2022). The same study showed that PS MPs and PS nanoparticles increased the solubilization of organics, but the latter inhibited hydrolysis to a level sufficiently severe to reduce hydrogen production. The former led to a higher hydrogen yield. Both PS MPs and nanoparticles could have a deleterious effect on methane production (Wang et al., 2022).

Similarly, the coexistence of PA and the antibiotic Ofloxacin during anaerobic sludge digestion reduces the harmful effect of Ofloxacin because PA adsorbs Ofloxacin partly (Fig. 1) (Xiang et al., 2023). This implies that MPs could provide the surfaces for adsorption of pollutants harmful to anaerobic digestion to a certain extent. With higher Ofloxacin, the beneficial effect of PA MPs on anaerobic sludge digestion wanes (Xiang et al., 2023). The study revealed increased solubilization of organic matter in the presence of PA, in line with the findings of Wang et al. (2022). This was likely attributed to the release of organic matter from extracellular polymers, which was inhibited by Ofloxacin. High concentrations of Ofloxacin resulted in the accumulation of humic acid, hence the inhibition of methane production. The co-existence of PA reversed the inhibitory effect of Ofloxacin partly and restored methane yield. With increasing concentrations of Ofloxacin, the antagonistic effect of PA diminished (Xiang et al., 2023).

A recent study by Liu et al. (2023) on how the co-existence of MPs and cadmium affected anaerobic sludge digestion revealed that when PVC MPs and cadmium were introduced separately to an aerobic sludge digester, methane production reduced by 45.5% with 10 mg/g total solids of cadmium added and only marginally decreased with 20–60 particles of MPs/g total solids added. This is in line with (Wei et al., 2019b) that the presence of PVC MPs at levels more than 10 particles/ g total solids resulted in lower methane production. However, the extent of methane decrease observed by Liu et al. (2023) as PVC MPs were added was significantly less than that of Wei et al. (2019b). When both PVC MPs and a fixed amount of cadmium (5 mg/g total solids) were added to the anaerobic sludge digester, methane production actually recovered in parallel to the concentrations of PVC MPs. PVC MPs of 1, 10, and 30 particles/g total solids led to a 66.9%, 86.8%, and 89.7% recovery of methane yield in comparison with the control, indicating the antagonistic effect of PVC MPs on cadmium toxicity (Liu et al., 2023). Liu et al. (2023) observed higher solubilization of organic matter, confirming the potential beneficial effects of MPs in raising soluble chemical oxygen demand at low concentrations (Wei et al., 2019b). PVC MPs alleviated the inhibitory effect of cadmium on dissolution of organics and formation of short-chain fatty acids by adsorbing cadmium substantially. This aligns with the findings of Chen et al. (2020) and Xiang et al. (2023) that MPs may antagonize the inhibitory effects of certain pollutants on biogas production during anaerobic sludge digestion.

4.2 Interactions with Microbial Community

An earlier experiment of Wei et al. (2019a) with PE MPs found negligible effects of the MPs on methane production at concentrations lower than 60 particles/g total solids but at 100 and 200 particles/g, up to a 27.5% reduction in methane yield was reported. Unlike PVC MPs, PE MPs did not alter soluble chemical oxygen demand significantly, hence solubilization of organics. High levels of PE MPs impeded the hydrolysis of dissolved organic matter in a manner depending on MP concentrations (Wei et al., 2019a). Since solubilization is a largely non-microbial process, the inhibitory effect of MPs on hydrolysis and subsequent acidification was most likely attributed to the impacts of MPs on the microbial profile. The study revealed exposure to MPs caused a decline in the relative abundances of hydrolytic bacteria of the *Rhodobacter* sp. (15% decline), acidification-related bacteria of the *Proteiniclasticum* sp. (24%), and *Proteiniborus* sp. (28%), as well as acetogenesis bacteria of the *Fonticella* sp. (19%) (Fig. 1) (Wei et al., 2019a). MPs-induced generation of reactive oxygen species with deleterious effects on microbes could be a reason behind the decline in the microbial populations (Fig. 2) (Wei et al., 2019a). Wang et al. (2022) noted a 15% decrease in the abundance of hydrogen-producing microbes, such as *Acetoanaerobium* sp., *Proteiniborus* sp., and *Proteiniclasticum* sp., taking part in anaerobic sludge digestion when PS nanoparticles were added, but this negative change was not observed with PS MPs. PS nanoparticles might have induced oxidative stress in the hydrogen-producing microbes exposed, causing the production of reactive oxygen species which disrupted intracellular redox conditions, membrane-associated electron transport system, and ultimately caused apoptosis (Fig. 2) (Wang et al., 2022). PS MPs were not observed to significantly increase the generation of reactive oxygen species. Meanwhile, PS MPs and nanoparticles were

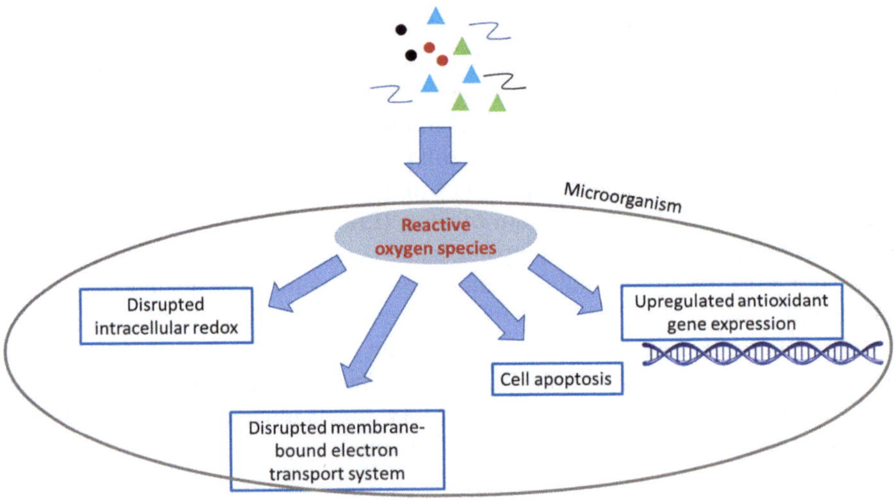

Fig. 2 Impacts of reactive oxygen species induced by nanoplastics and MPs on microbes

responsible for the decline in the key species in methane-producing sludge, comprising *Candidatus Cloacimonas* sp., *Diaphorobacter* sp., and *Gracilibacter* sp. Two methanogens were particularly affected by PS MPs and nanoparticles (Wang et al., 2022). The study of Chen et al. (2020) also pointed to the intracellular generation of reactive oxygen species induced by PS nanoparticles as the cause of reduced thermophilic bacteria responsible for hydrogen production. However, the co-existence of PFOS seemed to stimulate the synthesis of antioxidants and glutathione, which quenched the reactive oxygen species (Chen et al., 2020).

Similarly, the presence of PVC MPs affects the microbial profile of an anaerobic digester. They were found to adversely impact *Rhodobacter* sp., *Proteiniborus* sp., *Garciella* sp., and *Methanosaeta* sp. For PVC MPs, leaching of plastic additives was deemed a main contributing factor for the alteration in the microbial profile (Wei et al., 2019b). Nonetheless, not all types of MPs thwart anaerobic sludge digestion. Polyamide 6 (PA6) was reported to increase methane production by 39.5% at 10 particles/g total solids through enhanced acidification and methanogenesis (Fig. 1) (Chen et al., 2020). This was attained through increased activities of enzymes such as protease, butyrate kinase, and coenzyme F_{420}, which might result from positive changes in the microbial profile (Chen et al., 2020). The interactions of MPs with the microbial community in an anaerobic digester are highly dependent on the types and concentrations of MPs. Adding PS nanoparticles to a thermophilic anaerobic digester might have raised the generation of reactive oxygen species to a level beyond the protective effect of extracellular polymers, thus damaging thermophilic bacteria and reducing hydrogen yield by 54%. This concurs with the findings of Wang et al. (2022). The damage was likely due to negative changes in the composition, hydrolase, and surface charges of microbial extracellular polymers (Chen et al., 2020). It was suggested that nanoparticles could react with extracellular protein to form nanoparticle-protein complexes with altered toxicity and impacts on membrane functionality. PFOS and PS nanoparticles together induced structural damage to extracellular polymers through bond alterations of the macromolecules (Chen et al., 2020). While certain MPs could enhance enzymatic activities, hence biogas production, MPs could also inhibit enzymes involved in different stages of anaerobic sludge digestion. PS nanoparticles were observed to negatively affect the activity of α-amylase, a hydrolase of thermophilic bacteria crucial to the hydrolysis of sludge (Chen et al., 2020).

The deleterious effect of PS nanoparticles on anaerobic sludge digestion was also demonstrated by Fu et al. (2018) who reported a 40.7% decrease in the maximum daily methane production with 0.2 g/L PS nanoparticles added. The nanoparticles led to a decline in microbes in the families of *Anaerolinaceae*, *Cloacamonacea*, *Gracilibacteraceae*, and *Porphyromonadaceae*, possibly by altering microbial membrane permeability and biochemical activities (Fu et al., 2018). Nonetheless, as the effects of PS nanoparticles are likely to be different for different microbes, it was found that those in the families of *Clostridiaceae* and *Geobacteraceae*, for instance, had improved abundance (Fu et al., 2018). Chen et al. (2023) reported that the presence of 30 particles of polycarbonate (PC) MPs/g total solids could benefit anaerobic sludge digestion through the enrichment of certain hydrolytic bacteria (e.g., *Proteobacteria*) and methanogenic bacteria (e.g., *Methanosaeta* sp.). A study found PA MPs potentially assisted in the establishment of hydrolytic and acidogenic bacteria (e.g., *Hydrogenispora*) to enhance acetate formation (Xiang et al., 2023).

4.3 Leaching of Chemicals from MPs

Plastics usually contain a myriad of additives to confer them the desired properties and performance. The additives comprise plasticizers, antioxidants, light and thermal stabilizers, pigments, and flame retardants, among others (Hahladakis et al., 2018). These additives are retained in MPs when macroplastics break down, and the larger surface areas of MPs facilitate their release (Li & Tang, 2023). An examination of the leaching of stabilizers and plasticizers comprising bisphenol A (BPA), diethylhexyl phthalate (DEHP), diisodecyl phthalate (DiDP), and diisononyl phthalate (DiNP) from PVP MPs during anaerobic sludge digestion revealed the leaching of only BPA (Fig. 1) (Wei et al., 2019b). BPA leaching was high in the beginning and slowed down with time. BPA up to 3.6 μg/L was detected with 60 MPs particles/g total solids. A high level of BPA was deemed the main contributing factor of reduced methane yield as it could inhibit the hydrolysis-acidification process of anaerobic digestion. As for PE MPs, only the leaching of a plasticizer called acetyl tri-n-butyl citrate (ATBC) was detected, and it was not deemed to affect anaerobic sludge digestion significantly (Wei et al., 2019a).

PA6 was found to only leach out the monomer caprolactam (CPL). CPL in the anaerobic digester reached a maximum of 20.6 mg/L on day 45 (Fig. 1). The presence of CPL, in fact, led to an increase in methane yield (Chen et al., 2020). A recent study by Chen et al. (2023) on the effects of PC MPs on anaerobic sludge digestion revealed leaching of BPA, akin to PVC MPs, and more BPA was leached with increasing PC MPs. Similar to PVC MPs, low levels (10–60 particles/g total solids) of PC MPs enhanced methane yield by 25%, but adding 200 particles/g total solids PC MPs led to an 8% decrease in methane yield (Chen et al., 2023). Unlike PVC MPs, a higher level of PC MPs can be tolerated before a decrease in the performance of anaerobic digestion is observed. At low concentrations, BPA hampered the synthesis of reactive oxygen species, which in turn, promoted enzymatic activity and increased methanogen (*Methanobacterium* sp. and *Methanosarcina* sp.) populations (Chen et al., 2023). However, with increasing BPA leached from a higher number of PC MPs added, the synthesis of reactive oxygen species was enhanced (Chen et al., 2023). In this sense, the leaching of chemicals from MPs is linked to the interactions between MPs and anaerobic digesters' microbial communities (Fig. 1).

The effect of sludge pre-treatment on methane yield during anaerobic digestion has been examined. It was observed that the addition of untreated PET MPs enhanced methane production. Combined alkaline-thermal pretreatment of the MPs-contaminated sludge further increased methane yield (Hatinoglu & Sanin, 2022). Sludge pretreatment was supposed to weaken the MPs, making them more susceptible to leaching. In fact, pretreated sludge produced more reactive oxygen species than untreated sludge upon anaerobic digestion. However, the amount of reactive oxygen species produced did not seem to be high enough to affect methane production, hence not permitting the establishment of a correlation between reactive oxygen species and methane yield (Hatinoglu & Sanin, 2022).

4.4 Morphological and Chemical Changes

MPs in sludge treated with different methods have been found to undergo morphological and chemical changes. Hatinoglu and Sanin (2022) unveiled potential hydrolytic attack on ester bonds of PET after alkaline-thermal pretreatment, in addition to a reduction of carbonyl index in comparison to MPs in untreated sludge. There was also an increase in MP crystallinity after treatment. However, anaerobic digestion did not significantly change the crystallinity of MPs. Untreated sludge samples did not have the carbonyl index substantially reduced following anaerobic digestion, but a 12% reduction in the carbonyl index was reported for pre-treated sludge samples. After digestion, some spots on MPs had reduced surface thickness and seemed squashed. This might indicate some degradation of MPs (Fig. 1) (Hatinoglu & Sanin, 2022). Mahon et al. (2017) compared the changes to MPs in sludge subjected to different treatments and revealed that MPs in anaerobically digested sludge developed deep cleavage (Fig. 1). In comparison, thermally dried sludge had MPs showing blistering and fracturing, while MPs in lime-stabilized sludge tended to flake. Anaerobic digestion led to a reduction in MP abundance, and this could imply the ability of anaerobic digestion to degrade MPs, which could be enhanced through MP pretreatment (Mahon et al., 2017; Hatinoglu & Sanin, 2022). Therefore, it is likely that MPs also undergo biodegradation in anaerobic digesters, albeit slowly (Tang & Hadibarata, 2022).

5 Conclusion

Most of the MPs entering WWTPs are trapped in sludge during water treatment, suggesting sludge as a significant source of MPs. The reuse of sludge for agriculture often results in the re-entry of MPs into the environment. Sludge is treated by different means prior to disposal or reuse, one of which is anaerobic digestion. The presence of MPs in digested sludge means they have found their ways into anaerobic digesters. This is also confirmed by the detection of MPs in anaerobically digested sludge. The abundance of MPs in anaerobically digested sludge varies considerably from 0.02 MPs/g DW to 169,000 MPs/g DW. Influent quality and treatment capacity of WWTPs frequently determine the abundance of MPs in digested sludge. In anaerobic digesters, MPs could interact with the dissolved organic matter, causing decreased digestion or increased solubility, higher formation of volatile fatty acids, and acidification. MPs could also interact with other emerging pollutants through adsorption and may alleviate the negative impacts of the pollutants on anaerobic sludge digestion. MPs are known to alter the profiles of the microbial community in anaerobic digesters, resulting in the enhancement of certain microbes and inhibition of others, depending on the types and concentrations of MPs. The negative effects of MPs on microbes are possibly caused by the increased generation of reactive oxygen species by MPs and leaching of chemicals from MPs. Common chemicals leached from MPs are BPA, ATBC, and in some instances, MPs monomers like CPL. The morphology of MPs has been reported to change after anaerobic sludge digestion, causing the MPs to have reduced thickness and cleavage. The abundance of MPs reduces after anaerobic digestion, and this suggests degradation of MPs during the process. Bioaugmentation of anaerobic digestion and pretreatment of sludge, therefore, provide an important prospect for enhanced biodegradation of MPs through anaerobic sludge digestion.

References

Cesaro, A., Pirozzi, F., Zafırakou, A., & Alexandraki, A. (2022). Microplastics in sewage sludge destined to anaerobic digestion: The potential role of thermal pretreatment. *Chemosphere, 309*, 136669. https://doi.org/10.1016/j.chemosphere.2022.136669

Chand, R., Rasmussen, L. A., Tumlin, S., & Vollertsen, J. (2021). The occurrence and fate of microplastics in a mesophilic anaerobic digester receiving sewage sludge, grease, and fatty slurries. *Science of the Total Environment, 798*, 149287. https://doi.org/10.1016/j.scitotenv.2021.149287

Chen, H., Zou, Z., Tang, M., Yang, X., & Tsang, Y. F. (2023). Polycarbonate microplastics induce oxidative stress in anaerobic digestion of waste activated sludge by leaching bisphenol a. *Journal of Hazardous Materials, 443*, 130158. https://doi.org/10.1016/j.jhazmat.2022.130158

Chen, W., Yuan, D., Shan, M., Yang, Z., & Liu, C. (2020). Single and combined effects of amino polystyrene and perfluorooctane sulfonate on hydrogen-producing thermophilic bacteria and the interaction mechanisms. *Science of the Total Environment, 703*, 135015. https://doi.org/10.1016/j.scitotenv.2019.135015

Edo, C., González-Pleiter, M., Leganés, F., Fernández-Piñas, F., & Rosal, R. (2020). Fate of microplastics in wastewater treatment plants and their environmental dispersion with effluent and sludge. *Environmental Pollution, 259*, 113837. https://doi.org/10.1016/j.envpol.2019.113837

Franco, A. A., Martín-García, A. P., Egea-Corbacho, A., Arellano, J. M., Albendín, G., Rodríguez--Barroso, R., Quiroga, J. M., & Coello, M. D. (2023). Assessment and accumulation of microplastics in sewage sludge at wastewater treatment plants located in Cádiz, Spain. *Environmental Pollution, 317*, 120689. https://doi.org/10.1016/j.envpol.2022.120689

Fu, S.-F., Ding, J.-N., Zhang, Y., Li, Y.-F., Zhu, R., Yuan, X.-Z., & Zou, H. (2018). Exposure to polystyrene nanoplastic leads to inhibition of anaerobic digestion system. *Science of the Total Environment, 625*, 64–70. https://doi.org/10.1016/j.scitotenv.2017.12.158

Govani, J., Singh, E., Kumar, A., Zacharia, M., & Kumar, S. (2021). New generation technologies for solid waste management. In: Kumar, S., & Kumar, R., Pandey ABT-CD in B and B (reds). *Current developments in biotechnology and bioengineering* (pp. 77–106). Elsevier.

Hahladakis, J. N., Velis, C. A., Weber, R., Iacovidou, E., & Purnell, P. (2018). An overview of chemical additives present in plastics: Migration, release, fate and environmental impact during their use, disposal and recycling. *Journal of Hazardous Materials, 344*, 179–199. https://doi.org/10.1016/j.jhazmat.2017.10.014

Harley-Nyang, D., Memon, F. A., Jones, N., & Galloway, T. (2022). Investigation and analysis of microplastics in sewage sludge and biosolids: A case study from one wastewater treatment works in the UK. *Science of the Total Environment, 823*, 153735. https://doi.org/10.1016/j.scitotenv.2022.153735

Hatinoglu, M. D., & Sanin, F. D. (2022). Fate and effects of polyethylene terephthalate (PET) microplastics during anaerobic digestion of alkaline-thermal pretreated sludge. *Waste Management, 153*, 376–385. https://doi.org/10.1016/j.wasman.2022.09.016

Horton, A. A., Cross, R. K., Read, D. S., Jürgens, M. D., Ball, H. L., Svendsen, C., Vollertsen, J., & Johnson, A. C. (2021). Semi-automated analysis of microplastics in complex wastewater samples. *Environmental Pollution, 268*, 115841. https://doi.org/10.1016/j.envpol.2020.115841

Huang, W., Wang, Z., Zhou, Y., & Ng, W. J. (2015). The role of hydrogenotrophic methanogens in an acidogenic reactor. *Chemosphere, 140*, 40–46. https://doi.org/10.1016/j.chemosphere.2014.10.047

Iyare, P. U., Ouki, S. K., & Bond, T. (2020). Microplastics removal in wastewater treatment plants: A critical review. *Environmental Science: Water Research & Technology, 6*, 2664–2675.

Kebibeche, H., Khelil, O., Kacem, M., & Kaid Harche, M. (2019). Addition of wood sawdust during the co-composting of sewage sludge and wheat straw influences seeds germination. *Ecotoxicology and Environmental Safety, 168*, 423–430. https://doi.org/10.1016/j.ecoenv.2018.10.075

Kelessidis, A., & Stasinakis, A. S. (2012). Comparative study of the methods used for treatment and final disposal of sewage sludge in European countries. *Waste Management, 32*, 1186–1195. https://doi.org/10.1016/j.wasman.2012.01.012

Kothari, R., Pandey, A. K., Kumar, S., Tyagi, V. V., & Tyagi, S. K. (2014). Different aspects of dry anaerobic digestion for bio-energy: An overview. *Renewable and Sustainable Energy Reviews, 39*, 174–195. https://doi.org/10.1016/j.rser.2014.07.011

Lares, M., Ncibi, M. C., Sillanpää, M., & Sillanpää, M. (2018). Occurrence, identification and removal of microplastic particles and fibers in conventional activated sludge process and advanced MBR technology. *Water Research, 133*, 236–246. https://doi.org/10.1016/j.watres.2018.01.049

Lee, H., & Kim, Y. (2018). Treatment characteristics of microplastics at biological sewage treatment facilities in Korea. *Marine Pollution Bulletin, 137*, 1–8. https://doi.org/10.1016/j.marpolbul.2018.09.050

Li, C., & Tang, K. H. D. (2023). Effects of pH and temperature on the leaching of Di (2-Ethylhexyl) phthalate and Di-n-butyl phthalate from microplastics in simulated marine environment. *Biointerface Research in Applied Chemistry, 13*, 3.

Li, L., Geng, S., Li, Z., & Song, K. (2020). Effect of microplastic on anaerobic digestion of wasted activated sludge. *Chemosphere, 247*, 125874. https://doi.org/10.1016/j.chemosphere.2020.125874

Liong, R. M. Y., Hadibarata, T., Yuniarto, A., Tang, K. H. D., & Khamidun, M. H. (2021). Microplastic occurrence in the water and sediment of Miri River estuary, Borneo Island. *Water, Air, & Soil Pollution, 232*, 342. https://doi.org/10.1007/s11270-021-05297-8

Liu, X., Deng, Q., Du, M., Lu, Q., Zhou, W., & Wang, D. (2023). Microplastics decrease the toxicity of cadmium to methane production from anaerobic digestion of sewage sludge. *Sci Total Environ, 869*, 161780. https://doi.org/10.1016/j.scitotenv.2023.161780

Lusher, A., Hurley, R., Vogelsang, C., Nizzetto, L., & Olsen, M. (2018). *Mapping microplastics in sludge*. Norsk institutt for vannforskning.

Lv, X., Dong, Q., Zuo, Z., Liu, Y., Huang, X., & Wu, W.-M. (2019). Microplastics in a municipal wastewater treatment plant: Fate, dynamic distribution, removal efficiencies, and control strategies. *Journal of Cleaner Production, 225*, 579–586. https://doi.org/10.1016/j.jclepro.2019.03.321

Mahon, A. M., O'Connell, B., Healy, M. G., O'Connor, I., Officer, R., Nash, R., & Morrison, L. (2017). Microplastics in sewage sludge: Effects of treatment. *Environmental Science & Technology, 51*, 810–818. https://doi.org/10.1021/acs.est.6b04048

Meegoda, J. N., Li, B., Patel, K., & Wang, L. B. (2018). A review of the processes, parameters, and optimization of anaerobic digestion. *International Journal of Environmental Research and Public Health, 15*, 224.

Nagao, N., Tajima, N., Kawai, M., Niwa, C., Kurosawa, N., Matsuyama, T., Yusoff, F. M., & Toda, T. (2012). Maximum organic loading rate for the single-stage wet anaerobic digestion of food waste. *Bioresource Technology, 118*, 210–218. https://doi.org/10.1016/j.biortech.2012.05.045

Nielsen, J. L., Pedersen, N. K., Peydaei, A., Baudu, E., Fernando, W. E. Y., Gurevich, L., Fojan, P., Wimmer, R., & de Jonge, N. (2019). Potential for biodegradation of microplastics in thermophilic anaerobic digesters. In *Anaerobic digestion conference AD16*. Delft, Netherlands: Aalborg Universitet.

Pivokonsky, M., Cermakova, L., Novotna, K., Peer, P., Cajthaml, T., & Janda, V. (2018). Occurrence of microplastics in raw and treated drinking water. *Science of the Total Environment, 643*, 1644–1651. https://doi.org/10.1016/j.scitotenv.2018.08.102

Rasmussen, L. A., Iordachescu, L., Tumlin, S., & Vollertsen, J. (2021). A complete mass balance for plastics in a wastewater treatment plant – macroplastics contributes more than microplastics. *Water Research, 201*, 117307. https://doi.org/10.1016/j.watres.2021.117307

Shao, L., Wang, T., Li, T., Lü, F., & He, P. (2013). Comparison of sludge digestion under aerobic and anaerobic conditions with a focus on the degradation of proteins at mesophilic temperature. *Bioresource Technology, 140*, 131–137. https://doi.org/10.1016/j.biortech.2013.04.081

Tang, K. H. D. (2022). Abundance of microplastics in wastewater treatment sludge. *Journal of Human, Earth, and Future, 3*, 138–146.

Tang, K. H. D. (2023a). Microplastics in agricultural soils in China: Sources, impacts and solutions. *Environmental Pollution, 322*, 121235. https://doi.org/10.1016/j.envpol.2023.121235

Tang, K. H. D. (2023b). Bioaugmentation of anaerobic wastewater treatment sludge digestion: A perspective on microplastics removal. *Journal of Cleaner Production, 387*, 135864. https://doi.org/10.1016/j.jclepro.2023.135864

Tang, K. H. D., & Hadibarata, T. (2022). The application of bioremediation in wastewater treatment plants for microplastics removal: A practical perspective. *Bioprocess and Biosystems Engineering, 45*, 1865–1878. https://doi.org/10.1007/s00449-022-02793-x

Tang, K. H. D., & Hadibarata, T. (2021). Microplastics removal through water treatment plants: Its feasibility, efficiency, future prospects and enhancement by proper waste management. *Environmental Challenges, 5*, 100264. https://doi.org/10.1016/j.envc.2021.100264

Vollertsen, J., & Hansen, A. (2017). *Microplastic in Danish wastewater: Sources, occurrences and fate*. Danish Environmental Protection Agency.

Wainaina, S., Lukitawesa, Kumar Awasthi, M., & Taherzadeh, M. J. (2019). Bioengineering of anaerobic digestion for volatile fatty acids, hydrogen or methane production: A critical review. *Bioengineered, 10*, 437–458. https://doi.org/10.1080/21655979.2019.1673937

Wang, C., Wei, W., Chen, Z., Wang, Y., Chen, X., & Ni, B.-J. (2022). Polystyrene microplastics and nanoplastics distinctively affect anaerobic sludge treatment for hydrogen and methane production. *Science of the Total Environment, 850*, 158085. https://doi.org/10.1016/j.scitotenv.2022.158085

Wang, Z., Lin, T., & Chen, W. (2020). Occurrence and removal of microplastics in an advanced drinking water treatment plant (ADWTP). *Science of the Total Environment, 700*, 134520. https://doi.org/10.1016/j.scitotenv.2019.134520

Wei, W., Huang, Q.-S., Sun, J., Dai, X., & Ni, B.-J. (2019a). Revealing the mechanisms of polyethylene microplastics affecting anaerobic digestion of waste activated sludge. *Environmental Science & Technology, 53*, 9604–9613. https://doi.org/10.1021/acs.est.9b02971

Wei, W., Huang, Q.-S., Sun, J., Wang, J.-Y., Wu, S.-L., & Ni, B.-J. (2019b). Polyvinyl chloride microplastics affect methane production from the anaerobic digestion of waste activated sludge through leaching toxic Bisphenol-A. *Environmental Science & Technology, 53*, 2509–2517. https://doi.org/10.1021/acs.est.8b07069

Xiang, Y., Xiong, W., Yang, Z., Xu, R., Zhang, Y., Wu, M., Ye, Y., Peng, H., Tong, J., & Wang, D. (2023). Coexistence of microplastics alters the inhibitory effect of antibiotics on sludge anaerobic digestion. *Chemical Engineering Journal, 455*, 140754. https://doi.org/10.1016/j.cej.2022.140754

Zhang, J., Zhao, M., Li, C., Miao, H., Huang, Z., Dai, X., & Ruan, W. (2020). Evaluation the impact of polystyrene micro and nanoplastics on the methane generation by anaerobic digestion. *Ecotoxicology and Environmental Safety, 205*, 111095. https://doi.org/10.1016/j.ecoenv.2020.111095

Zhang, L., Liu, J., Xie, Y., Zhong, S., & Gao, P. (2021). Occurrence and removal of microplastics from wastewater treatment plants in a typical tourist city in China. *Journal of Cleaner Production, 291*, 125968. https://doi.org/10.1016/j.jclepro.2021.125968

Ziajahromi, S., Neale, P. A., Telles Silveira, I., Chua, A., & FDL, L. (2021). An audit of microplastic abundance throughout three Australian wastewater treatment plants. *Chemosphere, 263*, 128294. https://doi.org/10.1016/j.chemosphere.2020.128294

Zupančič, G. D., & Roš, M. (2008). Aerobic and two-stage anaerobic–aerobic sludge digestion with pure oxygen and air aeration. *Bioresource Technology, 99*, 100–109. https://doi.org/10.1016/j.biortech.2006.11.054

Micro-Nano-Plastics in Sewage Sludge: Sources, Occurrence, and Potential Environmental Risks

Deachen Angmo, Jaswinder Singh, Sartaj Ahmad Bhat, Babita Thakur, and Adarsh Pal Vig

1 Introduction

Plastics are a diverse category of synthetic polymers that have become a prominent representation of waste created by humans and the resulting environmental contamination (Chen et al., 2020; Yang et al., 2021; Weber & Opp, 2020). Plastics are categorized into four types based on their dimensions macroplastics (>25 mm), mesoplastics (5–25 mm), microplastics (0.1 μm–5 mm), and nanoplastics (<100 nm) (Alimi et al., 2018). Microplastics are pervasive specific contaminants that originate through deliberate manufacturing to achieve a millimeter or submillimeter size via grinding or extrusion (e.g., abrasives in cleaning products) (Primary) or the breakdown of larger plastic waste (secondary) due to physical, chemical, and biological processes (Browne et al., 2007, 2015; Cole et al., 2013; Dris et al., 2015). Furthermore, secondary microplastics are significantly more prevalent than primary microplastics (Hale et al., 2020). Their flexibility, stability, lightness, and cheap manufacturing costs have reduced the usage of naturally derived materials to a certain degree and in various spheres of social and economic growth (Hale et al., 2020; Yang et al., 2021; Lebreton & Andrady, 2019). PVC (polyvinyl chloride) has largely supplanted other materials, including metal and wood, in construction applications,

D. Angmo · A. P. Vig (✉)
Department of Botanical and Environmental Sciences, Guru Nanak Dev University, Amritsar, India

J. Singh
Department of Zoology, Khalsa College, Amritsar, Punjab, India

S. A. Bhat
River Basin Research Center, Gifu University, Gifu, Japan

B. Thakur
Department of Microbiology, School of Bioengineering and Biosciences, Lovely Professional University, Phagwara, Punjab, India

S. A. Bhat et al. (eds.), *Management of Micro and Nano-plastics in Soil and Biosolids*,
https://doi.org/10.1007/978-3-031-51967-3_14

accounting for about 20% of global production (Plastic Europe, 2017). Similarly, plastics have replaced natural fibers like cotton, wool, and silk used in textiles and carpets as comfort fibers (Lebreton & Andrady, 2019). Consequently, both the demand for and production of plastics are rising annually (Andrady, 2017). The insufficient and incorrect management of plastic waste has led to a noticeable buildup of plastic litter in the environment, endangering all life forms. Microplastic release in the environment via wastewater irrigation, application of sewage sludge as manure, plastic mulching, atmospheric deposition, car tire tearing, cosmetic products, etc. (Chen et al., 2020; Fendell & Sewell, 2009; Yang et al., 2021). The fragmentation of plastics is higher on land because of higher ambient temperatures, increased frictional forces, and exposure to UV radiation resulting in a higher accumulation of plastics in the soil (Andrady, 2017; Nizzetto et al., 2016; Rochman, 2018; van den Berg et al., 2020). Urban and agricultural soils are particularly prone to microplastic pollution due to their frequent exposure and contact with human activity (Möller et al., 2020). Once microplastics accumulate in the soil, they can be taken up by plants and transferred along the food chain. According to (de Souza Machado et al., 2018), there are indications that microplastic pollution could serve as a significant environmental pressure on terrestrial ecosystems, potentially leading to alterations in the soil ecosystem's physicochemical aspects. This influence encompasses changes in soil structure and chemical composition, such as the release of harmful substances through leaching. Additionally, the presence of microplastics caused alterations in the soil's microbial activity (de Souza Machado et al., 2018). Gradually, microplastics are increasingly permeating various environmental components, entering the food chain and even making their way into the human body. Considering all these aforementioned factors into account, the following objectives have been framed: (1) sources of microplastics in sewage sludge; (2) treatment and identification techniques of microplastics in sludge; (2.1) sampling collection; (2.2) extraction of microplastics from sewage sludge; (2.3) detection and identification methods of microplastics; (3) environmental risk of MPs during sludge soil amendment; (4) conclusions and future perspectives.

2 Sources of Microplastics in Sewage Sludge

The sewage treatment plant gathers microplastics from wastewater through multiple routes (Angmo et al., 2023). MPs are mostly produced by human activities, such as personal hygiene or cosmetic items, laundry, industrial operations, transportation, and other activities (Zhang & Chen, 2020). Wastewater and sewage sludge are regarded as the primary origin of microplastics found in agricultural land (Li et al., 2019). MPs are often concentrated in sludge throughout different wastewater treatment plant (WWTP) processes (Hatinoglu & Sanin, 2021; Jagadeesh & Sundaram, 2021; Crossman et al., 2020). Therefore, sewage serves as both a source and a sink of MPs (Ngo et al., 2019). Some microplastics discharge through the sewage system as it undergoes primary followed by secondary and sometimes tertiary treatment but, the majority of MPs settle in the solid product known as sludge (Gatidou et al., 2019). The primary treatment of a sewage treatment system is mechanical treatment, which removes larger-sized particles that are floatable and settleable. The majority of microplastics are removed by skimming the buoyant and lightweight microplastics through surface skimming in the primary clarifiers. The secondary treatment is the removal of pollutants using biological treatment, which often includes the sedimentation process. Tertiary treatment is an advanced treatment employed for the purpose of nutrient removal and sewage disinfectant (Freeman et al., 2020). Studies reveal that the retention rate of MPs in sewage sludge exceeds 90% (Zhang & Chen, 2020). As a result, microplastic concentration is significantly higher relative to effluent wastewater (Xu et al., 2020; Zhang et al., 2020). According to Sun et al. (2019), the concentration of MPs in sludge exhibited a wide range, ranging from 400 to 7000 particles/kg WW (wet weight). This level of concentration significantly exceeded the concentration present in the liquid phase of WWTPs. When assessed in terms of the sludge's dry weight, the concentration could vary from 1500 to 170,000 particles per kg of dry weight. The wide range in the concentration of microplastics in sludge highlights the need for a representative sample and a standardized detection technique in future research. Murphy et al. (2016) observed that the size of microplastic within sewage sludge exhibited notable distinctions from those in wastewater. Furthermore, their findings revealed that the average size of microplastics extracted from sewage sludge was significantly greater than those extracted from wastewater. Whereas, the concentration of microplastics in the sludge exhibit significant variation, ranging from 15,900 to 45,700 particles/kg dry weight (dw) of sludge. The wide difference in the quantity of microplastics in sludge in both studies may be related to the population served, climatic conditions, diversity in urbanization, plastics consumption, and treatment procedures (Hatinoğlu & Sanin, 2021). Pinpointing the source of MP contamination is difficult owing to the diversity of plastics and the pervasiveness of their probable origins. The bulk of textile materials is made of plastics like polyester and polyamide. As a result, it is suspected that effluents originating from household washing machines, industrial laundries, and textile manufacturing facilities play a significant role in introducing a substantial quantity of MP fibers into wastewater treatment systems (El Haynay et al., 2022). In terms of morphology, Li et al. (2018) evaluated the microplastic content within the sludge derived from 28 wastewater treatment plants (WWTPs) located in China. Their findings revealed that fibers accounted for 63% of the total microplastic composition. Additionally, Lares et al. (2018) found that the number of fibers in the Finland sludge sample might reach more than 80%. Seasonality may also influence the fluctuation of MP within sludge. Lee and Kim (2018) found an increase in MP levels in sludge throughout a three-month period of heavy precipitation. Population density also seems to have an effect on microplastic concentration in sludge (Rolsky et al., 2020). For instance, in China, there was a strong correlation between the expansion of the country's industrial and infrastructural sectors and the concentration level of MPs detected in sludge (Li et al., 2018). Several reports on microplastic concentration in sewage sludge and sources of microplastics are given in Table 1 and Fig. 1.

Table 1 Tabulated reports on microplastics concentrations in sewage sludge from various wastewater treatment plants worldwide

S.no	Sample type	Country	Concentration of MPs (particles kg – 1 d.w)	MPs composition	MPs shapes	MPs color	MPs size (μm)	References
1	Digested sludge	(Cádiz) Spain	0.02 ± 0.006 MP g DW^{-1} to 57.18 ± 20.69 MP g DW^{-1}	Acrylate, Polypropylene Polyethylene Polyethylene	Fibers	NM	100–355	Franco et al. (2023)
2	Sludge	Thailand	2.63 ± 1.26 × 10^{4}.	Polypropylene Polyethylene Polyethylene terephthalate	Fibers	NM	50–500	Tadsuwan & Babel. (2022).
3	Sludge	(Chengdu) China	44.4 n·kg^{-1} to 750.0 n·kg^{-1}.	Polypropylene Polyethylene Polystyrene	Fibers and debris	Green and translucent	>1000	Wei et al. (2022)
4	Sludge	Hong Kong	167 to 936 particles/g (d.w)	Polypropylene Polyethylene Polyethylene terephthalate	Fibers	NM	20–300	Zhang et al. (2022)
5	Sludge	China	6.74 × 10^{3}	Polypropylene Polyethylene Polyethylene terephthalate	Fragments, films, fibers, and granules	Transparent, white, blue, and black	<1000	Yuan et al. (2022)
6	Sludge	China	1448–11,125 n•kg^{-1} dw.	Polypropylene Polyethylene	Fibers	Transparent and white	NM	Li et al. (2022)
7	Sludge	China	320 ± 3 particles g^{-1} dw	Polypropylene Polyethylene	Pellet	Transparent and white	<500	Wang et al. (2022)
8	Sludge	Southwest Europe	12 and 39 MPs g^{-1} dw	Polyethylene terephthalate Polyamide Polystyrene	Fibers	NM	NM	Menéndez-Manjón et al. (2022)
9	Sludge cake	Turkey (Bursa)	17.9 ± 2.3 and 9.5 ± 2.3 n/g	Polypropylene Polyethylene Polyethylene terephthalate Polystyrene	Fragments	Black	500–1000	Üstün et al. (2022)

10	Sludge	Spain	1000 microparticles/g-dw	Polyethylene Polypropylene copolymer ethylene-ethyl-acrylate	Fragments and fibers	Blue, green, and orange	200–1000	Alvim et al. (2022)
11	Sludge	United Kingdom	301–10,380	Polypropylene Polyethylene Polyethylene terephthalate	NM	NM	25–178	Horton et al. (2021)
12	Waste activated sludge:	Italy Falconara Marittima	5.30×10^{3}	Polypropylene Polyethylene	Fragments and film	NM	30–500	Pittura et al. (2021)
13	Dewatered sludge	Morocco	$40.5 \pm 11.9 \times 10^{3}$	Polypropylene Polyethylene Polystyrene Aromatic polyamide	Fragments and fibers	NM	<500–2000	El Hayany et al. (2020)
14	Sludge	Spain	18,000 ± 15,940 LDP 32,070 ± 19,080 HDP	Polypropylene Polyvinylchloride	Fragments, fibers, and films	NM	50–1000	van den Berg et al. (2020)
15	Sludge	(Shaanxi) China	220×10^{3}	Polyvinylchloride Poly (1-butene) Polytetrafluoroethylene Polyethylene Polyacrylonitrile	Fragments, fibers, spheres, and films	White	8–1000	Ren et al. (2020)
16	Waste activated sludge	Australia	7.91 ± 0.4	Polyurethane Polyester Polyethylene Polyamide Polyethylene terephthalate Polyvinylchloride	Fragments, glitter and fibers	NM	5000	Raju et al. (2020)

(continued)

Table 1 (continued)

S.no	Sample type	Country	Concentration of MPs (particles kg – 1 d.w)	MPs composition	MPs shapes	MPs color	MPs size (μm)	References
17	Sludge	China	5553–13,460	Polyethylene Polyethylene terephthalate Polyacrylonitrile	Fibers	White	20–250	Li et al. (2019)
18	Sludge	India	$31 \times 10 \times 10^3$ MP/ kg dw	NM	Fiber, film, shaft, and sphere	NM	NA	Priyanka & Saravanakumar (2019)
19	Dewatered sludge	China	240.3 ± 31.4 n g – 1 dw	Polyamide	Microbeads Foams	NM	222.6	Liu et al. (2019)
20	Primary sludge	Canada	$(14.90 \pm 6.30) \times 10^3$	NM	Fibers	Pink, white	>1.0	Gies et al. (2018)
21	Sludge	Korea	14.9×10^3	NM	Fibers and fragments	NM	>106	Lee & Kim (2018)
22	Sludge	Mikkeli, Finland	23 ± 4.2 × 103	Polyethylene Polyamide (nylon)	Fibers and fragments	NM	250–5000	Lares et al. (2018)
23	Primary sludge	Australia	15.90–45.70 × 103	Polyethylene terephthalate	Fibers	NM	>25	Ziajaharaomi et al. (2017)
24	Dewatered sludge	Germany	$1.00–24.00) \times 10^3$	Polyethylene Polyester	Fibers	NM	> 500	Mintenig et al. (2017)
25	Sludge	Ireland	16,000	High-density polyethylene Polyamide Polyethylene terephthalate Acrylic Polyethylene	Fibers, fragments, spheres, and films	NM	>212	Mahon et al. (2017)

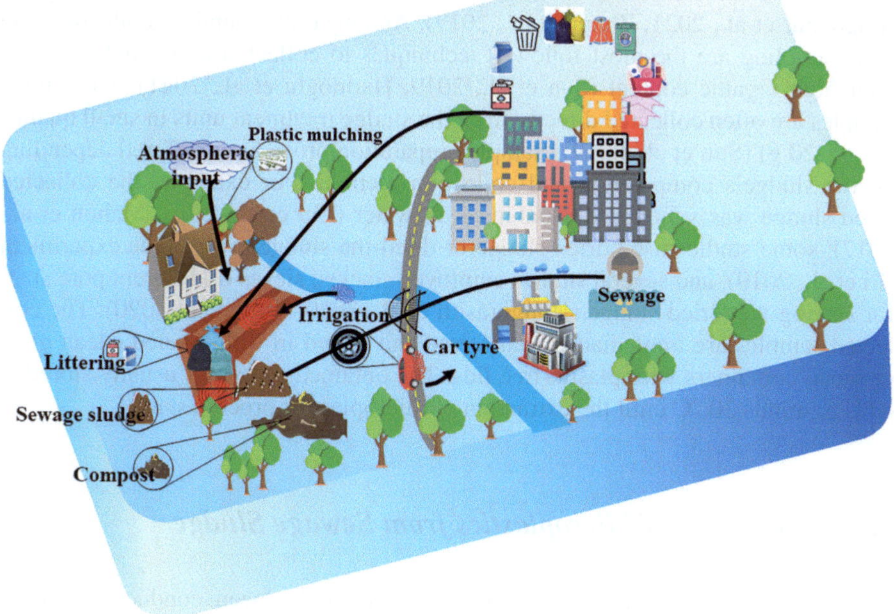

Fig. 1 Pictorial representation of microplastic origination from different sources in the environment

3 Treatment and Identification Techniques of Microplastics in Sludge

3.1 Sampling of Microplastics from Sewage Sludge

Every WWTP around the globe is designed based on a variety of characteristics, including wastewater type, population serviced, effluent discharge regulations, and investment cost. Due to variations in the methodologies employed for treating household wastewater, industrial wastewater, and stormwater runoff, the resulting composition varies accordingly (Koyuncuoğlu & Erden, 2021). Today, wastewater treatment plants (WWTPs) have the capacity to efficiently extract microplastics with a notable removal efficiency (99%), signifying that the MPs that are eliminated are subsequently transported into sewage sludge (Rolsky et al., 2020; Sun et al., 2019; van den Berg et al., 2020). Recently, Jiang et al. (2020) showed that in one WWTP in China, nearly 76% of microplastics were recovered from wastewater and transported into the sludge. As a result, higher concentration of microplastics in sludge than in effluents (Xu et al., 2020). There are various techniques to collect sludge samples including separate pumping and filtration (Mason et al., 2016; Minteing et al., 2017; Ziajahrohmi et al., 2017), surface filtration (Carr et al., 2016), container collection (Bayo et al., 2016; Lares et al., 2018; Murphy et al., 2016; Tagg et al., 2015), autosampler collection (Michielssen et al., 2016), and grab sampling (Ragoobur et al., 2021; Zhang et al., 2019). Amongst all, container collection and grab sampling are the most followed techniques to collect sewage sludge due to their high organic content (Sun et al., 2019; Hatinoğlu et al., 2021). The sludge samples are often collected directly from the sludge treatment units in small quantities (5–20 g) (Sun et al., 2019). Sample preparation procedures differed depending on the sludge's composition and collection method. For example, the collected dried sludge was softened by soaking it in water (Liu et al., 2019; Mahon et al., 2017); some studies used dried sludge or dried the sludge prior to the experiment (Li et al., 2019), and several studies employed stacked sieves of different pore sizes for sieving wet/dried sludge (Hongprasith et al., 2020; Ren et al., 2020). The collected samples are immediately transferred and stored in well-labeled, clean glass or metal containers with plastic-free lids. The obtained sample is usually stored at 4 °C or minus 20 °C until the extraction of microplastics process starts.

3.2 *Extraction of Microplastics from Sewage Sludge*

Extraction of MPs from sludge samples has typically been conducted using a density-based technique with a salt solution (usually NaCl), as is the case with other environmental samples (Hatinoğlu & Sanin, 2021; Koyuncuoğlu & Erden, 2021; Lusher et al., 2020).

3.2.1 Elutriation Column

The principle of elutriation was applied as the initial stage in the isolation of MPs from other samples. Elutriation, which divides lighter from heavier particles by an upward flow of liquid or gas, has been extensively utilized to separate the biota inside sediment samples (Mahon et al., 2017). Claessens et al. (2013) designed an apparatus based on the principle of elutriation. A PVC column (147 cm long with an internal diameter of 15 cm) is equipped with a 1 mm sieve on top and a 35 μm mesh screen (supported by a 1 mm mesh screen to support the sediment's weight) on the bottom. A 500 mL of sediment is transferred to the column after passing it through a 1 mm screen to eliminate any coarse particles. A sieve cover is used to avoid contamination caused by airborne particles or fibers. An upward water flow is produced by pumping tap water through the column from below. The silt gets fluidized at this phase. Aeration is given at the base of the column to facilitate the effective separation of plastic and silt particles. Mahon et al. (2017) performed a study to investigate the extraction efficiency of the elutriation column process using high-density polyethylene (HDPE). The author spiked the sediment with three different colors (red, blue, and black) HDPE and Polyvinyl chloride (PVC) particles. According to the findings of this study, the average extraction efficiency rate of the elutriation column for the spiked sediment samples was 90%, 94%, and 91% for the red, blue, and black HDPE particles, respectively, and for PVC particles, the elutriation column process is less effective, resulting in an average extraction efficiency of 80%.

3.2.2 Density Flotation Process Using Different Salts

Extraction of microplastics employing density salt solution is a widely used technique to isolate the microplastics from the sample matrix (Möller et al., 2020). Where the microplastics float to the surface of the high-density solutions of NaCl (1.2 g/cm^3), $ZnCl_2$ (1.5–1.7 g/cm^3), or NaI (1.6–1.8 g/cm^3) (Lv et al., 2019), while the denser sludge materials have a greater tendency to settle at the bottom of the solution. Subsequently, the remaining liquid, known as the supernatant, is subjected to filtration using a vacuum pump, resulting in the collection of microplastics (MPs) on a filter paper (Koyuncuoğlu & Erden, 2021) (Fig. 2). A majority of studies used NaCl as the density solution for extracting microplastics due to its inexpensive price and eco-friendly nature (Leslie et al., 2017). While NaCl serves as an affordable and ecologically viable salt, its density (1.2 g cm^3) is insufficient to effectively extract all types of microplastics, particularly PAN, PET, and PVC. Thus, NaI solution, sodium poly tungstate (1.4–1.5 g cm^{-3}), and $ZnCl_2$ solutions exhibited greater effectiveness in extracting a broader range of polymer types (Silva et al., 2018). The study conducted by Li et al. (2019) observed that the concentration of polyethylene terephthalate (PET) and polyacrylonitrile (PAN) in a solution of sodium iodide (NaI) was found to be greater compared to that in a solution of sodium chloride (NaCl). This difference was attributed to the lower density of saturated NaCl

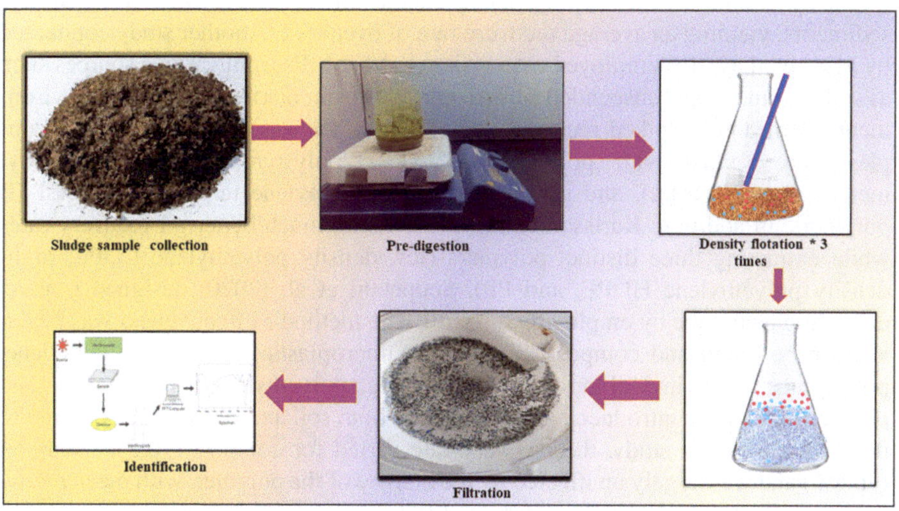

Fig. 2 Schematic diagram of microplastics extraction using density flotation method and identification from sludge sample

solution, which is measured as 1.2 g cm^{-3}, which falls below the densities of PAN and PET (Han et al., 2019) and further stated that the density of the salt solution and the quantity of MPs extracted are directly proportional. The findings by Alvim et al. (2022) demonstrated that a higher density solution, i.e., NaI, resulted in significantly higher retrieval rates of anthropogenic microparticles from sludge samples. Specifically, the recovery rates were approximately 1000 microparticles per gram of dry weight (g-dw) for dewatered sludge and 2000 MPs/ g-dw for digested sludge. In contrast, the lower-density solution (NaCl) resulted in lower retrieval rates, with 200 MPs and 600 MPs/g-dw for dewatered and digested sludge, respectively. However, these density separation reagents are expensive and environmentally dangerous. Whereas, Ren et al. (2020) detected microplastics with particle sizes ranging from 8 μm to 1000 μm, which accounted for 97.27%. In addition, 41.18% of the microplastics were PVC. Similarly, Rodrigues et al. (2020) reported that the recovery efficacy of $ZnCl_2$ was more than 95% when all polymer types were identified and recovered. Cutroneo et al. (2021) documented that NaCl stands as the predominant salt choice to extract MPs from sediment samples, constituting the common salt employed to extract MPs from sediment samples, constituting 45.6% of the articles reviewed by the researchers. In determining the most effective technique for extracting MPs, economic and environmental factors should also be taken into account.

3.2.3 Oil Extraction

The oil separation approach depends upon the intrinsic oleophilic properties exhibited by plastic polymers (He et al., 2021). The oil extraction technique was introduced by Crichton et al. (2017), using canola oil to remove MPs from aquatic sediments, yielding an average recovery rate of over 96%. Another study conducted by Mani et al. (2019) employed castor oil to extract MPs from various sources such as soil, sediment, and suspended surface particles from marine and riverine environments. A mean MP spiked-recovery rate of 99% ± 4% was obtained when the samples were spiked with polypropylene (PP), polystyrene (PS), polymethyl methacrylate (PMMA), and glycol-modified polyethylene terephthalate (PET-G) polymers. In addition, Karlsson et al. (2017) found much better oil recovery rates while extracting three distinct polymers (low-density polyethylene LDPE, high-density polyethylene HDPE, and PP). Scopetani et al. (2020) designed a novel extraction technique by employing olive oil. The method's effectiveness was evaluated on both soil and compost. Six distinct microplastics, namely polyethylene, polystyrene, polyvinyl chloride, polycarbonate, polyethylene terephthalate, and polyurethane, were introduced into samples of both soil and compost. According to the findings of this study, the recovery rate varied for compost (90.5%) and soil (96.5%) and was largely unaffected by the density of the polymer, with mean recovery rates of 90% ± 2%, 97% ± 5%, and 95% ± 4% for low, medium, and high-density polymers, respectively. Crichton et al. (2017) rinsed oil off the surface of

MPs using detergents and reagent alcohol (consisting of 90% ethanol, 5% methanol, and 5% isopropanol) and found that the treatment had no effect on the FTIR spectra. Lechthaler et al. (2020) stated that the oil extraction technique is a comparable and efficient separation medium and provides additional benefits for separating water and sediment samples owing to its density independence, easy and rapid application, as well as eco-friendliness after extracting microplastics using canola oil from sediment. Overall, the oil separation methodology is relatively low time-consuming and environmentally beneficial, and it deserves to be developed further as an alternative separation method.

3.3 Detection and Identification of Microplastics

To quantify and identify, the extracted MPs from the sample matrix are subjected to various physically and chemically based identification and quantification processes (Alvim et al., 2020). In physical analysis, MPs can be categorized according to their shape (fiber, film, foam, pellets, or fragments), color, size, and diameter, which are visualized using a stereo microscope or light microscope (Priyanka & Udayashankara, 2018). Among the microscopes, the stereomicroscope is the most commonly used microscope in studies because of its higher resolution than the light microscope (Koyuncuoğlu & Erden, 2021). The visualization of extracted MPs from the sample is a crucial step for the identification of microplastics (Hidalgo-Ruz et al., 2012). However, the ability to visually detect MPs is limited to bigger particles as a result of the stereomicroscope's relatively low magnification factor. Furthermore, the accuracy of the results heavily relies on the skills and expertise of the operator (Sun et al., 2019). According to Hidalgoruz et al. (2012), it was shown that a maximum error ratio of 70% could be identified, and this error rate increases as the particle size decreases and further suggests additional criteria for differentiating plastics from non-plastics during microscopic analysis, aiming to prevent misidentification and potential underestimation of MPs. The following criteria can be used to differentiate MP particles: (1) the object should be neither cellular nor organic in nature; (2) the fibers should have uniform thickness throughout their entire length; (3) the particles must be consistently colored throughout, and if they are translucent or white, they must be inspected with a high-power fluorescence microscope (Hidalgoruz et al. 2012). A considerable volume of microplastics makes it much more probable that counts will be doubled or missed. For instance, Shim et al. (2016) reported the incorrect identification of non-plastic particles under a microscope is frequently more than 20–70%, which was confirmed by further spectroscopic analysis, and the author suggests using additional analytical methods. Eriksen et al. (2013) reported the incorrect identification of around 20% of particles first recognized as microplastics by ocular inspection, later shown to be aluminum silicate derived from coal ash using scanning electron microscopy (SEM). However, SEM detection is slow, time-consuming, and relatively costly (Bläsing & Amelung, 2018). Additionally, SEM needs additional coatings during the initial preparation, which might lead to errors in determining the surface texture and color of microplastics (Shim et al., 2016). Liu et al. (2019) revealed that 81.4% of the suspicious particles were verified as microplastics. In general, infrared microscopy is one of the most accessible chemical identification methods for microplastics. Similar nondestructive vibrational methods include (micro-)Raman spectroscopy, attenuated total reflectance (ATR), and m-FT-IR. Among the employed spectroscopic methods, Fourier-transform infrared (FT-IR) spectroscopy is the most extensively utilized one. The FTIR technique is the most often described approach for evaluating microplastics extracted from sludge/soil (Rolsky et al., 2020). This method involves exposing microplastics to infrared light and obtaining a spectrum. These distinctive peaks correlate to particular chemical bonds between atoms. The obtained spectra can be utilized for the purpose of identifying the composition of a substance by comparing them to the library of reference spectra (Han et al., 2019). Micro-FTIR is the combination of FTIR spectroscopy with microscopy to generate high resolution for particles as small as 20 μm, with little or no sample preparation (Prata et al., 2019). The ATR-FTIR method utilizes a single beam configuration and measures percent transmission. Each sample is subjected to 40 scans, with a resolution of 0.4 cm^{-1} and a wavelength range spanning from 4000 to 350 cm^{-1} (Sun et al., 2019). Compared to FTIR spectroscopy, Raman methods are more sensitive instruments for identifying MPs with a spatial resolution as small as 1 μm. Raman spectroscopy is susceptible to fluorescence interference caused by microbial, organic, or inorganic components in samples. As a result, sample purification should be done carefully to minimize unwanted sample change before Raman analysis (Elert et al., 2017). The size reliance of the previously described approaches may result in an underestimation of the number of plastics, highlighting the need for a mass-based identification and quantification technique. GC-MS (gas chromatography coupled to mass spectrometry), which includes pyrolysis-GC-MS and thermal extraction desorption-GC-MS, has been rarely documented in studies analyzing sewage sludge samples. In addition to its rapid detection technique, the Pyr-GC/MS offers a quantitative estimate of MPs' mass independent of particle size (Mallow et al., 2020; Tian et al., 2023). Pyrolysis gas chromatography-mass spectrometry (Py-GC-MS) is a destructive approach based on the thermal breakdown of MP (up to 600 C) and its characterization using gas chromatography linked to mass spectrometry. MP types containing polar subunits (polyethers or polyesters) may be problematic for GCMS detection and separation depending on the GC column (El Hayany et al., 2022). In such circumstances, the use of a polar column may enhance detectability (Dignac et al., 2006). In addition, the Py-GC-MS approach requires a longer measurement time than Raman spectroscopy and FTIR spectroscopy (Käppler et al., 2018), and complicated data exploitation makes it more difficult; as a result, FTIR and Raman spectroscopy are frequently used for MP identification (El Hayany et al., 2022). Okoff et al. (2019) identified and quantified selected plastics (polystyrene (PS), polycarbonate (PC), poly-(methyl methacrylate) (PMMA), polypropylene (PP), polyethylene terephthalate (PET), polyethylene (PE), and polyvinyl chloride (PVC)) in biosolids samples (sewage sludge) using pressurized liquid extraction (PLE) combined with double-shot pyrolysis gas chromatography-mass spectrometry, which provides a basis for the uniform 333 reporting of results. However, because of its destructive nature, it precludes further microplastic analysis (Shim et al., 2016). Magnusson and Norén (2014) employed an alcohol burner to differentiate between synthetic and natural fibers, since plastics melted when exposed to heat while natural fibers did not. Furthermore, a staining methodology was employed to mitigate the potential overestimation of the putative MPs (Ziajahromi et al., 2017). Different detection techniques are presented in Fig. 3.

Fluorescence-staining techniques are straightforward and practical for detecting the amount of MP in environmental materials in bulk as they can increase the counting accuracy of tiny MPs relative to visual quantification (El Hayany et al., 2022). Fluorescence-staining techniques such as Nile Red have been used to detect MP in environmental matrices at optimum concentrations ranging from 1 to 1000 g mL and incubation times ranging from 5 min to 66 h. Increased Nile Red concentrations and prolonged incubation times enhanced fluorescence intensity, but also the background signal from the filters (Maes et al., 2017; Walia, 2018). Another study conducted by Liebezeit and Liebezeit (2014) used Rose-Bengal solution (4,5,6,7-tetrachloro-20,40,50,70-tetraiodo-fluorescein, Sigma-Aldrich, 95% dye concentration), natural microparticles, namely natural fibers, stained pink, enabling the visual differentiation of plastic and non-plastic particles. However, staining with rose bengal induced modest color alterations in PS and PVC fragments, particularly when staining was performed following WPO treatment (Lares et al., 2018). Fluorescence staining techniques have several disadvantages because of the co-staining of natural organics such as chitin; the fluorescent dye might provide false positive results. Furthermore, not all MP types can be stained with Nile Red (Shim et al., 2016). Therefore, we draw the conclusion that staining techniques cannot always be employed as an effective approach for counting and identifying MPs

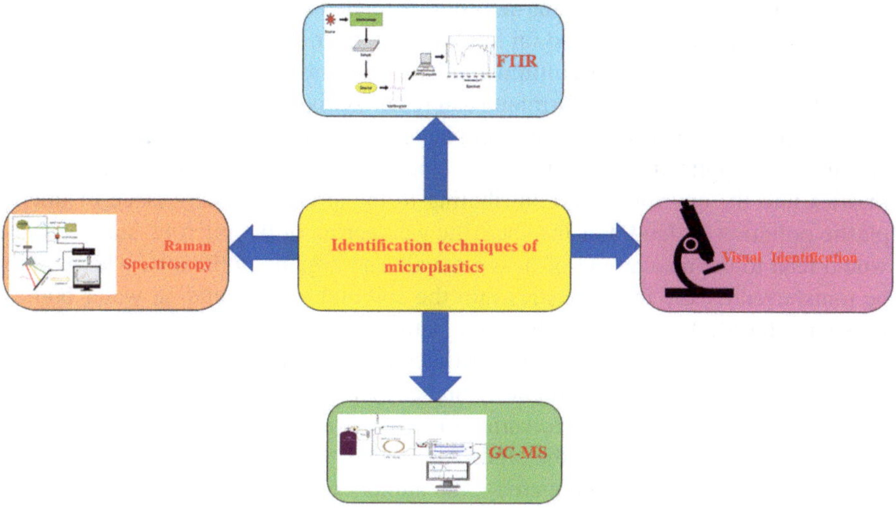

Fig. 3 Visual depiction of different identification techniques of microplastics

(Maes et al., 2017), but might be used as a fast and easy procedure to determine the potential MP content in sewage sludge (El Hayany et al., 2020). While there are several approaches for quantifying and characterizing MPs in environmental samples, we recommend that spectroscopic techniques be particularly developed for regular MP studies in sewage sludge and its compost. These approaches are quick, inexpensive, and appropriate for MP identification in environmental samples, implying that they might be particularly helpful for monitoring the MP contamination of sludge and its compost.

4 Environmental Risk of MPs During Sludge Soil Amendment

The presence of a substantial quantity of MPs in sewage sludge can lead to the buildup of these particles at levels that pose a potential risk to the environment (Koyuncuoğlu & Erden, 2021). The collected sewage sludge undergoes various treatment techniques, such as conditioning, dehydration, and burning, resulting in the conversion of the sludge into a mineral product that is both beneficial and cost-effective. Notably, the incineration process effectively eliminates microplastics from the final product. However, incineration of sewage sludge, including significant microplastics, results in airborne particle emission and solid residue. When treated sewage sludge is applied to agricultural land as an amendment and soil compost, it enhances soil fertility and provides benefits economically. Therefore, sludge amendments in land applications have gained a lot of attention. Rolsky et al. (2020) review reported that the percentage of sludge used in agricultural and soil/compost applications on land ranged from 0% to 94% across countries. Sewage sludge, the final product of sewage treatment facilities, is loaded with organic matter and trace elements; hence, it is often used on farms as fertilizer. Approximately 50% of sewage sludge is utilized for agricultural applications, while in Finland and Ireland, the proportion of sludge put into farmland can reach as high as 72% (Yang et al., 2021). Further, improper disposal of some sludge contributes to soil microplastic pollution. However, the application of sludge to the land might potentially release these microplastics into the environment, contributing to the already-existing issue of microplastic pollution on land. For instance, the weathering of the microplastics' surface would lead to the attachment of heavy metals and organic pollutants, which might be transferred by the microplastics into the soil and to surrounding water bodies. Zhang et al. (2020) discovered that soils with yearly 30 and 15 t/ha applications of sewage compost had average microplastic concentrations of 545.9 and 87.7 items/kg, respectively, which is significantly higher than soil without compost treatment. By evaluating fields subjected to varied rates of sewage sludge application, van den Berg et al. (2020) show that the sludge is loaded with microplastics. Subsequent applications of sewage sludge increase the number of microplastics in the soil from 280 to 430 per kilogram of soil by weight (van den Berg et al., 2020). Furthermore, according to Crossman et al. (2020), despite compliance with current regulations, biosolid applications may result in significant rates of MP export. According to studies, the yearly quantity of MPs entering the terrestrial ecosystem via land use of sludge is substantially larger than the number of MPs entering the seas and freshwater sediments from a variety of sources. Crossman et al. (2020) examined microplastic concentrations in biosolids from two distinct sources and in the soils of three agricultural areas in Ontario, Canada, where these biosolids were applied. In addition, as a control site, soils with no history of prior treatment of biosolids were investigated. Their findings indicated that significant levels of microplastic were present in biosolid samples, ranging from 8.7103 to 1.4104 MP/kg. When compared to the control field, significant levels of microplastic were found in all of the fields where biosolids were treated. In addition, they estimated that the yearly buildup of microplastics in agricultural soils is between 4.1×10^{11} and 1.3×10^{12} particles. Nevertheless, the use of sewage sludge as a biosolid opens up a route for microplastics to infiltrate agricultural soils, which may have undetermined effects on the long-term viability of agricultural practices and the safety of food supplies (Li et al., 2018). MPs entering the soil will be stored, translocated, eroded, degraded, and leached into groundwater, endangering organisms and subsequently affecting human health (He et al., 2018). Other potential routes include irrigation, plastic mulching, roadway runoff, tire deterioration, littering, and air deposition (Li et al., 2020). MPs pose a substantial threat to both flora and fauna like earthworms (Angmo et al., 2023; Cao et al., 2017; Jiang et al., 2020), *Triticum aestivum* (Liao et al., 2019), Oryza sativa L (Ma et al., 2022). They impede the growth and germination rates eventually resulting in their mortality.

5 Conclusion and Future Perspective

Microplastics have the potential to infiltrate terrestrial ecosystems via numerous pathways and originate from a diverse range of sources. Accumulation of microplastics in soil yields both beneficial and harmful effects, influencing its physical, chemical, and biological characteristics. At present, there is a lack of universally established procedures for extracting and quantifying microplastics in soil or biosolids. Therefore, it is crucial to develop accurate, achievable, and consistent techniques for extracting and measuring microplastics in soil ecosystems, encompassing a broad spectrum of particle sizes, polymer compositions, and physical states. Furthermore, limited research has been done on microplastic characteristics within terrestrial environments, as well as their ecological impacts and the fundamental mechanisms driving these effects. Additional research is required to comprehend the impact of microplastic pollution on soil organisms, plant growth, and agricultural crop yield in agricultural practices.

References

Alimi, O. S., Farner Budarz, J., Hernandez, L. M., & Tufenkji, N. (2018). Microplastics and nanoplastics in aquatic environments: Aggregation, deposition, and enhanced contaminant transport. *Environmental Science & Technology, 52*(4), 1704–1724.

Alvim, C. B., Mendoza-Roca, J. A., & Bes-Piá, A. (2020). Wastewater treatment plant as microplastics release source – Quantification and identification techniques. *Journal of Environmental Management, 255*, 109739.

Alvim, C. B., Valiente, S. N., Bes-Piá, M. A., & Mendoza-Roca, J. A. (2022). Methodology for removing microplastics and other anthropogenic microparticles from sludge dewatering system. *Journal of Environmental Management, 314*, 115010.

Andrady, A. L. (2017). The plastic in microplastics: A review. *Marine Pollution Bulletin, 119*(1), 12–22.

Angmo, D., Singh, J., Dutta, R., Chowdhary, A. B., Quadar, J., Sharma, M., et al. (2023). Earthworms modulate the toxicity effect of low-density polyethylene on plant development. *Journal of Soil Science and Plant Nutrition, 23*, 1–13.

Bayo, J., Olmos, S., López-Castellanos, J., & Alcolea, A. (2016). Microplastics and microfibers in the sludge of a municipal wastewater treatment plant. *International Journal of Sustainable Development and Planning, 11*(5), 812–821.

Bläsing, M., & Amelung, W. (2018). Plastics in soil: Analytical methods and possible sources. *Science of the Total Environment, 612*, 422–435.

Browne, M. A. (2015). Sources and pathways of microplastics to habitats. *Marine Anthropogenic Litter*, 229–244.

Browne, M. A., Galloway, T., & Thompson, R. (2007). Microplastic – An emerging contaminant of potential concern? *Integrated Environmental Assessment and Management, 3*(4), 559–561.

Cao, D., Wang, X., Luo, X., Liu, G., & Zheng, H. (2017). Effects of polystyrene microplastics on the fitness of earthworms in an agricultural soil. In *IOP conference series: Earth and environmental science* (Vol. 61, p. 12148). IOP Publishing.

Carr, S. A., Liu, J., & Tesoro, A. G. (2016). Transport and fate of microplastic particles in wastewater treatment plants. *Water Research, 91*, 174–182.

Chen, G., Feng, Q., & Wang, J. (2020a). Mini-review of microplastics in the atmosphere and their risks to humans. *Science of the Total Environment, 703*, 135504.

Chen, Y., Leng, Y., Liu, X., & Wang, J. (2020b). Microplastic pollution in vegetable farmlands of suburb Wuhan, Central China. *Environmental Pollution, 257*, 113449.

Claessens, M., Van Cauwenberghe, L., Vandegehuchte, M. B., & Janssen, C. R. (2013). New techniques for the detection of microplastics in sediments and field collected organisms. *Marine Pollution Bulletin, 70*(1–2), 227–233.

Cole, M., Lindeque, P., Fileman, E., Halsband, C., Goodhead, R., Moger, J., & Galloway, T. S. (2013). Microplastic ingestion by zooplankton. *Environmental Science & Technology, 47*(12), 6646–6655.

Crichton, E. M., Noël, M., Gies, E. A., & Ross, P. S. (2017). A novel, density-independent and FTIR-compatible approach for the rapid extraction of microplastics from aquatic sediments. *Analytical Methods, 9*(9), 1419–1428.

Crossman, J., Hurley, R. R., Futter, M., & Nizzetto, L. (2020). Transfer and transport of microplastics from biosolids to agricultural soils and the wider environment. *Science of the Total Environment, 724*, 138334.

Cutroneo, L., Reboa, A., Geneselli, I., & Capello, M. (2021). Considerations on salts used for density separation in the extraction of microplastics from sediments. *Marine Pollution Bulletin, 166*, 112216.

de Souza Machado, A. A., Lau, C. W., Till, J., Kloas, W., Lehmann, A., Becker, R., & Rillig, M. C. (2018a). Impacts of microplastics on the soil biophysical environment. *Environmental Science & Technology, 52*(17), 9656–9665.

de Souza Machado, A. A., Kloas, W., Zarfl, C., Hempel, S., & Rillig, M. C. (2018b). Microplastics as an emerging threat to terrestrial ecosystems. *Global Change Biology, 24*(4), 1405–1416.

Dignac, M. F., Houot, S., & Derenne, S. (2006). How the polarity of the separation column may influence the characterization of compost organic matter by pyrolysis-GC/MS. *Journal of Analytical and Applied Pyrolysis, 75*(2), 128–139.

Dris, R., Gasperi, J., Rocher, V., Saad, M., Renault, N., & Tassin, B. (2015). Microplastic contamination in an urban area: A case study in greater Paris. *Environmental Chemistry, 12*(5), 592–599.

El Hayany, B., El Fels, L., Quénéa, K., Dignac, M. F., Rumpel, C., Gupta, V. K., & Hafidi, M. (2020). Microplastics from lagooning sludge to composts as revealed by fluorescent staining-image analysis, Raman spectroscopy and pyrolysis-GC/MS. *Journal of Environmental Management, 275*, 111249.

El Hayany, B., Rumpel, C., Hafidi, M., & El Fels, L. (2022). Occurrence, analysis of microplastics in sewage sludge and their fate during composting: A literature review. *Journal of Environmental Management, 317*, 115364.

Elert, A. M., Becker, R., Duemichen, E., Eisentraut, P., Falkenhagen, J., Sturm, H., & Braun, U. (2017). Comparison of different methods for MP detection: What can we learn from them, and why asking the right question before measurements matters? *Environmental Pollution, 231*, 1256–1264.

Eriksen, M., Mason, S., Wilson, S., Box, C., Zellers, A., Edwards, W., et al. (2013). Microplastic pollution in the surface waters of the Laurentian Great Lakes. *Marine Pollution Bulletin, 77*(1–2), 177–182.

Fendall, L. S., & Sewell, M. A. (2009). Contributing to marine pollution by washing your face: Microplastics in facial cleansers. *Marine Pollution Bulletin, 58*(8), 1225–1228.

Franco, A. A., Martín-García, A. P., Egea-Corbacho, A., Arellano, J. M., Albendín, G., Rodríguez-Barroso, R., et al. (2023). Assessment and accumulation of microplastics in sewage sludge at wastewater treatment plants located in Cádiz, Spain. *Environmental Pollution, 317*, 120689.

Freeman, S., Booth, A. M., Sabbah, I., Tiller, R., Dierking, J., Klun, K., et al. (2020). Between source and sea: The role of wastewater treatment in reducing marine microplastics. *Journal of Environmental Management, 266*, 110642.

Gatidou, G., Arvaniti, O. S., & Stasinakis, A. S. (2019). Review on the occurrence and fate of microplastics in sewage treatment plants. *Journal of Hazardous Materials, 367*, 504–512.

Gies, E. A., LeNoble, J. L., Noël, M., Etemadifar, A., Bishay, F., Hall, E. R., & Ross, P. S. (2018). Retention of microplastics in a major secondary wastewater treatment plant in Vancouver, Canada. *Marine Pollution Bulletin, 133*, 553–561.

Hale, R. C., Seeley, M. E., La Guardia, M. J., Mai, L., & Zeng, E. Y. (2020). A global perspective on microplastics. *Journal of Geophysical Research: Oceans, 125*(1), e2018JC014719.

Han, X., Lu, X., & Vogt, R. D. (2019). An optimized density-based approach for extracting microplastics from soil and sediment samples. *Environmental Pollution, 254*, 113009.

Hatinoğlu, M. D., & Sanin, F. D. (2021). Sewage sludge as a source of microplastics in the environment: A review of occurrence and fate during sludge treatment. *Journal of Environmental Management, 295*, 113028.

He, D., Zhang, X., & Hu, J. (2021). Methods for separating microplastics from complex solid matrices: Comparative analysis. *Journal of Hazardous Materials, 409*, 124640.

Hidalgo-Ruz, V., Gutow, L., Thompson, R. C., & Thiel, M. (2012). Microplastics in the marine environment: A review of the methods used for identification and quantification. *Environmental Science & Technology, 46*(6), 3060–3075.

Hongprasith, N., Kittimethawong, C., Lertluksanaporn, R., Eamchotchawalit, T., Kittipongvises, S., & Lohwacharin, J. (2020). IR microspectroscopic identification of microplastics in municipal wastewater treatment plants. *Environmental Science and Pollution Research, 27*, 18557–18564.

Horton, A. A., Cross, R. K., Read, D. S., Jürgens, M. D., Ball, H. L., Svendsen, C., et al. (2021). Semi-automated analysis of microplastics in complex wastewater samples. *Environmental Pollution, 268*, 115841.

He, D., Luo, Y., Lu, S., Liu, M., Song, Y., & Lei, L. (2018). Microplastics in soils: Analytical methods, pollution characteristics and ecological risks. *TrAC Trends in Analytical Chemistry, 109*, 163–172.

Jagadeesh, N., & Sundaram, B. (2021). A review of microplastics in wastewater, their persistence, interaction, and fate. *Journal of Environmental Chemical Engineering, 9*(6), 106846.

Jiang, X., Chang, Y., Zhang, T., Qiao, Y., Klobučar, G., & Li, M. (2020). Toxicological effects of polystyrene microplastics on earthworm (Eisenia fetida). *Environmental Pollution, 259*, 113896.

Käppler, A., Fischer, M., Scholz-Böttcher, B. M., Oberbeckmann, S., Labrenz, M., Fischer, D., et al. (2018). Comparison of μ-ATR-FTIR spectroscopy and py-GCMS as identification tools for microplastic particles and fibers isolated from river sediments. *Analytical and Bioanalytical Chemistry, 410*, 5313–5327.

Karlsson, T. M., Vethaak, A. D., Almroth, B. C., Ariese, F., van Velzen, M., Hassellöv, M., & Leslie, H. A. (2017). Screening for microplastics in sediment, water, marine invertebrates and fish: Method development and microplastic accumulation. *Marine Pollution Bulletin, 122*(1–2), 403–408.

Koyuncuoğlu, P., & Erden, G. (2021). Sampling, pre-treatment, and identification methods of microplastics in sewage sludge and their effects in agricultural soils: A review. *Environmental Monitoring and Assessment, 193*(4), 1–28.

Lares, M., Ncibi, M. C., Sillanpää, M., & Sillanpää, M. (2018). Occurrence, identification and removal of microplastic particles and fibers in conventional activated sludge process and advanced MBR technology. *Water Research, 133*, 236–246.

Lebreton, L., & Andrady, A. (2019). Future scenarios of global plastic waste generation and disposal. *Palgrave Communications, 5*(1), 1–11.

Lechthaler, S., Hildebrandt, L., Stauch, G., & Schüttrumpf, H. (2020). Canola oil extraction in conjunction with a plastic free separation unit optimises microplastics monitoring in water and sediment. *Analytical Methods, 12*(42), 5128–5139.

Lee, H., & Kim, Y. (2018). Treatment characteristics of microplastics at biological sewage treatment facilities in Korea. *Marine Pollution Bulletin, 137*, 1–8.

Leslie, H. A., Brandsma, S. H., Van Velzen, M. J. M., & Vethaak, A. D. (2017). Microplastics en route: Field measurements in the Dutch river delta and Amsterdam canals, wastewater treatment plants, North Sea sediments and biota. *Environment International, 101*, 133–142.

Li, X., Chen, L., Mei, Q., Dong, B., Dai, X., Ding, G., & Zeng, E. Y. (2018). Microplastics in sewage sludge from the wastewater treatment plants in China. *Water Research, 142*, 75–85.

Li, Q., Wu, J., Zhao, X., Gu, X., & Ji, R. (2019). Separation and identification of microplastics from soil and sewage sludge. *Environmental Pollution, 254*, 113076.

Li, X., Chen, L., Ji, Y., Li, M., Dong, B., Qian, G., et al. (2020). Effects of chemical pretreatments on microplastic extraction in sewage sludge and their physicochemical characteristics. *Water Research, 171*, 115379.

Li, X., Wang, X., Chen, L., Huang, X., Pan, F., Liu, L., et al. (2022). Changes in physicochemical and leachate characteristics of microplastics during hydrothermal treatment of sewage sludge. *Water Research, 222*, 118876.

Liao, Y. C., Nazygul, J., Li, M., Wang, X. L., & Jiang, L. J. (2019). Effects of microplastics on the growth, physiology, and biochemical characteristics of wheat (*Triticum aestivum*). *Huan jing ke xue= Huanjing kexue, 40*(10), 4661–4667.

Liebezeit, G., & Liebezeit, E. (2014). Synthetic particles as contaminants in German beers. *Food Additives & Contaminants: Part A, 31*(9), 1574–1578.

Liu, X., Yuan, W., Di, M., Li, Z., & Wang, J. (2019). Transfer and fate of microplastics during the conventional activated sludge process in one wastewater treatment plant of China. *Chemical Engineering Journal, 362*, 176–182.

Lusher, A. L., Munno, K., Hermabessiere, L., & Carr, S. (2020). Isolation and extraction of microplastics from environmental samples: an evaluation of practical approaches and recommendations for further harmonization. *Applied Spectroscopy, 74*(9), 1049–1065.

Lv, X., Dong, Q., Zuo, Z., Liu, Y., Huang, X., & Wu, W. M. (2019). Microplastics in a municipal wastewater treatment plant: Fate, dynamic distribution, removal efficiencies, and control strategies. *Journal of Cleaner Production, 225*, 579–586.

Möller, J. N., Löder, M. G., & Laforsch, C. (2020). Finding microplastics in soils: A review of analytical methods. *Environmental Science & Technology, 54*(4), 2078–2090.

Ma, J., Aqeel, M., Khalid, N., Nazir, A., Alzuaibr, F. M., Al-Mushhin, A. A., et al. (2022). Effects of microplastics on growth and metabolism of rice (*Oryza sativa L.*). *Chemosphere, 307*, 135749.

Maes, T., Jessop, R., Wellner, N., Haupt, K., & Mayes, A. G. (2017). A rapid-screening approach to detect and quantify microplastics based on fluorescent tagging with Nile Red. *Scientific Reports, 7*(1), 1–10.

Magnusson, K., & Norén, F. (2014). Screening of microplastic particles in and down-stream a wastewater treatment plant. *Swedish Environmental Research Institute, 55*, 1–22.

Mahon, A. M., O'Connell, B., Healy, M. G., O'Connor, I., Officer, R., Nash, R., & Morrison, L. (2017). Microplastics in sewage sludge: Effects of treatment. *Environmental Science & Technology, 51*(2), 810–818.

Mallow, O., Spacek, S., Schwarzböck, T., Fellner, J., & Rechberger, H. (2020). A new thermoanalytical method for the quantification of microplastics in industrial wastewater. *Environmental Pollution, 259*, 113862.

Mani, T., Frehland, S., Kalberer, A., & Burkhardt-Holm, P. (2019). Using castor oil to separate microplastics from four different environmental matrices. *Analytical Methods, 11*(13), 1788–1794.

Mason, S. A., Garneau, D., Sutton, R., Chu, Y., Ehmann, K., Barnes, J., et al. (2016). Microplastic pollution is widely detected in US municipal wastewater treatment plant effluent. *Environmental Pollution, 218*, 1045–1054.

Menéndez-Manjón, A., Martínez-Díez, R., Sol, D., Laca, A., Laca, A., Rancaño, A., & Díaz, M. (2022). Long-term occurrence and fate of microplastics in WWTPs: A case study in Southwest Europe. *Applied Sciences, 12*(4), 2133.

Michielssen, M. R., Michielssen, E. R., Ni, J., & Duhaime, M. B. (2016). Fate of microplastics and other small anthropogenic litter (SAL) in wastewater treatment plants depends on unit processes employed. *Environmental Science: Water Research & Technology, 2*(6), 1064–1073.

Mintenig, S. M., Int-Veen, I., Löder, M. G., Primpke, S., & Gerdts, G. (2017). Identification of microplastic in effluents of waste water treatment plants using focal plane array-based micro-Fourier-transform infrared imaging. *Water Research, 108*, 365–337.

Murphy, F., Ewins, C., Carbonnier, F., & Quinn, B. (2016). Wastewater treatment works (WwTW) as a source of microplastics in the aquatic environment. *Environmental Science & Technology, 50*(11), 5800–5808.

Ngo, P. L., Pramanik, B. K., Shah, K., & Roychand, R. (2019). Pathway, classification and removal efficiency of microplastics in wastewater treatment plants. *Environmental Pollution, 255*, 113326.

Nizzetto, L., Futter, M., & Langaas, S. (2016). Are agricultural soils dumps for microplastics of urban origin? *Environmental Science & Technology, 50*, 10777.

Okoffo, E. D., O'Brien, S., O'Brien, J. W., Tscharke, B. J., & Thomas, K. V. (2019). Wastewater treatment plants as a source of plastics in the environment: A review of occurrence, methods for identification, quantification and fate. *Environmental Science: Water Research & Technology, 5*(11), 1908–1931.

Pittura, L., Foglia, A., Akyol, Ç., Cipolletta, G., Benedetti, M., Regoli, F., et al. (2021). Microplastics in real wastewater treatment schemes: Comparative assessment and relevant inhibition effects on anaerobic processes. *Chemosphere, 262*, 128415.

Plastics Europe. (2017). *Plastics—The facts*. Plastic Europe. https://www.plasticseurope.org/en/resources/publications/274-plastics-facts-2017

Prata, J. C., da Costa, J. P., Duarte, A. C., & Rocha-Santos, T. (2019). Methods for sampling and detection of microplastics in water and sediment: A critical review. *TrAC Trends in Analytical Chemistry, 110*, 150–159.

Priyanka, M., & Saravanakumar, M. P. (2019). Preliminary examination on isolation of microplastics (MPs) in sewage sludge from the local wastewater treatment plant. *International Journal of Engineering and Advanced Technology, 9*(1), 7514–7516.

Priyanka, G. M., & Udayashankara, T. H. (2018). Methodology for sampling, purification, extraction and identification of microplastic – A review. *International Journal of Science and Research, 7*(1), 1006–1009.

Ragoobur, D., Huerta-Lwanga, E., & Somaroo, G. D. (2021). Microplastics in agricultural soils, wastewater effluents and sewage sludge in Mauritius. *Science of the Total Environment, 798*, 149326.

Raju, S., Carbery, M., Kuttykattil, A., Senthirajah, K., Lundmark, A., Rogers, Z., et al. (2020). Improved methodology to determine the fate and transport of microplastics in a secondary wastewater treatment plant. *Water Research, 173*, 115549.

Ren, X., Sun, Y., Wang, Z., Barceló, D., Wang, Q., Zhang, Z., & Zhang, Y. (2020). Abundance and characteristics of microplastic in sewage sludge: A case study of Yangling, Shaanxi province, China. *Case Studies in Chemical and Environmental Engineering, 2*, 100050.

Rochman, C. M. (2018). Microplastics research—From sink to source. *Science, 360*(6384), 28–29.

Rodrigues, D., Antunes, J., Otero, V., Sobral, P., & Costa, M. H. (2020). Distribution patterns of microplastics in seawater surface at a Portuguese estuary and marine park. *Frontiers in Environmental Science, 8*, 582217.

Rolsky, C., Kelkar, V., Driver, E., & Halden, R. U. (2020). Municipal sewage sludge as a source of microplastics in the environment. *Current Opinion in Environmental Science & Health, 14*, 16–22.

Scopetani, C., Chelazzi, D., Mikola, J., Leiniö, V., Heikkinen, R., Cincinelli, A., & Pellinen, J. (2020). Olive oil-based method for the extraction, quantification and identification of microplastics in soil and compost samples. *Science of the Total Environment, 733*, 139338.

Shim, W. J., Song, Y. K., Hong, S. H., & Jang, M. (2016). Identification and quantification of microplastics using Nile red staining. *Marine Pollution Bulletin, 113*(1–2), 469–476.

Silva, A. B., Bastos, A. S., Justino, C. I., da Costa, J. P., Duarte, A. C., & Rocha-Santos, T. A. (2018). Microplastics in the environment: Challenges in analytical chemistry-A review. *Analytica chimica acta, 1017*, 1–19.

Sun, J., Dai, X., Wang, Q., van Loosdrecht, M. C., & Ni, B. J. (2019). Microplastics in wastewater treatment plants: Detection, occurrence and removal. *Water Research, 152*, 21–37.

Tadsuwan, K., & Babel, S. (2022). Microplastic abundance and removal via an ultrafiltration system coupled to a conventional municipal wastewater treatment plant in Thailand. *Journal of Environmental Chemical Engineering, 10*(2), 107142.

Tagg, A. S., Sapp, M., Harrison, J. P., & Ojeda, J. J. (2015). Identification and quantification of microplastics in wastewater using focal plane array-based reflectance micro-FT-IR imaging. *Analytical Chemistry, 87*(12), 6032–6040.

Tian, L., Skoczynska, E., van Putten, R. J., Leslie, H. A., & Gruter, G. J. M. (2023). Quantification of polyethylene terephthalate micro-and nanoplastics in domestic wastewater using a simple three-step method. *Science of the Total Environment, 857*, 159209.

Üstün, G. E., Bozdaş, K., & Can, T. (2022). Abundance and characteristics of microplastics in an urban wastewater treatment plant in Turkey. *Environmental Pollution, 310*, 119890.

van den Berg, P., Huerta-Lwanga, E., Corradini, F., & Geissen, V. (2020). Sewage sludge application as a vehicle for microplastics in eastern Spanish agricultural soils. *Environmental Pollution, 261*, 114198.

Walia, M. (2018). Identification and quantification of microplastics using Nile red staining. *Student Research Proceedings, 3*(1).

Wang, L., Shi, Y., Chai, J., Huang, L., Wang, Y., Wang, S., et al. (2022). Transfer of microplastics in sludge upon Fe (II)-persulfate conditioning and mechanical dewatering. *Science of the Total Environment, 156316*, 156316.

Weber, C. J., & Opp, C. (2020). Spatial patterns of mesoplastics and coarse microplastics in floodplain soils as resulting from land use and fluvial processes. *Environmental Pollution, 267*, 115390.

Wei, F., Xu, C., Chen, C., Wang, Y., Lan, Y., Long, L., et al. (2022). Distribution of microplastics in the sludge of wastewater treatment plants in Chengdu, China. *Chemosphere, 287*, 132357.

Xu, Q., Xing, R., Sun, M., Gao, Y., & An, L. (2020). Microplastics in sediments from an interconnected river-estuary region. *Science of the Total Environment, 729*, 139025.

Yang, L., Zhang, Y., Kang, S., Wang, Z., & Wu, C. (2021). Microplastics in soil: A review on methods, occurrence, sources, and potential risk. *Science of the Total Environment, 780*, 146546.

Yuan, F., Zhao, H., Sun, H., Sun, Y., Zhao, J., & Xia, T. (2022). Investigation of microplastics in sludge from five wastewater treatment plants in Nanjing. *China. Journal of Environmental Management, 301*, 113793.

Zhang, Z., & Chen, Y. (2020). Effects of microplastics on wastewater and sewage sludge treatment and their removal: A review. *Chemical Engineering Journal, 382*, 122955.

Zhang, Y., Gao, T., Kang, S., & Sillanpää, M. (2019). Importance of atmospheric transport for microplastics deposited in remote areas. *Environmental Pollution, 254*, 112953.

Zhang, K., Xu, S., Zhang, Y., Lo, Y., Liu, M., Ma, Y., et al. (2022). A systematic study of microplastic occurrence in urban water networks of a metropolis. *Water Research, 223*, 118992.

Zhang, X., Yan, B., & Wang, X. (2020). Selection and optimization of a protocol for extraction of microplastics from Mactra veneriformis. *Science of the Total Environment, 746*, 141250.

Ziajahromi, S., Neale, P. A., Rintoul, L., & Leusch, F. D. (2017). Wastewater treatment plants as a pathway for microplastics: Development of a new approach to sample wastewater-based microplastics. *Water Research, 112*, 93–99.

[illegible]

Part IV
Micro-Nano-plastics Remedies from Contaminated Areas

Cleaning Up the Smallest Pollutants: The Potential of Microbial Degradation in Tackling Micro- and Nano-Plastic Pollution

Ayushi Varshney

1 Introduction

Plastics are extensively used in various industries, agriculture, healthcare, and our daily lives, due to their affordability, convenience, and easy availability. Worldwide, according to reports, the generation of plastics has been increasing from 1.5 million tonnes in 1950 to an estimated 445.25 million tonnes in 2025 (https://www.statista.com/statistics/664906/plastics-production-volume-forecast-worldwide). Most plastic is released into the environment, with just 71% being recycled, 12% being burned for potential energy recovery, 8% being buried in landfills, and the other 12% being burnt (Enfrin et al., 2019; Jahnavi et al., 2021). Over the past few decades, plastic has significantly raised the quality of our lives. However, due to its extremely low biodegradability, it typically ends up in the trash after use, having a negative influence on the environment (Ali et al., 2021). The broad dispersion of plastic waste and the ensuing environmental and ecological implications have sparked global concerns regarding the challenges posed by plastic debris (Zhang et al., 2020). The concern and interest in plastic pollution have been growing due to its potential to accumulate and remain in the environment for many centuries, especially in conditions with low-light and low-oxygen levels (Horton et al., 2017).

Among all plastic debris, microplastics (MPs) are frequently referred to as plastic particles between 1 μm and 5 mm in size, whereas nanoplastics (NPs) are defined as plastic particles smaller than 1 μm (Andrady, 2017). Plastic waste, once introduced into the environment, can remain for prolonged periods and subsequently disintegrate into smaller fragments, also referred to as mesoplastics or microplastics (MPs), and even nanoplastics (NPs), before eventually undergoing complete decomposition (Kumar et al., 2021). These MPs and NPs, collectively known as MNPs,

A. Varshney (✉)
Anantaa GSK Innovations Pvt Ltd., Faridabad, Haryana, India
e-mail: dr.ayushi@agskipl.com

S. A. Bhat et al. (eds.), *Management of Micro and Nano-plastics in Soil and Biosolids*,
https://doi.org/10.1007/978-3-031-51967-3_15

have become a significant environmental concern, commonly referred to as "white pollution" (Naik et al., 2020). The impact of nanoplastics (NPs) on living tissue is more severe due to their smaller size (<100 nm). They can easily penetrate cell membranes and have adverse effects on cells and tissues, as compared to microplastics (MPs), which are larger in size and easier to handle and can easily be removed but still have a significant impact on living organisms. MPs can release toxicants and accumulate in living organisms, acting as carriers of toxicants. They may also interact with other environmental pollutants, both organic and inorganic, to facilitate their transit to environmental compartments and increase the bioaccumulation of further pollutants present in the environment. The accumulation and persistence of nanoplastic waste can create a major danger to both the environment and human health. These waste products have the potential to significantly deform cells and agglomerate nanoplastics in organs like the liver, kidney, and gut (Deng et al., 2017; Tiwari et al., 2020). The ubiquitous presence of micro- and nano-sized plastics in environmental matrices is an increasing cause of concern because of their unintended impacts on the biodiversity of terrestrial and aquatic ecosystems. The presence of MNPs in the soil ecosystem inhibits the growth of earthworms, lowers soil pH, prevents seed germination, and lowers plant shoot height (Mohanty et al., 2022). Ingesting micro- and nano-sized plastics obstructs the digestive tract, stunts growth, disrupts reproduction, and increases mortality of aquatic life forms such as marine bivalves, microalgae, amphipods, etc. (Rout et al., 2022). A number of new pollutants, including Bisphenol A, atrazine, perfluoroalkylates, etc., can have detrimental consequences on biodiversity through their adsorption and desorption on MNPs. The growing contaminant-loaded MNPs could pose risks to the ecosystem and human health if they enter the food chain through biota consumption (Zhang et al., 2021). Therefore, in order to reduce the amount of micro- and nanoplastics entering various environmental components, the development of advanced treatment technologies is required to set restricted discharge standards for MNPs.

In recent years, studies have focused on using microbial biotechnology as a potential tool to degrade these plastics (Pathak & Navneet, 2017). A large number of microbes, including bacteria, algae, and fungi, are linked to the breakdown of plastics. These microbes include *Pseudomonas aeruginosa*, *Streptomyces badius*, *Streptomyces setonii*, *Rhodococcusruber*, *Pseudomonasstutzeri*, cyanobacteria, *Chlamydomonas reinhardtii*, *Phaeodactylum tricornutum*, *Aspergillus niger*, *Fusarium lini*, and *Aspergillus flavus* (Pathak & Navneet, 2017). In response to environmental stress (such as MNPs), microorganisms can modify their growth, metabolic rate, and biosynthesize noble microbial bioproducts to combat environmental stress (Guan & Liu, 2020). These microbes have evolved defense mechanisms against the harmful effects of MNPs present in the natural environment (Yang et al., 2020). In their cultivation media, these bacteria can utilize plastic polymers as a source of energy, which reduces the dry weight of the polymers and causes physico-chemical changes, including alterations in surface morphologies.

The present chapter highlights the biological and toxicological effects of MNPs, the contribution of different microbes, their enzymes, and various biodegradation mechanisms in wastewater, municipal solid waste (MSW), and composting. The chapter begins by exploring the ecotoxicological impacts of MNPs in agricultural soils. In addition, it discusses the enzymatic, advanced molecular, and biomembrane techniques utilized in MNPs remediation and their benefits and drawbacks. Finally, it involves the gaps in current research that require further investigation to enhance the efficiency and sustainability of MNPs biodegradation.

2 Origins of Microplastic Contamination in Agricultural Soils

The agricultural and horticultural industries have emerged as major consumers of plastics in the form of film sheets, foam, pipes, and other materials (Andrady, 2017). These plastic pollutants are used further for the transportation of fertilizer, the management of weeds and pests, storage, crop preservation, and construction of buildings and structures. However, the use of plastic materials in the agricultural sector is necessary from the nursery stage to the post-harvest stage; as a result, 2,250,000 tonnes of plastics are consumed annually by the agricultural and agriculture-related sectors (Sa'adu & Farsang, 2023). Furthermore, plastic pollutants can come in a variety of shapes, including films, fibers, pieces, beads, and foam. In agricultural soils, the primary sources of these contaminants include sewage sludge, coated fertilizers, irrigation water, agrochemicals, etc., while secondary sources come from the gradual decomposition of more substantial plastic components, such as mulch and greenhouse films. Contaminants made of macro- and microplastic are moved horizontally, vertically, and within soil profiles. Vertical transfer occurs through leaching, the actions of soil and microbiological organisms, and agricultural practices, including mulching, irrigation, and greenhouse farming while horizontal transfer is facilitated by the action of wind and water (O'Connor et al., 2019).

MNPs are generally classified into two categories, primary and secondary MNPs. Household goods, cosmetics, and polymeric raw materials used in the plastics industry, such as polyethylene (PE), polystyrene (PS), and polypropylene (PE), are the primary sources of MNPs. Primary MNPs can be found in plastic pellets and personal care goods and are plastic particles that have been purposefully created for a variety of uses, such as film-forming agents, functionalized polymers, hydrophilic agents, and silicones (Amobonye et al., 2021). On the contrary, secondary MNPs come up as a result of abiotic factors like temperature, UV radiation, microbial degradation, and atmospheric pressure fragmenting large amounts of plastic debris (Geyer et al., 2017). The oxidation of plastic materials can lead to secondary MNPs. Due to the composition of their molecules, polymeric materials including polypropylene (PP), polyvinyl chloride (PVC), and polybutadiene (PBD) are especially prone to oxidative-thermal deterioration even at ambient temperature (Hu et al., 2021). Due to the extensive generation of plastic in the 1990s and earlier, secondary microplastics are currently being produced in the oceans throughout the globe. Although efforts are being made to decrease environmental pollution, specialized knowledge is required to remove MNPs.

MPs and NPs can enter soils through a variety of natural processes and man-made activities. Agricultural practices, such as using plastic mulch, applying soil amendments containing plastics, and irrigating with plastic-polluted wastewater, are the main human sources of these particles. The presence of polymers in agricultural soils contributed significantly to the amount of microplastics, which can accumulate in crops as they break down (Fig. 1). Plastic mulch films are an important producer of microplastics and are generally used in arid and cold climates to regulate temperature and increase crop yield. These films break down into microplastics upon exposure to light and mechanical forces like tillage and are then transported into agricultural soils, particularly in areas where the films are not effectively recovered (Fig. 1).

Despite being overlooked in recent years, atmospheric deposition is another potential pathway for microplastic pollution of terrestrial ecosystems. Weather patterns, especially precipitation and snowfall, can have an impact on the consequences of microplastics from the earth's atmosphere to soil ecosystems, while wind and atmospheric circulation can influence remote transport to different locations (Mohajerani & Karabatak, 2020). Agricultural soils and human health may be at risk from airborne microplastics from metropolitan areas that settle on fields and plant leaves, even if the direct harmful effects of these particles on soils are not fully understood. Compost, which is also frequently used to enhance soil fertility, contains a high proportion of plastic particles, suggesting it is a significant route for microplastics entering agricultural soils. The rapid disintegration of larger plastics into smaller pieces and microbial activity during composting processes can increase the concentration of microplastics in agricultural soils.

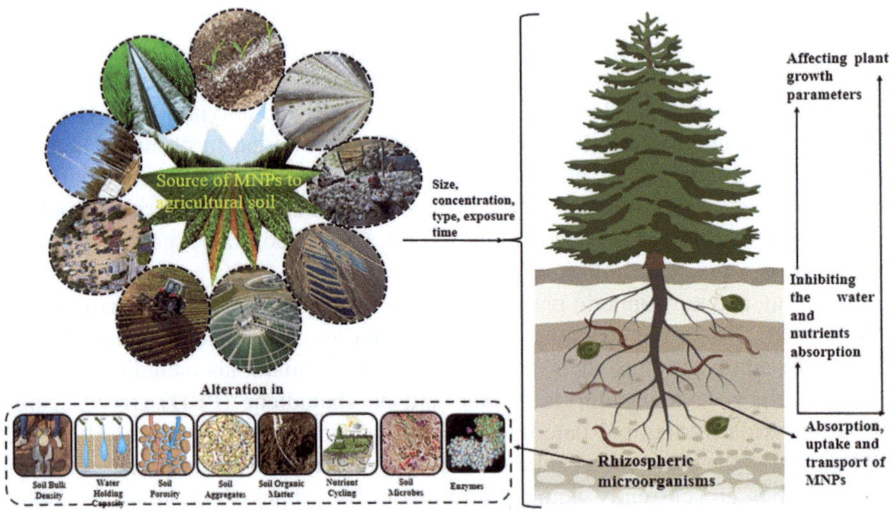

Fig. 1 Source and impact of MNPs in agricultural soil

In addition to having an immediate effect on soil microbes and crops, the presence of microplastics in soil can modify a number of soil properties, including soil bulk density, soil aggregate size fraction, and evapotranspiration (Fig. 1). These changes can have indirect effects on plant performance, such as altered water movement routes leading to increased soil water evaporation and changes in the soil microbial population, particularly in terms of root colonization microorganisms. Furthermore, research has shown that microplastics can mitigate the harmful effects of sulfamethazine on plant growth by modifying microbial populations. Microplastics interact with other pollutants, including polycyclic aromatic hydrocarbons (PAHs), organochlorine insecticides, and toxic metals (such as Zn, Cd, and Pb), in agricultural soil due to their huge surface area, which can lead to complex environmental consequences. As microplastics move through the soil profile via various means, including bioturbation, surface runoff, water penetration, irrigation canals, and even the atmosphere, they may adsorb and attach to sorbed organic contaminants. Therefore, microplastics and their co-existing contaminants pose significant environmental hazards in agricultural soil systems, acting as vectors for other pollutants. Further research is needed to thoroughly examine these risks.

3 Microbial Strategies to Remediate Micro-Nano Plastics (MNPs)

Micro-nano plastics (MNPs) pollutants can be eliminated in an efficient, economical, and sustainable manner using microbial degradation. Several biochemical reactions take place during the process, and they differ depending on the chemical composition of the pollutants. It is essential to thoroughly comprehend these mechanisms to safeguard the environment (Velez et al., 2018). Biostimulation (the supply of nutrients that stimulate microbial growth) and bioaugmentation (the addition of living cells capable of breakdown) are essential to increase the biodegradation rate of pollutants in environments with limited nutrition (Tiwari et al., 2020). The microbial degradation of polymers is influenced by numerous factors that can alter the physicochemical properties of the polymers. The process of microplastic degradation is affected by biotic and abiotic factors (Fig. 2). The type and concentration of microorganisms, the availability of single or mixed cultures, cell hydrophobicity, diversity of enzymatic systems, production of extracellular enzymes, and the potential for biosurfactant formation are biotic factors that influence the biodegradation rate. Abiotic factors are environmental variables that include factors like temperature and pH of the media, UV radiation, salinity, oxygen concentration, and moisture content. Each of these parameters should be adjusted to the employment of the particular microorganisms to achieve the highest biodegradation rate. Additionally, the critical attributes of MNPs, such as their additive and hydrophobic nature, along with the presence of persistent organic pollutants on their surfaces, play a vital role in the remediation process (Miloloža et al., 2022).

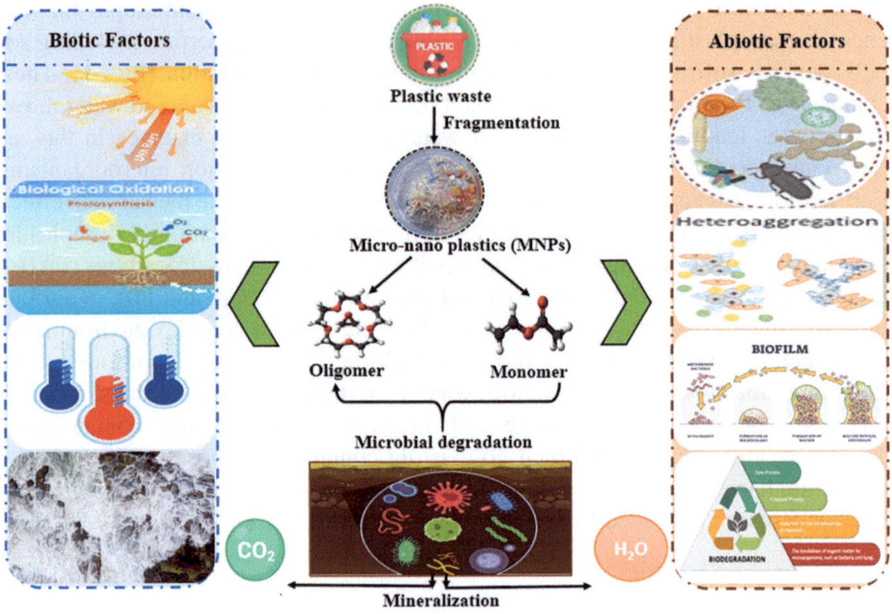

Fig. 2 The degradation of microplastics through the combined influence of abiotic and biotic factors

3.1 *Mechanism of MNPs Biodegradation*

Microplastics can undergo degradation via physical, chemical, and biological means, with the latter being linked to a wide variety of enzymes (Bacha et al., 2021). Microbial plastic degradation proceeds through a series of consecutive steps: Assimilation is the integration of molecules by microorganisms; biodeterioration is the alteration of the physical and chemical properties of polymers by biological agents; biofragmentation is the lytic cleavage of complex polymers into simpler forms through enzymes and acids; and mineralization is the process of oxidation of metabolites generated by degradation, such as CO_2, CH_4, and H_2O (Tiwari et al., 2020). In general, the degradation starts with the breakdown of polymers into smaller particles, followed by further degradation into oligomers, dimers, and monomers (Fig. 3). Finally, microbes facilitate the mineralization steps to complete the degradation process. When microplastics undergo complete mineralization, they are converted into intermediate compounds that serve as an energy source for promoting microbial growth, with carbon dioxide being released in the process. It was observed that abiotic factors like UV radiation and photo-oxidation can increase microbial degradation (Min et al., 2020). However, the microbial degradation process can be hindered by the presence of large molecular fragments in higher molecular weight plastic polymers that are not easily absorbed by cells. Microorganisms use two mechanisms to break down these large particles: intracellular and

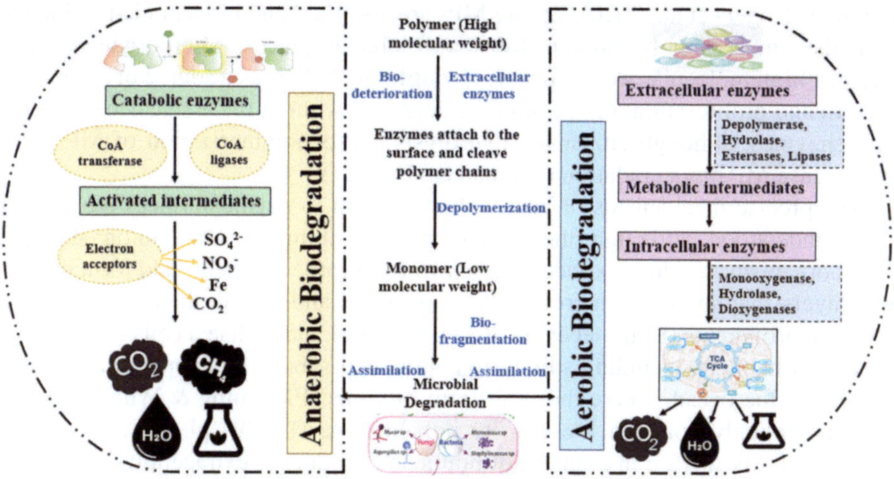

Fig. 3 Mechanism of biological degradation of plastics using aerobic and anaerobic conditions

extracellular degradation (Fig. 3) (Yuan et al., 2020). In intracellular degradation, microbes acquire on the surface of MNPs to break down the plastic into shorter chains. In extracellular degradation, bacteria release extracellular enzymes like hydrolases to break down complex polymers into simpler units that can be metabolized by different metabolic pathways (Yuan et al., 2020). Then, through aerobic or anaerobic metabolism, microorganisms transform these shorter chains into final products (such as CH_4, CO_2, or H_2O) in a process called mineralization. Finally, biological natural attenuation leads to the consumption of these end products as carbon and energy sources (Tiwari et al., 2020).

Several extracellular enzymes are associated with the degradation of microplastics, including esterases, lipases, lignin peroxides, laccases, and manganese peroxides, which enhance the hydrophilicity of the plastics and convert them into carbonyl or alcohol residues (Taniguchi et al., 2019). Bacteria release extracellular hydrolase enzymes that are capable of polymerizing complex compounds into hydroxyl acid monomers. Polyhydroxy butyrate (PHB) hydrolysis yields R-3-hydroxybutyric acid (HBA), while PHBV degradation yields 3-hydroxybutyrate and 3-hydroxy valerate. Aerobic conditions lead to the production of CO_2 and H_2O, while anaerobic conditions lead to the production of methane without any detrimental molecules produced during the PHB degradation (Fig. 3) (Kovalcik et al., 2020). Hydrolase enzymes, such as lipases, esterases, and cutinases, work on the surface of the plastic to enhance chain cleavage reactions and cause cracking (Lear et al., 2021). Although these enzymes do not penetrate the polymer, the monomers they produce are assimilated by microbes and enter different metabolic pathways. Chitinases are enzymes synthesized by bacteria, such as *Achromobacter*, *Flavobacterium*, *Micrococcus*, *Pseudomonas*, and *Vibrio* sp., that can break down polymers. Although the precise mechanism of MPs particle degradation by *Pseudomonas* sp. remains ambiguous, chitinase enzyme has been found to be useful in the enzymatic breakdown of plastic polymer (Rogers et al., 2020). Before MPs are absorbed and biodegraded inside the cell, they must first be broken down into smaller pieces. Microorganisms can degrade high molecular weight plastics by utilizing MNPs as a source of carbon for growth and proliferation. MPs cannot be degraded without the assistance of microbial enzymes, although enzymatic hydrolysis and depolymerization of MP polymers cause MP biodegradation (Othman et al., 2021).

The precise mechanisms of bacterial attachment to plastic surfaces are not yet fully understood, although the formation of biofilms is a common strategy. Microorganisms produce extracellular enzymes to gather nutrients and form protective layers that shield them from toxic substances and enable them to survive in harsh environments (Rummel et al., 2017). Bacterial and archaeal cells can live in a variety of habitats, including soil, deep continental subsurface, upper marine sediment, and the ocean, where they can form biofilms (Flemming & Wuertz, 2019). Bacteria quickly colonize low-density polyethylene (LDPE) plastic in seawater, attaching to its surface in a matter of hours and producing extracellular polymeric substances and other bacterial metabolites that can break down debris, such as microplastic-related compounds, and improve polymer/microbe adsorption (Harrison et al., 2011). Changes in physicochemical properties, such as the identification of degraded plastic products and corresponding cellular responses, can help understand the role of microorganisms in the degradation/accumulation of MNPs.

3.2 Microbial Degradation of Micro-Nano Plastics in Wastewater, Compost, Sewage Sludge and Municipal Solid Waste

MNPs were found to be present in significant quantities in various water sources, including filtered and contaminated water, oceanic water, streams, and rainwater (Kelly et al., 2021). The pace of plastic degradation is influenced by several factors, including molecular weight, crystal structure, organic functional groups, and additives (Arpia et al., 2021). Bacterial species can create biofilms on the surface of plastic fragments by biosynthesizing extracellular polysaccharides (EPS), which can eventually lead to the fragmentation of MNPs into smaller sizes and eventual biodegradation. The treatment of wastewater by certain bacterial taxa, such as Vibrio, Campylobacter, and Arcobacter, can also lead to the degradation of MNPs (Kelly et al., 2021). To eliminate MNPs from wastewater streams, it is advisable to examine the tiny molecules that can influence the production of enzymes such as PETase and MHETase, which bacteria use to degrade MNPs (Dey et al., 2021).

Numerous bacterial groups, such as Alteromonadaceae and Burkholderiales, have been linked to the biodegradation of poly (3-hydroxybutyrate-co-3-hydroxyhe xanoate) (PHBH). It was demonstrated that *Erythrobacter* Alcanivorax borkumensis, growing on MNPs biofilms, can eliminate low-density PE in aquatic environments (Yang et al., 2020). Recent studies have reported MNPs removal percentages from wastewater through primary (50–98%), secondary (0.2–14%), and tertiary (0.2–2%) treatments (Dey et al., 2021). However, there is little information available about the microbial breakdown of MNPs in wastewater via biological/secondary treatment in the majority of research. Future research should, therefore, look into how bacteria contribute to the biodegradation of MNPs in wastewater streams.

For effective plastic treatment, a variety of remediation technologies and biotechnologies have been proposed so far, including membrane separation, coagulation, and biodegradation. MNP removal is frequently carried out in wastewater treatment plants (WWTPs) during the coagulation process, where Al and Fe salts attach to plastic particles and make it easier for their remediation through complexation (Chorghe et al., 2017). The effectiveness of MNP removal by coagulation has been confirmed by multiple studies (Ma et al., 2019; Novotna et al., 2019). For instance, a study had assessed the effectiveness of coagulation using ferric chloride and aluminium chloride to remove PE particles of various sizes. While particles of less than 5 mm were larger than the typical colloidal particles removed by coagulation, the results still supported the effectiveness of plastic removal by coagulation in the case of relatively small particles (Ma et al., 2019). While studies on the conventional membrane separation process and membrane bioreactors (MBRs) for efficient wastewater treatment and energy production have increased significantly in recent years, the removal of MNPs using membrane technology is still limited (Poerio et al., 2019). Additionally, ultrafiltration employing membrane separation has been suggested as a successful technique for removing high MP concentrations, enabling the attainment of high-quality drinking water with relatively low energy consumption, high separation efficiency, and a small plant size (Abomohra & Hanelt, 2022). In recent years, biodegradation has been referred to as a cutting-edge, environmentally friendly method of cleaning up plastic waste. There has been a lot of research on the ability of microbes (mostly bacteria and fungi) to break down plastic by enzymatic hydrolysis (Zhu et al., 2020). Further mechanistic elucidations and the investigation of novel microbial enzymes are essential for improved MNP remediation via biodegradation. Thus, a new line of research that is still in its earliest stages aims to identify plastic-active enzymes for further application in biotechnological processes and to characterize their actual purpose in nature (Chow et al., 2023). The four primary stages of MNP degradation are as follows: the establishment of a biofilm on plastispheres, which reduces the hydrophobicity and buoyancy of plastic particles (Rai et al., 2021). The next two steps are biodeterioration via exopolysaccharides and enzymatic action (endo−/exoenzymes). The destabilization of the carbon skeleton in MNP particles through enzymatic depolymerization using oxidases, amidases, peroxidases, and laccases is the third stage (Gómez-Méndez et al., 2018). Then, microbial biomass assimilates the monomers. According to reports, hydrolases are crucial to the hydrolysis of plastic polymers, which determines how quickly MNP degrades.

In general, most of the microplastic particles (MNPs) in wastewater are removed during treatment, but they tend to accumulate in the sludge (Iyare et al., 2020). Studies have shown that the sludge contains more concentration of MNPs than wastewater (Aslam et al., 2020; Sun et al., 2019). Additionally, primary sludge contains a greater number of MNPs than secondary sludge. Gies et al. (2018) found that a significant number of MNPs (0.54–1.28 trillion) were present in primary sludge, whereas a wastewater treatment plant (WWTP) in Vancouver, Canada, found a lower concentration (0.22–0.36 trillion) of MNPs in activated/secondary sludge. The remediation of MNPs with microbial aid in secondary treatment is, therefore, essential. Municipal solid waste (MSW), in addition to sludge, is regarded as a possible source of MNPs in the environment. These particles could be linked to a variety of microorganic and organic contaminants, which, when they enter the food chain, can have detrimental impacts on the environment and human health (Mahesh et al., 2023).

Plastic can be broken down by the native microbial populations that live in municipal solid waste (MSW) and sewage sludge. Recently, Ru et al. (2020) found that the mesophilic bacterium, *Stenotrophomonas panacihumi*, can break down polypropylene (PP) into low and high molecular weight forms (with Mn values of 2800, 3600 Da and 44,000 Da, respectively) after 90 days. Similarly, it was reported that the bacterial strain Rhodococcus was able to degrade 6.4% of the PP polymer mass in just 40 days (Camila Ariza-Tarazona et al., 2018). A study showed that the bacterium *Ideonella sakaiensis* was able to completely reduce a polyethylene terephthalate (PET) film in 42 days (Yoshida et al., 2016), while wild bacterial strains *Bacillus* sp. and *Paenibacillus* sp. retrieved from a landfill location were able to reduce particle diameter by 28% after 60 days (Park & Kim, 2019). Microbial consortia have been found to be more successful at degrading plastic polymers in the environment than pure microbial strains. For example, the use of two *Pseudomonas* sp. in a microbial consortium demonstrated better degrading abilities than established species. Actinomycetes and fungi can also create microbial consortia, which could result in better plastic and polymer degradation outcomes (Skariyachan et al., 2018).

The compost industry is facing a serious challenge of contamination due to smaller plastic waste fragments. In order to tackle this problem, the in situ biodegradation using hyperthermophilic composting (hTC) approach has been proposed and tested. Results indicate that 43.7% of MNPs were eliminated from wastewater sludge after 6–7 weeks of hTC, the highest proportion ever attained through MNP biodegradation. According to high-throughput sequencing, the most frequent bacteria involved in the biodegradation of MNPs during hTC are Thermus, Bacillus, and Geobacillus, demonstrating the future potential of this approach for MNP removal (Wang & Wu, 2021). The goal of scientific study is rapidly changing towards establishing guidelines that permit the biological method of bioplastic degradation in both aerobic and anaerobic natural waste treatment measures (Ruggero Federica et al., 2019). Thermophilic composting is an appealing method to eliminate unwanted plastic particles from the environment. Small plastic fragments are naturally decomposed by utilizing a variety of techniques. *Streptomyces* sp., a thermophilic microbe that is naturally occurring in the soil and has been documented to degrade poly(D-3-hydroxybutyrate) at 50 °C, has also demonstrated the ability to degrade PE succinate. Similarly, the thermophilic bacterial strain *Bacillus* sp. TT96 has shown the potential to degrade PE succinate during composting (Pathak & Navneet, 2017).

PUR foam, commonly known as polyurethane foam, is a serious environmental pollutant that poses a major challenge for waste disposal. In vitro experiments have shown that *Pseudomonas chlororaphis*, a Gram-negative multi-species aerobic proteobacterium commonly found in soil, is capable of breaking down polyester PUR foam (Pathak & Navneet, 2017). *Brevibacillus borstelensis* and *Rhodococcus ruber*, two more Gram-positive bacteria, have also been discovered to utilize polymers as a carbon source and break them down. A consortium that enhances the biodegradation of plastic pieces through a syntrophic mechanism can be formed by fungi (*Aspergillus niger*) and bacteria (Pseudomonas and Vibrio) (Skariyachan et al., 2018). Additionally, certain thermophilic bacteria, including *Flavobacterium* sp., *Micrococcus* sp., *Staphylococcus* sp., *Bacillus* sp., *Chelatococcus* sp., are well known for their capability to metabolize plastic in compost (Skariyachan et al., 2015, 2018).

4 Recent Developments in the Breakdown of Micro-Nano Plastics

Several cutting-edge methods have recently been implemented to improve the biodegradation of plastic particles, including enzymatic/enzyme-assisted degradation, sophisticated molecular tools and technologies, membrane bioreactor (MBR)-assisted remediation, and remediation based on nanotechnologies.

4.1 Enzyme-Based Degradation

It is extremely difficult to break down plastic using microbial enzymes because of the lack of hydrolysable groups in the carbon-carbon backbone. The involvement of microbial diversity in the alteration of plastic polymers and the subsequent process of depolymerization has been widely acknowledged. This is due to the diverse array of genes, proteins, enzymes, and metabolic pathways that are involved (Fig. 4) (Priya et al., 2022). However, the first step that can be accomplished by the action of both biotic and abiotic forces is the reduction of molecular weight. For instance, exposure to UV radiation can make microbial enzymes more likely to attack the carbonyl group of the polymer (Kopecká et al., 2022; Kour et al., 2023). Several enzymes, including laccase, manganese-dependent enzymes (enzymes that degrade lignin), urease, lipase, and protease, are used to break down polymers. Different types of microbial enzymes, including cutinases, lipases, and esterases, as well as *Streptomyces* sp., have demonstrated their effectiveness in degrading polyethylene (PE) succinate. Thermostable laccase, for instance, can degrade PE in just 48 h of

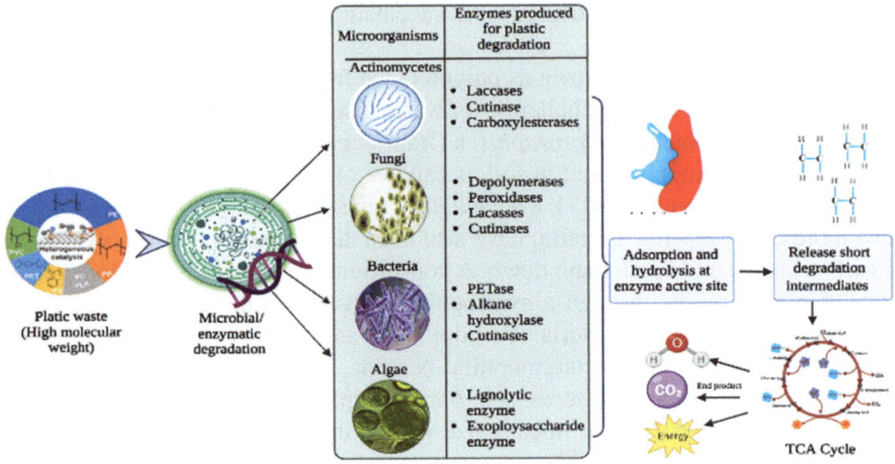

Fig. 4 Biodegradation of plastic waste using enzyme-based technology

incubation at 37 °C (Jaiswal et al., 2020). Additionally, it has been discovered that the composting may cause the thermophilic bacterial strain (*Bacillus* sp. TT96) to degrade PE succinate (Pathak & Navneet, 2017).

Ideonella sakaiensis, a bacterium, has been found to be able to use polyethylene terephthalate (PET) as its main source of energy and carbon (Yoshida et al., 2016). Mainly two enzymes (PETase and MHETase) in this bacteria are responsible for breaking down PET into its monomers, terephthalic acid, and ethylene glycol (Palm et al., 2019). Despite the partial understanding of a few enzymes and their impact on plastic degradation, the issue has not been fully addressed. This is primarily due to a lack of technologies that would improve the effectiveness and industrial production of these enzymes that degrade plastics.

Recent studies have focused on increasing enzyme activity by protein and enzyme engineering in order to better understand the enzymatic breakdown of plastics. For instance, an engineered PETase mutant from *Ideonella sakaiensis* showed increased activity with three mutations (R61A, L88F, 303, and I179F) compared to the wild-type strain. This increase in enzyme activity becomes possible by effectively altering proteins and altering major hydrophobic grooves of substrate binding sites (Ma et al., 2016). A recent study has demonstrated the effective use of protein-engineered enzymes in the proper breakdown of microplastics, such as polyurethane. The study suggests that synthetic polymer-binding peptides like anchor peptides act as catalysts for the breakdown of these microplastics (Islam et al., 2019). The impressive outcomes of protein/enzyme engineering highlight its potential as a technique for improving the removal of MNPs. The engineering or enhancement of microbial strains is another modern technique. In a recent study, it was discovered that, in comparison to enzymes under neutral pH settings, an alkaline microbial strain when exposed to high pH conditions reduced the size of PET polymer and increased by-product concentration. These microorganisms can proliferate by using the products produced in the medium, and the feedback inhibition process can be avoided (Gong et al., 2018).

It is interesting to note that considerable effort has been expended on creating techniques for enzyme immobilization and their extensive relevance in environmental cleanup. Environmental contaminants such as dyes, phenolic endocrine disruptors, medicines, and other micropollutants can be removed from the environment using these eco-friendly methods (Shakerian et al., 2020). Bisphenol A (BPA), a monomer of polycarbonate plastics, is one of the most frequently produced compounds in the world (Hacıosmanoğlu et al., 2019). The enzyme laccase has been discovered as one of the most frequently used enzymes in immobilized systems for BPA degradation (Piao et al., 2019). Oxidative enzymes like laccase and horseradish peroxidase have been demonstrated to be more cost-effective, reusable, stable, and long-lasting when immobilized (Shakerian et al., 2020). Therefore, it is expected that integrating membrane and enzyme/microbial technology would offer an encouraging future for the remediation of environmental MNPs. Therefore, new techniques for immobilizing plastic-degrading enzymes on a large scale, or at least at the pilot scale, are required for water filtration. Additionally, new techniques that synthesize a large amount of enzymes in an easy and affordable manner must be developed to address this challenge.

4.2 Advanced Molecular Techniques

Metabolic engineering and synthetic biology have led to the development of resistant microbial strains with superior eco-friendly conversion and recycling capabilities for synthetic plastics. The development of biotechnological strategies for plastic degradation is facilitated by the discovery of novel DNA sequences, enzymes, metabolic pathways, and active compounds with higher metabolic capacities (De et al., 2018). The systems biology approach also uses several omics technologies like transcriptomics, metabolomics, proteomics, and genomics to improve the monitoring of the disintegration of various environmental pollutants. These advancements have the potential to significantly benefit the environment (Fig. 5).

In recent years, a variety of different metabolic engineering methods have been developed that can be used either alone or in conjunction with genetically modified constructs that are intended to degrade persistent pollutants. One effective approach is systems metabolomic engineering, which helps improve the growth of engineered microbes and enhances their ability to degrade plastic more efficiently. To improve the metabolic activity of catalysts, high-throughput screening methods have been developed to analyze the activities of polyester hydrolases in PET hydrolysis. However, discovering novel microbial polyester hydrolases and assembling a group of metabolically active enzymes are still significant challenges in producing a metabolic process for removing polyethylene waste (Wei & Zimmermann, 2017). Researchers have utilized the genetic engineering approach to enhance the biodegradation potential of native microorganisms by manipulating their genetic pathways

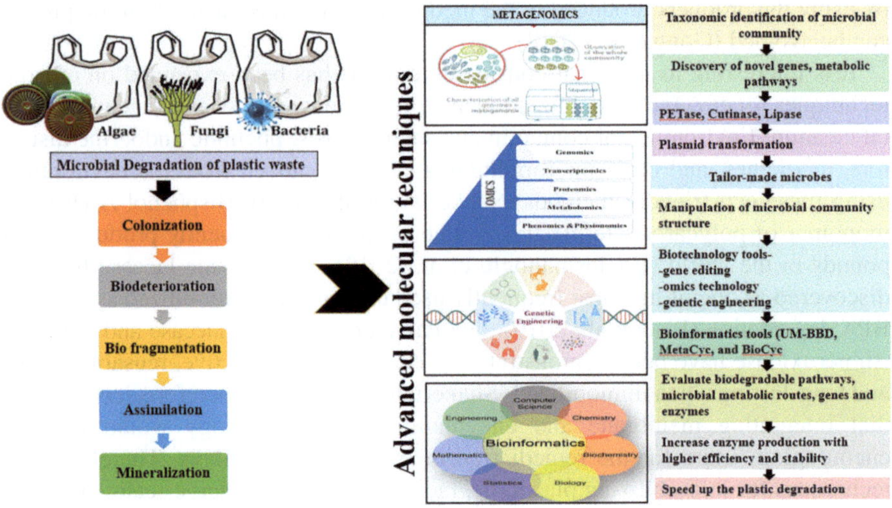

Fig. 5 Biodegradation of plastic waste using advanced molecular techniques

or by introducing plastic degradation genes into non-native microorganisms. Advances in gene editing tools and technologies, including TALENs, CRISPR/Cas9, electroporation, and chemical transformation, can be utilized to enhance or alter the genetic structure of microorganisms (Fig. 5). These methods can be used practically to alter bacterial genomes by adding genes that encode different plastic-degrading enzymes including PETase, esterase, depolymerase, laccase, and others.

It was observed that naturally occurring bacteria degrade plastic waste considerably more slowly when compared to artificial methods (engineered constructs) (Gu, 2021). For instance, a synthetic enzyme construct called cutinase has been demonstrated to accelerate the degradation of PUR in 4.8–6.2 h compared to the time taken by wild microorganisms (Islam et al., 2019). The effectiveness of plastic biodegradation has also been confirmed in manipulated marine microbial consortia (Syranidou et al., 2019). Such altered microorganisms that break down plastic can be used to depolymerize plastic polymers. Despite several successes of genetic engineering of microorganisms in the lab, the majority of the engineered microbes have not produced good results when tested in the field.

Bioinformatics, which has been shown to be a successful method for improving the biodegradation of pollutants like MNPs, can be used to improve understanding of the biodegradation of plastic polymers (Purohit et al., 2020). A number of databases have been created to assess biodegradation pathways and toxicities, including UM-BBD, MetaCyc, and BioCyc. These databases provide important details about the microbial metabolic pathways, genes, and enzymes, as well as their enzymatic reactions, involved in the biodegradation of recalcitrant substances such as plastic polymers (Priya et al., 2022). These computational techniques are particularly useful for detecting the relevant enzymes and anticipating the biodegradation pathways of hazardous substances that were not previously recognized. This bioinformatics approach offered a common foundation to the fields of metabolic engineering and synthetic biology for developing an innovative approach for plastic biodegradation (Ali et al., 2021). However, it is essential to note that bioinformatics and experimental data have some significant disadvantages, including lack of availability, accessibility problems, and validation concerns that need to be taken into account in future studies.

The application of synthetic biology, which incorporates "omic" studies, computational biology, and high throughput sequencing, has significantly advanced our understanding of the interactions between bacteria and plastic polymers and the potential degradation of these polymers (Bouhajja et al., 2016). An important component of synthetic biology is creating a metabolic pathway for the biodegradation of synthetic polymers (Purohit et al., 2020). However, little is currently known about the different kinds of microbes that may break down synthetic polymers and the enzymes they employ. Therefore, future research should focus on developing techniques to identify and characterize robust microorganisms that can degrade polymers and their associated enzymes. An extensive study is required at all tiers of environmental microbiology and biotechnology, as well as gene modification and protein engineering, to address the issues related to the biodegradation of plastic polymers. Genetics, metabolic engineering, molecular biology, and systems biology-based combinatorial strategies may ultimately offer innovative insights into the biodegradation of plastic polymers.

4.3 Membrane Technology

Membrane bioreactor (MBR) technology, also known as biomembrane technology, employs a film-based approach such as microfiltration or ultrafiltration to facilitate a partition method that utilizes biological catalysts such as enzymes and microbes (Dey et al., 2021). This cutting-edge technology has received recognition globally as a viable strategy for the effective treatment of both industrial and municipal wastewater (Fig. 6) (Baker, 2012). Additionally, the use of ultrafiltration membranes is a proven method for the elimination of several pollutants from wastewater treatment, including microorganisms, tiny organic acids, and soluble microbial metabolites (Wu et al., 2019). It is important to understand that the porosity of membranes used in water treatment processes falls within the same range of sizes as microplastics. As a result, the membrane displays outstanding durability and improved selectivity for removing microplastic contaminants and their byproducts (Baker, 2012; Parks et al., 2017). The development of membrane technology has led to the creation of membrane bioreactor (MBR) technology, which offers superior contaminant removal efficiency compared to conventional methods (Lares et al., 2018). MBR technology, when combined with traditional biological sludge treatment with a micro-ultrafiltration system, provides significant benefits over conventional approaches, such as high pollutant removal rates, low energy usage, and decreased sludge production (Arias et al., 2020). MBR technology reportedly removes

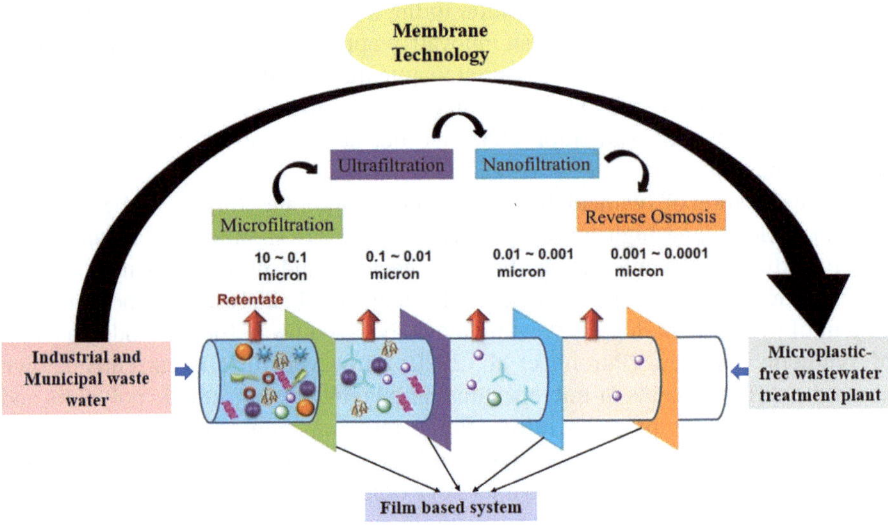

Fig. 6 Biodegradation of plastic waste using membrane technology

microplastic contamination from wastewater treatment plants more prominently than conventional technology (Talvitie et al., 2017). Even after ultrafiltration of pollutants, nanoscale particles (polyvinyl chloride, polyethylene terephthalate, polystyrene, and polyethylene) are still detected (Halle et al., 2017). For instance, MBR increased the removal efficiency of MPs by up to 99.4% in comparison to traditional activated-sludge-based treatment (Pico et al., 2019). The use of MBRs for MNPs remediation in WWTPs indicated their superiority to oxidation ditches (Lv et al., 2019), which can be related to their dual action for anti-fouling and separation performance (Sun et al., 2020). Although ultrafiltration in combination with coagulation is already utilized in WWTPs and allows the considerable removal of organic matter, this technology has not yet been successfully developed for the removal of MNPs that remain in the final effluents.

In the past, the elimination of nano-sized pollutants was a difficult task, but advancements in technology, such as membrane technology, have provided viable solutions. Currently, there is a rising trend in the development of nanofiltration membranes, which are a combination of reverse osmosis and ultrafiltration. The efficiency of contaminant removal may differ based on the constituents of the membranes. A promising and effective approach for membrane technology is the immobilization of nanomaterials either on or within the membrane matrix. The selection of appropriate nanomaterials, based on the type of pollutant, can result in improved performance and increased stability. Zeolite-coated ceramic membranes, inorganic and organic thin film nanocomposite (TFN) membranes, and hybrid protein polymer biomimetic membranes are a few examples of the novel membrane materials using nanotechnology being developed for water purification (Ma et al., 2016). These composite materials represent the next generation of membrane technology, by using novel techniques that take advantage of interactions between pollutants and membrane dynamics. In the removal of boron from saltwater and desalination processes, nanofiltration, a membrane-based technology that uses membranes containing nanopores, has shown outstanding results (Yang et al., 2019). It is noteworthy that nanoporous membranes hold great potential as fine filters that selectively permit the passage of water molecules through their structure (Li et al., 2017). While the use of membrane technology has proven to be a successful method for removing MNPs from water bodies, it is essential to pursue additional approaches for degrading or utilizing these toxic contaminants in order to achieve a pollution-free environment.

5 Conclusion and Future Scope

Micro-nano plastic (MNPs) pollution is considered to be a prominent form of pollution that has a long-lasting impact on the environment. Utilizing microbial processes and enzymes to remediate this pollution can serve as a means to break down or mitigate the effects of MNPs. The biodegradation of plastic materials involves the identification of several breakdown pathways, which requires further studies. It is also a significant challenge to use microbial enzymes to improve the breakdown and restructuring of plastics for improved biodegradation. Future studies will be greatly facilitated by the identification of organisms that can affect a variety of plastics, including high molecular weight polymers. After a polymer degrades, monomers, dimers, and oligomers synthesize that can be recycled to make new materials. However, identifying highly active and functional enzymes is a difficult task, but the metagenomics approach could aid in identifying non-cultivated microorganisms and establishing function-based assays. Although combining membrane and enzyme technology, nanoparticle technology, and metagenomics aids in the remediation of MNPs, it is crucial to investigate the individual enzymes and processes that degrade micro-nano plastics. A comprehensive study in this field is required for a considerable reduction of micro- and nano-sized plastics so that the promotion of long-term health could be achieved.

References

Abomohra, A., & Hanelt, D. (2022). Recent advances in micro-/nanoplastic (MNPs) removal by microalgae and possible integrated routes of energy recovery. *Microorganisms, 10*(12), 2400. https://doi.org/10.3390/MICROORGANISMS10122400

Ali, I., Ding, T., Peng, C., Naz, I., Sun, H., Li, J., & Liu, J. (2021). Micro- and nanoplastics in wastewater treatment plants: Occurrence, removal, fate, impacts and remediation technologies – A critical review. *Chemical Engineering Journal, 423*, 130205. https://doi.org/10.1016/J.CEJ.2021.130205

Amobonye, A., Bhagwat, P., Raveendran, S., Singh, S., & Pillai, S. (2021). Environmental impacts of microplastics and nanoplastics: A current overview. *Frontiers in Microbiology, 12*, 3728. https://doi.org/10.3389/FMICB.2021.768297/BIBTEX

Andrady, A. L. (2017). The plastic in microplastics: A review. *Marine Pollution Bulletin, 119*(1), 12–22. https://doi.org/10.1016/J.MARPOLBUL.2017.01.082

Arias, D. M., García, J., & Uggetti, E. (2020). Production of polymers by cyanobacteria grown in wastewater: Current 1 status, challenges and future perspectives. *New Biotechnology, 55*, 46–57.

Arpia, A. A., Chen, W. H., Ubando, A. T., Naqvi, S. R., & Culaba, A. B. (2021). Microplastic degradation as a sustainable concurrent approach for producing biofuel and obliterating hazardous environmental effects: A state-of-the-art review. *Journal of Hazardous Materials, 418*, 1–11. https://doi.org/10.1016/J.JHAZMAT.2021.126381

Aslam, H., Ali, T., Mortula, M. M., & Attaelmanan, A. G. (2020). Evaluation of microplastics in beach sediments along the coast of Dubai, UAE. *Marine Pollution Bulletin, 150*, 110739. https://doi.org/10.1016/J.MARPOLBUL.2019.110739

Bacha, A.-U.-R., Nabi, I., & Zhang, L. (2021). Mechanisms and the engineering approaches for the degradation of microplastics. *ACS ES&T Engineering, 1*(11), 1481–1501. https://doi.org/10.1021/ACSESTENGG.1C00216

Baker, R. W. (2012). Membrane technologies and applications. In K. Mohanty & M. K. Purkait (Eds.), *Membrane technologies and applications* (pp. 1–485). Wiley. https://doi.org/10.1002/0470020393

Bouhajja, E., Agathos, S. N., & George, I. F. (2016). Metagenomics: Probing pollutant fate in natural and engineered ecosystems. *Biotechnology Advances, 34*(8), 1413–1426. https://doi.org/10.1016/J.BIOTECHADV.2016.10.006

Camila Ariza-Tarazona, M., Francisco Villarreal-Chiu, J., Barbieri, V., Siligardi, C., & Cedillo-González, E. I. (2018). New strategy for microplastic degradation: Green photocatalysis using a protein-based porous N-TiO2 semiconductor. *Ceramics International, 45*(7), 9618–9624. https://doi.org/10.1016/j.ceramint.2018.10.208

Chorghe, D., Sari, M. A., & Chellam, S. (2017). Boron removal from hydraulic fracturing wastewater by aluminum and iron coagulation: Mechanisms and limitations. *Water Research, 126*, 481–487. https://doi.org/10.1016/J.WATRES.2017.09.057

Chow, J., Perez-Garcia, P., Dierkes, R., & Streit, W. R. (2023). Microbial enzymes will offer limited solutions to the global plastic pollution crisis. *Microbial Biotechnology, 16*(2), 195–217. https://doi.org/10.1111/1751-7915.14135

De, L., Alves, F., Westmann, C. A., Lovate, G. L., Viana De Siqueira, G. M., Cabral Borelli, T., & Guazzaroni, M.-E. (2018). Review article metagenomic approaches for understanding new concepts in microbial science. *Hindawi International Journal of Genomics, 2018*, 1–15. https://doi.org/10.1155/2018/2312987

Deng, Y., Zhang, Y., Lemos, B., & Ren, H. (2017). Tissue accumulation of microplastics in mice and biomarker responses suggest widespread health risks of exposure. *Scientific Reports, 7*, 46687. https://doi.org/10.1038/SREP46687

Dey, T. K., Uddin, M. E., & Jamal, M. (2021). Detection and removal of microplastics in wastewater: Evolution and impact. *Environmental Science and Pollution Research, 28*, 16925–16947. https://doi.org/10.1007/s11356-021-12943-5/Published

Enfrin, M., Dumée, L. F., & Lee, J. (2019). Nano/microplastics in water and wastewater treatment processes – Origin, impact and potential solutions. *Water Research, 161*, 621–638. https://doi.org/10.1016/J.WATRES.2019.06.049

Federica, R., Riccardo, G., & Claudio, L. (2019). Methodologies to assess biodegradation of bioplastics during aerobic composting and anaerobic digestion: A review. *Waste Management and Research, 37*(10), 959–975. https://journals.sagepub.com/doi/pdf/10.1177/0734242X19854127

Flemming, H. C., & Wuertz, S. (2019). Bacteria and archaea on Earth and their abundance in biofilms. *Nature Reviews. Microbiology, 17*(4), 247–260. https://doi.org/10.1038/S41579-019-0158-9

Geyer, R., Jambeck, J. R., & Law, K. L. (2017). Production, use, and fate of all plastics ever made. *Science Advances, 3*(7). https://doi.org/10.1126/SCIADV.1700782

Gies, E. A., LeNoble, J. L., Noël, M., Etemadifar, A., Bishay, F., Hall, E. R., & Ross, P. S. (2018). Retention of microplastics in a major secondary wastewater treatment plant in Vancouver, Canada. *Marine Pollution Bulletin, 133*, 553–561. https://doi.org/10.1016/J.MARPOLBUL.2018.06.006

Gómez-Méndez, L. D., Moreno-Bayona, D. A., Poutou-Piñales, R. A., Salcedo-Reyes, J. C., Pedroza-Rodríguez, A. M., Vargas, A., & Bogoya, J. M. (2018). Biodeterioration of plasma pretreated LDPE sheets by Pleurotus ostreatus. *PLoS One, 13*(9), e0203786. https://doi.org/10.1371/JOURNAL.PONE.0203786

Gong, J., Kong, T., Li, Y., Li, Q., Li, Z., & Zhang, J. (2018). Biodegradation of microplastic derived from poly(ethylene terephthalate) with bacterial whole-cell biocatalysts. *Polymers, 10*(12). https://doi.org/10.3390/POLYM10121326

Gu, J.-D. (2021). Biodegradability of plastics: The issues, recent advances, and future perspectives. *Environmental Science and Pollution Research, 28*, 1278–1282. https://doi.org/10.1007/s11356-020-11501-9/Published

Guan, N., & Liu, L. (2020). Microbial response to acid stress: Mechanisms and applications. *Applied Microbiology and Biotechnology, 104*(1), 51–65. https://doi.org/10.1007/S00253-019-10226-1

Hacıosmanoğlu, G. G., Doğruel, T., Genç, S., Oner, E. T., & Can, Z. S. (2019). Adsorptive removal of bisphenol A from aqueous solutions using phosphonated levan. *Journal of Hazardous Materials, 374*, 43–49. https://doi.org/10.1016/J.JHAZMAT.2019.04.015

Halle, A. T., Jeanneau, L., & Martignac, M. et al. (2017). Nanoplastic in the North Atlantic Subtropical Gyre. *Environmental Science and Technologies*, https://www.cordouan-tech.com/wp-content/uploads/2018/02/Nanoplastic-in-the-North-Atlantic-Subtropical-Gyre-Ter-Halle-2017.pdf.

Harrison, J. P., Sapp, M., Schratzberger, M., & Osborn, A. M. (2011). Interactions between microorganisms and marine microplastics: A call for research. *Marine Technology Society Journal, 45*, 12–20.

Horton, A. A., Walton, A., Spurgeon, D. J., Lahive, E., & Svendsen, C. (2017). Microplastics in freshwater and terrestrial environments: Evaluating the current understanding to identify the knowledge gaps and future research priorities. *Science of the Total Environment, 586*, 127–141. https://doi.org/10.1016/J.SCITOTENV.2017.01.190

Hu, K., Tian, W., Yang, Y., Nie, G., Zhou, P., Wang, Y., Duan, X., & Wang, S. (2021). Microplastics remediation in aqueous systems: Strategies and technologies. *Water Research, 198*, 117144. https://doi.org/10.1016/J.WATRES.2021.117144

Islam, S., Apitius, L., Jakob, F., & Schwaneberg, U. (2019). Targeting microplastic particles in the void of diluted suspensions. *Environment International, 123*, 428–435. https://doi.org/10.1016/J.ENVINT.2018.12.029

Iyare, P. U., Ouki, S. K., & Bond, T. (2020). Microplastics removal in wastewater treatment plants: A critical review. *Environmental Science: Water Research & Technology, 6*(10), 2664–2675. https://doi.org/10.1039/D0EW00397B

Jahnavi, N., Kanmani, K., Kumar, P. S., & Varjani, S. (2021). Conversion of waste plastics into low emissive hydrocarbon fuel using catalyst produced from biowaste. *Environmental Science and Pollution Research, 28*(45), 63638–63645. https://doi.org/10.1007/S11356-020-11398-4/METRICS

Jaiswal, S., Sharma, B., & Shukla, P. (2020). Integrated approaches in microbial degradation of plastics. *Environmental Technology and Innovation, 17*, 100567. https://doi.org/10.1016/J.ETI.2019.100567

Kelly, J. J., London, M. G., McCormick, A. R., Rojas, M., Scott, J. W., & Hoellein, T. J. (2021). Wastewater treatment alters microbial colonization of microplastics. *PLoS One, 16*(1), e0244443. https://doi.org/10.1371/JOURNAL.PONE.0244443

Kopecká, R., Kubínová, I., Sovová, K., Mravcová, L., Vítěz, T., & Vítězová, M. (2022). Microbial degradation of virgin polyethylene by bacteria isolated from a landfill site. *SN Applied Science, 4*(302), 1–12. https://doi.org/10.1007/s42452-022-05182-x

Kour, H., Shareif Khan, S., Kour, D., Rasool, S., Sharma, Y. P., Rai, P. K., Singh, S., Chaubey, K. K., Rai, A. K., & Yadav, N. (2023). Microbes mediated plastic degradation: A sustainable approach for environmental sustainability. *Journal of Applied Biology and Biotechnology, 11*(3), 9–19. https://doi.org/10.7324/JABB.2023.110515

Kovalcik, A., Obruca, S., Kalina, M., Machovsky, M., Enev, V., Jakesova, M., Sobkova, M., & Marova, I. (2020). Enzymatic hydrolysis of poly(3-Hydroxybutyrate-co-3-Hydroxyvalerate) scaffolds. *Materials, 13*(13), 1–21. https://doi.org/10.3390/MA13132992

Kumar, M., Chen, H., Sarsaiya, S., Qin, S., Liu, H., Awasthi, M. K., Kumar, S., Singh, L., Zhang, Z., Bolan, N. S., Pandey, A., Varjani, S., & Taherzadeh, M. J. (2021). Current research trends on micro- and nano-plastics as an emerging threat to global environment: A review. *Journal of Hazardous Materials, 409*, 124967. https://doi.org/10.1016/J.JHAZMAT.2020.124967

Lares, M., Ncibi, M. C., Sillanpää, M., & Sillanpää, M. (2018). Occurrence, identification and removal of microplastic particles and fibers in conventional activated sludge process and advanced MBR technology. *Water Research, 133*, 236–246. https://doi.org/10.1016/J.WATRES.2018.01.049

Lear, G., Kingsbury, J. M., Franchini, S., Gambarini, V., Maday, S. D. M., Wallbank, J. A., Weaver, L., & Pantos, O. (2021). Plastics and the microbiome: Impacts and solutions. *Environmental Microbiomes, 16*(1), 1–19. https://doi.org/10.1186/S40793-020-00371-W

Li, M., Feng, Y., Wang, K., Yong, W. F., Yu, L., & Chung, T. S. (2017). Novel hollow fiber air filters for the removal of ultrafine particles in PM2.5 with repetitive usage capability. *Environmental Science and Technology, 51*(17), 10041–10049. https://doi.org/10.1021/ACS.EST.7B01494/SUPPL_FILE/ES7B01494_SI_001.PDF

Lv, X., Dong, Q., Zuo, Z., Liu, Y., Huang, X., & Wu, W. M. (2019). Microplastics in a municipal wastewater treatment plant: Fate, dynamic distribution, removal efficiencies, and control strategies. *Journal of Cleaner Production, 225*, 579–586. https://doi.org/10.1016/J.JCLEPRO.2019.03.321

Ma, Y., Huang, A., Cao, S., Sun, F., Wang, L., Guo, H., & Ji, R. (2016). Effects of nanoplastics and microplastics on toxicity, bioaccumulation, and environmental fate of phenanthrene in fresh water. *Environmental Pollution, 219*, 166–173. https://doi.org/10.1016/J.ENVPOL.2016.10.061

Ma, B., Xue, W., Ding, Y., Hu, C., Liu, H., & Qu, J. (2019). Removal characteristics of microplastics by Fe-based coagulants during drinking water treatment. *Journal of Environmental Sciences, 78*, 267–275. https://doi.org/10.1016/J.JES.2018.10.006

Mahesh, S., Gowda, N. K., & Mahesh, S. (2023). Identification of microplastics from urban informal solid waste landfill soil; MP associations with COD and chloride. *Water Science and Technology, 87*(1), 115–129. https://doi.org/10.2166/WST.2022.412

Miloloža, M., Cvetnić, M., Kučić Grgić, D., Ocelić Bulatović, V., Ukić, Š., Rogošić, M., Dionysiou, D. D., Kušić, H., & Bolanča, T. (2022). Biotreatment strategies for the removal of microplastics from freshwater systems. A review. *Environmental Chemistry Letters, 20*(2), 1377–1402. https://doi.org/10.1007/S10311-021-01370-0

Min, K., Cuiffi, J. D., & Mathers, R. T. (2020). Ranking environmental degradation trends of plastic marine debris based on physical properties and molecular structure. *Nature Communications, 11*(1). https://doi.org/10.1038/S41467-020-14538-Z

Mohajerani, A., & Karabatak, B. (2020). Microplastics and pollutants in biosolids have contaminated agricultural soils: An analytical study and a proposal to cease the use of biosolids in farmlands and utilise them in sustainable bricks. *Waste Management, 107*, 252–265. https://doi.org/10.1016/J.WASMAN.2020.04.021

Mohanty, A., Rout, P. R., Dubey, B., Meena, S. S., Pal, P., & Goel, M. (2022). A critical review on biogas production from edible and non-edible oil cakes. *Biomass Conversion and Biorefinery, 12*(3), 949–966. https://doi.org/10.1007/S13399-021-01292-5/FIGURES/5

Naik, R. A., Rowles, L. S., Hossain, A. I., Yen, M., Aldossary, R. M., Apul, O. G., Conkle, J., & Saleh, N. B. (2020). Microplastic particle versus fiber generation during photo-transformation in simulated seawater. *Science of the Total Environment, 736*, 139690. https://doi.org/10.1016/J.SCITOTENV.2020.139690

Novotna, K., Cermakova, L., Pivokonska, L., Cajthaml, T., & Pivokonsky, M. (2019). Microplastics in drinking water treatment – Current knowledge and research needs. *The Science of the Total Environment, 667*, 730–740. https://doi.org/10.1016/J.SCITOTENV.2019.02.431

O'Connor, D., Pan, S., Shen, Z., Song, Y., Jin, Y., Wu, W. M., & Hou, D. (2019). Microplastics undergo accelerated vertical migration in sand soil due to small size and wet-dry cycles. *Environmental Pollution, 249*, 527–534. https://doi.org/10.1016/J.ENVPOL.2019.03.092

Othman, A. R., Hasan, H. A., Muhamad, M. H., Ismail, N. 'Izzati, & Abdullah, S. R. S. (2021). Microbial degradation of microplastics by enzymatic processes: A review. Environmental Chemistry Letters, 19(4), 3057–3073. https://doi.org/10.1007/S10311-021-01197-9.

Palm, G. J., Reisky, L., Böttcher, D., Müller, H., Michels, E. A. P., Walczak, M. C., Berndt, L., Weiss, M. S., Bornscheuer, U. T., & Weber, G. (2019). Structure of the plastic-degrading Ideonella sakaiensis MHETase bound to a substrate. *Nature Communications, 10*(1717), 1–10. https://doi.org/10.1038/s41467-019-09326-3

Park, S. Y., & Kim, C. G. (2019). Biodegradation of micro-polyethylene particles by bacterial colonization of a mixed microbial consortium isolated from a landfill site. *Chemosphere, 222*, 527–533. https://doi.org/10.1016/J.CHEMOSPHERE.2019.01.159

Parks, D. H., Rinke, C., Chuvochina, M., Chaumeil, P. A., Woodcroft, B. J., Evans, P. N., Hugenholtz, P., & Tyson, G. W. (2017). Recovery of nearly 8,000 metagenome-assembled genomes substantially expands the tree of life. *Nature Microbiology, 2*(11), 1533–1542. https://doi.org/10.1038/s41564-017-0012-7

Pathak, V. M., & Navneet. (2017). Review on the current status of polymer degradation: A microbial approach. *Bioresources and Bioprocessing, 4*(1), 1–31. https://doi.org/10.1186/S40643-017-0145-9

Piao, M., Zou, D., Ren, X., Gao, S., Qin, C., & Piao, Y. (2019). High efficiency biotransformation of bisphenol A in a fluidized bed reactor using stabilized laccase in porous silica. *Enzyme and Microbial Technology, 126*, 1–8. https://doi.org/10.1016/J.ENZMICTEC.2019.03.006

Pico, Y., Alfarhan, A., & Barcelo, D. (2019). Nano- and microplastic analysis: Focus on their occurrence in freshwater ecosystems and remediation technologies. *TrAC Trends in Analytical Chemistry, 113*, 409–425. https://doi.org/10.1016/J.TRAC.2018.08.022

Poerio, T., Piacentini, E., & Mazzei, R. (2019). Membrane processes for microplastic removal. *Molecules, 24*(22). https://doi.org/10.3390/MOLECULES24224148

Priya, A., Dutta, K., & Daverey, A. (2022). A comprehensive biotechnological and molecular insight into plastic degradation by microbial community. *Journal of Chemical Technology and Biotechnology, 97*(2), 381–390. https://doi.org/10.1002/JCTB.6675

Purohit, J., Chattopadhyay, A., & Teli, B. (2020). Metagenomic exploration of plastic degrading microbes for biotechnological application. *Current Genomics, 21*(4), 270. https://doi.org/10.2174/1389202921999200525155711

Rai, P. K., Lee, J., Brown, R. J. C., & Kim, K. H. (2021). Micro- and nano-plastic pollution: Behavior, microbial ecology, and remediation technologies. *Journal of Cleaner Production, 291*, 125240. https://doi.org/10.1016/J.JCLEPRO.2020.125240

Rogers, K. L., Carreres-Calabuig, J. A., Gorokhova, E., & Posth, N. R. (2020). Micro-by-micro interactions: How microorganisms influence the fate of marine microplastics. *Limnology and Oceanography Letters, 5*(1), 18–36. https://doi.org/10.1002/LOL2.10136

Rout, P. R., Goel, M., Mohanty, A., Pandey, D. S., Halder, N., Mukherjee, S., Bhatia, S. K., Sahoo, N. K., & Varjani, S. (2022). Recent advancements in microalgal mediated valorisation of wastewater from hydrothermal liquefaction of biomass. *BioEnergy Research, 16*(1), 45–60. https://doi.org/10.1007/S12155-022-10421-5

Ru, J., Huo, Y., & Yang, Y. (2020). Microbial degradation and valorization of plastic wastes. *Frontiers in Microbiology, 11*, 442. https://doi.org/10.3389/FMICB.2020.00442/BIBTEX

Rummel, C. D., Jahnke, A., Gorokhova, E., Kü, D., & Schmitt-Jansen, M. (2017). Impacts of biofilm formation on the fate and potential effects of microplastic in the aquatic environment. *Environmental Science and Technology Letters, 4*, 258–267. https://doi.org/10.1021/acs.estlett.7b00164

Sa'adu, I., & Farsang, A. (2023). Plastic contamination in agricultural soils: A review. *Environmental Sciences Europe, 35*(1), 1–11. https://doi.org/10.1186/S12302-023-00720-9/TABLES/1

Shakerian, F., Zhao, J., & Li, S. P. (2020). Recent development in the application of immobilized oxidative enzymes for bioremediation of hazardous micropollutants – A review. *Chemosphere, 239*, 124716. https://doi.org/10.1016/J.CHEMOSPHERE.2019.124716

Skariyachan, S., Megha, M., Kini, M. N., Mukund, K. M., Rizvi, A., & Vasist, K. (2015). Selection and screening of microbial consortia for efficient and ecofriendly degradation of plastic garbage collected from urban and rural areas of Bangalore, India. *Environmental Monitoring and Assessment, 187*(1), 1–14. https://doi.org/10.1007/S10661-014-4174-Y/METRICS

Skariyachan, S., Manjunath, M., Shankar, A., Bachappanavar, N., & Patil, A. A. (2018). Application of novel microbial consortia for environmental site remediation and hazardous waste management toward low- and high-density polyethylene and prioritizing the cost-effective, eco-friendly, and sustainable biotechnological intervention. *Handbook of Environmental Materials Management, 1–48*. https://doi.org/10.1007/978-3-319-58538-3_9-1

Sun, J., Dai, X., Wang, Q., van Loosdrecht, M. C. M., & Ni, B. J. (2019). Microplastics in wastewater treatment plants: Detection, occurrence and removal. *Water Research, 152*, 21–37. https://doi.org/10.1016/J.WATRES.2018.12.050

Sun, Z. K., Zhou, Y., Jiao, Y., Cheng, X. Q., Zhang, Y., Wang, P., Liang, H., Yang, X., Drioli, E., Figoli, A., Ma, J., & Shao, L. (2020). Multi-hydrophilic functional network enables porous membranes excellent anti-fouling performance for highly efficient water remediation. *Journal of Membrane Science, 608*, 118191. https://doi.org/10.1016/J.MEMSCI.2020.118191

Syranidou, E., Karkanorachaki, K., Amorotti, F., Avgeropoulos, A., Kolvenbach, B., Zhou, N. Y., Fava, F., Corvini, P. F. X., & Kalogerakis, N. (2019). Biodegradation of mixture of plastic films by tailored marine consortia. *Journal of Hazardous Materials, 375*, 33–42. https://doi.org/10.1016/J.JHAZMAT.2019.04.078

Talvitie, J., Mikola, A., Koistinen, A., & Setälä, O. (2017). Solutions to microplastic pollution – Removal of microplastics from wastewater effluent with advanced wastewater treatment technologies. *Water Research, 123*, 401–407. https://doi.org/10.1016/J.WATRES.2017.07.005

Taniguchi, I., Yoshida, S., Hiraga, K., Miyamoto, K., Kimura, Y., & Oda, K. (2019). Biodegradation of PET: Current status and application aspects. *ACS Catalysis, 9*(5), 4089–4105. https://doi.org/10.1021/ACSCATAL.8B05171/ASSET/IMAGES/MEDIUM/CS-2018-05171U_0012.GIF

Tiwari, N., Santhiya, D., & Sharma, J. G. (2020). Microbial remediation of micro-nano plastics: Current knowledge and future trends. *Environmental Pollution, 265*, 115044. https://doi.org/10.1016/J.ENVPOL.2020.115044

Velez, J. F. M., Shashoua, Y., Syberg, K., & Khan, F. R. (2018). Considerations on the use of equilibrium models for the characterisation of HOC-microplastic interactions in vector studies. *Chemosphere, 210*, 359–365. https://doi.org/10.1016/J.CHEMOSPHERE.2018.07.020

Wang, S., & Wu, Y. (2021). Hyperthermophilic composting Technology for Organic Solid Waste Treatment: Recent research advances and trends. *PRO, 9*(675), 1–11. https://doi.org/10.3390/pr9040675

Wei, R., & Zimmermann, W. (2017). Microbial enzymes for the recycling of recalcitrant petroleum-based plastics: How far are we? *Microbial Biotechnology, 10*(6), 1308–1322. https://doi.org/10.1111/1751-7915.12710

Wu, M., Liu, W., & Liang, Y. (2019). Probing size characteristics of disinfection by-products precursors during the bioavailability study of soluble microbial products using ultrafiltration fractionation. *Ecotoxicology and Environmental Safety, 175*, 1–7. https://doi.org/10.1016/J.ECOENV.2019.02.077

Yang, Z., Zhou, Y., Feng, Z., Rui, X., Zhang, T., & Zhang, Z. (2019). Polymers A review on reverse osmosis and nanofiltration membranes for water purification. *Polymers, 11*(1252), 1–22. https://doi.org/10.3390/polym11081252

Yang, Y., Liu, W., Zhang, Z., Grossart, H. P., & Gadd, G. M. (2020). Microplastics provide new microbial niches in aquatic environments. *Applied Microbiology and Biotechnology, 104*(15), 6501–6511. https://doi.org/10.1007/S00253-020-10704-X/FIGURES/2

Yoshida, S., Hiraga, K., Takehana, T., Taniguchi, I., Yamaji, H., Maeda, Y., Toyohara, K., Miyamoto, K., Kimura, Y., & Oda, K. (2016). A bacterium that degrades and assimilates poly(ethylene terephthalate). *Science, 351*(6278), 1196–1199. https://doi.org/10.1126/SCIENCE.AAD6359/SUPPL_FILE/AAD6359-YOSHIDA-SM.PDF

Yuan, J., Ma, J., Sun, Y., Zhou, T., Zhao, Y., & Yu, F. (2020). Microbial degradation and other environmental aspects of microplastics/plastics. *The Science of the Total Environment, 715*, 136968. https://doi.org/10.1016/J.SCITOTENV.2020.136968

Zhang, X., Chen, J., & Li, J. (2020). The removal of microplastics in the wastewater treatment process and their potential impact on anaerobic digestion due to pollutants association. *Chemosphere, 251*, 126360. https://doi.org/10.1016/J.CHEMOSPHERE.2020.126360

Zhang, Y., Jiang, H., Bian, K., Wang, H., & Wang, C. (2021). A critical review of control and removal strategies for microplastics from aquatic environments. *Journal of Environmental Chemical Engineering, 9*(4), 105463. https://doi.org/10.1016/J.JECE.2021.105463

Zhu, L., Zhao, S., Bittar, T. B., Stubbins, A., & Li, D. (2020). Photochemical dissolution of buoyant microplastics to dissolved organic carbon: Rates and microbial impacts. *Journal of Hazardous Materials, 383*, 121065. https://doi.org/10.1016/J.JHAZMAT.2019.121065

Enzyme-Assisted Biodegradation of Micro-Nanoplastics: Advances and Future Outlook on the Management of Plastic Pollution

Arun Dhanasekaran and Kannabiran Krishnan

1 Introduction

Plastics, versatile polymers, do not undergo degradation naturally due to their unique physico-chemical properties. They persist for a very long time and cause severe damage to the environment, ecosystems and living systems. Though they undergo degradation by abiotic variables such as UV irradiation, temperature and climatic conditions, they degrade into micro- and nanoplastics, remaining a major pollutant affecting the globe. They also possess hydrophobicity as their parent compounds. Major sources of MPs are single-use plastics, which further degrade into NPs, with wastewater treatment plants being the major source of NPs. These micro-nanoplastics (MNPs) end up with bioaccumulation and biomagnification in aquatic and terrestrial organisms. They interfere with the biogeochemical cycle and affect soil microbiome and plant growth in terrestrial ecosystem. It is reported that polystyrene nanoplastics accumulate in wheat leaves and roots and onion root cells as well. In aquatic system, fish, rotifers, bivalves and algae are affected by NPs. In human beings, NPs penetrate through skin and blood and is distributed throughout the body, thereby creating stress and associated health hazards. They also enter into food chain of higher trophic levels. MNPs are not only pollutants but also serve as carriers for many toxic chemicals and pathogens. They are polychlorinated biphenyls, bisphenol A, heavy metals, including copper (Cu) and lead (Pb), and human pathogens (Wang et al., 2021). These MNPs are sourced from single-use plastics, wastewater discharged from industries, cosmetics and washing of clothes (Ahmaditabatabaei et al., 2021). Several physico-chemical techniques are applied for plastic degradation; however, they have many disadvantages such as the release

A. Dhanasekaran · K. Krishnan (✉)
Department of Biomedical Sciences, School of Biosciences and Technology, Vellore Institute of Technology, Vellore, Tamil Nadu, India
e-mail: kkb@vit.ac.in

S. A. Bhat et al. (eds.), *Management of Micro and Nano-plastics in Soil and Biosolids*,
https://doi.org/10.1007/978-3-031-51967-3_16

of toxic gases and requirement of high energy and costly chemicals. Therefore, they are inappropriate for circular economy and are not environmentally friendly. Biodegradation of plastics by microbes and their associated metabolites were seemed to be an effective, eco-friendly and cost-effective approach. Despite the existence of multiple reports on the microbial breakdown of plastics, pathogenicity and horizontal gene transfer being the bottlenecks make this approach argumentative. Microbial enzymes that biodegrade plastics have no harmful effects and play a crucial role in the breakdown of plastics. Yoshida et al. (2016) discovered an enzyme that can degrade polyethylene terephthalate (PET) effectively, called as PETase, brought the research on plastic biodegradation to the limelight. More research on plastic-degrading enzymes has since been performed. With the available interdisciplinary knowledge and biotechnological tools, it is possible to design and engineer enzymes for plastic biodegradation. Not many reports are available on discovery, mechanism, design and production of plastic-biodegrading enzymes; hence, an overview of enzymes involved in plastic biodegradation, current advancements, challenges and the future outlook on plastic biodegradation are discussed.

2 Plastic-Biodegrading Enzymes (PBEs): Discovery Methods

A surge to discover plastic-biodegrading enzymes (PBEs) from plastic biodegrading microorganisms (PDMs) has been increasing in the recent past. The evolution of PDMs leads to the progression of PBEs for the majority of common plastics in a very short period of time (Viljakainen & Hug, 2021). While a few enzymes have been identified that work on high-molecular weight polymers like polyethylene (PE), polypropylene (PP) and polystyrene (PS), many PET-degraders have been identified in the last decade (Danso et al., 2019). PBEs have traditionally been discovered using culture-based methodologies in which microbes are cultivated from plastic-polluted environmental samples and tested for plastic-biodegrading activity by visual inspection, clear zone formation, weight reduction and measuring carbon dioxide evolution (Viljakainen & Hug, 2021). Traditionally, biochemical and molecular biology methods have been used to identify PBEs (Gangoiti et al., 2012). Plastic-contaminated sites are a niche for PDMs from where they can be isolated and identified (Yoshida et al., 2016). Once proteins exhibiting plastic biodegradation are isolated from PDMs, mass spectrometry (MS) can be used for their identification by extracting PBE sequence information (Gangoiti et al., 2012). Alternatively, after finding out a PDM, the presence of PBE genes can be predicted by using whole genome sequencing and annotation (Yoshida et al., 2016). Predicted PBEs can be tested for activity by cloning them into a heterologous host. Yoshida et al. (2016) used a culture enrichment method to isolate a PET-degrading bacterium from more than 200 soil samples of a plastic water-bottle recycling unit. Initially, screening for plastic biodegradation was performed with PET film as a predominant carbon source that was morphologically altered by the bacterium. On 16S rDNA sequencing, it was found to be *Ideonella sakaiensis*, and on whole genome sequencing, a reputed lipase was identified similar to the hydrolase from *Thermobifida fusca* that was reported to have PET-degrading ability. It was also found that *I. sakaiensis* uses a two-enzyme system (PETase and MHETase) to biodegrade PET. Culture-based techniques are useful to identify and confirm the PDMs that in turn also help to identify significant PBEs. However, due to the lack of cultured candidates of bacterial phyla, the culture-based techniques are time-consuming and limiting (Hug et al., 2016). This has prompted PBE bioprospecting attempts to concentrate on unexplored microbial diversity (Danso et al., 2019). Scientists are exploring uncultured microbial ancestries from a variety of habitats by computational biology approaches for annotating large sequence data. This has allowed for high throughput PBE metagenomic screening (Zhu et al., 2022). The metagenomic approach is based mainly on homology and functional screening. They do not follow conventional culture-dependent methods. BLAST search tools are used for homology-based screening that identifies PBEs based on their similarities of query sequences to known PBEs (Danso et al., 2019).

Although in silico homology search provides a quick and economical approach to screen enormous sequence data and significantly increases projected PBE families, this method may bypass novel enzymes that do not have enough sequence matching with recognised enzymes. Because sequence matching does not ensure similar activity, and mutation can alter or delay the activity, biochemical evaluation is essential to establish the activity of predicted enzymes. Furthermore, PBE expression needs to be validated by protein or RNA analysis. Due to the efficient and rapid nature of homology sequence screening, it frequently surpasses the time-consuming process of functional confirmation and protein characterisation. As a consequence, there exists a disparity between the number of predicted PBEs and those that have been experimentally validated (Viljakainen & Hug, 2021). The application of bioinformatic analysis in the directed exploration of the vast number of projected PBEs enables the identification of potential novel PBE clades. This approach offers the opportunity to discover enzymes with unique properties or originating from microbial lineages that have not been previously associated with plastic biodegradation (Viljakainen & Hug, 2021). Future research will rely on the discovery of effective computational tools to detect PBEs, as well as the development of high-throughput technology which is crucial in validating computationally predicted outcomes (Danso et al., 2019). Cloning of DNA obtained from an environmental sample into a fosmid or cosmid clone library, followed by transformation into a vector for expression and functional screening, is the strategy currently followed (Lam et al., 2015). Different approaches that are useful for finding out plastic biodegrading microbes and enzymes are shown in Fig. 1.

Functional metagenomic screening, in contrast to homology screening, validates plastic-biodegrading characteristics of enzyme families that were previously unidentified, which do not depend on pre-existing sequence information (Zhu et al., 2022). Also, it is a high-throughput technique for studying gene function. Although expensive and technically difficult, library building can frequently result in significant DNA loss or difficulties with heterologous expression (Lam et al., 2015). However, functional metagenomic screening has proven to be an effective method

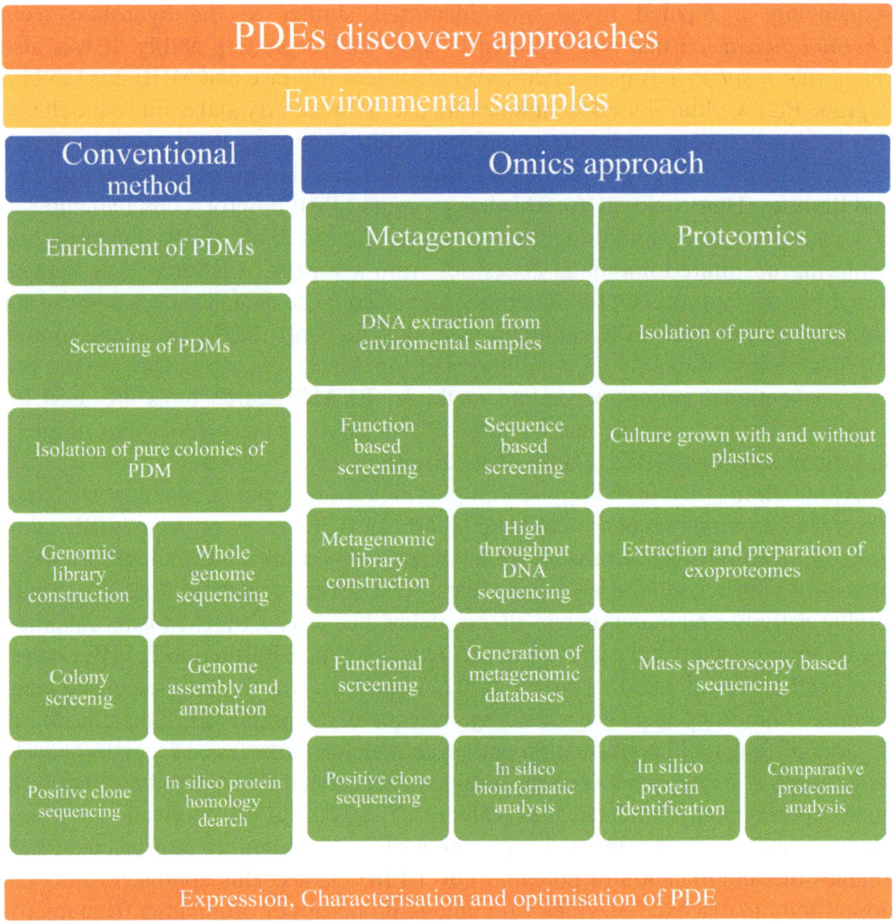

Fig. 1 Different approaches to discover plastic-degrading microbes/enzymes

for discovering new PBEs. Cutinase enzyme derived from the compost of leaf branch has the ability to degrade polyethylene terephthalate (PET) (Sulaiman et al., 2012). Tchigvintsev et al. (2015) used functional metagenomics to identify polyesterases. The discovery of novel enzymes for biotechnological uses by proteomics has enormous potential. Experiments involving the comparison of microorganisms in the medium with and without plastics can elucidate the metabolic pathways and enzymes associated with interactions between plastics and microbes, particularly those involved in the decomposition of plastics (Hug et al., 2016). Exoproteome screening is especially promising for identifying PBEs being released from cells and can react with plastic polymers (Zhu et al., 2022). Currently, comparative proteomics is used, but investigations involving metaproteomics or metatranscriptomics in plastic biodegradation experiments are uncommon at the moment. The potential of multi-omics techniques and advanced computer analysis in microbe-plastic research has yet to be realised. Marques et al. (2021) discussed the web server required to discover and engineer enzymes. For enzyme discovery, tools such as HEC-net, Bio2Rxn, GSP4PDB, LIBRA-WA, EnzymeMiner could be used. Enzymes with an affinity to bind plastic substrate can be screened using molecular docking tools and can be visualised using AutoDock Vina, AutoDock, Glide and Gold (Ebert & Pelletier, 2017).

3 PETase as a Model Enzyme

The enzyme PETase (IsPETase), which is derived from the bacterium *I. sakaiensis* 201–F6, has been recognised as the first characterised PET-degrading enzyme. It catalyses the conversion of PET into primary metabolites known as mono-(2-hydroxyethyl) terephthalate (MHET), as well as secondary metabolites including ethylene glycol (EG), terephthalic acid (TPA) and bis(2-hydroxyethyl terephthalate) (BHET) (Magalhães et al., 2021). The enzyme possesses a canonical α/β--hydrolase shape, consisting of a core β-sheet composed of nine mixed β-strands, which is encircled by seven α helices and has 290 amino acid residues. According to Joo et al. (2018), the active site has a triangle consisting of Serine (Ser160), Histidine (His237) and Aspartate (Asp206). The study revealed that PETase exhibits a conventional mechanism characteristic of the serine hydrolase family, which is akin to that of cutinase. However, the sole distinction lies in the function of Trp185, the unstable tryptophan located in close proximity to the catalytic centre. The residue's mobility is believed to have a crucial role in both substrate binding and pocket release, as suggested by Knott et al. (2020) and Papadopoulou et al. (2019).

The presence of a shallow crevice on the apo-form enzyme facilitates the binding of substrates on the surface. The protein is bound by four moieties, with one moiety binding to the first subsite, where catalysis occurs, and the remaining three moieties binding to the second subsite. The initial stage involves the establishment of a charge relay circuit between the catalytic residues Asp206 and His237. The present system facilitates the deprotonation of the hydroxyl group located on Ser160 in His237, resulting in the enhancement of its nucleophilicity. Consequently, the deprotonated hydroxyl group proceeds to engage in an attack on the carbonyl group of PET ester. The catalytic process is initiated by this assault, resulting in the production of the first tetrahedral intermediate. This intermediate is stabilised by the presence of the oxyanion hole, which consists of the amino acid residues Met161 and Tyr87. The intermediate possessing a negative charge exhibits a high degree of instability, leading to its subsequent decomposition and the consequent liberation of the initial product, as well as the formation of an acyl-enzyme intermediate. The second intermediate is formed through the nucleophilic attack of a water molecule, which has been deprotonated by the Asp206 and His237 charge relay system, on the carbonyl carbon. This intermediate is also characterised by its instability and subsequently undergoes breakdown, leading to the release of the second product. The aforementioned attacks induce the creation of nicks and subsequently lead to the degradation of PET chains, resulting in the emergence of unique terminals from each subsite. For instance, a TPA terminal arises from subsite I, whereas an HE terminal emerges from subsite II. During the final stage of digestion, the two cleaved chains undergo different processes to be converted into MHET (as well as BHET, TPA and residual quantities of EG). These processes are determined by the terminal structure of each molecule. In the case of both the HE and TPA terminals, after the four moieties have been bonded to subsite I and II, the previously mentioned process of ester bond cleavage occurs, leading to the formation of MHET and TPA, respectively. Additionally, the ester bond cleavage also leads to the synthesis of HE-PET. The various interactions between binding agents and terminals give rise to a sequence of polyethylene terephthalate (PET) monomers and dimers, which are subsequently metabolised into the ultimate products by a multimodal process. This process leads to the accumulation of four specific molecules, namely MHET, TPA, EG and BHET. Furthermore, it is worth noting that MHET and EG can undergo additional degradation. However, several arguments regarding catalytic functions and mechanisms are ongoing (Wei et al., 2019; Seo et al., 2019).

4 Enzymes Involved in Biodegradation of Plastics

Plastics are hydrophobic, non-bioavailable and recalcitrant materials. The surface properties of plastic restrict enzymatic adsorption and activity. Therefore, pre-treatment may be required to reduce the size of the plastics, resulting in an increase in surface area, which facilitates accessibility and hence biodegradation. While crystalline plastics are resistant, semi-crystalline plastics are prone to microbial and enzymatic biodegradation. Though several plastic degradation techniques are available, biodegradation using enzymes is a non-toxic and eco-friendly approach. Biodegradation of non-hydrolysable plastics, including PP, PS, PE and polyvinyl chloride (PVC), is tough because of the C-C backbone. Any pre-treatment method involving UV irradiation or oxidising agent is required for creating carbonyl groups that can be accessed by microbes/enzymes (Wei & Zimmermann, 2017). Rarely, PBEs are reported for such plastics, namely alkane hydroxylase, laccase and hydroquinone peroxidase. Interestingly, laccases, one of the oxidoreductive enzymes, are also involved in the biodegradation of the natural polymer called lignin containing a carbon backbone and hydrophobicity similar to plastics (Inderthal et al., 2021).

4.1 Biodegradation of PE

Polyethylene (PE) is proposed to degrade similarly to alkane because of the linear arrangement of monomers without any side functional groups. Therefore, alkane hydroxylase (AH), which degrades hydrocarbons by terminal or subterminal oxidation, can be used for PE biodegradation. Evidently, recombinant AH from

Pseudomonas sp. E4 and *Pseudomonas aeruginosa* E7 expressed in *E. coli* biodegraded 20% and 30% of PE samples respectively (Wei & Zimmermann, 2017; Inderthal et al., 2021). Apart from these enzymes, lignolytic enzymes were shown to have significant effects on plastic biodegradation. Due to the homogenous C-C linkage among lignin and plastics such as PE and PS, the oxidoreductases that degrade lignin may also be involved in plastic degradation. Manganese peroxidase (MnP) from the lignin-biodegrading fungus, *Trarnetes versicolor* and *Phanerochaete chrysosporium* are reported to biodegrade PE (Iiyoshi et al., 1998). Laccases derived from *Pleurotus ostreatus*, *Aspergillus flavus* and *Rhodococcus ruber* are capable of biodegrading PE (Santo et al., 2013; Gomez-Mendez et al., 2018; Zhang et al., 2020). Lignin peroxidase (LiP) from *Streptomyces* species degraded heat-treated PE within 15 days of incubation (Pometto et al., 1992). *Phanerochaete chrysosporium* MTCC-787's extracellular LiP and MnP were reported to break down 70% of a preoxidised high molecular weight PE sample (Wei & Zimmermann, 2017).

4.2 Biodegradation of PS

PS biodegradation involves main chain cleavage (β-carbon) and side chain cleavage (phenyl group). While the former is more likely to be attacked by enzymes because of C-H and C-C linkage, the latter is resistant because of the C=C bond (Kang et al., 2023). The proposed enzymes and their degrading mechanism may be helpful for the primary understanding of PS biodegradation. Some of the oxidoreductases, such as cytochrome P450, monooxygenase, alkane hydroxylase and aromatic ring hydroxylase, seem to be involved in PS biodegradation. Initial cleavage of β-carbon on main chain may be due to alkane hydroxylase, monooxygenase and cytochrome P450. As discussed earlier in PE degradation, alkane hydroxylase may involve terminal or subterminal oxidation of PS. The main chain of PS is actually a substituted alkane in which an atom of oxygen is incorporated by monooxygenase for degradation. Monoxygenases from *Geobacillus thermodenitrificans* NG80-2 and *Pusillimonas* sp. strain T7-7 were able to oxidise lengthy hydrocarbons. Cytochrome P450 produced by *Bacillus subtilis* may be involved in the hydroxylation reaction of ethylbenzene and styrene epoxidation in the presence of hydrogen peroxide. It may also be involved in the initial and subterminal oxidation of PS, resulting in both the formation and degradation of single-ring aromatic compounds. In addition, 4-hydroxybenzoate and 3-hydroxybenzoate produced during main aromatic chain cleavage are catalysed by respective monooxygenase enzymes. The phenyl group of PS may be degraded by aromatic ring hydroxylases by adding two oxygen atoms to the styrene. For instance, ring-hydroxylating dioxygenase from *Rhodococcus* sp. P14 and *Sphingobium* sp. FB3 can degrade benzo[a]anthracene. Benzoate compound 2-hydro-1,2-dihydroxybenzoate can be degraded by *Microbacterium esteraromaticum* SBS1-7 containing benzoate 1,2-dioxygenase (Hou & Majumder, 2021). Recently reported plastic-biodegrading enzymes are listed in Table 1.

Table 1 Microbial enzymes involved in biodegrading plastics and their sources

Plastics	PDEs	Producers	References
PET	Esterase	*Thermobifida halotolerans*	Viljakainen and Hug (2021)
	Carboxylesterase	*Thermobifida fusc* KW3	Gao et al. (2021)
	Cutinase	*Aspergillus oryzae, Aspergillus nidulans, Fusarium solani, Humicola insolens, Fusarium solani pisi, Fusarium oxysporum, Penicillium citrinum,*	Ahmaditabatabaei et al. (2021)
		Thermobifida fusca, T. cellulolysitica	Viljakainen and Hug (2021)
		Pseudomonas mendocina, Saccharomonospora viridis AHK190	Gao et al. (2021)
	PETase	*Ideonella sakaiensis*	Urbanek et al. (2021)
	MHETase	*Ideonella sakaiensis*	Urbanek et al. (2021)
	Hydrolase	*Penicillium funiculosum, Fusarium oxysporum* LCH I, *Fusarium solani*	Ahmaditabatabaei et al. (2021)
	Polyesterase	*Penicillium citrinum*	Ahmaditabatabaei et al. (2021)
	Lipase	*Candida polymerase, Candida rugosa, Penicillium simplicissimum, Thermomyces lanuginosus*	Ahmaditabatabaei et al. (2021)
		Pseudomonas aestusnigri, Oleispira antarctica RB-8, *Polyangium brachysporum, Vibrio gazogenes, Thermomonospora curvata* DSM 43183	Gao et al. (2021)
	Suberinase	*Streptomyces scabies*	Amoboyne et al. (2021)
PUR	Polyurethanase	*Pseudomonas chlororaphis, Serratia marcescens*	Mohanan et al. (2020)
	Aryl acylamidase	*Rhodococcus equi*	Iram et al. (2019)
	Polyamidase	*Nocardia farcinia*	Zhu et al. (2019)
PE	Laccase	*Rhodococcus ruber*	Zhang et al. (2022)
	Alkane monooxygenase	*Pseudomonas aeruginosa*	Zhang et al. (2022)
	Rubredoxin reductase	*Pseudomonas putida*	Zhang et al. (2022)
	Manganese peroxidase	*Phanerochaete chrysosporium*	Zhang et al. (2022)
	Lignin peroxidase	*Phanerochaete chrysosporium*	Zhang et al. (2022)
	Alkane hydroxylase	*Pseudomonas* sp. E4	Amoboyne et al. (2021)

(continued)

Table 1 (continued)

Plastics	PDEs	Producers	References
PS	Oxidative (exo)enzymes	*Gloeophyllum trabeum* DSM 1398	Hou and Majumder (2021)
	Extracellular depolymerase enzymes	*Citrobacter sedlakii*, *Alcaligenes* sp., *Enterobacter* sp., *Brevundimonas diminuta*	Hou and Majumder (2021)
	Hydroquinone peroxidase	*Azotobacter beijerinckii* HM121	Moog et al. (2019)
	Styrene monooxygenase	*Pseudomonas putida*, *Bacillus cereus*, *Pseudomonas* sp. VLB120, *Rhodocccus* sp. ST-5	Mohanan et al. (2020)
PVC	Lignin peroxidase	Phanerochaete chrysosporium	Zhang et al. (2022)
	PVC esterase	*Methylocystis* sp. MJC1	Mohanan et al. (2020)

The comprehension of PBE kinetics facilitates the process of plastic biodegradation. A limited number of research has been conducted to examine the degradation kinetics of PBEs responsible for the breakdown of PP, PE, PS and PVC. Further investigation is necessary to ascertain the order and rate of the reaction and the rate of enzyme deactivation in the biodegradation of plastics (Hou & Majumder, 2021).

5 Synthetic Biology in PBEs Production

5.1 Role of Genetic Engineering

Recently, environmental microbes have been widely modified using genetic engineering techniques. Amino acid sequences are altered by genetic engineering to develop recombinant DNA-based mutations (Kumar et al., 2023). Techniques that allow high-level manufacturing of plastic-degrading enzymes are necessary for industrial applications. Most of the studies use the intracellular expression system of *E. coli* for experimentation (Chen et al., 2023). However, due to the high expense of the purifying techniques, it is difficult in commercial applications. Since PETase is a well-studied PBE, production of PETase at a large scale requires a potential chassis (Chen et al., 2023). Some of the mostly used hosts for genetic modifications for PET biocatalyst production are discussed as follows.

In recent times, *E. coli* has emerged as a prominent model organism utilised extensively in various industries for the purpose of over-expressing industrial enzymes and facilitating large-scale biocatalyst manufacturing (Olajuyin et al., 2016). With a high cell density, *E. coli* can cut the doubling time to 24 h in recombinant protein production (Samak et al., 2018). *E. coli* requires inexpensive and readily accessible compounds for expression. Also, the DNA transformation protocol in *E. coli* is fast and handy with a high success rate (Rosano & Ceccarelli, 2014). Recently, *E. coli* has been used more frequently for the creation of PET biocatalysts to solve the crystal structures of enzymes, including leaf compost cutinase (LCC), PETase and MHETase. It also provides great insight into the PET substrate affinity and degradation mechanism (Almeida et al., 2019; Tournier et al., 2020). Two genes encoding esterases derived from *Clostridium botulinum* ATCC3502 were expressed in *E. coli*. The hydrolysis of PET was carried out at a temperature of 50 °C. Notably, catalysis was observed to be increased by the introduction of a mutation in the zinc-binding domain (Perz et al., 2016). *E. coli* was also used for the whole-cell catalytic extracellular expression of PETase. In heterologous *E. coli* expression, Sec-dependent and SRP-dependent signal peptides were fused with IsPETase using Pet22B-SPMalE. PET films were successfully biodegraded after 72 h of incubation at 30 °C using extracellular secreted IsPETase directly in a culture medium (Seo et al., 2019; Samak et al., 2020).

Expression of PET biocatalyst using *Pichia pastoris* yielded increased cell density, profound growth and also facilitates the extraction of PET biocatalyst directly from the culture medium using extracellular expression (Samak et al., 2020). *P. pastoris* also performs post-translation modification, including glycosylation, acylation and methylation. Recombinant cutinase LCC expression in *P. pastoris* results in producing glycosylated LCC with 10 °C higher thermostability than the native enzyme is reported (Shirke et al., 2018). For PET biocatalyst production, microalgae are more promising cell factory. In recent times, the availability of algal genomic data and transformation tools and procedures has facilitated enhanced biocatalyst production (Samak et al., 2020). Till now, for genetic manipulation, over 40 different microalgae strains were used (Gangl et al., 2015). *Phaeodactylum tricornutum* is used for PETase extracellular gene expression. Extracellular expression of PETase in microalgae *P. tricornutum* demonstrated 80-fold faster development of polyethylene terephthalate glycol (PETG) with low crystallinity (Samak et al., 2020). High saline conditions and mesophilic temperature of 21 °C with TPA and MHET, increase the activity of expressed PETase against shredded pieces of PET (Moog et al., 2019). Recently the heterologous production of PETase was reported from the green algae, *Chlamydomonas reinhardtii* CC-124. Production of biocatalyst using microalgae has many advantages over other organisms. In a study conducted by Mayfield et al. (2007), it was observed that cell disruption was achieved rapidly and effectively, with a timeframe of 24 h, and even as little as 3 h in the case of *Chlorella sorokiniana*. Gong et al. (2019) showed that the utilisation of photobioreactors employing phototrophic growth can effectively impede the dissemination of transgenes into the surrounding environment. The recombinant biocatalysts can be expressed via chloroplast genomes (Lauersen et al., 2013a). Extracellular production of glycosylated biocatalysts is achieved by post-translational modification (Lauersen et al., 2013b). The future holds promising prospects for the field of microalgae engineering, as the advent of data tools will facilitate the cost-effective manufacturing of improved PET biocatalysts.

Clostridium thermocellum got attention in recent years, because of its thermophilic and anaerobic characteristics for heterologous expression chassis. It also has the optimum growth conditions and can reach up to 60 °C. A great effort has been put into *C. thermocellum* for genetic manipulation by developing an efficient transformation tool to prevent gene disruption. The transformation efficiency of *C. thermocellum* has been enhanced by using the electrophoresis method (Olson & Lynd, 2012). This method of transformation allows *C. thermocellum* to utilise whole-cell biocatalyst for lignocellulose saccharification and biofuel production. Yan et al. (2021) studied cutinase expression (LCC) enzyme at 60 °C for the degradation of PET films using *C. thermocellum* as a microbial chassis. This study also showed the conversion of 60% of PET films into monomers after 2 weeks of incubation by extracellular expression of the LCC enzyme.

5.2 Whole-Cell Catalyst

Immobilised enzymes used in industry as free enzymes are not stable and reusable. However, high cost for enzyme purification and immobilisation, utilisation of chemicals that may denature and reduce the activity of enzymes, and high application cost lead to recession (Sheldon & Van Pelt, 2013). Expressing the target enzymes on the cell surface by tagging them to the anchor protein thereby the obstructions of enzyme immobilisation can be reduced (Dwevedi & Kayastha, 2012). This technique can be implemented for whole-cell biocatalysis that helps in the production of PBEs. For reactions involving substantial substrates, such as PET, which are unable to translocate to the interior of the cell, surface display is the optimal setting. Furthermore, surface display reduces the need for purification of enzyme for catalysis, facilitates catalyst reuse, eliminates product contamination and can improve the stability of catalyst (Loll-Krippleber et al., 2022).

Cell surface display can be either prokaryotic or eukaryotic. However, due to the necessity for post-translational modifications in prokaryotic cell system to express complex proteins, the eukaryotic system is preferred mostly (Park, 2020). Researchers employed anchored proteins Sed1p, YlPir1p and Lpp'OmpA to display lipase produced from *Rhizopus oryzae*, *Yarrowia lipolytica* and *Bacillus licheniformis* respectively on cell surface of *Pichia pastoris*, *Saccharomyces cerevisiae* and *E. coli* enhancing the ideal temperature, pH, and other features of lipase (Han et al., 2017; Li et al., 2015; Yuzbasheva et al., 2015; Chen et al., 2021). A whole-cell biocatalyst expressing recombinant LCC has been shown to be important for industry in PET biodegradation (Wei et al., 2020; Chen et al., 2021). Chen et al. (2021) produced many engineered yeast cells displaying IsPETase for degradation of high crystalline PET (hcPET). They also developed a whole-cell biocatalyst for the first time by employing surface display of PETase on *P. pastoris* (Chen et al., 2021). It showed 36-fold enzymatic activity compared to normal PETase. In addition, the biocatalyst showed significant activity even after several times of reuse. MHET being an enzyme (PETase) inhibitor produced in large amounts when only PETase is expressed, Loll-Krippleber et al. (2022) considered MHETase as an important part of PET degradation as it can convert MHET into EG and TPA. They also engineered MHETase from *Ideonella sakaiensis* to be displayed on *S. cerevisiae*. Significant results showing activity up to 12 days at 45 °C were obtained.

Although nature remains the richest source of novel enzymes, most do not exhibit adequate availability, stability and productivity to be employed in commercial applications. Furthermore, a few chemical or environmental stimuli or signals may be required to initiate the formation of molecules genetically that are not created in laboratories. The utilisation of host organisms for the purpose of biosynthesis continues to face challenges, such as post-transcriptional alterations and metabolic changes. Additionally, the extraction of improved proteins, especially enzymes, may prove challenging if they are produced in the periplasm or develop inclusion bodies (de Carvalho, 2017). However, microbial hosts such as *Corynebacterium glutamicum, E. coli, Pseudomonas* sp. and *Saccharomyces cerevisiae* have been effectively employed in many applications (Pearsall et al., 2015).

6 Enzyme Optimisation

6.1 Role of Protein Engineering

Modification in the amino acids may result in enhanced thermal stability and activity of the PETase protein for a long time. Most of the enzyme improvements are focused on site-directed mutagenesis. Joo et al. (2018) used site-directed mutagenesis to create 14 mutants, among which the variant IsPETase R280A (replacement of arginine with alanine at position 280) showed increased activity by 22.4% and 32.4% in 18 h and 36 h, respectively, by producing TPA and MHET from PET films. Han et al. (2017) constructed 12 mutants, of which Y58A (replacement of tyrosine with alanine at position 58) exhibited 80.73% MHET production. Liu et al. (2018) developed two mutants, such as W130H and S209F, which showed increased activity towards PET bottles in comparison with unmodified PETase. The introduction of two mutations may establish additional hydrogen bonds that stabilise the protein. Accordingly, an IsPETase variant S121D/D186H and S121E/D186H (aspartic acid or glutamic acid in the place of serine and histidine in the place of aspartate) was developed. Consequently, triple mutant was created by introducing R280A mutant, previously created by Joo et al. (2018), to IsPETase S121E/D186H, and it was developed as IsPETase S121E/D186H/R280A. This mutant showed 13.9- and 2.3-fold better activity than the native protein and mutant R280A, respectively (Son et al., 2019). Ma et al. (2018) made modifications to six distinct residues located in the vicinity of the substrate-binding groove of PETase. They obtained three mutants, namely R61A, L88F and I179F, by a rapid cell-free screening method. The three mutants exhibited 1.4-, 2.1- and 2.5-fold increased activity than the wild-type PETase, respectively. On SEM (Scanning Electron Microscope) analysis, the morphology of PET film was found roughed and eroded along with many holes. Austin et al. (2018) improved PET degradation by narrowing the binding cleft of active

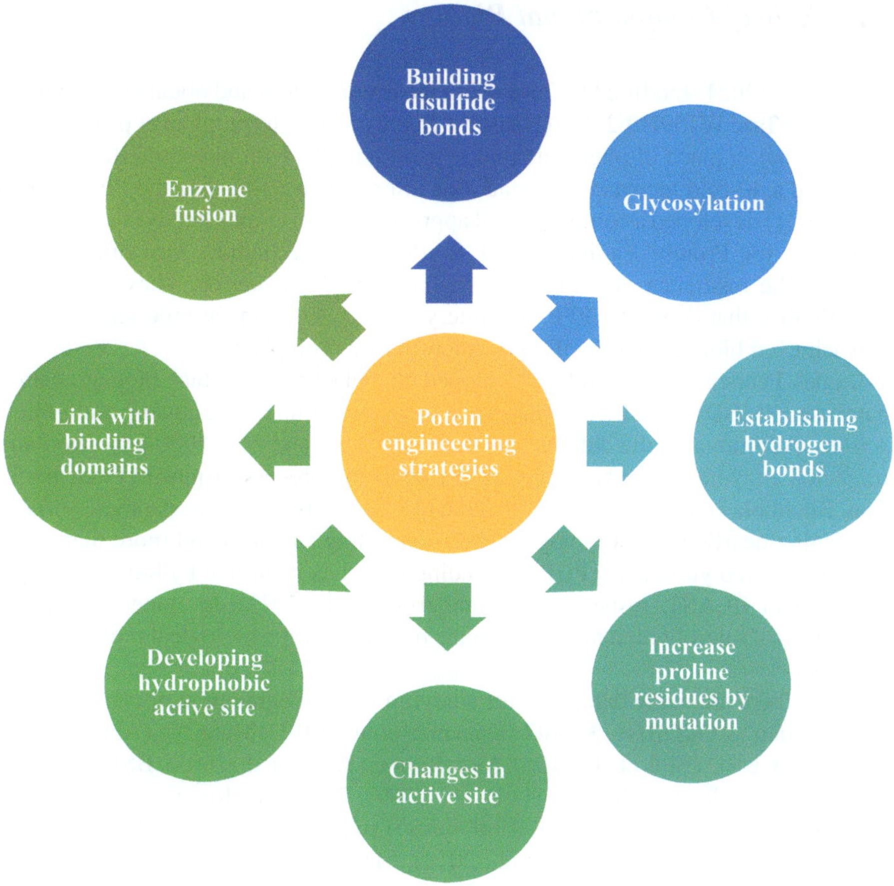

Fig. 2 Protein engineering strategies for PDEs

sites of PETase. They developed a double mutant S238F/W159H (change of serine to phenylalanine and tryptophan to histidine) through site-directed mutagenesis. The introduction of the S238F mutation resulted in the formation of more p-stacking and hydrophobic contacts, leading to the close proximity of terephthalate moieties. On the other hand, the W159H mutation greatly facilitated the binding of PET molecules to the active channel. Dai et al. (2021) changed the structure of protein by the addition of a hydrophobic substrate binding domain such as cellulose binding domain (CBM), poly (3-hydroxybutyrate) binding domain (PBM) or hydrophobin binding domain (HBM) to the C-terminus of PETase. Among them, only CBM-added protein showed a 2.28-fold increase in the activity compared to the original mutant. The protein engineering-based strategies followed for making enzymes as effective plastic graders are shown in Fig. 2.

6.2 Role of Computational Biology

Meng et al. (2021) modified PETase using the Premuse tool and obtained the double mutant PETase W159H/F229Y (Histidine in the place of tryptophan at position 159 and tyrosine in place of phenylalanine at position 229) with increased thermal stability and a 40-fold increase in degradation after 24 h at 40 °C than the PETase. Cui et al. (2021) developed a computational approach known as the Greedy Accumulated technique for Protein Engineering (GRAPE). This technique was employed to enhance the resilience of a PETase enzyme from *I. sakaiensis*. They redesigned DuraPETase that degrades 30% of semicrystalline PET films at mild temperatures and achieved biodegradation of 2 g/L microplastics completely.

James-Pearson et al. (2023) developed the plastics microbial biodegradation database (PMBD), which encompasses a total of 949 microorganism-plastic interactions. Additionally, the database includes 79 genes and over 8000 annotated and projected enzyme sequences that play a role in the process of plastic biodegradation. Buchholz et al. (2022) developed the plastics-active enzymes database (PAZy) containing nearly 3000 homologs of PET-degrading enzymes and more than 2000 homologs of polyurethane (PUR) degrading enzymes identified by hidden Markov models and BLAST, respectively. Gambarini et al. (2022) developed PlasticDB consisting of microorganisms and proteins reported for the biological breakdown of plastics.

Also, there are many databases for studying the biodegradation of various xenobiotics, such as OxDBase, KEGG (Kanehisa et al., 2017), MetaCyc (Ahmed et al., 2018), UM-BBD (Ellis et al., 2006), MetaRouter (Pazos et al., 2005), BSD (Urbance et al., 2003), Bionemo (Carbajosa et al., 2009), PAHbase (Kessner et al., 2008), BioRadBase (Reena et al., 2012), BiofOmics (Lourenco et al., 2012), KBase (Arkin et al., 2016), NCLASS (Ghosh et al., 2013), MetaboLights (Kale et al., 2016) and BioCyc, ESIS (León-Zayas et al., 2019).

Enzymes in 3D structure will provide a better understanding of catalytic mechanisms. For this purpose, PDB, structural classification of proteins (SCOP), class-architecture-topology and homology (CATH), molecular modelling databases (MMDB) and visualisation of molecular tools such as RasMol, PyMOL, UCSF chimera, Cn3D, visual molecular dynamics (VMD) and molecular graphic library (MGL) can be used (Skariyachan et al., 2022). We can also compare the structure of our desired compound with the PDB database using Dali, FATCAT, MATRAS and PDBeFold (Ebert & Pelletier, 2017). There are computational pipelines facilitating the analysis, design and engineering of enzymes (Scherer et al., 2021). In order to gain insights into the structure and functioning of enzymes, researchers employ the utilisation of trRosetta, a method for producing enzyme structure models by protein structure prediction. Additionally, GROMACS is employed to investigate the mechanism of catalysis and flexibility in the conformation of enzymes through the application of molecular dynamics simulations. The GALigandDock software was utilised to generate the enzyme-substrate complex through the process of molecular docking. ConSurf is employed for the purpose of identifying design positions by studying the evolutionary conservation of individual residues within an enzyme. The PROSS method is employed for stabilising enzyme-substrate complexes. FuncLib, a tool utilised for the purpose of enhancing the activity and specificity of enzymes, is employed in the field of enzyme engineering. DUET and Rosetta flex-ddG are computational tools commonly employed for the assessment of stability, affinity and activity in enzymes. DUET is specifically designed to predict stability changes resulting from mutations in enzymes, whereas Rosetta flexddG focuses on predicting changes in substrate affinity caused by mutations (Scherer et al., 2021).

Damborsky and Brezovsky (2014) addressed various computational tools for designing enzymes. Selection of active binding sites for enzyme redesign (SABER) examines the catalytic sites of proteins added to the PDB, adaptive substituent reordering algorithm (ASRA) predicts the properties of uncharacterised proteins, POCKETOPTIMIZER designs the binding pockets of proteins for small ligands, and combinatorial co-evolving site saturation mutagenesis (CCSM) identifies hotspots for modification. Enzyme engineering can be performed for the betterment in solubility, stability, selectivity, catalytic activity and dynamics of proteins. There are tools for each case, such as AGGRESCAN3D 2.0, SOLart, AggreRATE-pred, solubility-weighted index, SoluProt for protein solubility, FuncLib, CaverDock and Caver Web, DaReUS-Loop, LoopGrafter, nAPOLI for enzyme activity and selectivity, FireProt, TKSA-MC, pStab, ProTSPoM, Yosshi, SSbondPr, DenseCPD for protein stability and DynaMut2, CABS-flex 2.0, ProSNEx, AlloSigMA 2, LARMD for protein dynamics (Marques et al., 2021).

7 Enzyme Immobilisation

7.1 Enzyme Nano-immobilisation

Traditional immobilisation techniques are influenced by carrier choices for increased stability, reuse and quick extraction from the reaction mixture, as compared to the unbound forms. Owing to tiny size and a huge surface area, nanostructures were found to stabilise enzymes for a long time. Biomimetic mineralisation has caught attention recently for ease of separation, high enzyme catalytic activity, increased stability and gentle reaction conditions. In this process, a dilute buffer solution is employed to mitigate the kinetics of inorganic salt generation, while enzymes are incorporated into the reaction to facilitate the gradual adhesion of the inorganic salt. This study involves the targeted chelation of the imidazole group present in the His-tag region exposed on the surface of the protein, in conjunction with the use of metal ions, to achieve the immobilisation of the protein (enzyme). This technique was used to immobilise lipase and transaminase successfully. Nucleation of cobalt ions with PETase enzyme produces cobalt nanoparticles. The enzyme–inorganic nanoflowers, PETase@$Co_3(PO)_4$ exhibit advantageous characteristics such as stacking of more enzymes and promotion of mass transfer. The production efficiency of TPA was found to be 3.5 times greater compared to that of free enzymes. The PETase@$Co_3(PO)_4$ exhibited a retention of 75% of its initial activity even after a period of 12 days, when compared to the activity of free enzymes. This immobilisation technique is suitable for industrial PET depolymerisation (Jia et al., 2021).

Multienzymatic complex was created using SpyTag/SpyCatcher (Spy) chemistry. The SpyTag exhibits the potential to undergo spontaneous formation of stable iso-peptide bonds with SpyCatcher, displaying a notably accelerated reaction rate, increased yield, and remarkable flexibility to various reaction conditions. Under biomimetic mineralisation process, a bienzymatic cascade system was developed by co-immobilising glucose-oxidase and glucoamylase. They exhibited 92% conversion efficiency of starch to gluconic acid with 1.5 times increase in the activity of enzyme within 80 min. Direct immobilisation via biomimetic mineralisation would render the PETase enzyme encapsulated within the crystals inaccessible for PET degradation. Therefore, with the help of Spy chemistry and biomimetic mineralisation techniques, DuraPETase and MHETase was co-immobilised on calcium phosphate nanocrystals. The fusion enzyme MHETase-SpyCatcher was initially incorporated into $Ca_3(PO4)_2$ by biomimetic mineralisation, then DuraPETase-SpyTag was attached to MHETase@$Ca_3(PO4)_2$ nanoparticles by means of SpyCatcher-SpyTag interactions. The DuraPETase was found to be situated on the outer layer of the co-immobilised bienzyme system. The bienzyme compartmentalisation has led to an enhanced interaction between DuraPETase and the solid substrate. Additionally, the formation of nanocrystals has provided a protective environment for the enzymes, resulting in improved stability. Thereby the PET degradation efficiency of DuraPETase-MHETase@$Ca_3(PO4)_2$ was increased significantly over the free bienzyme system with 10 days of incubation at 40 °C (Chen et al., 2022).

However, the temperature required for PETase maximal activity is around 50 °C. More energy is required to maintain such temperature. It is confirmed that Fe_3O_4 nanoparticles can increase system temperature to 46 °C by converting solar energy into thermal energy. Examination of DuraPETase, free DuraPETase + Fe_3O_4 and DuraPETase immobilised on Fe_3O_4 was conducted. Comparatively, DuraPETase+ Fe_3O_4 and DuraPETase@Fe_3O_4 showed higher activity and stability than free DuraPETase. Even after six cycles, DuraPETase@Fe_3O_4 showed 57% of their initial activity. However, DuraPETase + Fe_3O_4 were much more effective than DuraPETase@Fe_3O_4 (Li et al., 2022). This implied that DuraPETase immobilisation is unnecessary and mere DuraPETase + Fe_3O_4 is recommended for solar-driven enzymatic PET biodegradation. Enzymatic and nanoparticle technologies encompass a range of methodologies, one of which involves the fusion of enzymes with hydrophobins. Additionally, advancements have been made in the development of novel binding modules involving nanoparticles and polymers (Kumar et al., 2023).

8 Enzyme-Based Bioreactors

Though PET materials can be degraded by PETase, complete biodegradation is hampered by the formation of MHET as intermediates. Ultrafiltration was shown to be effective in removing soluble products that inhibit enzymatic hydrolysis thereby promoting higher yields. The utilisation of immobilised enzymes in a membrane-associated bioreactor facilitates the uninterrupted and continuous application of the enzyme. Also, a membrane separation unit is associated with the bioreactor in order to filter out the soluble products after catalytic reaction. In this reactor, biocatalysis and separation of intermediates take place in parallel. There are many advantages of using a membrane-based reactor over free enzyme biochemical reactions, such as better yield, enzyme reusability and continuous operation. In addition, membranes are a convenient matrix for enzyme immobilisation and thus provides high surface area for loading enzymes (Nath et al., 2014).

Numerous research have also demonstrated the efficacious utilisation of a membrane reactor in the context of the heterogeneous hydrolysis of intractable cellulose. The intermediate cellobiose, which acts as an inhibitor of cellulose, was effectively eliminated, resulting in increased hydrolysis and yield. Cutinase enzyme TfCut2, derived from *Thermobifida fusca* KW3, was utilised to catalyse the hydrolysis of PET films in an aqueous medium housed in an ultrafiltration membrane reactor. The effectiveness of the biocatalytic hydrolysis of PET was enhanced by 70% during a 24-h reaction period through the regular elimination of inhibitory compounds from hydrolysis using the membrane, as compared to a batch mode hydrolysis process. (Barth et al., 2015; Tournier et al., 2023).

9 Biotransformation of Plastics

Industrialising plastic biodegradation is necessary to sustainably manage the huge accumulation of plastic trash. Value-added products from plastic biodegradation can help scale up plastic biodegradation economically. To promote the plastics circular economy, research has focused on repurposing plastic biodegradation products. PET can be broken down into TPA and EG by PETase and MHETase. In the last few years, soil microbes like *Pseudomonas putida* have been used for the conversion of EG into bioproducts with additional products, namely rhamnolipids, glycolic acid (GLA) and glyoxylic acid. GLA is used in cosmetics as an exfoliant and is obtained from EG by a two-step conversion process. *Gluconobacter oxydans* KCCM 0109 can convert EG from PET hydrolysate into glycolic acid (Hee et al., 2020). Polyhydroxyalkanoate (PHA), intracellularly produced by microbes, has similar chemical and physical properties as synthetic plastics and is biodegradable. *Pseudomonas putida* KT2440 can increase medium chain length (mcl) PHA under a limited nitrogen source and is not able to use EG as a carbon source. Franden et al. (2018) showed that in engineered *Pseudomonas putida* KT2440, the overexpression of glyoxylate carboligase operon with glycolate oxidase operon results in the use of EG as the primary source of carbon.

Conversion of TPA to valuable products like gallic acid (GA), protocatechuic acid (PCA), catechol, vanillic acid (VA) and muconic acid (MA) can enhance the economic viability of PET recycling (Hee et al., 2020). *Comamonas* sp., *Rhodococcus* sp., *Delftia tsuruhatensis* can be capable of using terephthalic acid as the sole carbon source (Liu et al., 2022). TPA degradation pathway involves PCA as an intermediate, which is then transformed into highly valued aromatic compounds. Three microorganisms isolated by Kenny et al. (2008), namely *Pseudomonas frederiksbergensis* GO23, *Pseudomonas putida* GO19 and *Pseudomonas putida* GO16, can utilise TPA for growth and accumulation of mclPHA. Enzymes involved in converting terephthalic acid to protocatechuic acid are DCD dehydrogenase and TPA1,2-dioxygenase. These enzymes are reported to be found in *Comamonas* sp. E6, *Rhodococcus* sp. DK17 and *Delftia tsuruhatensis* T7. GA is of medicinal value used to produce trimethoprim, propyl gallate and antioxidants. Increased expression of PobA obtained from *Pseudomonas putida* KT2440, engineered *E.coli* HBH-1 can produce 1.4 mM of gallic acid (Kim et al., 2019; Liu et al., 2022). Pyrogallol is a high-value chemical that is used in the oil industry as an antioxidant. Catechol is subjected to catechol hydroxylation and converted to pyrogallol by PhKLMNOPQ (Hee et al., 2020). Bioconversion of TPA to VA has also been reported. VA is used in the pharmaceutical industry as a direct precursor of vanillin. Bioconversion of TPA to muconic acid has also been reported (Bai et al., 2013).

As discussed earlier, PVC, PP, PE and PS cannot be biodegraded easily. They should be degraded prior by any physico-chemical methods, and the intermediates would be used as a substrate for bioconversion. Zhang et al. (2018) reported that phenol, one of the major products of PS, can transform into PHA using a microbial consortium. PE degradation product hexadeconoate was convertible using *P. aeruginosa*, *Ralstonia eutropha* and *Y. lipolytica* (Tiso et al., 2022). The biotransformation of styrene, a PS monomer, by a four-step enzyme cascade reaction yielded up to 70% mandelic acid (Lukito et al., 2021). Furthermore, this biological mandelic acid manufacturing removes the use of poisonous cyanide, which is required in conventional synthesis, providing a greener and more sustainable production approach while boosting the plastics circular economy (Lee et al., 2023). Ward et al. (2005) converted styrene, phenyl acetic acid into PHA using *P. putida* CA-3. PP pyrolysis also produces alkenes and branched-chain fatty alcohols which can be fermented into fatty acids using *Yarrowia lipolytica* strain 78-003 (Mihreteab et al., 2019; Lee et al., 2023). A hybrid system involving chemical and biological methods is recommended for non-hydrolysable plastics. Solvolysis and pyrolysis are chemical methods usually employed. The product of pyrolysis, fatty acid obtained from PE and thermally oxidised PP, were reported to produce PHA using *Pseudomonas putida* and *Cupriavidus necator*, respectively (Lee et al., 2023). Not much reports are there on PVC degradation and utilisation of their metabolites for making value-added products. The biodegradation and biotransformation process of plastics are shown in Fig. 3.

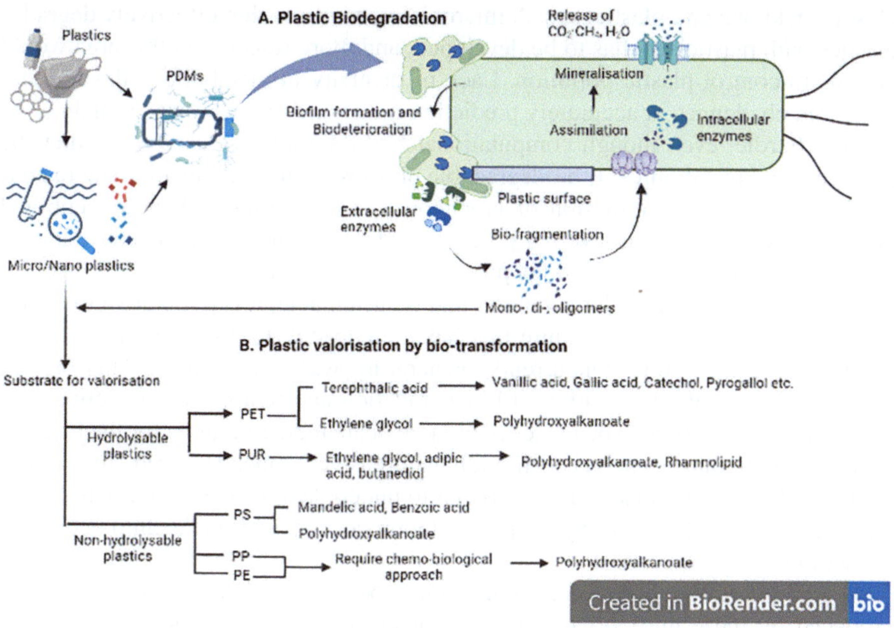

Fig. 3 Biodegradation and biotransformation of plastics

These researches have demonstrated that plastic biodegradation byproducts have potential value; hence stimulating more research would be beneficial to the society. The effective decomposition of synthetic plastics, along with the conversion of degradation byproducts into valuables, is of paramount importance in the sustainable handling of plastic trash in the natural surroundings.

10 Future Outlook

Plastics have been in use for more than 100 years by humans. Biocatalysts would have evolved to degrade the plastics by adaptation and/or natural evolution (Lee et al., 2023). PETase discovery led to effective biorecycling of plastics. Catalytic promiscuity of enzymes may extend their applicability. There is a need for comprehensive exploration of key enzymes that exhibit promiscuity and possess activity towards synthetic polymers, in order to assess their potential for enhancing biodegradation efficiency. Catalytic performances of computer-designed enzymes are more theoretically, but practically not, effective for the degradation of commercial plastics. Metabolic engineering of plastic-degrading microbes for the development of microbial cell factories may lead to valorisation of plastic waste for plastic circular economy (Lee et al., 2023). One of the widely used plastics, PVC, is still understudied because of their recalcitrance nature due to plasticisers. Degradation of plasticisers is also understudied, insisting research on the degradation of both plastic backbone and plasticisers. A microbial consortium that effectively degrades plastics with plasticisers has to be developed, and more research in this area would help us to control plastic pollution. Lack of diversity in metabolic pathway databases and challenges in accurately predicting metabolic intermediates remain to be a major hurdle, even though computational biology tools are used to predict the pathways and metabolites. If biodegradable plastics are used as an alternate to conventional plastics, the emission of greenhouse gases is higher than conventional plastics. Therefore, the up-cycling process needs to be encouraged because it converts plastic waste into value-added products rather than the release of CO_2. In order to construct a carbon circular economy that is sustainable, it is imperative to devise an innovative industrial procedure that has the capacity to diminish emissions of greenhouse gases while concurrently generating waste into value. This can be achieved through the integration of biodegradation and biological up-cycling processes (Lee et al., 2023). The presence of large-scale industrial units for the biological recycling of plastics in an eco-friendly nature is very much required. To enhance the pace of commercialisation, it is crucial to undertake thorough empirical studies on a significant scale, focusing on specific bioprocess designs and techno-economic evaluations.

In addition to the direct utilisation of microbes, the application of exoenzymes produced by these microorganisms, such as esterases, lipases and cutinases, has been investigated for the purpose of biodegrading MPs (Iram et al., 2019). While the microbial method has demonstrated numerous benefits, it is important to note that its full potential has not yet been fully explored (Endeshaw et al., 2017). The potential occurrence of horizontal gene transfer among the genetically modified microorganisms and the indigenous microbes may result in inadvertent gene transfer, resulting in the acquisition of antibiotic resistance traits (Kumar et al., 2023). This phenomenon poses environmental risks and hinders the unrestricted utilisation of these microbes in waste treatment applications (Jong et al., 2020; Pant et al., 2021). Enhancing the biodegradation technique can be facilitated by gaining a comprehensive knowledge of the shape and size characteristics of MPs that need to be degraded (Kumar et al., 2023). Lack of rigorous policies and stringent regulations regarding the use and safe disposal of plastics and the accumulation of MPs in developing economies has been identified as the primary underlying pollutant (Iroegbu et al., 2021; Kumar et al., 2023). Further exploration is necessary to further our understanding of plastic-biodegrading enzymes and the underlying molecular mechanisms. In order to enhance the speed of biodegradation by plastic-biodegrading microbial enzymes, a comprehensive and interdisciplinary approach is needed to save the ecosystem and the environment (Garrido et al., 2021).

11 Conclusions

Indeed, plastics are a favourite material for industrial and other domestic applications. But their detrimental effects pose a serious environmental threat. Several measures have been taken in many countries, such as banning single-use plastics.

MNPs remain to be a major threat to the ecosystem due to their free distribution and accumulation, affecting all living systems. Biodegradation of plastics using microbes is an environmental-friendly method, but most of the microbes involved are pathogenic in nature. However, their enzymes and biosurfactants are effective in plastic biodegradation. Recent advancements in biotechnology and interdisciplinary research provide us with various tools and techniques to enhance and improve the enzymatic biodegradation of plastics. Protein engineering techniques and computational biology tools are useful in designing efficient PDEs, but the safety aspects need to be studied well. Despite the ongoing efforts for effective biodegradation of plastics, non-hydrolysable plastics such as PP and PVC that are used in everyday applications remain a major challenge and threat to the environment and human health and well-being. More research on these plastics and their biodegradation will help us to understand their biodegradation and scale up of the process without affecting the ecosystem. Stringent rules and regulations and a ban on plastic production and use would make the world free from plastic pollution and save the environment, ecosystem and all the living systems from plastic-associated health hazards.

References

Ahmaditabatabaei, S., Kyazze, G., Iqbal, H. M., & Keshavarz, T. (2021). Fungal enzymes as catalytic tools for polyethylene terephthalate (PET) degradation. *Journal of Fungi, 7*(11), 931.

Ahmed, T., Shahid, M., Azeem, F., Rasul, I., Shah, A. A., Noman, M., Hameed, A., Manzoor, N., Manzoor, I., & Muhammad, S. (2018). Biodegradation of plastics: Current scenario and future prospects for environmental safety. *Environmental Science and Pollution Research, 25*, 7287–7298.

Almeida, E. L., Rincón, A. F. C., Jackson, S. A., & Dobson, A. D. (2019). In silico screening and heterologous expression of a polyethylene terephthalate hydrolase (PETase)-like enzyme (SM14est) with polycaprolactone (PCL)-degrading activity, from the marine sponge-derived strain Streptomyces sp. SM14. *Frontiers in Microbiology, 10*, 2187.

Amobonye, A., Bhagwat, P., Singh, S., & Pillai, S. (2021). Plastic biodegradation: Frontline microbes and their enzymes. *Science of the Total Environment, 759*, 143536.

Arkin, A. P., Stevens, R. L., Cottingham, R. W., Maslov, S., Henry, C. S., Dehal, P., Ware, D., Perez, F., Harris, N. L., Canon, S., & Sneddon, M. W. (2016). The DOE systems biology knowledgebase (KBase). *BioRxiv*, 096354.

Austin, H. P., Allen, M. D., Donohoe, B. S., Rorrer, N. A., Kearns, F. L., Silveira, R. L., Pollard, B. C., Dominick, G., Duman, R., El Omari, K., & Mykhaylyk, V. (2018). Characterization and engineering of a plastic-degrading aromatic polyesterase. *Proceedings of the National Academy of Sciences, 115*(19), 4350–4357.

Bai, Y. Y., Xiao, L. P., Shi, Z. J., & Sun, R. C. (2013). Structural variation of bamboo lignin before and after ethanol organosolv pretreatment. *International Journal of Molecular Sciences, 14*(11), 21394–21413.

Barth, M., Oeser, T., Wei, R., Then, J., Schmidt, J., & Zimmermann, W. (2015). Effect of hydrolysis products on the enzymatic degradation of polyethylene terephthalate nanoparticles by a polyester hydrolase from Thermobifida fusca. *Biochemical Engineering Journal, 93*, 222–228.

Buchholz, P. C., Feuerriegel, G., Zhang, H., Perez-Garcia, P., Nover, L. L., Chow, J., Streit, W. R., & Pleiss, J. (2022). Plastics degradation by hydrolytic enzymes: The plastics-active enzymes database—PAZy. *Proteins: Structure, Function, and Bioinformatics, 90*(7), 1443–1456.

Carbajosa, G., Trigo, A., Valencia, A., & Cases, I. (2009). Bionemo: Molecular information on biodegradation metabolism. *Nucleic Acids Research, 37*(1), 598–602.

Chen, Z., Xiao, Y., Weber, G., Wei, R., & Wang, Z. (2021). Yeast cell surface display of bacterial PET hydrolase as a sustainable biocatalyst for the degradation of polyethylene terephthalate. *Methods in Enzymology, 648*, 457–477.

Chen, K., Dong, X., & Sun, Y. (2022). Sequentially co-immobilized PET and MHET hydrolases via spy chemistry in calcium phosphate nanocrystals present high-performance PET degradation. *Journal of Hazardous Materials, 438*, 129517.

Chen, C. C., Li, X., Shen, P., Yang, Y., Min, J., Zhang, Q., Liu, Y., Huang, J. W., & Guo, R. T. (2023). The future for biochemical recycling of poly (ethylene terephthalate): From lab to process. Current opinion in green and sustainable. *Chemistry*, 100822.

Cui, Y., Chen, Y., Liu, X., Dong, S., Tian, Y. E., Qiao, Y., Mitra, R., Han, J., Li, C., Han, X., & Liu, W. (2021). Computational redesign of a PETase for plastic biodegradation under ambient condition by the GRAPE strategy. *ACS Catalysis, 11*(3), 1340–1350.

Dai, L., Qu, Y., Huang, J. W., Hu, Y., Hu, H., Li, S., Chen, C. C., & Guo, R. T. (2021). Enhancing PET hydrolytic enzyme activity by fusion of the cellulose–binding domain of cellobiohydrolase I from Trichoderma reesei. *Journal of Biotechnology, 334*, 47–50.

Damborsky, J., & Brezovsky, J. (2014). Computational tools for designing and engineering enzymes. *Current Opinion in Chemical Biology, 19*, 8–16.

Danso, D., Chow, J., & Streit, W. R. (2019). Plastics: Environmental and biotechnological perspectives on microbial degradation. *Applied and Environmental Microbiology, 85*(19), 1095–1019.

de Carvalho, C. C. (2017). Whole cell biocatalysts: Essential workers from nature to the industry. *Microbial Biotechnology, 10*(2), 250–263.

Dwevedi, A., & Kayastha, A. M. (2012). Enzyme immobilization: A breakthrough in enzyme technology and boon to enzyme based industries. *Proteomics Research Journal, 3*, 333.

Ebert, M. C., & Pelletier, J. N. (2017). Computational tools for enzyme improvement: Why everyone can–and should–use them. *Current Opinion in Chemical Biology, 37*, 89–96.

Ellis, L. B., Roe, D., & Wackett, L. P. (2006). The University of Minnesota biocatalysis/biodegradation database: The first decade. *Nucleic Acids Research, 34*(1), 517–521.

Endeshaw, A., Gizaw, B., Tsegaye, Z., & Wassie, M. (2017). The role of microorganisms in bioremediation–A review. *Open Journal of Environmental Biology, 2*, 038–046.

Franden, M. A., Jayakody, L. N., Li, W. J., Wagner, N. J., Cleveland, N. S., Michener, W. E., Hauer, B., Blank, L. M., Wierckx, N., Klebensberger, J., & Beckham, G. T. (2018). Engineering Pseudomonas putida KT2440 for efficient ethylene glycol utilization. *Metabolic Engineering, 48*, 197–207.

Gambarini, V., Pantos, O., Kingsbury, J. M., Weaver, L., Handley, K. M., & Lear, G. (2022). PlasticDB: A database of microorganisms and proteins linked to plastic biodegradation. *Database*, baac008.

Gangl, D., Zedler, J. A., Rajakumar, P. D., Martinez, E. M. R., Riseley, A., Włodarczyk, A., Purton, S., Sakuragi, Y., Howe, C. J., Jensen, P. E., & Robinson, C. (2015). Biotechnological exploitation of microalgae. *Journal of Experimental Botany, 66*(22), 6975–6990.

Gangoiti, J., Santos, M., Prieto, M. A., de la Mata, I., Serra, J. L., & Llama, M. J. (2012). Characterization of a novel subgroup of extracellular medium-chain-length polyhydroxyalkanoate depolymerases from actinobacteria. *Applied and Environmental Microbiology, 78*(20), 7229–7237.

Gao, R., Pan, H., & Lian, J. (2021). Recent advances in the discovery, characterization, and engineering of poly (ethylene terephthalate)(PET) hydrolases. *Enzyme and Microbial Technology, 150*, 109868.

Garrido, R., Cabeza, L. F., & Falguera, V. (2021). An overview of bioplastic research on its relation to national policies. *Sustainability, 13*(14), 7848.

Ghosh, S. K., Pal, S., & Ray, S. (2013). Study of microbes having potentiality for biodegradation of plastics. *Environmental Science and Pollution Research, 20*, 4339–4355.

Gómez-Méndez, L. D., Moreno-Bayona, D. A., Poutou-Pinales, R. A., Salcedo-Reyes, J. C., Pedroza-Rodríguez, A. M., Vargas, A., & Bogoya, J. M. (2018). Biodeterioration of plasma pretreated LDPE sheets by Pleurotus ostreatus. *PLoS One, 13*(9), 203786.

Gong, W., Xie, B., Deng, S., Fan, Y., Tang, X., & Liang, H. (2019). Enhancement of anaerobic digestion effluent treatment by microalgae immobilization: Characterized by fluorescence excitation-emission matrix coupled with parallel factor analysis in the photobioreactor. *Science of the Total Environment, 678*, 105–113.

Han, X., Liu, W., Huang, J. W., Ma, J., Zheng, Y., Ko, T. P., Xu, L., Cheng, Y. S., Chen, C. C., & Guo, R. T. (2017). Structural insight into catalytic mechanism of PET hydrolase. *Nature Communications, 8*(1), 2106.

Hee, J., Schlögel, K., Lechthaler, S., Plaster, J., Bitter, K., Blank, L. M., & Quicker, P. (2020). Comparative analysis of the behaviour of marine litter in thermochemical waste treatment processes. *Processes, 9*(1), 13.

Hou, L., & Majumder, E. L. W. (2021). Potential for and distribution of enzymatic biodegradation of polystyrene by environmental microorganisms. *Materials, 14*(3), 503.

Hug, L. A., Thomas, B. C., Sharon, I., Brown, C. T., Sharma, R., Hettich, R. L., Wilkins, M. J., Williams, K. H., Singh, A., & Banfield, J. F. (2016). Critical biogeochemical functions in the subsurface are associated with bacteria from new phyla and little studied lineages. *Environmental Microbiology, 18*(1), 159–173.

Iiyoshi, Y., Tsutsumi, Y., & Nishida, T. (1998). Polyethylene degradation by lignin-degrading fungi and manganese peroxidase. *Journal of Wood Science, 44*, 222–229.

Inderthal, H., Tai, S. L., & Harrison, S. T. (2021). Non-hydrolyzable plastics–an interdisciplinary look at plastic bio-oxidation. *Trends in Biotechnology, 39*(1), 12–23.

Iram, D., Riaz, R., & Iqbal, R. K. (2019). Usage of potential micro-organisms for degradation of plastics. *Open Journal of Environmental Biology, 4*(1), 7–15.

Iroegbu, A. O. C., Ray, S. S., Mbarane, V., Bordado, J. C., & Sardinha, J. P. (2021). Plastic pollution: A perspective on matters arising: Challenges and opportunities. *ACS Omega, 6*(30), 19343–19355.

James-Pearson, L. F., Dudley, K. J., Te'o, V. S. J., & Patel, B. K. (2023). A hot topic: Thermophilic plastic biodegradation. *Trends in Biotechnology, 41*(9), 1117–1126.

Jia, Y., Samak, N. A., Hao, X., Chen, Z., Yang, G., Zhao, X., Mu, T., Yang, M., & Xing, J. (2021). Nano-immobilization of PETase enzyme for enhanced polyethylene terephthalate biodegradation. *Biochemical Engineering Journal, 176*, 108205.

Jong, M. C., Harwood, C. R., Blackburn, A., Snape, J. R., & Graham, D. W. (2020). Impact of redox conditions on antibiotic resistance conjugative gene transfer frequency and plasmid fate in wastewater ecosystems. *Environmental Science & Technology, 54*(23), 14984–14993.

Joo, S., Cho, I. J., Seo, H., Son, H. F., Sagong, H. Y., Shin, T. J., Choi, S. Y., Lee, S. Y., & Kim, K. J. (2018). Structural insight into molecular mechanism of poly (ethylene terephthalate) degradation. *Nature Communications, 9*(1), 382.

Kale, N. S., Haug, K., Conesa, P., Jayseelan, K., Moreno, P., Rocca-Serra, P., & Steinbeck, C. (2016). MetaboLights: an open-access database repository for metabolomics data. *Current Protocols in Bioinformatics, 53*(1), 14.13.1–14.13.18.

Kanehisa, M., Furumichi, M., Tanabe, M., Sato, Y., & Morishima, K. (2017). KEGG: New perspectives on genomes, pathways, diseases and drugs. *Nucleic Acids Research, 45*, 353–361.

Kang, M. G., Kwak, M. J., & Kim, Y. (2023). Polystyrene microplastics biodegradation by gut bacterial Enterobacter hormaechei from mealworms under anaerobic conditions: Anaerobic oxidation and depolymerization. *Journal of Hazardous Materials*, 132045.

Kenny, S. T., Runic, J. N., Kaminsky, W., Woods, T., Babu, R. P., Keely, C. M., Blau, W., & O'Connor, K. E. (2008). Up-cycling of PET (polyethylene terephthalate) to the biodegradable plastic PHA (polyhydroxyalkanoate). *Environmental Science & Technology, 42*(20), 7696–7701.

Kessner, D., Chambers, M., Burke, R., Agus, D., & Mallick, P. (2008). ProteoWizard: Open source software for rapid proteomics tools development. *Bioinformatics, 24*(21), 2534–2536.

Kim, H. T., Kim, J. K., Cha, H. G., Kang, M. J., Lee, H. S., Khang, T. U., Yun, E. J., Lee, D. H., Song, B. K., Park, S. J., & Joo, J. C. (2019). Biological valorization of poly (ethylene terephthalate) monomers for upcycling waste PET. *ACS Sustainable Chemistry & Engineering, 7*(24), 19396–19406.

Knott, B. C., Erickson, E., Allen, M. D., Gado, J. E., Graham, R., Kearns, F. L., Pardo, I., Topuzlu, E., Anderson, J. J., Austin, H. P., & Dominick, G. (2020). Characterization and engineering of a two-enzyme system for plastics depolymerization. *Proceedings of the National Academy of Sciences, 117*(41), 25476–25485.

Kumar, V., Singh, E., Singh, S., Pandey, A., & Bhargava, P. C. (2023). Micro-and nano-plastics (MNPs) as emerging pollutant in ground water: Environmental impact, potential risks, limitations and way forward towards sustainable management. *Chemical Engineering Journal, 459*, 141568.

Lam, K. N., Cheng, J., Engel, K., Neufeld, J. D., & Charles, T. C. (2015). Current and future resources for functional metagenomics. *Frontiers in Microbiology, 6*, 1196.

Lauersen, K. J., Berger, H., Mussgnug, J. H., & Kruse, O. (2013a). Efficient recombinant protein production and secretion from nuclear transgenes in Chlamydomonas reinhardtii. *Journal of Biotechnology, 167*(2), 101–110.

Lauersen, K. J., Vanderveer, T. L., Berger, H., Kaluza, I., Mussgnug, J. H., Walker, V. K., & Kruse, O. (2013b). Ice recrystallization inhibition mediated by a nuclear-expressed and-secreted recombinant ice-binding protein in the microalga Chlamydomonas reinhardtii. *Applied Microbiology and Biotechnology, 97*, 9763–9772.

Lee, S., Lee, Y. R., Kim, S. J., Lee, J. S., & Min, K. (2023). Recent advances and challenges in the biotechnological upcycling of plastic wastes for constructing a circular bioeconomy. *Chemical Engineering Journal, 454*, 140470.

León-Zayas, R., Roberts, C., Vague, M., & Mellies, J. L. (2019). Draft genome sequences of five environmental bacterial isolates that degrade polyethylene terephthalate plastic. *Microbiology Resource Announcements, 8*(25), 0–1128.

Li, W., Shi, H., Ding, H., Wang, L., Zhang, Y., Li, X., & Wang, F. (2015). Cell surface display and characterization of Rhizopus oryzae lipase in Pichia pastoris using Sed1p as an anchor protein. *Current Microbiology, 71*, 150–155.

Li, Z., Chen, K., Yu, L., Shi, Q., & Sun, Y. (2022). Fe_3O_4 nanoparticles-mediated solar-driven enzymatic PET degradation with PET hydrolase. *Biochemical Engineering Journal, 180*, 108344.

Liu, M., Yang, S., Long, L., Cao, Y., & Ding, S. (2018). Engineering a chimeric lipase-cutinase (Lip-Cut) for efficient enzymatic deinking of waste paper. *BioResources, 13*(1), 981–996.

Liu, H., Xu, L., Bao, X., Zhou, J., Qian, X., Dong, W., & Jiang, M. (2022). Microbial fermentation for biodegradation and biotransformation of waste plastics into high value–added chemicals. *Biomass, Biofuels, Biochemicals*, 395–412.

Loll-Krippleber, R., Sajtovich, V. A., Ferguson, M. W., Ho, B., Burns, A. R., Payliss, B. J., Bellissimo, J., Peters, S., Roy, P. J., Wyatt, H. D., & Brown, G. W. (2022). Development of a yeast whole-cell biocatalyst for MHET conversion into terephthalic acid and ethylene glycol. *Microbial Cell Factories, 21*(1), 1–15.

Lourenco, A., Ferreira, A., Veiga, N., Machado, I., Pereira, M. O., & Azevedo, N. F. (2012). BiofOmics: A web platform for the systematic and standardized collection of high-throughput biofilm data. *PLoS One, 7*(6), 39960.

Lukito, B. R., Wang, Z., Sundara Sekar, B., & Li, Z. (2021). Production of (R)-mandelic acid from styrene, L-phenylalanine, glycerol, or glucose via cascade biotransformations. *Bioresources and Bioprocessing, 8*(1), 1–11.

Ma, Y., Yao, M., Li, B., Ding, M., He, B., Chen, S., Zhou, X., & Yuan, Y. (2018). Enhanced poly (ethylene terephthalate) hydrolase activity by protein engineering. *Engineering, 4*(6), 888–893.

Magalhães, R. P., Cunha, J. M., & Sousa, S. F. (2021). Perspectives on the role of enzymatic biocatalysis for the degradation of plastic PET. *International Journal of Molecular Sciences, 22*(20), 11257.

Marques, S. M., Planas-Iglesias, J., & Damborsky, J. (2021). Web-based tools for computational enzyme design. *Current Opinion in Structural Biology, 69*, 19–34.

Mayfield, S. P., Manuell, A. L., Chen, S., Wu, J., Tran, M., Siefker, D., Muto, M., & Marin-Navarro, J. (2007). Chlamydomonas reinhardtii chloroplasts as protein factories. *Current Opinion in Biotechnology, 18*(2), 126–133.

Meng, X., Yang, L., Liu, H., Li, Q., Xu, G., Zhang, Y., Guan, F., Zhang, Y., Zhang, W., Wu, N., & Tian, J. (2021). Protein engineering of stable IsPETase for PET plastic degradation by Premuse. *International Journal of Biological Macromolecules, 180*, 667–676.

Mihreteab, M., Stubblefield, B. A., & Gilbert, E. S. (2019). Microbial bioconversion of thermally depolymerized polypropylene by Yarrowia lipolytica for fatty acid production. *Applied Microbiology and Biotechnology, 103*, 7729–7740.

Mohanan, N., Montazer, Z., Sharma, P. K., & Levin, D. B. (2020). Microbial and enzymatic degradation of synthetic plastics. *Frontiers in Microbiology, 11*, 580709.

Moog, D., Schmitt, J., Senger, J., Zarzycki, J., Rexer, K. H., Linne, U., Erb, T., & Maier, U. G. (2019). Using a marine microalga as a chassis for polyethylene terephthalate (PET) degradation. *Microbial Cell Factories, 18*(1), 1–15.

Nath, A., Chakraborty, S., & Bhattacharjee, C. (2014). Bioreactor and enzymatic reactions in bioremediation. In *Microbial biodegradation and bioremediation* (pp. 455–495).

Olajuyin, A. M., Yang, M., Liu, Y., Mu, T., Tian, J., Adaramoye, O. A., & Xing, J. (2016). Efficient production of succinic acid from Palmaria palmata hydrolysate by metabolically engineered Escherichia coli. *Bioresource Technology, 214*, 653–659.

Olson, D. G., & Lynd, L. R. (2012). Transformation of Clostridium thermocellum by electroporation. *Methods in Enzymology, 510*, 317–330.

Pant, G., Garlapati, D., Agrawal, U., Prasuna, R. G., Mathimani, T., & Pugazhendhi, A. (2021). Biological approaches practised using genetically engineered microbes for a sustainable environment: A review. *Journal of Hazardous Materials, 405*, 124631.

Papadopoulou, A., Hecht, K., & Buller, R. (2019). Enzymatic PET degradation. *Chimia, 73*(9), 743.

Park, M. (2020). Surface display technology for biosensor applications: A review. *Sensors, 20*(10), 2775.

Pazos, F., Guijas, D., Valencia, A., & De Lorenzo, V. (2005). MetaRouter: Bioinformatics for bioremediation. *Nucleic Acids Research, 33*(1), 588–592.

Pearsall, S. M., Rowley, C. N., & Berry, A. (2015). Advances in pathway engineering for natural product biosynthesis. *ChemCatChem, 7*(19), 3078–3093.

Perz, V., Bleymaier, K., Sinkel, C., Kueper, U., Bonnekessel, M., Ribitsch, D., & Guebitz, G. M. (2016). Substrate specificities of cutinases on aliphatic–aromatic polyesters and on their model substrates. *New Biotechnology, 33*(2), 295–304.

Pometto, A. L., 3rd, Lee, B. T., & Johnson, K. E. (1992). Production of an extracellular polyethylene-degrading enzyme (s) by Streptomyces species. *Applied and Environmental Microbiology, 58*(2), 731–733.

Reena, R., Majhi, M. C., Arya, A. K., Kumar, R., & Kumar, A. (2012). BioRadBase: A database for bioremediation of radioactive waste. *African Journal of Biotechnology, 11*(35), 8718–8721.

Rosano, G. L., & Ceccarelli, E. A. (2014). Recombinant protein expression in Escherichia coli: Advances and challenges. *Frontiers in Microbiology, 5*, 172.

Samak, N. A., Hu, J., Wang, K., Guo, C., & Liu, C. (2018). Development of a novel microaerobic cultivation strategy for high potential CotA laccase production. *Waste and Biomass Valorization, 9*, 369–377.

Samak, N. A., Jia, Y., Sharshar, M. M., Mu, T., Yang, M., Peh, S., & Xing, J. (2020). Recent advances in biocatalysts engineering for polyethylene terephthalate plastic waste green recycling. *Environment International, 145*, 106144.

Santo, M., Weitsman, R., & Sivan, A. (2013). The role of the copper-binding enzyme–laccase– in the biodegradation of polyethylene by the actinomycete Rhodococcus ruber. *International Biodeterioration & Biodegradation, 84*, 204–210.

Scherer, M., Fleishman, S. J., Jones, P. R., Dandekar, T., & Bencurova, E. (2021). Computational enzyme engineering pipelines for optimized production of renewable chemicals. *Frontiers in Bioengineering and Biotechnology, 9*, 673005.

Seo, H., Kim, S., Son, H. F., Sagong, H. Y., Joo, S., & Kim, K. J. (2019). Production of extracellular PETase from Ideonella sakaiensis using sec-dependent signal peptides in E. coli. *Biochemical and Biophysical Research Communications, 508*(1), 250–255.

Sheldon, R. A., & van Pelt, S. (2013). Enzyme immobilisation in biocatalysis: Why, what and how. *Chemical Society Reviews, 42*(15), 6223–6235.

Shirke, A. N., White, C., Englaender, J. A., Zwarycz, A., Butterfoss, G. L., Linhardt, R. J., & Gross, R. A. (2018). Stabilizing leaf and branch compost cutinase (LCC) with glycosylation: Mechanism and effect on PET hydrolysis. *Biochemistry, 57*(7), 1190–1200.

Skariyachan, S., Taskeen, N., Kishore, A. P., & Krishna, B. V. (2022). Recent advances in plastic degradation–From microbial consortia-based methods to data sciences and computational biology driven approaches. *Journal of Hazardous Materials, 426*, 128086.

Son, H. F., Cho, I. J., Joo, S., Seo, H., Sagong, H. Y., Choi, S. Y., Lee, S. Y., & Kim, K. J. (2019). Rational protein engineering of thermo-stable PETase from Ideonella sakaiensis for highly efficient PET degradation. *ACS Catalysis, 9*(4), 3519–3526.

Sulaiman, S., Yamato, S., Kanaya, E., Kim, J. J., Koga, Y., Takano, K., & Kanaya, S. (2012). Isolation of a novel cutinase homolog with polyethylene terephthalate-degrading activity from leaf-branch compost by using a metagenomic approach. *Applied and Environmental Microbiology, 78*(5), 1556–1562.

Tchigvintsev, A., Tran, H., Popovic, A., Kovacic, F., Brown, G., Flick, R., Hajighasemi, M., Egorova, O., Somody, J. C., Tchigvintsev, D., & Khusnutdinova, A. (2015). The environment shapes microbial enzymes: Five cold-active and salt-resistant carboxylesterases from marine metagenomes. *Applied Microbiology and Biotechnology, 99*, 2165–2178.

Tiso, T., Winter, B., Wei, R., Hee, J., de Witt, J., Wierckx, N., Quicker, P., Bornscheuer, U. T., Bardow, A., Nogales, J., & Blank, L. M. (2022). The metabolic potential of plastics as biotechnological carbon sources–Review and targets for the future. *Metabolic Engineering, 71*, 77–98.

Tournier, V., Topham, C. M., Gilles, A., David, B., Folgoas, C., Moya-Leclair, E., Kamionka, E., Desrousseaux, M. L., Texier, H., Gavalda, S., & Cot, M. (2020). An engineered PET depolymerase to break down and recycle plastic bottles. *Nature, 580*(7802), 216–219.

Tournier, V., Duquesne, S., Guillamot, F., Cramail, H., Taton, D., Marty, A., & André, I. (2023). Enzymes' power for plastics degradation. *Chemical Reviews, 123*(9), 5612–5701.

Urbance, J. W., Cole, J., Saxman, P., & Tiedje, J. M. (2003). BSD: The biodegradative strain database. *Nucleic Acids Research, 31*(1), 152–155.

Urbanek, A. K., Kosiorowska, K. E., & Mirończuk, A. M. (2021). Current knowledge on polyethylene terephthalate degradation by genetically modified microorganisms. *Frontiers in Bioengineering and Biotechnology, 9*, 771133.

Viljakainen, V. R., & Hug, L. A. (2021). New approaches for the characterization of plastic-associated microbial communities and the discovery of plastic-degrading microorganisms and enzymes. *Computational and Structural Biotechnology Journal, 19*, 6191–6200.

Wang, L., Wu, W. M., Bolan, N. S., Tsang, D. C., Li, Y., Qin, M., & Hou, D. (2021). Environmental fate, toxicity and risk management strategies of nanoplastics in the environment: Current status and future perspectives. *Journal of Hazardous Materials, 401*, 123415.

Ward, P. G., de Roo, G., & O'Connor, K. E. (2005). Accumulation of polyhydroxyalkanoate from styrene and phenylacetic acid by Pseudomonas putida CA-3. *Applied and Environmental Microbiology, 71*(4), 2046–2052.

Wei, R., & Zimmermann, W. (2017). Microbial enzymes for the recycling of recalcitrant petroleum-based plastics: How far are we? *Microbial Biotechnology, 10*(6), 1308–1322.

Wei, R., Breite, D., Song, C., Gräsing, D., Ploss, T., Hille, P., Schwerdtfeger, R., Matysik, J., Schulze, A., & Zimmermann, W. (2019). Biocatalytic degradation efficiency of postconsumer polyethylene terephthalate packaging determined by their polymer microstructures. *Advanced Science, 6*(14), 1900491.

Wei, R., Tiso, T., Bertling, J., O'Connor, K., Blank, L. M., & Bornscheuer, U. T. (2020). Possibilities and limitations of biotechnological plastic degradation and recycling. *Nature Catalysis, 3*(11), 867–871.

Yan, F., Wei, R., Cui, Q., Bornscheuer, U. T., & Liu, Y. J. (2021). Thermophilic whole-cell degradation of polyethylene terephthalate using engineered Clostridium thermocellum. *Microbial Biotechnology, 14*(2), 374–385.

Yoshida, S., Hiraga, K., Takehana, T., Taniguchi, I., Yamaji, H., Maeda, Y., Toyohara, K., Miyamoto, K., Kimura, Y., & Oda, K. (2016). A bacterium that degrades and assimilates poly (ethylene terephthalate). *Science, 351*(6278), 1196–1199.

Yuzbasheva, E. Y., Yuzbashev, T. V., Perkovskaya, N. I., Mostova, E. B., Vybornaya, T. V., Sukhozhenko, A. V., Toropygin, I. Y., & Sineoky, S. P. (2015). Cell surface display of Yarrowia lipolytica lipase Lip2p using the cell wall protein YlPir1p, its characterization, and application as a whole-cell biocatalyst. *Applied Biochemistry and Biotechnology, 175*, 3888–3900.

Zhang, Y., Wusiman, A., Liu, X., Wan, C., Lee, D. J., & Tay, J. (2018). Polyhydroxyalkanoates (PHA) production from phenol in an acclimated consortium: Batch study and impacts of operational conditions. *Journal of Biotechnology, 267*, 36–44.

Zhang, J., Gao, D., Li, Q., Zhao, Y., Li, L., Lin, H., Bi, Q., & Zhao, Y. (2020). Biodegradation of polyethylene microplastic particles by the fungus Aspergillus flavus from the guts of wax moth Galleria mellonella. *Science of the Total Environment, 704*, 135931.

Zhang, N., Ding, M., & Yuan, Y. (2022). Current advances in biodegradation of polyolefins. *Microorganisms, 10*(8), 1537.

Zhu, B., Chen, Y., & Wei, N. (2019). Engineering biocatalytic and biosorptive materials for environmental applications. *Trends in Biotechnology, 37*(6), 661–676.

Zhu, B., Wang, D., & Wei, N. (2022). Enzyme discovery and engineering for sustainable plastic recycling. *Trends in Biotechnology, 40*(1), 22–37.

Microbial Nanobioremediation of Micro-Nanoplastics: Current Strategies, Challenges, and Future Prospects

Jyothirmayee Kola Pratap and Kannabiran Krishnan

1 Introduction

Man-made plastics are widely used by humans in daily life due to their durability and affordable cost. Plastics and their products are extensively utilized in several industries for multiple applications, and because of their extensive use, they become an environmental menace globally (Cheng et al., 2021). Asia is the largest user and producer of plastics. Tiwari et al. (2020) reported that China contributes the most (30%) around the globe and followed by Europe (17%), Canada, Mexico, and the United States (18%). The improper usage of waste management techniques is the reason for the increase in MNPs contamination (Geyer et al., 2017). Around 12.8 million tonnes of plastics are deposited every year in the oceans, rivers, lakes, etc., without any proper waste management treatment, which leads to an increase in plastic pollution globally (Lebreton et al., 2017). Water sources in Asia are polluted with high levels of MNPs. In China, the Yangtze River is the most contaminated river, which generates around 310,000 tonnes of plastic trash, contributing to greater than 4% of plastic pollution that occurs every year. The Ganges in India is one of the most polluted rivers, which carry around 115,000 tonnes of plastic trash. The most used plastic materials are polyvinyl chloride (PVC), polyethylene terephthalate (PET), Polyethylene (PE), Polypropylene (PP), and polystyrene (PS). The usage of plastic on a large scale leads to the accumulation of huge amounts of plastic waste, which becomes a major environmental pollutant affecting the ecosystem (agricultural, freshwater, and marine ecosystems) that cannot be degraded for decades (Ricardo et al., 2021). According to Brahney et al. (2020), by 2025,

J. K. Pratap · K. Krishnan (✉)
Department of Biomedical Sciences, School of Biosciences & Technology, Vellore Institute of Technology, Vellore, Tamil Nadu, India
e-mail: kkb@vit.ac.in

S. A. Bhat et al. (eds.), *Management of Micro and Nano-plastics in Soil and Biosolids*,
https://doi.org/10.1007/978-3-031-51967-3_17

around 11 billion tonnes of plastic pollutants will be accumulated in the environment. Macroplastics (MPs) are converted into MNPs by natural climatic conditions. MPs are less than 5 mm in diameter, and nanoplastics (NPs) are smaller than 100 nm in size. When compared to MPs, NPs are more dangerous to the living species because of their size (Wu et al., 2021). NPs are difficult to identify, quantify, and remove from the source and are capable of releasing organic-inorganic toxins (Bradney et al., 2019). A huge population of microbes such as algae, fungi, and bacteria are involved in the plastic degradation process and adapts or responds to biotic and abiotic conditions. In order to overcome environmental stress, microbes respond in several ways, like increasing or lowering growth, metabolism, and production of novel microbial bioproducts.

This review focuses on microbial nanobioremediation of MNPs using advanced technologies such as membrane technology, enzymatic technology, molecular technology, and nanotechnology using nanoparticles.

2 Biodegradation

Biodegradation refers to the disintegration of environmental pollutants by microbes such as bacteria, fungi and viruses (Sarkar et al., 2021). Bio-related strategies for plastic degradation are gaining momentum due to their environmentally friendly nature (Talvitie et al., 2017). Genetic engineering techniques exploit the plastic degrading potential of microbes with the help of enzymatic hydrolysis. *Ideonella sakaiensis*, a novel bacterium, has the ability to degrade PET into ethylene glycol and terephthalic acid, both harmless to the environment (Yoshida et al., 2016). Microplastic biodegradation has four steps: (1) formation of biofilm – decreases plastic hydrophobicity and buoyancy; (2) biodeterioration – catalysed by exopolysaccharides, exo-enzymes and endo-enzymes; (3) biofragmentation – destabilization of the carbon skeleton by enzymatic depolymerization with enzymes, oxidases, amidases, peroxidases, and laccases; and (4) Mineralization – it is the final and rate-determining step, where complete degradation of polymer occurs with the release of H_2O, CO_2, and CH_4 as final products. In unfavourable environmental conditions, multiple biotic stress factors affect the rate of biodegradation of plastics (Rai et al., 2021). Microplastic biodegradation is influenced by many physicochemical factors, including high MW (molecular weight), hydrophobicity, recalcitrant nature, 3D structure, and chemical complexity of additives or plasticizers (Hadad et al., 2005). Environmental factors, including many abiotic factors like temperature, pH, UV, and salinity, affect the hydrolysis rate by exerting control over catalytic site bioavailability (Ho et al., 1999). Many plastic biodegradation studies reveal that bacterial phyla, including *Firmicutes, Actinobacteria,* and *Proteobacteria*, originated from landfills, have the capability to utilize the plastics (Matjašič et al., 2021). Strains of *Bacillus* isolated from mangrove environments showed the ability to degrade polypropylene in 40 days. A marine fungus, *Z.*

maritimum, can utilize polyethylene as a substrate for growth and demonstrated a significant decrease in PE particle proportion in 28 days. Degradation of the plastic surface with green algae, blue-green algae, *Scenedesmus dimorphus,* diatoms, and *Navicula pupula* has been reported. The rate of PE microplastic decomposition was increased with a hybrid bacterial culture comprised primarily of landfill-isolated *Bacillus* sp. and *Paenibacillus* sp. (Peng et al., 2022).

2.1 Chemical Degradation Techniques

Chemical removal and degradation are chemical methods used for managing microplastic pollution in water. In chemical removal techniques, microplastics are agglomerated or deposited, which is then removed from treated water. These techniques alter the interparticle interactions and surface chemistry, thus leading to efficient removal. Chemical degradation methods involve MP decomposition or degradation. These methods alter the physicochemical properties of MPs (Hashmi et al., 2022). Many factors, such as chemical structure, type, size, and process conditions, significantly affect changes in microplastics. The coagulation process is based on charge, which is driven by electrostatic interactions between the MPs. The addition of chemical coagulants causes the repulsive forces to reduce, which in turn destabilizes MPs and forms an appropriate aqueous medium needed for effective removal (Bhandari & Ranade, 2014). Inorganic coagulants such as polyaluminum chloride and alum, and organic coagulants such as polyacrylamide and poly-diallyl dimethylammonium chloride are widely used (Golmohammadi et al., 2022). Nearly 54.7% removal of PET with activated silicic acid as a coagulant and 91.45% removal with anionic polyacrylamide was reported (Zhang et al., 2021). The hydrolysis method is inexpensive and produces products that are environmentally friendly during degradation. In general, hydrolysis is performed in three forms: acid hydrolysis with concentrated sulfuric acid, neutral hydrolysis with steam or water, and alkaline hydrolysis with sodium hydroxide (Golmohammadi et al., 2022). Products of different types of hydrolysis depend on the targeted MPs. Temperature has a major role in hydrolysis as it affects the degradation rate (Zhang et al., 2021). Advanced oxidation processes cause the mineralization of MPs by employing reactive oxygen species (ROS). Key factors that control the efficiency are oxidant dosage, temperature, plastic concentration, and catalyst (Rout et al., 2022). ROS have a high potential for degradation, and hydroxyl radicals, peroxide, and superoxide anions are produced through multiple routes such as hydrothermal carbonization, Fenton, heat-activated persulfate treatments, chlorination, ozonation, and UV-induction (Golmohammadi et al., 2022). Multiple studies suggest that Fenton treatment can cause effective decomposition of plastic wastes and results in useful products (Du et al., 2021).

3 Bioremediation (BR)

BR has developed as an effective method for reducing the harmful consequences of environmental pollution in several ways. The United States Environmental Protection Agency (USEPA) stated that waste management techniques depend upon microorganisms, which use their biological activity to interact with the environment and factors like sunlight, air, water, oxygen, and carbon dioxide, etc., for changing the physical and chemical state of toxic waste by breaking down into non-toxic or less toxic compounds such as H_2O and CO_2 (Sharma et al., 2017).

3.1 Nanobioremediation (NBR)

In NBR, both BR and nanotechnology (NT) approaches are explored for the effective removal of heavy metals, hydrocarbons, MNPs, etc. There are no consistent reports to prove the efficiency of this method; there have been several studies that have incorporated this approach for the effective removal/ degradation of contaminants. Although this looks like a promising strategy for the removal of MNPs since BR uses organic and living organisms, several parameters must be considered before combining the two strategies. It is known that certain NPs are toxic to microbial growth while some are not (Vazquez-Núnez et al., 2020). The nanomaterials (NMs)/ nanoparticles interaction with biota is dependent on the material used, nanoparticle size, surface coating, the organism, the contaminant type, pH, temperature, etc. (Tan et al., 2018).

4 Degradation of MNPs by Microrganisms

The degradation of MNPs using microbes is an environmentally friendly approach. Plastic pollution can be controlled by microbial degradation procedures because it is completely based on biotic and abiotic factors like pH, temperature, sunlight, oxygen, etc. (Zhou et al., 2022). MNPs can be used as carbon sources by microbes, and the degradation can be measured using various omics approaches (Tiwari et al., 2020; Knott et al., 2020). Microbial biodegradation of MNPs involves the following four stages: biodeterioration, biofragmentation, assimilation, and mineralization (Zhou et al., 2022; Dhanasekar & Krishnan, 2023). Biodeterioration is the first phase of the degradation process during which the physical (size, shape) and chemical properties of plastics are altered by microorganisms. The microbes attach to the MNPs and form a biofilm, changing the pH by creating an acidic environment that causes external damage. The high MW of plastics can be degraded to lower MW that enters the cell through the cell membrane and gets absorbed and degraded within the cell with the help of enzymes resulting in biofragmentation.

The broken fragments are consumed as a carbon source for their growth and development by the microorganisms. The MNP particles are absorbed through the membrane and transported to the cytoplasm of the cell. In the final step of mineralization, the MNPs are oxidized, resulting in H_2O and CO_2 as a result of degradation (Dhanasekarn and Krishnan, 2023; Zhou et al., 2022).

4.1 Bacterial Bioremediation of MNPs

According to Matjašič et al. (2021), hundreds of bacterial strains are involved in the biodegradation of plastics. Among these, 48% are proteobacteria, 37.4% are firmicutes, and 9.8% are actinobacteria. The plastic-degrading bacteria use MPs, NPs, organic compounds, etc., as a carbon and energy source for their growth and development. Plastic-degrading bacteria reported recently are given in Table 1. PE films were degraded by *Rhodococcus ruber* strain C208 after incubation for 56 days under aerobic conditions and showed a 7.5% gravimetric weight reduction (Sivan et al., 2006). PP was degraded after 90 days at 37 °C incubation under composting conditions by *Stenotrophomonas panacihumi* PA3-2 (Jeon & Kim, 2016). PVC films were degraded by *Bacillus flexus* DSM 1320 and *Pseudomonas citronellolis* DSM 50332 between 30 and 90 days of incubation at 37 °C in an orbital shaker and observed a reduction of gravimetric weight between 13.7% and 18.58% after 40 days (Giacomucci et al., 2019). *Exiguobacterium* sp. strain YT2 degrades styrofoam PS films within 28 days with 11% weight reduction (Yang et al., 2015). The mesophilic strain, *Stenotrophomonas panacihumi*, takes 90 days to degrade 12.7–20.3% of the high molecular weight of MPs (3600 Da) and low molecular weight of PP (2800 Da) (Ru et al., 2020). Microbial consortia of two *Pseudomonas* sp. strains degrade 28% and 7% w/w of PP (Judy et al., 2019). The consortia of *Rhizopus delemar, R. arrhizus; P. aeruginosa, Penicillium simplicissimum,* and *Pseudomonas fluorescence* capable of hydrolysing the polymers like PEA (polyethyleneadipate) and PCL (polycaprolactone) (Rana 2019). The co-culture *Bacillus amyloliquefaciens* and *A. clavatus* degrade LDPE in 90 days by altering pH; the degradation efficiency of the polymer can be measured by weight reduction and evaluation of CO_2 (Pathak & Navneet, 2017). LDPE was degraded by many actinobacterial species like *Rhodococcus ruber* C208, *Streptomyces badius; Streptomyces setnii, Streptomyces viridosporus, Arthrobacter oxydans, Arthrobacter globiformis Microbacterium paaraoxydans, Cellulosimicrobium flunkei, Arthrobacter paraffineus, Kocuria palustris* (Dhanasekarn and Krishnan, 2023). *Pseudomonas* sp. degrades low molecular weight MPs after 40 days of incubation (Wilkes and Aristilde 2017). *Streptomyces* sp. (thermophilic) degrades poly (D-3-hydroxybutyrate) (P3HB) at 50 °C (Pathak & Navneet, 2017).

Table 1 Bacterial degradation of micro-nanoplastics

Microbes	Type of plastic	Percentage of degradation	References
Bacillus flexus DSM 1320 and *Pseudomonas citronellolis* DSM 50332	PVC	13.7–18.58% weight reduction	Giacomucci et al. (2019)
Pseudomonas spp.	PPs and MPs	28% and 7% w/w	Judy et al. (2019)
P. aeruginosa, *P. fluorescence*, *R. delemar*, *P. simplicissimum*, *R. arrhizus*, *C. cylindracea*, and *Achromobacter* sp.	Microplastics	PEA and PCL hydrolysis	Rana (2019)
Geobacillus and *Thermus*, *Bacillus*	MNPs	43.7%	Chen et al. (2019)
Vibrio, *Campylobacter*, and *Arcobacter*	Microplastics	Breakage of MPs	Kelly et al. (2021)
Rhodococcus ruber C208, *Mycobacterium* sp., *Aquihabitans* sp., and *Paenibacillus urinalis*	PS	–	Matjašič et al. (2021), Raddadi and Fava (2019)
Rhodococcus equi, *A. globiformis* SBI-5, and *Corynebacterium* sp. AF14	PUR	–	Iram et al. (2019), Matjašič et al. (2021)
Arthrobacter sp., *Micrococcus* sp., *Arthrobacter* sp. GMB5, and *Leucobacter* sp.	HDPE	–	Jaiswal et al. (2020), Matjašič et al. (2021)
Rhodococcus ruber C208, *Streptomyces badius*, *Streptomyces setnii*, *Streptomyces viridosporus*, *Arthrobacter oxydans*, *Arthrobacter globiformis*, *Microbacterium paaraoxydans*, *Cellulosimicrobium flunkei*, *Arthrobacter paraffineus*, and *Kocuria palustris*	LDPE	–	Dhanasekar and Krishnan (2023)
Micrococcus luteus, *Brevibacterium* sp., and *Streptomyces albogriseolus* LBX-2	PE	–	Dhanasekarn and Krishnan (2023)

PP polypropylene; *PS* polystyrene; *PEA* poly ethylene adipate; *PCL* poly caprolactone; *PHB* poly D-3-hydroxybutyrate; *HDPE* high density polyethylene; *LDPE* low density polyethylene; *PUR* polyurathane; *PVC* polyvinayl chloride

4.2 Fungal Bioremediation of MNPs

MNPs are used by fungi as a carbon and energy source. Fungi can use non-substratum-specific enzymes to detoxify contaminants and infiltrate substrates. Fungal mycelium binds to the hydrophobic materials present in plastics and even penetrates the three-dimensional surfaces (Peng et al., 2022). PE was degraded by *Penicillium, Aspergillus,* and *Fusarium* (Ghosh et al., 2019). The enzyme

LC-cutinase from *Saccharomonospora viridis; Thermobifida fusca cutinase* hydrolyses 8.4% of PET packing film at 50 °C causing a weight loss of up to 50% over 7 days (Ru et al., 2020). After 16 weeks of incubation, 0.16% and 0.04% of low and high molecular weight PS film were degraded by *Penicillium variabile* CCF3219 (Tian et al., 2017). *Penicillium oxalicum* and *Penicillium chrysogenum* (Ascomycete) cause morphological damage on LDPE and HDPE sheets after 60 days of incubation (Ojha et al., 2017). *Monascus ruber, Monascus sanguineus* forms a complex between the PPU (Polyester polyurethane) and hyphae after 14 days (El-Morsy et al., 2017). *Gloeophyllum trabeum* (Ascomycete) degrades 50% by weight of polystyrene sulfonate (microplastic) after 20 days (Krueger et al., 2015). *Xepiculopsis graminea, Penicillium griseofulvum,* and *Cladosporium cladosporioides* break down the polyester polyurethane (PPU) into small fragments by an enzymatic degradation process and the cultured plates show clear zones around them between 6 and 14 days (Brunner et al., 2018). Some of the fungal microorganisms that degrade micro-nanoplastics (MNPs) are listed in the given Table 2.

4.3 Algal Bioremediation of MNPs

Identifying the algal species and their toxins, which could be used for the breaking down of polymeric substances, is one of the strategies to remove white pollution. Plastics can be utilized as a carbon source for their growth and development. Algae migrate on the surface of plastics in sewage water by producing ligninolytic enzymes and exopolysaccharide enzymes to initiate the degradation process (Sarmah & Rout, 2018). Algae involved in MNPs degradation are listed in Table 3. LDPE was degraded by many microalgae, including *Cyanobacteria* (blue-green algae), *Scenedesmus dimorphus*, *Navicula pupula,* and *Anabaena spiroides* (Kumar et al., 2017; Chia et al., 2020). *Oscillatoria subbrevis* and *Phormidium lucidum* degrade PE and LDPE without any addition or pretreatment with additives (Sarmah & Rout, 2018). *Chlamydomonas reinhardtii* is capable of producing the plastic-degrading enzyme PETase, which forms holes on the surface of PET for complete degradation (Kim et al., 2020). Microalgae have been extensively used for the removal of MNPs (Abomohra & Hanelt, 2022).

5 Bioremediation Using Various Other Technologies

The conventional method of MNP degradation produces highly toxic compounds, posing environmental and health risks. Bioremediation is the best alternative method for degrading micro-nanoplastics by using various technologies.

Table 2 Fungal degradation of micro-nanoplastics

Microorganism	Type of plastic	Degradation (%)	References
Gloeophyllum trabeum (Ascomycete)	PS Sulfonate MPs	Degradation and 50% weight reduction	Krueger et al. (2015)
P. chrysogenum (Ascomycete) and *P. oxalicum*	LDPE and HDPE sheets	Causes damage to PE sheets	Ojha et al. (2017)
M. sanguineus, *M. ruber*	PPU	Degradation after 14 days	El-Morsy et al. (2017)
P. variabile CCF3219	C-PS polymer film	0.04–0.16% degradation	Tian et al. (2017)
Zalerion maritium ATTC 34329 (marine fungus)	LLDPE MPs	Degradation after 28 days at 25 °C	Paço et al. (2017)
Cladosporium cladosporioides, *Penicillium griseofulvum*, *and Xepiculopsis graminea*	PPU	Degradation after 6–14 days of growth	Brunner et al. (2018)
Aspergillus terreus	PE	Weight and tensile strength reduction after 60 days	Sangale et al. (2019)
Chaetomium globosum	PVC	Degradation after 28 days	Vivi et al. (2019)
Penicillium, Aspergillus, and Fusarium	PE	Degradation of MPs	Ghosh et al. (2019)
Saccharomonospora viridis, *Thermobifida fusca cutinas*	MPs, PET	8.4% at 50 °C and 50% weight loss; 7–8.4% of PET hydrolysed at 63 °C	Ru et al. (2020)
Microbial consortium: *Bacillus subtilis*, *Sarcina aurantiaca*, *Aspergillus niger*, and *Aspergillus flavus*	PET	Weight reduction 28.78	Maheswaran et al. (2023)
Aspergillus niger	PS, PMMA	Weight reduction	Wang et al. (2023)
Fusarium spp., *Lecanicillium* spp.	PE and PP	Weight reduction	Wróbel et al. (2023)

PS polystyrene; *HDPE* high density polyethylene; *LDPE* low density polyethylene; *PPU* polyester polyurethane; *PET* polyethylene terephthalate; *LLDPE* linear low-density polyethylene; *PMMA* poly (methyl methacrylate); *PVC* polyvinyl chloride; *PE* polyethylene; *PP* polypropylene

5.1 Nanoremediation

MNPs are extensively generated from various sources of macroplastics due to climatic conditions. The continuous increase of MNPs in the ecosystem, as well as the toxicological impact on many species of animals, warns us to limit plastic manufacturing and consumption. MNPs in the environment are eliminated by making them eco-friendly plastics that degrade naturally. Nanoparticles and nanomaterials can be explored for cleaning up MNPs in the environment. Nanotechnology is used for making inorganic nanosensors to trace abiotic particles, including NPs (Mitrano et al., 2019). Nanoparticles can be added to microbial culture for plastic

Table 3 Algal degradation of micro-nanoplastics

Microalgae	Type of plastic	Degradation capability	References
Chlorella fusca var. vacuolate (Green algae)	BPA	Degraded without any estrogenic activity	Hirooka et al. (2005)
Chlamydomonas reinhardtii	PP	18%	Lagarde et al. (2016)
Chlorella vulgaris (microalgae) and *Aeromonas hydrophilia* (bacteria)	BPA	Degrade after 7 days of incubation	Gulnaz and Dincer (2009)
Cyanobacteria Anabaena spiroides	LDPE	8.18% weight reduction	Kumar et al. (2017)
Navicula pupula	LDPE	4.44% weight loss	Kumar et al. (2017)
Oscillatoria subbrevis and Phormidium lucidum	LDPE	Degrade without additives	Sarmah and Rout (2018)
Scenedesmus dimorphus	LDPE	3.74% weight reduction	Chia et al. (2020)
Oscillatoria subbrevis, Phormidium lucidum, Lyngbya diguetii, Cylindrospermum muscicola, and *Nostoc carneum*	WPBs	–	Sarmah and Rout (2018b)
Oscillatoria limentica, P. calcicole, Lyngbya cinerascens, O. earlei, Spirulina major, N. linckia, Hydrocoleum sp., *Pithophora* sp., and *Chlorella* sp.	PCBs	–	Sarmah and Rout (2018a)
Calothrix fusca, Scenedesmus quadricauda, Stigeoclonium tenue, Navicula minuta, Nitzschia sp., *N. intermedia, N. dicephala, Spirogyra* sp., and *Synedra tabulata*	PCBs	–	Sarmah and Rout (2018a)
Chlamydomonas reinhardtii (Genetically modified microorganism)	PET	Degrades completely	Kim et al. (2020)
Chlorella sp.	LDPE	Weight reduction	Govindan et al. (2021)

LDPE low density polyethylene; *BPA* bisphenol A; *PET* polyethylene terephthalate; *WPB* waste polythene bags; *PCB* polythene carry bags

degradation (Tiwari et al., 2020). Immobilized microbial nanoparticles can be used for the biosorption of NPs. The SiO_2 (silicon dioxide) nanoparticles with different concentrations influence the growth of bacteria that are capable of degrading plastics (Pathak & Navneet, 2017). SiO_2 reduces the lag phase for better growth and increases the rate of plastic degradation by bacteria drastically through nano-plastic interaction. A high concentration of fullerene 60 nanoparticles is considered harmful, though a low concentration promotes plastic degradation by promoting bacterial growth and LDPE degradation by breaking the hydrocarbon chain (Sah et al., 2010). Microbial NR is used as an effective tool to eliminate emerging pollutants macro, micro-nanoplastics.

5.2 *Membrane Technology*

The high concentration of MPs present in polymeric debris is recovered efficiently by membrane technology. PET, PP, and PE are some of the major MPs. Wastewater treatment plants are the primary source of MNPs contamination in the environment (Tiwari et al., 2020). Conventional methods such as activated charcoal, coagulation, precipitation, flocculation, sedimentation, and filtration are adopted for the elimination of MNPs from fresh and drinking water (Pivokonsky et al., 2018). The disadvantages of traditional methods for the removal of MNPs are chemical sludge formation in large quantities, more turbidity, and low disposal potential (Zinicovscaia, 2016). Different filtration techniques, nanofiltration, reverse osmosis, and ultrafiltration, are employed for the purification of wastewater using membrane technology. The membrane ultrafiltration technology was established to eliminate wastes completely from wastewater (Wu et al., 2019). The main priority is treating water free from MNPs, and unfortunately, MPs and membranes have similar porosities. Membranes with high stability and increased selectivity can eliminate impurities made of MPs and their by-products (Parks et al., 2017). Membrane bioreactor technology is the advanced method in membrane technology that shows better results in the removal of contaminants than traditional methods. The development of membrane technology contributed to the development of a membrane bioreactor technology (MBT), which is more efficient at removing pollutants (Lares et al., 2018). The major advantage of MBT is the low rate of sludge production, low energy consumption, and high removal rate of impurities (Zinicovscaia, 2016).

5.3 *Enzymatic Degradation*

Biodegradation of various polymers is recognized as an effective, safe, and environmentally friendly approach to promote MNPs waste reduction. Different microbial proteins, genes, enzymes, and their roles in many metabolic pathways have led to the degradation of plastic polymers (Yuan et al., 2020; Priya et al., 2021). Esterases, lipases, carboxylesterases, cutinases, and other enzymes produced by microbes either modify or break down a wide range of plastic components (Zhang et al., 2020). Monooxygenase enzymes have the responsibility for hydroxylation, dealkylation, deamination, dehalogenation, sulphoxidation, N-oxide reduction, and desulfuration methods and play a key role in aliphatic, aromatic, and alicyclic hydrocarbon degradation (Shin et al., 2018). Hydrolytic enzymes such as hydrolases, class-II peroxidases, lignin peroxidase, peroxidases, manganese peroxidase, laccases, and peroxygenases can be used in plastic biodegradation (Dey et al., 2023).

Hydrolase, an extracellular enzyme released by bacteria, can transform complex molecules into polymers and convert them to hydroxylic acid monomers (Lear et al., 2021). Hydrolase, aminotransferase, and dehydrogenase degrade polyamide produced by *Bacillus, Vibrio, Anoxybacillus, Pseudomonas, Fusarium, Caulobacter,* and *Aspergillus* degrade polyurethane by lipase, cutinase, hydrolase, and esterase (Zhou et al., 2022). *Corynebacterium, Aspergillus*, and *Pseudomonas* produce hydrolase, oxygenase, styrene, and oxide isomerase which degrade PS potentially. *Ideonella, Thermobifida*, and *Humicola* degrade PET by hydrolase, lipase, and depolymerase. *Candida, Aspergillus*, and *Pseudomonas* release lipase, esterase, and tannase enzymes, which degrade polyvinyl chloride (PVC). *Penicillium simplicissimum, Phanerochaete chrysosporium,* and *Rhodococcus ruber* release lipase and manganese peroxidase that degrade PE (Yamada-Onodera et al., 2001). Nylon was degraded by *Agromyces* sp. by producing a nylon hydrolase enzyme (Negoro et al., 2012).

5.4 Molecular Degradation

Genetic modification techniques are effective tools for altering the properties of microbes, which increases their ability to biodegrade plastic waste materials (Gu, 2021). Identification of DNA sequences, novel enzymes, biochemical pathways, active compounds, and new metabolic functions has improved the degradation of polymers for biotechnological applications (de Alves et al., 2018). The strategies used for identifying novel DNA sequences and genes can be separated into screening methods based on sequences and functions (Schmeisser et al., 2007), with function-based screening methods able to reveal these distinctive characteristics of enzymes (Uchiyama & Miyazaki, 2009). The biological screening of metagenomics is investigated using LC cutinase (altered through protein engineering) activity on tributyrin agar plates. It is an excellent model for the investigation of pathways and also plays a critical role in surface alteration and PET degradation (Sulaiman et al., 2012; Tiwari et al., 2020).

The genetically modified enzyme cutinase has the capability of reducing the degradation period from 41.8 to 6.2 hours, better than naturally produced enzymes (Islam et al., 2019). The modified marine consortia have greater plastic biodegradation ability and are also used in the depolymerization of MNPs (Syranidou et al., 2019). Different omics technologies (genomics, transcriptomics, proteomics, metabolomics, etc.) have been applied to standardize and study the degradation process in many environmental pollutants (Kumar et al., 2019). Bioinformatics is employed to improve pollutant biodegradation, such as MNPs (Purohit et al., 2020). Various databases such as UM-BBD, MetaCyc, and BioCyc are linked to plastic biodegradation processes (Priya et al., 2021).

6 Plastic Polymers and the Ecosystem

MNPs are divided into two categories: primary MNPs, which include cosmetics, household products, and polymeric raw materials (nylon, PE, PP, PS, etc.), and secondary MNPs formed by extensive plastic waste fragmentation due to exposure to abiotic factors like UV radiation, temperature, atmospheric pressure, and microbial degradation. These are the major sources of MNPs. Polymer degradation is also influenced by stress and reactive molecules like ozone. Surprisingly, physical damage to plastic polymers can lead to additional MNPs production (Roex et al., 2013; Arthur et al., 2009). Due to their chemical composition, even at room temperature, PP, polybutadiene (PBD), and PVC can undergo oxidative degradation; hence, antioxidants and stabilizers are added to make them withstand deterioration. The strong chemical bonds and molecular chains weaken and undergo oxidative thermal degradation only on physical abrasion. The extensive usage of single-use plastic food packaging materials such as PE, PP, and PET leads to plastic contaminants (Brooks et al., 2018). MNPs used in the electronic, textile industry, paint industry, and automobile industries have the ability to be released directly into rivers and watersheds, adversely affecting the marine ecosystem (Kay et al., 2018; da Costa et al., 2016). Household items like scrubs, toothpaste, cosmetics, and cleaning liquids used in daily life can also contribute to MNPs contamination at the basal level.

According to Carr et al. (2016), tertiary treatment is not a source of MPs pollution because these pollutants have been eliminated by the settling and skimming process. The extensive usage of plastic materials like water bottles, disposal bags, pens, refills, caps, disposable tea cups, pharmaceuticals, etc. for packing, storing, and distributing leads to major sources of MNPs pollutants in drinking water (Krishnan, 2023). The other source of MNPs is the decomposition of synthetic fibres while washing clothes; items used during agricultural activities can release compounds like polyethylene, polyurethane, polystyrene foils, and films. It may be present in biochar, sulfur, vermiculite, hydro-absorbent polymers, business activities like carving of thermal devices from polystyrene, ship-induced waste in the sea, drug delivery, etc. (Liebezeit & Liebezeit, 2014; Zhang et al., 2012). PVC distribution pipes may be a source of MNPs in drinking water. These pipelines deteriorate and dissolve with time, becoming a source of MNPs (Krishnan, 2023). Secondary MNPs are rapidly released into various ecosystems like marine, terrestrial, and the atmosphere, resulting in massive plastic pollution all around the world (Lebreton et al., 2019). These secondary micro-nanoplastics affect public health either directly or indirectly, as shown in Fig. 1.

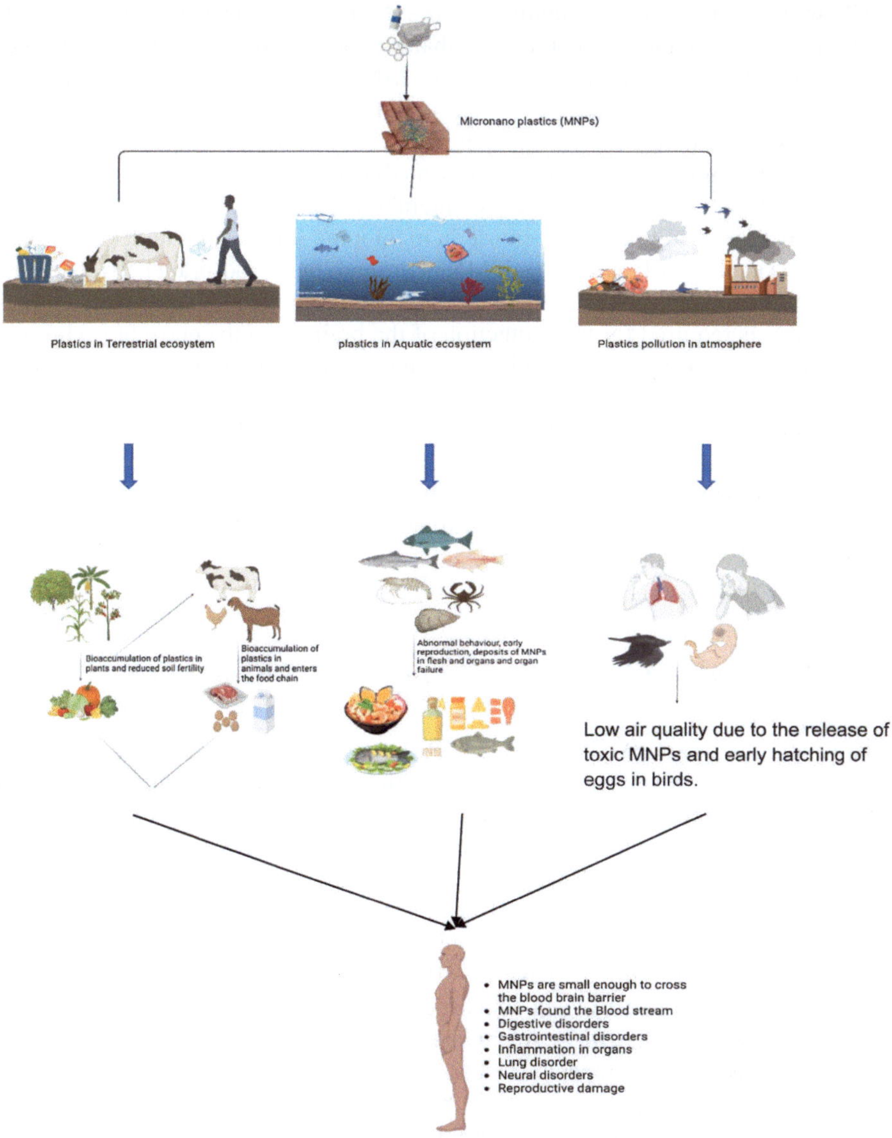

Fig. 1 Sources of micro-nanoplastics and their effect on the ecosystem

6.1 Effect of MNPs on Public Health

MNPs enter the human body via inhalation, direct skin contact, food chain, and ingestion. MNPs block the digestive tract and cause irritation and erosion in the mucosa (Rehse et al., 2016; Barnes et al., 2009). Inhalation is the major means of exposure of MNPs to humans. An individual ingests 26-130 airborne MNPs while walking through a plastic-contaminated area (Prata, 2018). The use of plastic medical equipment, including nebulizers, inhalers, therapeutic substances, and other respiratory aids, may lead to the release of MNPs entering the lungs through the inhalation process (Krishnan, 2023). The cellular uptake of MNPs and their toxicity is still being studied. PS nanoparticles can easily penetrate lipid membranes via phagocytosis, pinocytosis, and clathrin-mediated pathways, affecting cellular activity and disrupting the integrity of membrane proteins. PS affects the membrane structure, resulting in a decrease in membrane fluidity and molecular diffusion (Rossi et al., 2014). Crossing of MNPs through the blood-brain barrier and the presence of PS nanoplastics in the brain have been reported. A selectively permeable barrier is required for the protection of the brain from the entry of toxins and the maintenance of neuronal activity (Almutairi et al., 2016). According to Yadav et al. (2022), MNPs identified in food samples are PET, PS, polynorbornene, nylon, polychloroprene, and copolymer polyacrylamide. Exposure of MNPs to macrophages results in the production of more IL-6 and IL-8, leading to an enormous rise in the inflammatory response, thereby affecting the immunity. TNF and IL-1 mediated inflammatory responses after exposure to polyethylene MNPs have been reported by Lehner et al. (2020).

6.2 Effect of MNPs on Marine and Coastal Ecosystems

The accumulation of 39–500 nm of MNPs in the gills, testis, intestine, liver, and blood of the organism was observed, which decreases the survival rate of *Oryziaslatipes* (Japanese rice fish) (Manabe et al., 2011). In *chlorella* species and *Scenedermus* species, the exposure of PS particles affects photosynthesis and also induces ROS production (Bhattacharya et al., 2010). In *Crassostrea virginica* (Eastern oyster) and *Mytilus edulis* (Blue mussel), the accumulation of PS particles in the digestive tract was reported earlier (Ward & Kach, 2009). According to a recent study, LDPE particles have no effect on Zebra fish after three weeks of exposure compared to PVC and PS, which develop abnormalities in tissues and structure (Tiwari et al., 2020). Accidental intake of PE particles as food by fish leads to a decrease in rapacious acts and productivity in the common goby (*Pomatoschistus microps*) (de Sá et al., 2015). Susceptibility of microbes to MNPs can cause major consequences like proinflammatory responses, early death, slow growth and development, decreased energy, blocked feeding capacity, decreased immunity, behavioural distortion, etc., in many microorganisms (Ziajahromi et al., 2018).

6.3 Effect of MNPs on Agriculture

The effect of MNPs on agricultural soil has a huge influence on the environment and human well-being. According to Cao et al. (2017), MPs of PE and biodegradable plastic residues have been proven to have a negative impact on wheat vegetation and reproduction. Excessive amounts of leftover plastics in the soil can damage the texture, water-retaining capacity, nutrient transportation, minimizing crop production, and enhancing carbon/nitrogen metabolism (Steinmetz et al., 2016). MNPs' accumulation in terrestrial soil affects the diversity of soil organisms, leading to a severe ecological impact on the ecosystem (Ng et al., 2018). MNPs carried through the oral route exhibit severe toxicity on the gastrointestinal tract of organisms (Bouwmeester et al., 2015).

6.4 Effect of MNPs on the Atmosphere

MNPs affect agricultural diversity, aquatic ecosystems, and environmental health. MNPs in both indoor and outdoor air severely affect humans and their well-being. The most fascinating aspect is the link between microplastic strength, exposure, and breakdown mechanisms. Almost 48,000 outdoor MNPs are exposed each day through inhalation (Vianello et al., 2019). According to Dris et al. (2017), the concentration of synthetic plastics (fibres) in indoor air needs to be investigated further. These synthetic and plastic fibres degrade photo-oxidatively before being fragmented by wind shear (Krishnan, 2023). The release of MNPs in the air causes lung cancer, inflammatory disorders, gastrointestinal disorders, organ failure and abnormal behaviour in humans and animals, early ripening or early maturation of fruits in plants, reproductive disorders in birds, and early hatching of eggs in hens and other severe pathophysiological problems (Fig. 2).

7 Application of Nanobioremediation

In NBR, sorption plays a pivotal role and can be classified into two types, namely adsorption and absorption. During adsorption, the pollutant interacts with the NPs at the surface level, and the converse occurs during absorption, i.e. the pollutant penetrates deeply into the sorbent material. During these processes, the pollutants can be immobilized, sequestered, and concentrated at the end (Vieira & Volesky, 2000).

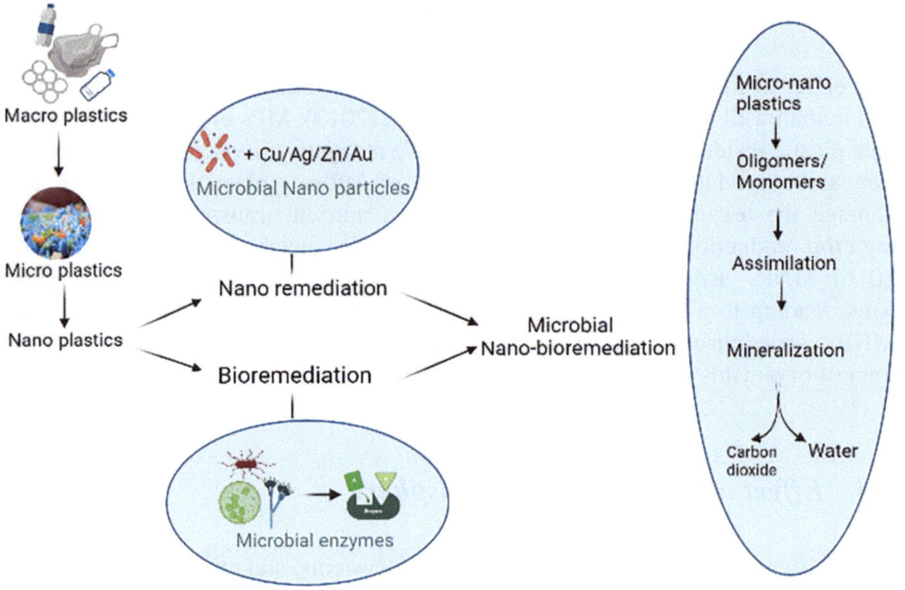

Fig. 2 Nanobioremediation of micro-nanoplastics

7.1 Adsorption-Based Processes

MPs from water are removed by adsorption-based processes. The nanoparticles are prepared with different functional groups for targeting specific pollutants in water. Also, they have a greater pore size and surface area, which are desirable features for sorption processes.

7.1.1 Metal-Organic Frameworks (MOFs)

Their unique attributes, such as increased surface area, easy customization, porous structures, and tenable architecture, have made them excellent candidates for sorption methods. In recent times, MOFs have been used in several studies for the effective removal of MNPs from water due to their efficiency. ZIF-based MOFs have also been widely used to remove both organic and inorganic contaminants from water through hydrogen bonding, π-π, and electrostatic interactions (Li et al., 2021). A MOF known as ZIF-67 was also reported to be used for the removal of PS MPs. Under optimum conditions, it could remove about 92.1% of PS MPs (Wan et al., 2022). MOFs are generally in powder form and need to be mixed with aerogels or foams. Zirconium MOF-foam possesses pores, high uniformity, and robustness. It was used in various simulated water and seawater filtration processes

(You et al., 2022). UiO-66-OH@MF-3 was capable of removing up to 95.5 ± 1.2% of MNPs and had high performance in recycling and large quantity filtration experiments (Chen et al., 2020).

7.1.2 Carbon-Based Nanoparticles

Carbon nanoparticles have been used widely in the removal of MPs from seawater. Graphenes, carbon nanotubes (CNTs), carbon powder, and fly ash have been exploited as carbon-based NPs. It has a high surface area, numerous active sites, hydrophobicity, and functionalization that can be tailored according to the pollutant's nature (Sajid et al., 2022). Graphene-based nanoparticles can remove MPs under varying pH conditions, due to their special hexagonal honeycomb structure, which gives them great stability under acidic and alkaline environments (Peng et al., 2022). Combining CNTs with magnetic nanoparticles can be used to collect the adsorbed MPs. Magnetic CNTs were also used to remove PET, PA, and PE MPs adsorbed (Tang et al., 2021). Fly ash amended with Fe ions was utilized as an adsorbent to remove MNPs from water. By using it for water treatment, its negative impacts can be reduced. Modified fly ash is extensively used for water treatment due to its large surface area, high affinity, and porous structure. CNTs magnetically altered fly ash can be easily collected and can be used many times (Zhao et al., 2022). A major problem in using carbon nanoparticles is their complete recovery from water and their toxicity.

7.1.3 Biobased Nanoparticles

Biobased nanomaterials help to combat the problem of toxicity associated with most adsorbents. Generally, adsorbents are not reusable, but biobased nanomaterials are cheap, non-toxic, and stable. Chitin sponge composites with graphene oxide or carbon nitride doped with oxygen were used for the adsorptive removal of charged MPs (Baig et al., 2019). MPs are often found trapped in coral reefs due to their structure. Coral-reef-inspired aerogels were synthesized and used for the removal of microplastics from water (Zheng et al., 2022). Biochar is another application of biobased nanoparticles; they are cheap, environment-friendly, have more surface area, and functionalization properties (Chen et al., 2022). Plant protein-derived sponges are another great candidate for bioremediation of MPs. This does not cause secondary pollution. In a recent study, chemical cross-linking was used to make a natural sponge from plant protein. It was biodegradable and showed excellent mechanical properties, capable of removing 81.2% of MPs (1 mg/L) at pH-9 in water systems (Wang et al., 2021). Even though there are a lot of options available for biobased NPs, only a few have been utilized and shown positive results. They are advantageous over other materials as they do not cause secondary pollution and are biodegradable.

7.1.4 Double-Layered Hydroxides and Double-Layered Oxides

They have negatively charged anions and positively charged metal hydroxide layers, and the inside is filled with water molecules to balance the overall charge. G@LDO showed ≥80% removal of MPs at pH 3–11, and 60% of PS at pH = 1 and 13. It maintained its removal efficiency even after 5 cycles of adsorption-desorption (Peng et al., 2022). Zn-Al LDHs were used for the elimination of NP debris from the water. Electrostatic interactions between Zn-Al-LDHs and nanoplastic debris ensured that they stuck to Zn-Al-LDHs (Tiwari et al., 2020). The only disadvantage associated with LDHs/LDOs is their instability in adsorption under harsh conditions. This can be overcome by compositing them with other materials (Saraji & Ghani, 2014).

7.1.5 Magnetic Nanomaterials

MPs can be efficiently magnetized by using Fe_3O_4 NPs. In a recent study, a magnet was used for removing magnetized MPs of various types and sizes in water and seawater. A fluid removal rate of 81.33 ± 16.7%, 82.28 ± 5.77%, and 80.56 ± 17.35% in river water, household sewage and seawater respectively was achieved (Shi et al., 2022). In another study, MPs were magnetically removed from environmental samples. MPs were removed with the help of hydrophobic Fe nanoparticles. About 92% of 10–20 μm PES and PS beads from seawater were removed effectively (Grbic et al., 2019).

8 Current Challenges and Future Aspects

MNPs represent a significant environmental contaminant that exhibits prolonged persistence within the ecosystem. The accumulation of large amounts MNPs leads to environmental pollution and affects the whole ecosystem. A lack of adequate disposal methods and recycling procedures results in unmanageable MNP pollution in the environment. Disposing of plastics into soil, water, and the environment is creating a health hazard both directly and indirectly (Krishnan, 2023). Nanoremediation and bioremediation of plastics and the combination of both, nanobioremediation, are more advantageous with respect to the big surface area to volume ratio of nanoparticles and specificity of microorganisms used for bioremediation to enhance reactivity. The synergistic effect of both would raise the effectiveness of the process for the efficient removal of pollutants.

As a possible alternative for in situ or ex situ bioremediation, genetic engineering, synthetic biology, microbial consortia, biomembrane, enzymatic, and nanoparticle technologies have been designed and utilized as new biotechnological approaches (Silva, 2021). Genome editing using genetic engineering tools to modify the enzyme (dehalogenase, polyethylene terephthalate hydrolase, laccase, depolymerase, and esterase) coding gene or incorporating genes coding for specific enzymes of interest for effective degradation of MNPs. Oxygenase of *Streptomyces albogriseolus* LBX-2 was genetically modified for PE (Shao et al., 2019).

Although numerous research studies have been carried out to decrease MNPs by bioremediation and thereby decrease their eco-toxicological effects, more research on MNP pollution, consequences, and possibilities for mitigation is still needed (Zhou et al., 2022). To avoid continuous remediation and removal activities, innovative remediation techniques must be facilitated with efficient mitigation approaches that concentrate on reducing the source of MNPs contamination. It will be possible only by enacting a suitable law that emphasizes controlled usage of plastics and improving waste management facilities (Syranidou et al., 2017). Furthermore, future studies should focus more on NPs, which have a higher penetrating capacity than MPs. Only a few fungal species have been identified that can degrade PVC (thermoplastic) (Zhou et al., 2022). Bacterial and enzyme-assisted PVC breakdown has not been reported. Most personal care products (PCPs) and cosmetics contain PP. Plenty of microbial groups are reported with efficient degradation of the PP polymers by using different degradation techniques, but there are no findings on enzymatic degradation and their mechanism. Nanotechnologies for pollutant removal are still limited to the laboratory, and industrialization of the process for application at the field level is still challenging for researchers.

Bioinformatics databases such as the enviPath (Environmental Contaminant Biotransformation Pathway Resource), the University of Minnesota Biocatalysis/Biodegradation Database (UM-BBD), MetaCyc database, and BioCyc database provide details on genes, metabolic pathways, and biocatalytic enzymes involved in the biodegradation of MNPs (Purohit et al., 2020; Caspi et al., 2020). The computational tools and databases help in identifying the specific enzymes catalysing the metabolic pathway and also predicting the biodegradation pathways and toxic chemicals produced (Ali et al., 2021). More validated experimental data are required to support the bioinformatics/systems biology approach for identifying a suitable organism and metabolic pathways for synthetic polymer degradation. All these approaches, including genetics, bioinformatics tools, systems biology, metabolic engineering, and molecular biology, would certainly provide a sustainable option for the effective biodegradation of MNPs.

9 Conclusion

Plastic has been the major pollutant of the world for decades. MNPs are leaching out into the air, soil, and water and also entering the food chain, thereby affecting the environment, ecosystem, and all living systems, including humans, causing severe health hazards. Recently, reports of MNPs being found in fish and human bloodstreams as an evidence. The conventional methods that deal with plastic pollution, such as chemical and physical methods, often lead to secondary pollution and are not eco-friendly. They often prove toxic to the flora and fauna and also to the microbial community. Bioremediation helps to detoxify toxic metabolites; at the same time, it is cheap, eco-friendly, and non-toxic. The only drawback is the time it takes for remediation. Hence, the focus has to be shifted to improving bioremediation to yield faster and better results; nanobioremediation combines both bioremediation and nanotechnology to remediate MNPs. Although several studies have been reported with this approach, there is no conclusive evidence of its efficiency. Hence extensive studies are needed to understand the nanobioremediation process and to improve further. It is also important to mitigate the usage of plastics and plastic-based products; on the other hand, the use of bioplastics should be encouraged. A global level of awareness, strict policies, and stringent punishment for violators have to be followed to combat the rising problem of MNPs.

References

Abomohra, A., & Hanelt, D. (2022). Recent advances in micro-/Nanoplastic (MNPs) removal by microalgae and possible integrated routes of energy recovery. *Microorganisms, 10*(12), 2400.

Ali, S. S., Elsamahy, T., Koutra, E., Kornaros, M., El-Sheekh, M., Abdelkarim, E. A., & Sun, J. (2021). Degradation of conventional plastic wastes in the environment: A review on current status of knowledge and future perspectives of disposal. *Science of the Total Environment, 771*, 144719.

Almutairi, M. M., Gong, C., Xu, Y. G., Chang, Y., & Shi, H. (2016). Factors controlling permeability of the blood–brain barrier. *Cellular and Molecular Life Sciences, 73*, 57–77.

Arthur, C., Baker, J., & Bamford, H. (2009). *Proceedings of the international research workshop on the occurrence, effects, and fate of microplastic marine debris*, Sept 9–11, 2008.

Baig, N., Sajid, M., & Saleh, T. A. (2019). Graphene-based adsorbents for the removal of toxic organic pollutants: A review. *Journal of Environmental Management, 244*, 370–382.

Barnes, D. K., Galgani, F., Thompson, R. C., & Barlaz, M. (2009). Accumulation and fragmentation of plastic debris in global environments. *Philosophical Transactions of the Royal Society B Biological Sciences, 364*(1526), 1985–1998.

Bhandari, V. M., & Ranade, V. V. (2014). Advanced physico-chemical methods of treatment for industrial wastewaters. In *Industrial wastewater treatment, recycling and reuse* (p. 81–140).

Bhattacharya, P., Lin, S., Turner, J. P., & Ke, P. C. (2010). Physical adsorption of charged plastic nanoparticles affect algal photosynthesis. *Journal of Physical Chemistry C, 114*(39), 16556–16561.

Bouwmeester, H., Hollman, P. C., & Peters, R. J. (2015). Potential health impact of environmentally released micro-and nano plastics in the human food production chain: experiences from nanotoxicology. *Environmental Science & Technology, 49*, 8932–8947.

Bradney, L., Wijesekara, H., Palansooriya, K. N., Obadamudalige, N., Bolan, N. S., Ok, Y. S., Rinklebe, J., Kim, K. H., & Kirkham, M. B. (2019). Particulate plastics as a vector for toxic trace-element uptake by aquatic and terrestrial organisms and human health risk. *Environment International, 131*, 104937.

Brahney, J., Hallerud, M., Heim, E., Hahnenberger, M., & Sukumaran, S. (2020). Plastic rain in protected areas of the United States. *Science, 368*(6496), 1257–1260.

Brooks, A. L., Wang, S., & Jambeck, J. R. (2018). The Chinese import ban and its impact on global plastic waste trade. *Science Advances, 4*(6), eaat0131.

Brunner, I., Fischer, M., Rüthi, J., Stierli, B., & Frey, B. (2018). Ability of fungi isolated from plastic debris floating in the shoreline of a lake to degrade plastics. *PLoS One, 13*(8), e0202047.

Cao, D., Wang, X., Luo, X., Liu, G., & Zheng, H. (2017). Effects of polystyrene microplastics on the fitness of earthworms in agricultural soil, in: IOP Conference Series: Earth and Environmental Science. *IOP Publishing, 61*(1), 012148.

Carr, S. A., Liu, J., & Tesoro, A. G. (2016). Transport and fate of microplastic particles in wastewater treatment plants. *Water Research, 91*, 174–182.

Caspi, R., Billington, R., Keseler, I. M., Kothari, A., Krummenacker, M., Midford, P. E., Ong, W. K., Paley, S., Subhraveti, P., & Karp, P. D. (2020). The MetaCyc database of metabolic pathways and enzymes-a 2019 update. *Nucleic Acids Research, 48*(D1), D445–DD45.

Chen, Z., Zhao, W. Q., Xing, R. Z., Xie, S. J., Yang, X. G., Cui, P., Lü, J., Liao, H. P., Yu, Z., Wang, S. H., & Zhou, S. G. (2019). Enhanced in situ biodegradation of micro-plastics in sewage sludge using hyper-thermophilic composting technology. *Journal of Hazardous Materials, 384*, 121271.

Chen, Y. J., Chen, Y., Miao, C., Wang, Y. R., Gao, G. K., Yang, R. X., Zhu, H. J., Wang, J. H., Li, S. L., & Lan, Y. Q. (2020). Metal–organic framework-based foams for efficient microplastics removal. *Journal of Materials Chemistry A, 8*(29), 14644–14652.

Chen, W. H., Hoang, A. T., Nižetić, S., Pandey, A., Cheng, C. K., Luque, R., Ong, H. C., Thomas, S., & Nguyen, X. P. (2022). Biomass-derived biochar: From production to application in removing heavy metal-contaminated water. *Process Safety and Environmental Protection, 160*, 704–733.

Cheng, Y. L., Kim, J. G., Kim, H. B., Choi, J. H., Tsang, Y. F., & Baek, K. (2021). Occurrence and removal of micro-plastics in wastewater treatment plants and drinking water purification facilities: A review. *Chemical Engineering Journal, 410*, 128381.

Chia, W. Y., Tang, D. Y. Y., Khoo, K. S., Lup, A. N. K., & Chew, K. W. (2020). Nature's fight against plastic pollution: Algae for plastic biodegradation and bioplastics production. *Environmental Science and Ecotechnology, 4*, 100065.

da Costa, J. P., Santos, P. S., Duarte, A. C., & Rocha-Santos, T. (2016). (Nano) plastics in the environment–sources, fates and effects. *Science of the Total Environment, 566*, 15–26.

de Alves, L. F., Westmann, C. A., Lovate, G. L., de Siqueira, G. M. V., Borelli, T. C., & Guazzaroni, M. E. (2018). Metagenomic approaches for understanding new concepts in microbial science. *International Journal of Genomics, 2018*, 2312987.

de Sá, L. C., Luís, L. G., & Guilhermino, L. (2015). Effects of microplastics on juveniles of the common goby (Pomatoschistus microps): Confusion with prey, reduction of the predatory performance and efficiency, and possible influence of developmental conditions. *Environmental Pollution, 196*, 359–362.

Dey, S., Anand, U., Kumar, V., Kumar, S., Ghorai, M., Ghosh, A., Kant, N., Suresh, S., Bhattacharya, S., Bontempi, E., & Bhat, S. A. (2023). Microbial strategies for degradation of microplastics generated from COVID-19 healthcare waste. *Environmental Research, 216*, 114438.

Dhanasekar, A., & Krishnan, K. (2023). Plastic associated environmental pollution: A systematic review on biodegradation methods, challenges, and future prospects. *Research Journal of Chemistry and Environment, 27*(2), 122–134.

Dris, R., Gasperi, J., Mirande, C., Mandin, C., Guerrouache, M., Langlois, V., & Tassin, B. (2017). A first overview of textile fibers, including microplastics, in indoor and outdoor environments. *Environmental Pollution, 221*, 453–458.

Du, H., Xie, Y., & Wang, J. (2021). Microplastic degradation methods and corresponding degradation mechanism: Research status and future perspectives. *Journal of Hazardous Materials, 418*, 126377.

El-Morsy, E. M., Hassan, H. M., & Ahmed, E. (2017). Biodegradative activities of fungal isolates from plastic contaminated soils. *Mycosphere, 8*(8), 1071–1087.

Geyer, R., Jambeck, J. R., & Law, K. L. (2017). Production, use, and fate of all plastics ever made. *Science Advances, 3*(7), e1700782.

Ghosh, S., Qureshi, A., & Purohit, H. J. (2019). Microbial degradation of plastics: Biofilm and degradation pathways. Contaminants in agriculture and environment: Health risks and remediation. *Agro and Environment, 1*, 184–199.

Giacomucci, L., Raddadi, N., Soccio, M., Lotti, N., & Fava, F. (2019). Polyvinyl chloride biodegradation by *Pseudomonas citronellolis* and *Bacillus flexus*. *New Biotechnology, 52*, 35–41.

Golmohammadi, M., Musavi, S. F., Habibi, M., Maleki, R., Golgoli, M., Zargar, M., Dumée, L. F., Baroutian, S., & Razmjou, A. (2022). Molecular mechanisms of microplastics degradation: A review. *Separation and Purification Technology, 309*, 22906.

Govindan, N., Bhuyar, P., Sundararaju, S., Feng, H. X., Rahim, M. H. A., Muniyasamy, S., & Maniam, G. P. (2021). Evaluation of microalgae's plastic biodeterioration property by a consortium of *Chlorella* sp. and *Cyanobacteria* sp. *Environmental Research, Engineering and Management, 77*(3), 86–98.

Grbic, J., Nguyen, B., Guo, E., You, J. B., Sinton, D., & Rochman, C. M. (2019). Magnetic extraction of microplastics from environmental samples. *Environmental Science & Technology Letters, 6*(2), 68–72.

Gu, J. D. (2021). Biodegradability of plastics: The issues, recent advances, and future perspectives. *Environmental Science and Pollution Research, 28*(2), 1278–1282.

Gulnaz, O., & Dincer, S. (2009). Biodegradation of bisphenol A by Chlorella vulgaris and Aeromonas hydrophilia. *Journal of Applied Biological Sciences, 3*(2), 79–84.

Hadad, D., Geresh, S., & Sivan, A. (2005). Biodegradation of polyethylene by the thermophilic bacterium Brevibacillus borstelensis. *Journal of Applied Microbiology, 98*(5), 1093–1100.

Hashmi, M. Z., Nasrullah, A., Basharat, H., & Ashfaq, M. (2022). *Chemical technologies to degrade microplastic pollution. Microplastic pollution: Environmental occurrence and treatment technologies* (p. 487–510). Springer international Publishing.

Hirooka, T., Nagase, H., Uchida, K., Hiroshige, Y., Ehara, Y., Nishikawa, J. I., Nishihara, T., Miyamoto, K., & Hirata, Z. (2005). Biodegradation of bisphenol A and disappearance of its estrogenic activity by the green alga Chlorella fusca var. vacuolata. *Environmental Toxicology and Chemistry: An International Journal, 24*(8), 1896–1901.

Ho, K. L. G., Pometto, A. L., Gadea-Rivas, A., Briceno, J. A., & Rojas, A. (1999). Degradation of polylactic acid (PLA) plastic in Costa Rican soil and Iowa University compost rows. *Journal of Polymers Degradation, 7*, 173–177.

Iram, D., Riaz, R., & Iqbal, R. K. (2019). Usage of potential microorganisms for degradation of plastics. *Open Journal of Environmental Biology, 4*(1), 7–15.

Islam, S., Apitius, L., Jakob, F., & Schwaneberg, U. (2019). Targeting micro-plastic particles in the void of diluted suspensions. *Environment International, 123*, 428–435.

Jaiswal, S., Sharma, B., & Shukla, P. (2020). Integrated approaches in microbial degradation of plastics. *Environmental Technology & Innovation, 17*, 100567.

Jeon, H. J., & Kim, M. N. (2016). Isolation of mesophilic bacterium for biodegradation of polypropylene. *International Biodeterioration & Biodegradation, 115*, 244–249.

Judy, J. D., Williams, M., Gregg, A., Oliver, D., Kumar, A., Kookana, R., & Kirby, J. K. (2019). Micro-plastics in municipal mixed-waste organic outputs induce minimal short to long-term toxicity in key terrestrial biota. *Environmental Pollution, 252*(1), 522–531.

Kay, P., Hiscoe, R., Moberley, I., Bajic, L., & McKenna, N. (2018). Wastewater treatment plants as a source of microplastics in river catchments. *Environmental Science and Pollution Research, 25*, 20264–20267.

Kelly, J. J., London, M. G., McCormick, A. R., Rojas, M., Scott, J. W., & Hoellein, T. J. (2021). Wastewater treatment alters microbial colonization of micro-plastics. *PLoS One, 16*(1), e0244443.

Kim, J. W., Park, S. B., Tran, Q. G., Cho, D. H., Choi, D. Y., Lee, Y. J., & Kim, H. S. (2020). Functional expression of polyethylene terephthalate-degrading enzyme (PETase) in green microalgae. *Microbial Cell Factories, 19*(1), 1–9.

Knott, B. C., Erickson Mark, D., Allen, E., Gado, J. E., Graham, R., Kearns, F. L., Pardo, I., Topuzlu, E., Anderson, J. J., Austin, H. P., Dominick, G., Johnson, C. W., Rorrer, N. A.,

Szostkiewicz, C. J., Copié, V., Payne, C. M., Woodcock, H. L., Donohoe, B. S., Beckham, G. T., & McGeehan, J. E. (2020). Characterization and engineering of a two-enzyme system for plastics depolymerization. *Proceedings of the National Academy of Sciences, 117*(41), 25476–25485.

Krishnan, K. (2023). A systematic review on the impact of micro-nanoplastics exposure on human health and diseases. *Biointerface Research in Applied Chemistry, 13*, 1–12.

Krueger, M. C., Hofmann, U., Moeder, M., & Schlosser, D. (2015). Potential of wood-rotting fungi to attack polystyrene sulfonate and its depolymerisation by Gloeophyllum trabeum via hydroquinone-driven Fenton chemistry. *PLoS One, 10*(7), e0131773.

Kumar, R. V., Kanna, G. R., & Elumalai, S. (2017). Biodegradation of polyethylene by green photosynthetic microalgae. *Journal of Bioremediation & Biodegradation, 8*(381), 2.

Kumar, M., Kumar, M., Pandey, A., & Thakur, I. S. (2019). Genomic analysis of carbon dioxide sequestering bacterium for exopolysaccharides production. *Scientific Reports, 9*(1), 1–12.

Lagarde, F., Olivier, O., Zanella, M., Daniel, P., Hiard, S., & Caruso, A. (2016). Microplastic interactions with freshwater microalgae: Hetero-aggregation and changes in plastic density appear strongly dependent on polymer type. *Environmental Pollution, 215*, 331–339.

Lares, M., Ncibi, M. C., Markus, S., & Mika, S. (2018). Occurrence, identification and removal of microplastic particles and fibers in conventional activated sludge process and advanced MBR technology. *Water Research, 133*, 236–246.

Lear, G., Kingsbury, J. M., Franchini, S., Gambarini, V., Maday, S. D. M., Wallbank, J. A., Weaver, L., & Pantos, O. (2021). Plastics and the microbiome: Impacts and solutions. *Environmental Microbiology, 16*(1), 1–19.

Lebreton, L. C., Van Der Zwet, J., Damsteeg, J. W., Slat, B., Andrady, A., & Reisser, J. (2017). River plastic emissions to the world's oceans. *Nature Communications, 8*(1), 15611.

Lebreton, L., Egger, M., & Slat, B. (2019). A global mass budget for positively buoyant macroplastic debris in the ocean. *Scientific Reports, 9*(1), 12922.

Lehner, R., Wohlleben, W., Septiadi, D., Landsiedel, R., Petri-Fink, A., & Rothen- Rutishauser, B. (2020). A novel 3D intestine barrier model to study the immune response upon exposure to microplastics. *Archives of Toxicology, 94*, 2463–2479.

Li, Z., Guo, Z., Zhang, T., Li, Q., Chen, J., Ji, W., Liu, C., & Wei, Y. (2021). Fabrication of in situ ZIF-67 grown on alginate hydrogels and its application for enhancing Cu (II) adsorption from aqueous solutions. *Colloids and Surfaces B: Biointerfaces, 207*, 112036.

Liebezeit, G., & Liebezeit, E. (2014). Synthetic particles as contaminants in German beers. *Food Additives & Contaminants: Part A, 31*(9), 1574–1578.

Maheswaran, B., Al-Ansari, M., Al-Humaid, L., Raj, J. S., Kim, W., Karmegam, N., & Rafi, K. M. (2023). In vivo degradation of polyethylene terephthalate using microbial isolates from plastic polluted environment. *Chemosphere, 310*, 136757.

Manabe, M., Tatarazako, N., & Kinoshita, M. (2011). Uptake, excretion and toxicity of nano-sized latex particles on medaka (Oryzias latipes) embryos and larvae. *Aquatic Toxicology, 105*(3–4), 576–581.

Matjašič, T., Simčič, T., Medveščček, N., Bajt, O., Dreo, T., & Mori, N. (2021). Critical evaluation of biodegradation studies on synthetic plastics through a systematic literature review. *Science of the Total Environment, 752*, 141959.

Mitrano, D. M., Beltzung, A., Frehland, S., Schmiedgruber, M., Cingolani, A., & Schmidt, F. (2019). Synthesis of metal-doped nanoplastics and their utility to investigate fate and behaviour in complex environmental systems. *Nature Nanotechnology, 14*(4), 362–368.

Negoro, S., Shibata, N., Tanaka, Y., Yasuhira, K., Shibata, H., Hashimoto, H., Lee, Y. H., Oshima, S., Santa, R., Mochiji, K., & Goto, Y. (2012). Three-dimensional structure of nylon hydrolase and mechanism of nylon-6 hydrolysis. *The Journal of Biological Chemistry, 287*(7), 5079–5090.

Ng, E. L., Lwanga, E. H., Eldridge, S. M., Johnston, P., Hu, H. W., Geissen, V., & Chen, D. (2018). An overview of microplastic and nanoplastic pollution in agroecosystems. *Science of the Total Environment, 627*, 1377–1388.

Ojha, N., Pradhan, N., Singh, S., Barla, A., Shrivastava, A., Khatua, P., Rai, V., & Bose, S. (2017). Evaluation of HDPE and LDPE degradation by fungus, implemented by statistical optimization. *Scientific Reports, 7*(1), 39515.

Paço, A., Duarte, K., da Costa, J. P., Santos, P. S., Pereira, R., Pereira, M. E., Freitas, A. C., Duarte, A. C., & Rocha-Santos, T. A. (2017). Biodegradation of polyethylene microplastics by the marine fungus Zalerion maritimum. *Science of the Total Environment, 586*, 10–15.

Parks, D. H., Rinke, C., Chuvochina, M., Chaumeil, P. A., Woodcroft, B. J., Evans, P. N., Hugenholtz, P., & Tyson, G. W. (2017). Recovery of nearly 8,000 metagenome-assembled genomes substantially expands the tree of life. *Nature Microbiology, 2*(11), 1533–1542.

Pathak, V. M., & Navneet. (2017). Review on the current status of polymer degradation: A microbial approach. *Bioresources and Bioprocessing, 4*(1), 15.

Peng, L., Mehmood, T., Bao, R., Wang, Z., & Fu, D. (2022). An overview of micro (nano) plastics in the environment: Sampling, identification, risk assessment and control. *Sustainability, 14*(21), 14338.

Pivokonsky, M., Cermakova, L., Novotna, K., Peer, P., Cajthaml, T., & Janda, V. (2018). Occurrence of microplastics in raw and treated drinking water. *Science of the Total Environment, 643*, 1644–1651.

Prata, J. C. (2018). Airborne microplastics: Consequences to human health? *Environmental Pollution, 234*, 115–126.

Priya, A., Dutta, K., & Daverey, A. (2021). A comprehensive biotechnological and molecular insight into plastic degradation by microbial community. *Journal of Chemical Technology and Biotechnology, 97*(2), 381–390.

Purohit, J., Chattopadhyay, A., & Teli, B. (2020). Metagenomic exploration of plastic degrading microbes for biotechnological application. *Current Genomics, 21*(4), 253–270.

Raddadi, N., & Fava, F. (2019). Biodegradation of oil-based plastics in the environment: Existing knowledge and needs of research and innovation. *Science of the Total Environment, 679*, 148–158.

Rai, P. K., Lee, J., Brown, R. J., & Kim, K. H. (2021). Micro-and nano-plastic pollution: Behavior, microbial ecology, and remediation technologies. *Journal of Cleaner Production, 291*, 125240.

Rana, K. I. (2019). Usage of potential micro-organisms for degradation of plastics. *Journal of Environmental Biology, 4*, 7–15.

Rehse, S., Kloas, W., & Zarfl, C. (2016). Short-term exposure with high concentrations of pristine microplastic particles leads to immobilization of Daphnia magna. *Chemosphere, 153*, 91–99.

Ricardo, I. A., Alberto, E. A., Júnior, A. H. S., Macuvele, D. L. P., Padoin, N., Soares, C., Riella, H. G., Starling, M. C. V. M., & Trovó, A. G. (2021). A critical review on micro-plastics, interaction with organic and inorganic pollutants, impacts and effectiveness of advanced oxidation processes applied for their removal from aqueous matrices. *Chemical Engineering Journal, 424*, 130282.

Roex, E., Vethaak, D., Leslie, H., & Kreuk, M. D. (2013). *Potential risk of microplastics in the freshwater environment*. STOWA Amersfoort.

Rossi, G., Barnoud, J., & Monticelli, L. (2014). Polystyrene nanoparticles perturb lipid membranes. *Journal of Physical Chemistry Letters, 5*(1), 241–246.

Rout, P. R., Mohanty, A., Sharma, A., Miglani, M., Liu, D., & Varjani, S. (2022). Micro-and nanoplastics removal mechanisms in wastewater treatment plants: A review. *Journal of Hazardous Materials Advances, 6*, 100070.

Ru, J., Huo, Y., & Yang, Y. (2020). Microbial degradation and valorization of plastic wastes. *Frontiers in Microbiology, 11*, 442.

Sah, A., Kapri, A., Zaidi, M. G. H., Negi, H., & Goel, R. (2010). Implications of fullerene-60 upon in-vitro LDPE biodegradation. *Journal of Microbiology and Biotechnology, 20*(5), 908–916.

Sajid, M., Asif, M., Baig, N., Kabeer, M., Ihsanullah, I., & Mohammad, A. W. (2022). Carbon nanotubes-based adsorbents: Properties, functionalization, interaction mechanisms, and applications in water purification. *Journal of Water Process Engineering, 47*, 102815.

Sangale, M. K., Shahnawaz, M., & Ade, A. B. (2019). Potential of fungi isolated from the dumping sites mangrove rhizosphere soil to degrade polythene. *Scientific Reports, 9*(1), 5390.

Saraji, M., & Ghani, M. (2014). Dissolvable layered double hydroxide coated magnetic nanoparticles for extraction followed by high performance liquid chromatography for the determination of phenolic acids in fruit juices. *Journal of Chromatography A, 1366*, 24–30.

Sarkar, D. J., Das Sarkar, S., Das, B. K., Praharaj, J. K., Mahajan, D. K., Purokait, B., Mohanty, T. R., Mohanty, D., Gogoi, P., Kumar, S., & Behera, B. K. (2021). Microplastics removal efficiency of drinking water treatment plant with pulse clarifier. *Journal of Hazardous Materials, 413*, 125347.

Sarmah, P., & Rout, J. (2018). Efficient biodegradation of low-density polyethylene by cyanobacteria isolated from submerged polyethylene surface in domestic sewage water. *Environmental Science and Pollution Research, 25*, 33508–33520.

Sarmah, P., & Rout, J. (2018a). Algal colonization on polythene carry bags in a domestic solid waste dumping site of Silchar town in Assam. *Phykos, 48*(67), e77.

Sarmah, P., & Rout, J. (2018b). Biochemical profile of five species of cyanobacteria isolated from polythene surface in domestic sewage water of Silchar town, Assam (India). *Current Trends In Biotechnology And Pharmacy, 12*(1), 7–15.

Schmeisser, C., Steele, H., & Streit, W. R. (2007). Metagenomics, biotechnology with nonculturable microbes. *Applied Microbiology and Biotechnology, 75*, 955–962.

Shao, H., Chen, M., Fei, X., Zhang, R., Zhong, Y., Ni, W., & Tan, X. (2019). Complete genome sequence and characterization of a polyethylene biodegradation strain, Streptomyces albogriseolus LBX-2. *Microorganisms, 7*(10), 379.

Sharma, B., Rawat, H., & Pooja, S. R. (2017). Bioremediation-A progressive approach toward reducing plastic wastes. *International Journal of Current Microbiology and Applied Sciences, 6*(12), 1116–1131.

Shi, X., Zhang, X., Gao, W., Zhang, Y., & He, D. (2022). Removal of microplastics from water by magnetic nano-Fe3O4. *Science of the Total Environment, 802*, 149838.

Shin, Y. H., Jung, I., Park, H., Pyeon, J. J., Son, J. G., Koo, C. M., Kim, S., & Kang, C. Y. (2018). Mechanical fatigue resistance of piezoelectric PVDF polymers. *Micromachines, 9*(10), 503.

Silva, A. L. P. (2021). New frontiers in remediation of (micro) plastics. *Current Opinion in Green and Sustainable Chemistry, 28*, 100443.

Sivan, A., Szanto, M., & Pavlov, V. (2006). Biofilm development of the polyethylene-degrading bacterium Rhodococcus ruber. *Applied Microbiology and Biotechnology, 72*, 346–352.

Steinmetz, Z., Wollmann, C., Schaefer, M., Buchmann, C., David, J., Tröger, J., Muñoz, K., Frör, O., & Schaumann, G. E. (2016). Plastic mulching in agriculture. Trading short-term agronomic benefits for long-term soil degradation? *Science of the Total Environment, 550*, 690–705.

Sulaiman, S., Yamato, S., Kanaya, E., Kim, J. J., Koga, Y., Takano, K., & Kanaya, S. (2012). Isolation of a novel cutinase homolog with polyethylene terephthalate-degrading activity from leaf-branch compost by using a metagenomic approach. *Applied and Environmental Microbiology, 78*(5), 1556–1562.

Syranidou, E., Karkanorachaki, K., Amorotti, F., Repouskou, E., Kroll, K., Kolvenbach, B., Corvini, P. F., Fava, F., & Kalogerakis, N. (2017). Development of tailored indigenous marine consortia for the degradation of naturally weathered polyethylene films. *PLOS One, 12*(8), e0183984.

Syranidou, E., Karkanorachaki, K., Amorotti, F., Avgeropoulos, A., Kolvenbach, B., Zhou, N. Y., Fava, F., Corvini, P. F. X., & Kalogerakis, N. (2019). Biodegradation of mixture of plastic films by tailored marine consortia. *Journal of Hazardous Materials, 375*, 33–42.

Talvitie, J., Mikola, A., Koistinen, A., & Setala, O. (2017). Solutions to microplastic pollution-removal of microplastics from wastewater effluent with advanced wastewater treatment technologies. *Water Research, 123*, 401–407.

Tan, W., Peralta-Videa, J. R., & Gardea-Torresdey, J. L. (2018). Interaction of titanium dioxide nanoparticles with soil components and plants: Current knowledge and future research need–a critical review. *Environmental Science: Nano, 5*(2), 257–278.

Tang, Y., Zhang, S., Su, Y., Wu, D., Zhao, Y., & Xie, B. (2021). Removal of microplastics from aqueous solutions by magnetic carbon nanotubes. *Chemical Engineering Journal, 406*, 126804.

Tian, L., Kolvenbach, B., Corvini, N., Wang, S., Tavanaie, N., Wang, L., Ma, Y., Scheu, S., Corvini, P. F. X., & Ji, R. (2017). Mineralisation of 14C-labelled polystyrene plastics by Penicillium variabile after ozonation pre-treatment. *New Biotechnology, 38*, 101–105.

Tiwari, N., Santhiya, D., & Sharma, J. G. (2020). Microbial remediation of micro-nano plastics: Current knowledge and future trends. *Environmental Pollution, 265*, 115044.

Uchiyama, T., & Miyazaki, K. (2009). Functional metagenomics for enzyme discovery: Challenges to efficient screening. *Current Opinion in Biotechnology, 20*(6), 616–622.

Vazquez-Núnez, E., Molina-Guerrero, C. E., Pena-Castro, J. M., Fernandez-Luqueno, F., & de la Rosa-Alvarez, M. (2020). Use of nanotechnology for the bioremediation of contaminants: A review. *PRO, 8*(7), 826.

Vianello, A., Jensen, R. L., Liu, L., & Vollertsen, J. (2019). Simulating human exposure to indoor airborne microplastics using a Breathing Thermal Manikin. *Scientific Reports, 9*(1), 8670.

Vieira, R. H., & Volesky, B. (2000). Biosorption: A solution to pollution? *International Microbiology, 3*(1), 17–24.

Vivi, V. K., Martins-Franchetti, S. M., & Attili-Angelis, D. (2019). Biodegradation of PCL and PVC: Chaetomium globosum (ATCC 16021) activity. *Folia Microbiologica, 64*, 1–7.

Wan, H., Wang, J., Sheng, X., Yan, J., Zhang, W., & Xu, Y. (2022). Removal of polystyrene microplastics from aqueous solution using the metal–organic framework material of ZIF-67. *Toxics, 10*(2), 70.

Wang, Z., Sun, C., Li, F., & Chen, L. (2021). Fatigue resistance, re-usable and biodegradable sponge materials from plant protein with rapid water adsorption capacity for microplastics removal. *Chemical Engineering Journal, 415*, 129006.

Wang, P., Liu, J., Han, S., Wang, Y., Duan, Y., Liu, T., Hou, L., Zhang, Z., Li, L., & Lin, Y. (2023). Polyethylene mulching film degrading bacteria within the plastisphere: Co-culture of plastic degrading strains screened by bacterial community succession. *Journal of Hazardous Materials, 442*, 30045.

Ward, J. E., & Kach, D. J. (2009). Marine aggregates facilitate ingestion of nanoparticles by suspension-feeding bivalves. *Marine Environmental Research, 68*(3), 137–142.

Wilkes, R. A., & Aristilde, L. (2017). Degradation and metabolism of synthetic plastics and associated products by Pseudomonas sp.: capabilities and challenges. *Journal of Applied Microbiology, 123*(3), 582–593.

Wróbel, M., Szymańska, S., Kowalkowski, T., & Hrynkiewicz, K. (2023). Selection of microorganisms capable of polyethylene (PE) and polypropylene (PP) degradation. *Microbiological Research, 267*, 127251.

Wu, M., Liu, W., & Liang, Y. (2019). Probing size characteristics of disinfection by-products precursors during the bioavailability study of soluble microbial products using ultrafiltration fractionation. *Ecotoxicology and Environmental Safety, 175*, 1–7.

Wu, X., Lu, J., Du, M., Xu, X., Beiyuan, J., Sarkar, B., Bolan, N., Xu, W., Xu, S., Chen, X., & Wu, F. (2021). Particulate plastics-plant interaction in soil and its implications: A review. *Science of the Total Environment, 792*, 148337.

Yadav, H., Sethulekshmi, S., & Shriwastav, A. (2022). Estimation of microplastic exposure via the composite sampling of drinking water, respirable air, and cooked food from Mumbai, India. *Environmental Research, 214*, 113735.

Yamada-Onodera, K., Mukumoto, H., Katsuyaya, Y., Saiganji, A., & Tani, Y. (2001). Degradation of polyethylene by a fungus, penicillium simplicissimum YK. *Polymer Degradation and Stability, 72*(2), 323–327.

Yang, Y., Yang, J., Wu, W. M., Zhao, J., Song, Y., Gao, L., Yang, R., & Jiang, L. (2015). Biodegradation and mineralization of polystyrene by plastic-eating mealworms: Part 2. Role of gut microorganisms. *Environmental Science & Technology, 49*(20), 12087–12093.

Yoshida, S., Hiraga, K., Takehana, T., Taniguchi, I., Yamaji, H., Maeda, Y., Toyohara, K., Miyamoto, K., Kimura, Y., & Oda, K. (2016). A bacterium that degrades and assimilates poly(ethylene terephthalate). *Science, 351*(6279), 1196–1199.

You, D., Zhao, Y., Yang, W., Pan, Q., & Li, J. (2022). Metal-organic framework-based wood aerogel for effective removal of micro/nano plastics. *Chemical Research in Chinese Universities, 38*(1), 186–191.

Yuan, J., Ma, J., Sun, Y., Zhou, T., Zhao, Y., & Yu, F. (2020). Microbial degradation and other environmental aspects of micro-plastics/plastics. *Science of the Total Environment, 715*, 136968.

Zhang, H., Kuo, Y. Y., Gerecke, A. C., & Wang, J. (2012). *Environmental Science & Technology, 46*(20), 10990–10996.

Zhang, B., He, Y., Zhu, H., Huang, X., Bai, X., Kannan, K., & Zhang, T. (2020). Concentrations of bisphenol A and its alternatives in paired maternal–fetal urine, serum and amniotic fluid from an e-waste dismantling area in China. *Environment International, 136*, 105407.

Zhang, Y., Zhou, G., Yue, J., Xing, X., Yang, Z., Wang, X., Wang, Q., & Zhang, J. (2021). Enhanced removal of polyethylene terephthalate microplastics through polyaluminum chloride coagulation with three typical coagulant aids. *Science of the Total Environment, 800*, 149589.

Zhao, H., Huang, X., Wang, L., Zhao, X., Yan, F., Yang, Y., Li, G., Gao, P., & Ji, P. (2022). Removal of polystyrene nanoplastics from aqueous solutions using a novel magnetic material: Adsorbability, mechanism, and reusability. *Chemical Engineering Journal, 430*, 133122.

Zheng, B., Li, B., Wan, H., Lin, X., & Cai, Y. (2022). Coral-inspired environmental durability aerogels for micron-size plastic particles removal in the aquatic environment. *Journal of Hazardous Materials, 431*, 128611.

Zhou, Y., Kumar, M., Sarsaiya, S., Sirohi, R., Awasthi, S. K., Sindhu, R., Binod, P., Pandey, A., Bolan, N. S., Zhang, Z., & Singh, L. (2022). Challenges and opportunities in bioremediation of micro-nano plastics: a review. *Science of the Total Environment, 802*, 149823.

Zhou, Y., Kumar, M., Sarsaiya, S., Sirohi, R., Awasthi, S. K., Sindhu, R., Binod, P., Pandey, A., Bolan, N. S., Zhang, Z., & Singh, L. (2022). Challenges and opportunities in bioremediation of micro-nano plastics: A review co-release of hexabromocyclododecane (HBCD) and nano-and microparticles from thermal cutting of polystyrene foams. *Science of the Total Environment, 802*, 149823.

Ziajahromi, S., Kumar, A., Neale, P. A., & Leusch, F. D. (2018). Environmentally relevant concentrations of polyethylene microplastics negatively impact the survival, growth and emergence of sediment-dwelling invertebrates. *Environmental Pollution, 236*, 425–431.

Zinicovscaia, I. (2016). Conventional methods of wastewater treatment. In *Cyanobacteria for bioremediation of wastewaters* (pp. 17–25). Springer.

Book Description

"Management of Micro and Nano-plastics in Soil and Biosolids: Fate, Occurrence, Monitoring, and Remedies" encompasses a broad and updated body of knowledge concerning the fate, occurrence, monitoring, and remediation of micro and nano-plastics in contaminated soil and biosolids. It presents innovative applications and approaches to reduce micro and nano-plastics pollution. The book aids in understanding the fate of micro and nano-plastics, their transition from biosolids to soil, and their impacts on fauna and flora.

Key Features

- Discusses studies that explore the utilization of the nexus approach in the management of micro and nano-plastics.
- Highlights and identifies emerging technologies that facilitate the remediation of micro and nano-plastics from soil and biosolids.
- Compiles authoritative information on recent advancements in the fate, monitoring, and remedies of micro and nano-plastics in contaminated soil and biosolids.
- Provides the latest developments and current research perspectives, technological advancements, and critical thinking related to the management of micro and nano-plastics.

S. A. Bhat et al. (eds.), *Management of Micro and Nano-plastics in Soil and Biosolids*,
https://doi.org/10.1007/978-3-031-51967-3

Index

S. A. Bhat et al. (eds.), *Management of Micro and Nano-plastics in Soil and Biosolids*,
https://doi.org/10.1007/978-3-031-51967-3

MIX
Papier aus verantwortungsvollen Quellen
Paper from responsible sources
FSC® C105338

If you have any concerns about our products, you can contact us on
ProductSafety@springernature.com

In case Publisher is established outside the EU, the EU authorized representative is:
Springer Nature Customer Service Center GmbH
Europaplatz 3, 69115 Heidelberg, Germany

Printed by Libri Plureos GmbH
in Hamburg, Germany